Mechanical Properties of Solid Polymers

Mechanical Properties of Solid Polymers

Fourth Edition

John Sweeney
School of Computing and Engineering
University of Bradford
Bradford, UK

Peter Hine
School of Physics and Astronomy
University of Leeds
UK

WILEY

This edition first published 2026

Edition History
John Wiley & Sons, Ltd (3e, 2012; 2e, 2004).

Registered Offices
John Wiley & Sons, Inc., 111 River Street, Hoboken, NJ 07030, USA
John Wiley & Sons Ltd. New Era House, 8 Oldlands Way, Bognor Regis, West Sussex, PO22 9NQ, UK

For details of our global editorial offices, customer services, and more information about Wiley products visit us at www.wiley.com.

The manufacturer's authorized representative according to the EU General Product Safety Regulation is Wiley-VCH GmbH, Boschstr. 12, 69469 Weinheim, Germany, e-mail: Product_Safety@wiley.com.

Wiley also publishes its books in a variety of electronic formats and by print-on-demand. Some content that appears in standard print versions of this book may not be available in other formats.

Library of Congress Cataloging-in-Publication Data
Names: Sweeney, J. (John), author. | Hine, Peter (Peter J.), author.
Title: Mechanical properties of solid polymers / John Sweeney, University of Bradford, United Kingdom, Peter Hine, University of Leeds, United Kingdom.
Description: Fourth edition. | Hoboken, NJ, USA : Wiley, [2026] | Earlier editon by I.M. Ward, School of Physics and Astronomy, Leeds University, Leeds, UK, J. Sweeney, School of Engineering, Design and Technology, University of Bradford, Bradford, UK. | Includes bibliographical references and index.
Identifiers: LCCN 2025024600 (print) | LCCN 2025024601 (ebook) | ISBN 9781394202065 (cloth) | ISBN 9781394202072 (adobe pdf) | ISBN 9781394202096 (epub)
Subjects: LCSH: Polymers–Mechanical properties.
Classification: LCC TA455.P58 W37 2026 (print) | LCC TA455.P58 (ebook)
LC record available at https://lccn.loc.gov/2025024600
LC ebook record available at https://lccn.loc.gov/2025024601

Cover Design: Wiley
Cover Image: © hiroyuki nakai/Getty Images

Set in 9.5/12.5pt STIXTwoText by Straive, Pondicherry, India

SKY10129270_102225

Contents

Preface

This book is the fourth edition of *Mechanical Properties of Solid Polymers* and follows the third edition that was published in 2013. This series of textbooks was instigated by the late Ian M. Ward, FRS, in 1971 and includes two editions of the student-oriented work *An Introduction to the Mechanical Properties of Solid Polymers.*

Many of the changes in material in this edition are driven by progress over time in polymer technology, experimental practice and analysis. Thus, this edition includes examples of the science and application of naturally occurring polymers and those from renewable resources. The availability of commercially produced machines such as highly versatile dynamic mechanical testers has reduced the need for specialized dedicated apparatus. Similarly, the availability of software packages in areas such as finite element analysis means that experiments involving complex stress fields can be routinely modelled, reflected by more reports of numerical analysis in this edition. The same packages allow complex material models to be implemented for realistic problems and in effect make such models applicable. In addition, the final chapter on fracture has been expanded to include anisotropic material, most applicable to oriented polymers.

The previous edition included a chapter on polymer composites. This field has greatly expanded in recent years and is the subject of several of dedicated textbooks. We acknowledge (with thanks to a reviewer) that attempting to cover this area is no longer appropriate here. We have replaced this chapter with a new one, Morphology and Structural Effects, where we give an account of morphology development and its structural consequences. We believe that this improves the focus of the book.

13 March 2025

John Sweeney
University of Bradford, Bradford, UK

Peter Hine
University of Leeds, UK

1

Structure of Polymers

The mechanical properties that form the subject of this book are a consequence of the chemical composition of the polymer and also of its structure at the molecular and supermolecular levels. We shall therefore introduce a few elementary ideas concerning these aspects.

1.1 Chemical Composition

1.1.1 Polymerization

Linear polymers consist of long molecular chains of covalently bonded atoms, each chain being a repetition of much smaller chemical units. One of the simplest polymers is polyethylene, which is an addition polymer made by polymerizing the monomer ethylene, $CH_2{=}CH_2$, to form the polymer.

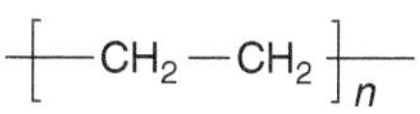

Note that the double bond is removed during the polymerization (Figure 1.1). The well-known vinyl polymers are made by polymerizing compounds of the form

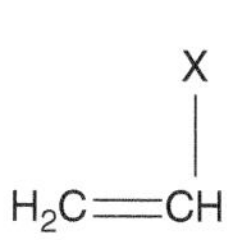

where X represents a chemical group; examples are as follows:

Polypropylene

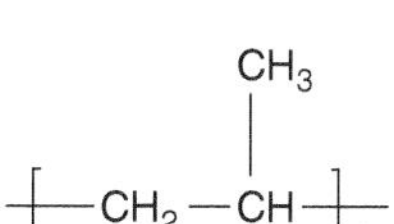

Mechanical Properties of Solid Polymers, Fourth Edition. John Sweeney and Peter Hine.

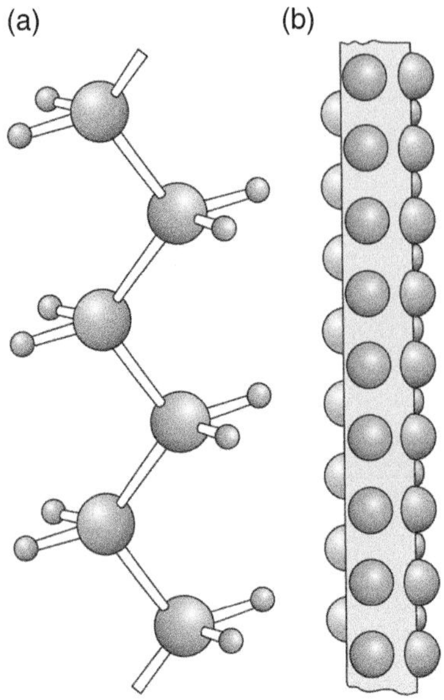

Figure 1.1 (a) The polyethylene chain $(CH_2)_n$ in schematic form (larger spheres, carbon; smaller spheres, hydrogen) and (b) sketch of a molecular model of a polyethylene chain.

Polystyrene

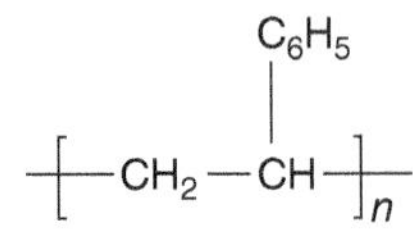

and

poly(vinyl chloride)

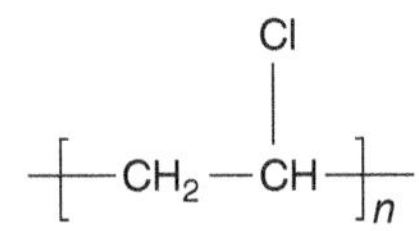

Natural rubber, polyisoprene, is a diene, and its repeat unit

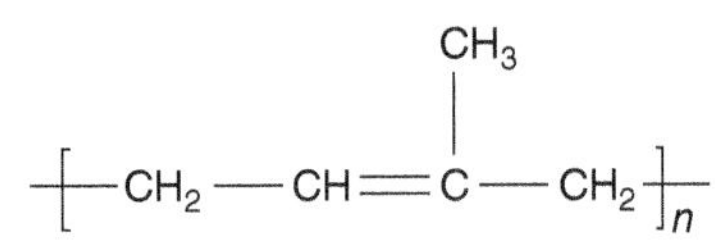

contains a double bond.

A condensation reaction is one in which two or more molecules combine into a larger molecule with or without the loss of a small molecule (such as water). One example is the formation of polyethylene terephthalate (the polyester used for terylene and Dacron fibres and transparent films and bottles) from ethylene glycol and terephthalic acid:

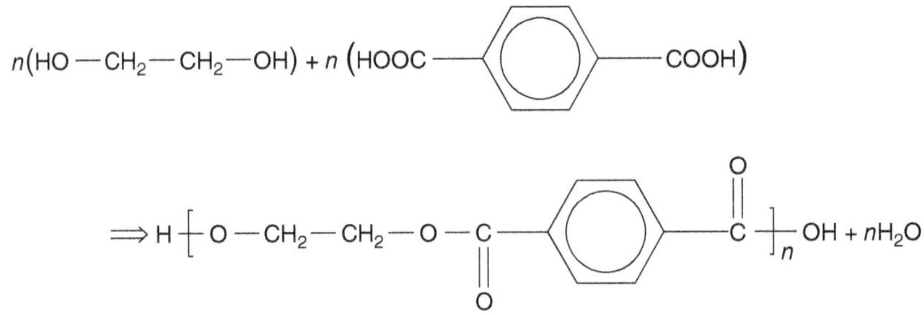

Another common condensation polymer is nylon 6,6.

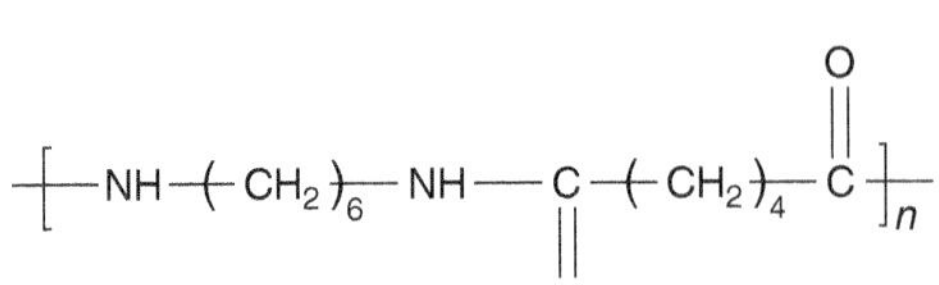

1.1.2 Cross-Linking and Chain-Branching

Linear polymers can be joined by other chains at points along their length to make a cross-linked structure (Figure 1.2). Chemical cross-linking produces a thermosetting polymer, so called because the cross-linking agent is normally activated by heating, after which the material does not soften and melt when heated further, for example Bakelite and epoxy resins. A small amount of cross-linking through sulphur bonds is needed to give natural rubber its characteristic feature of rapid recovery from a large extension.

Very long molecules in linear polymers can entangle to form temporary physical cross-links, and we shall show later that a number of the characteristic properties of solid polymers are explicable in terms of the behaviour of a deformed network.

A less extreme complication is chain-branching, where a secondary chain initiates from a point on the main chain, as is illustrated for polyethylene in Figure 1.3. Low-density

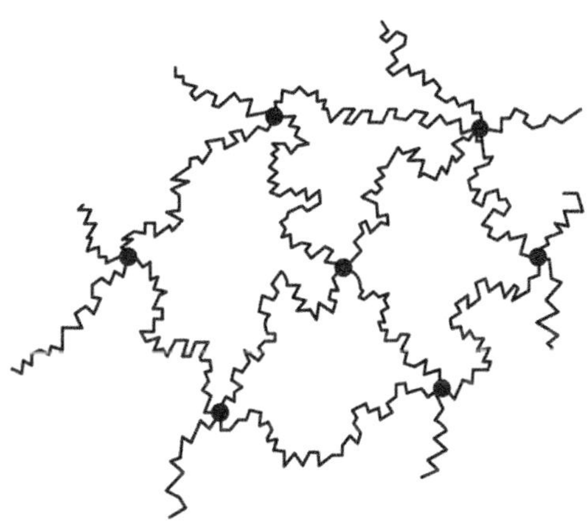

Figure 1.2 Schematic diagram of a cross-linked polymer.

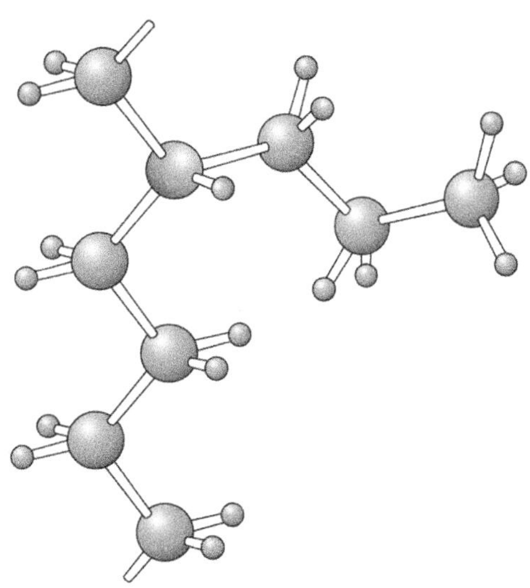

Figure 1.3 A chain branch in polyethylene.

polyethylene, as distinct from the high-density linear polyethylene shown in Figure 1.1, possesses on average one long branch per molecule and a larger number of small branches, mainly ethyl ($-CH_2-CH_3$) or butyl ($-(CH_2)_3-CH_3$) side groups. The presence of these branch points leads to considerable differences in mechanical behaviour compared with linear polyethylene.

1.1.3 Average Molecular Mass and Molecular Mass Distribution

Each sample of a polymer contains molecular chains of varying lengths, that is of varying molecular mass (Figure 1.4). The mass (length) distribution is of importance in determining the properties of the polymer, but until the advent of gel permeation chromatography (Vaughan 1960; Moore 1964), it could be determined only by tedious fractionation procedures. Most investigations therefore quoted different types of average molecular mass, the commonest being the number average $\overline{M}_n$ and the weight average $\overline{M}_n$, defined as

$$\overline{M}_n = \frac{\sum N_i M_i}{\sum N_i} \qquad \overline{M}_w = \frac{\sum (N_i M_i) M_i}{\sum N_i M_i},$$

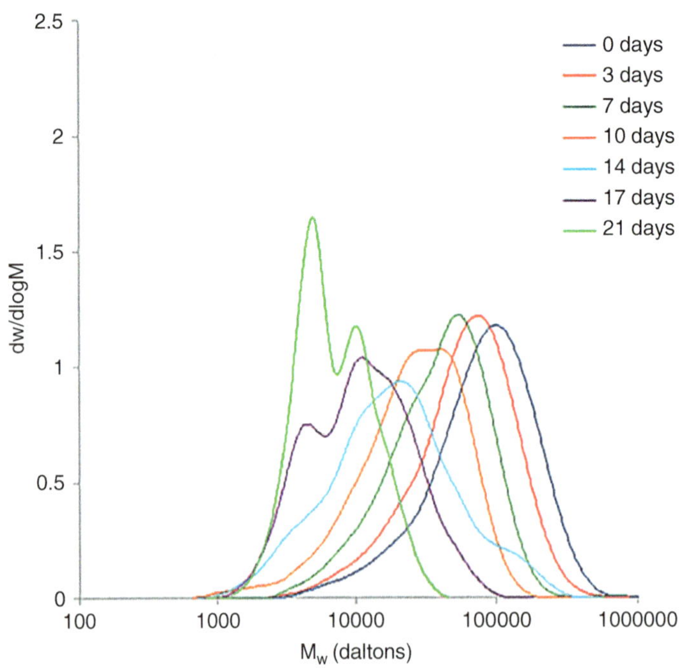

Figure 1.4 The gel permeation chromatograph traces give a direct indication of the molecular mass distribution. Results are for polylactic acid (PLA) undergoing degradation for periods up to 21 days; note the generally decreasing molecular mass with time. Stloukl et al., 2016 / with permission of Elsevier.

where N_i is the number of molecules of molecular mass M_i and Σ denotes summation over all i molecular masses.

The weight average molecular mass is always higher than the number average, as the former is strongly influenced by the relatively small number of very long (massive) molecules. The ratio of the two averages gives a general idea of the width of the molecular mass distribution.

Fundamental measurements of average molecular mass must be performed on solutions so dilute that intermolecular interactions can be ignored or compensated for. The commonest techniques are osmotic pressure for the number average and light scattering for the weight average. Both methods are rather lengthy, so in practice an average molecular mass was often deduced from viscosity measurements of either a dilute solution of the polymer (which relates to M_n) or a polymer melt (which relates to M_w). Each method yielded a different average value, which made it difficult to correlate specimens characterized by different groups of workers.

The molecular mass distribution is important in determining flow properties and so may affect the mechanical properties of a solid polymer indirectly by influencing the final physical state. Direct correlations of molecular mass to viscoelastic behaviour and brittle strength have also been obtained.

1.1.4 Chemical and Steric Isomerism and Stereoregularity

A further complication of the chemical structure of polymers lies in the possibility of different chemical isomeric forms within a repeat unit or between a series of repeat units. Both natural rubber and gutta percha are chemically polyisoprene, but the former is the *cis* form and the latter is the *trans* form (see Figure 1.5). The characteristic properties of rubber are a consequence of the loose packing of molecules (i.e. large *free volume*) that arises from its structure.

Vinyl monomer units

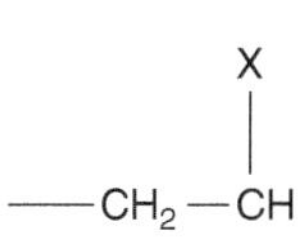

can be added to a growing chain either head-to-tail:

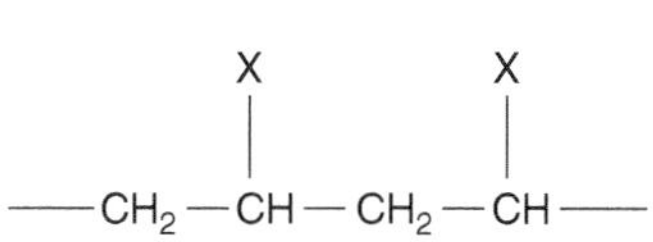

or head-to-head:

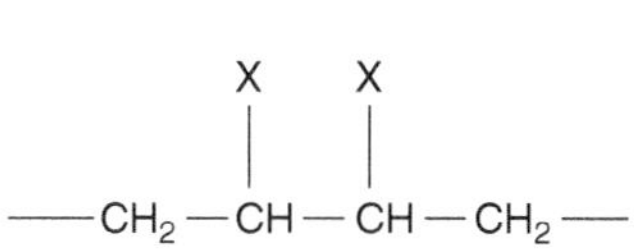

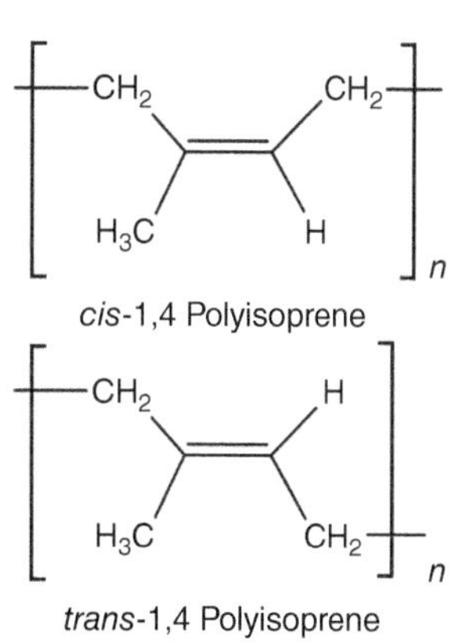

Figure 1.5 cis-1,4-Polyisoprene and trans-1,4-polyisoprene.

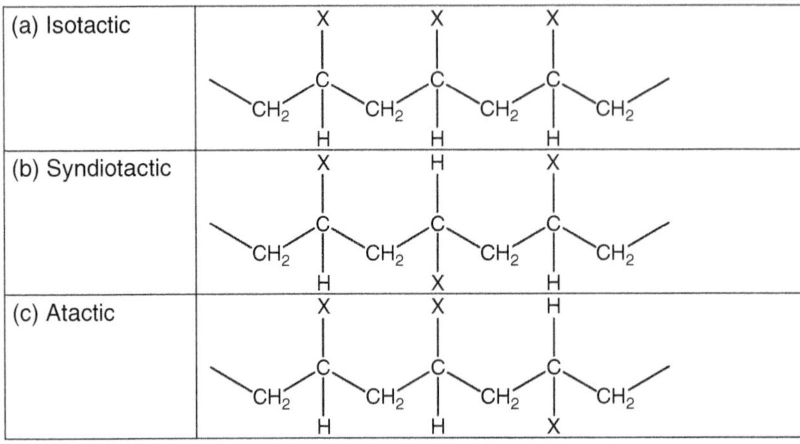

Figure 1.6 A substituted α-olefin can take three stereosubstituted forms.

Head-to-tail substitution is usual, and only a small proportion of head-to-head linkages can produce a reduction in the tensile strength because of the loss of regularity.

Stereoregularity provides a more complex situation, which we will examine in terms of the simplest type of vinyl polymer (Figure 1.6) and for which we shall suppose that the polymer chain is a planar zigzag. Two very simple regular polymers can be constructed. In the first (Figure 1.6(a)) the substituent groups are all added in an identical manner to give an *isotactic* polymer. In the second regular polymer (Figure 1.6(b)) there is an inversion of the manner of substitution between consecutive units, giving a *syndiotactic* polymer for which the substituent groups alternate regularly on opposite sides of the chain. The regular sequence of units is called *stereoregularity*, and stereoregular polymers are crystalline and can possess high melting temperatures. The working range of a polymer is thereby extended compared with the amorphous atactic form, whose range is limited by the lower softening point. The final alternative structure is formed when the orientation of successive substituents takes place randomly (Figure 1.6(c)) to give an irregular *atactic* polymer that is incapable of crystallizing. Polypropylene $(\text{—}CH_2CHCH_3\text{—})_n$ was for many years obtainable only as an atactic polymer, and its widespread use began only when stereospecific catalysts were developed to produce the isotactic form. Even so, some faulty substitution occurs and atactic chains can be separated from the rest of the polymer by solvent extraction.

1.1.5 Liquid Crystalline Polymers

Liquid crystals (or plastic crystals as they are sometimes called) are materials that show molecular alignment in one direction but not three-dimensional crystalline order. During the last 20 years, liquid crystalline polymers have been developed where the polymer chains are so straight and rigid that small regions of almost uniform orientation (domains) separated by distinct boundaries are produced. In the case where these domains occur in solution, polymers are termed *lyotropic*. Where the domains occur in the melt, the polymers are termed *thermotropic*.

An important class of lyotropic liquid crystal polymers are the aramid polymers such as polyparabenzamide

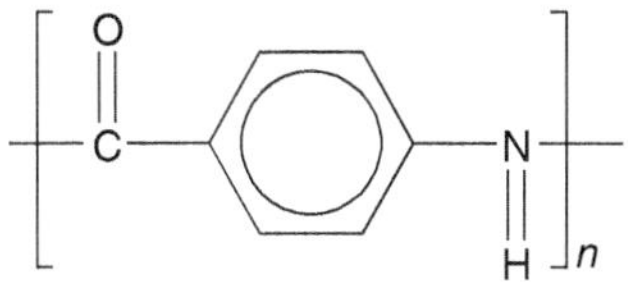

and polyparaphenylene terephthalamide

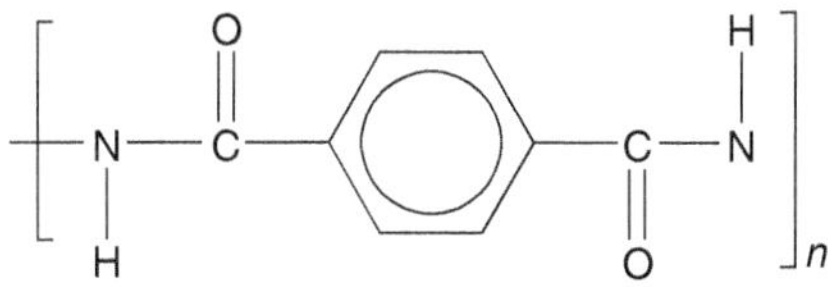

better known as Kevlar, which is a commercially produced high stiffness and high strength fibre. It is important to emphasize that although Kevlar fibres are prepared by spinning a lyotropic liquid crystalline phase, the final fibre shows clear evidence of three-dimensional order.

Important examples of thermotropic liquid crystalline polymers are copolyesters produced by condensation of hydroxybenzoic acid (HBA)

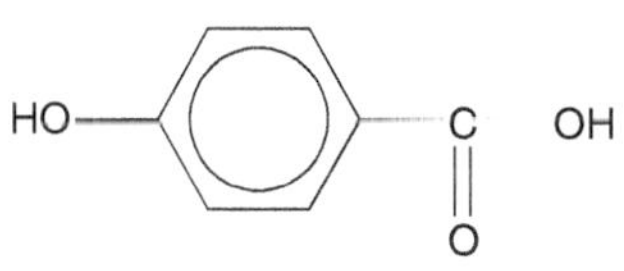

and 2,6-hydroxynaphthoic acid (HNA)

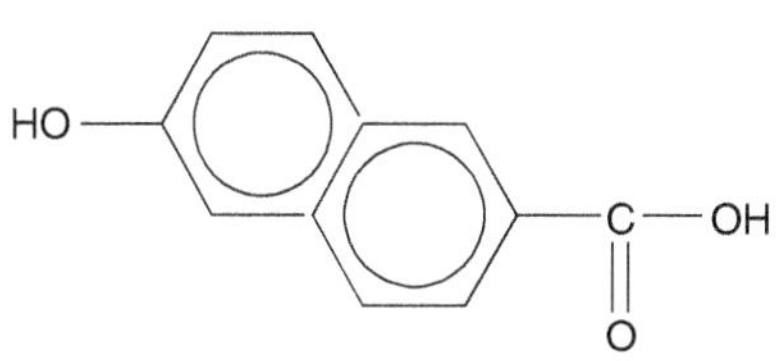

most usually in the proportions HBA:HNA = 73:27.

In addition to these main-chain liquid crystalline polymers, there are also side-chain liquid crystalline polymers, where the liquid crystalline nature arises from the presence of rigid straight side-chain units (called the mesogens) chemically linked to an existing polymer backbone either directly or via flexible spacer units.

The review by Noël and Navard (1991) gives further information on liquid crystalline polymers, including methods of preparation.

1.1.6 Naturally Occurring Polymers

While a significant number of published works and interest over the last 90 years (since polyethylene was invented in 1933) have concentrated on what are termed synthetic

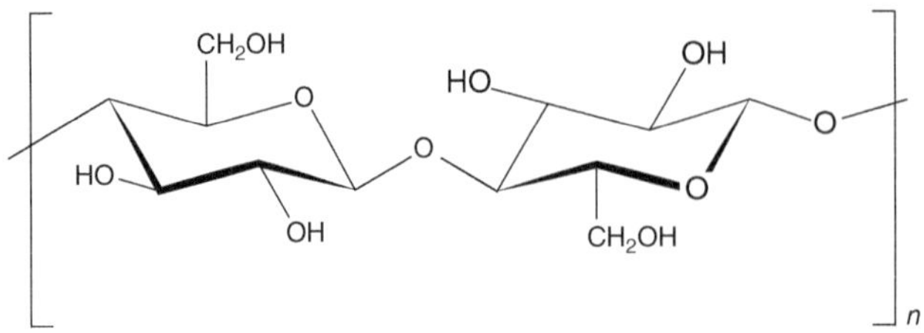

Figure 1.7 The chemical structure of cellulose.

polymers, there is also another class of polymers which are termed natural polymers. These are defined as being found naturally in the environment and produced via a biological process, as opposed to the synthetic polymers that are human created. This class of natural polymers includes two major groups: polysaccharides (long-chain carbohydrate-based molecules) such as cellulose and starch; and proteins (based on amino acids) such as keratin (wool, leather), silk and collagen. The renewed interest in these natural polymers (which were used long before synthetic polymers were invented – for example cotton and paper) comes from their combination of deriving from renewable sources, biocompatibility and their ability to biodegrade (in contrast to most synthetic polymers). Hence these natural polymers are finding increasingly wide usage in a range of fields such as construction, biomedical and general engineering (automotive, etc.).

Of all the naturally produced polymers, cellulose is the most abundant and most widely used. Cellulose is the main structural component of all plant walls and so forms a major part of the world's biomass. The repeat unit of cellulose is shown in Figure 1.7 and comprises two glucose monomers linked by an oxygen atom which is termed a glycosidic linkage. For each repeat unit, one glucose unit is rotated by 180° leading to a straight linear chain when these are connected together by nature to form a natural polymer.

An important aspect is that the cellulose chains form a dense network of hydrogen bonds (both intramolecular and intermolecular) which then allows the molecules to stack and form strong and stiff meso-structures (with a significant crystallinity) that are resistant to most solvents. The hydrogen bond network has the result that cellulose polymer does not melt like a synthetic thermoplastic polymer, and so chemical process is required if further processing is necessary, but giving excellent temperature resistance.

A further subset of natural cellulosic polymers is termed semi-synthetic natural polymers which are formed by the chemical modification of cellulose. In these polymers, acid treatment substitutes some or all of the OH groups of cellulose, leading to increased solubility and thermal processing depending on the level of substitution. Two important polymers resulting from this are cellulose acetate (from treating natural cellulose with acetic acid) and cellulose nitrate (by treating with nitric acid).

1.1.7 Polymers from Renewable Sources

A third group of polymers that fits between the two groups of synthetic and naturally occurring polymers are those that are human produced but derived from renewable biomass sources such as sugar cane or corn starch; these are often termed biopolymers. Poly lactic acid (PLA) is one of the most common of these and is currently the most widely used polymer made from biomass.

PLA is a linear thermoplastic polyester with a repeat unit as shown in Figure 1.8. It has a glass transition temperature around 60°C and a melting temperature around 160°C. PLA refers to a family of polymers with differing isomeric forms. PLLA is a polymer of L-lactide and PDLA is a polymer of D-lactide, and there is a co-polymer form PLDLA and other variations. PLAs are biocompatible and degrade hydrolytically, i.e. in response to the ingress of water molecules; PLLA is semi-crystalline and therefore least permeable to water and less degradable than the effectively amorphous PLDA and PLDLA. These characteristics have resulted in the widespread use of PLAs in medical applications, such as implants (pins, screws, plates, etc.), due to the ability to degrade over the timescale of a year. It is non-toxic and biocompatible with good mechanical properties at body temperature. They are also used to make disposable items such as drinking cups, as they are biodegradable. However, this capability is only realized via an industrial composting process, and if released into the environment PLA may persist over decades and form microparticles in seawater (Haider et al. 2019).

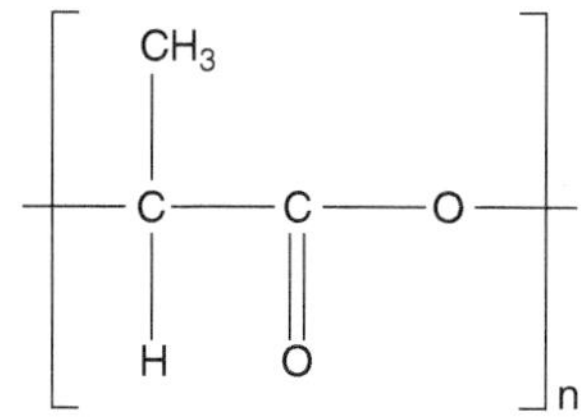

Figure 1.8 The chemical structure of poly lactic acid.

1.1.8 Blends, Grafts and Copolymers

A *blend* is a physical mixture of two or more polymers. A *graft* is formed when long side chains of a second polymer are chemically attached to the base polymer. A *copolymer* is formed when chemical combination exists in the main chain between two or more polymers, $_n$, $[B]_n$ and so on. The two principal forms are block copolymers ([AAAA. . .] [BBB. . .]) and *random* copolymers, the latter having no long sequences of A or B units.

All these processes are commonly used to enhance the ductility and toughness of brittle homopolymers or increase the stiffness of rubbery polymers. An example of a blend is acrylonitrile–butadiene–styrene copolymer (ABS), where the separate rubber phase gives much improved impact resistance.

The basic properties of polymers may be enhanced by physical as well as chemical means. An important example is the use of finely divided carbon black as a filler in rubber compounds. Polymers may be combined with stiffer filaments, such as glass and carbon fibres, to form a composite. We shall show later that some semi-crystalline polymers may be treated as composites at a molecular level.

It must not be forgotten that all useful polymers contain small quantities of additives to aid processing and increase the resistance to degradation. The physical properties of the base polymer may be modified by the presence of such additives.

1.2 Physical Structure

The physical properties of a polymer of a given chemical composition are dependent on two distinct aspects of the arrangement of the molecular chains in space.

1) The arrangement of a single chain without regard to its neighbour: rotational isomerism.
2) The arrangement of chains with respect to each other: orientation and crystallinity.

1.2.1 Rotational Isomerism

Rotational isomerism arises because of the alternative conformations of a molecule that can result from the possibility of hindered rotation about the many single bonds in the structure. Spectroscopic techniques (Mizushima 1954) developed in small molecules have been extended to polymers, and as an example we illustrate (Figure 1.9) the alternative trans and gauche conformations in the glycol residue of polyethylene terephthalate (Grime and Ward 1958): the former is a crystalline conformation, but the latter is present in amorphous regions.

To pass from one rotational isomeric form to another requires that an energy barrier be surmounted (Figure 1.10), so that the possibility of the chain molecules changing their conformations depends on the relative magnitude of the energy barrier compared with thermal energies and the perturbing effects of applied stress. Hence, arises the possibility of linking molecular flexibility to deformation mechanisms, a theme to which we will return on several occasions.

1.2.2 Orientation and Crystallinity

When we consider the arrangement of molecular chains with respect to each other, there are again two largely separate aspects, those of molecular orientation and crystallinity. In semi-crystalline polymers, this distinction may at times be an artificial one.

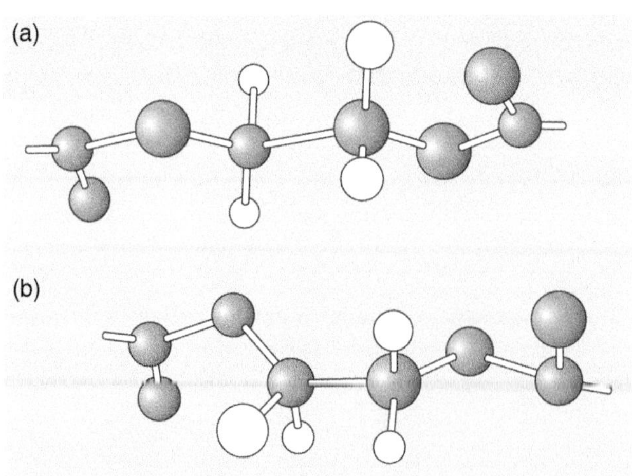

Figure 1.9 Polyethylene terephthalate in the crystalline trans conformation (a) and in the gauche conformation present in 'amorphous' regions (b). Grime, D. and Ward, I.M. (1958) / Royal Society of Chemistry.

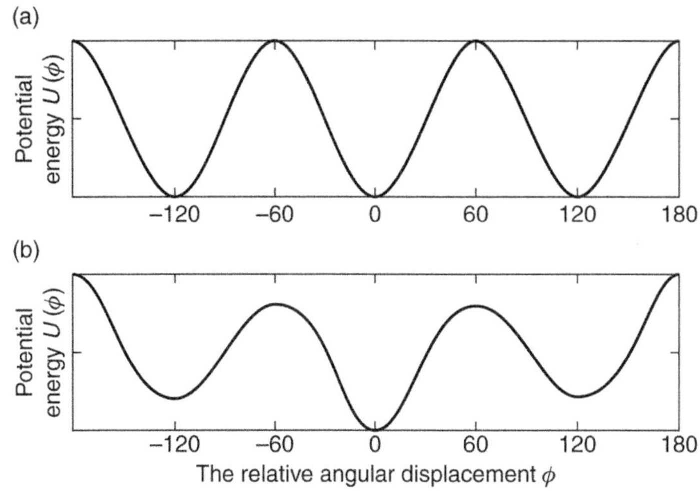

Figure 1.10 Potential energy for rotation (a) around the C—C bond in ethane and (b) around the central C—C bond in n-butane. Adapted from McCrum et al., 1991.

When cooled from the melt, many polymers form a disordered structure called the amorphous state. Some of these materials, such as polymethyl methacrylate, polystyrene and rapidly cooled (melt-quenched) polyethylene terephthalate, have a comparatively high modulus at room temperature, but others, such as natural rubber and atactic polypropylene, have a low modulus. These two types of polymer are often termed glassy and rubberlike, respectively, and we shall see that the form of behaviour exhibited depends on the temperature relative to a glass–rubber transition temperature (T_g) that is dependent on the material and the test method employed. Although an amorphous polymer may be modeled as a random tangle of molecules (Figure 1.11(a)), features such as the comparatively high density (Robertson 1965) show that the packing cannot be completely random. X-ray diffraction techniques indicate no distinct structure, rather a broad diffuse maximum (the amorphous halo) that indicates a preferred distance of separation between the molecular chains.

When an amorphous polymer is stretched, the molecules may be preferentially aligned along the stretch direction. In polymethyl methacrylate and polystyrene, such molecular orientation may be detected by optical methods, which measure the small difference between the refractive index in the stretch direction and that in the perpendicular direction. X-ray diffraction methods for relaxed amorphous polymers still reveal no evidence of three-dimensional order, so the structure may be regarded as a somewhat oriented tangled skein (Figure 1.11(b)) that is oriented amorphous but not crystalline.

In polyethylene terephthalate, however, stretching produces both molecular orientation and small regions of three-dimensional order, termed crystallites, because the orientation processes have brought the molecules into adequate juxtaposition for regions of three-dimensional order to form.

Many polymers, including polyethylene terephthalate, also crystallize if they are cooled slowly from the melt. In this case, we may say that they are crystalline but unoriented.

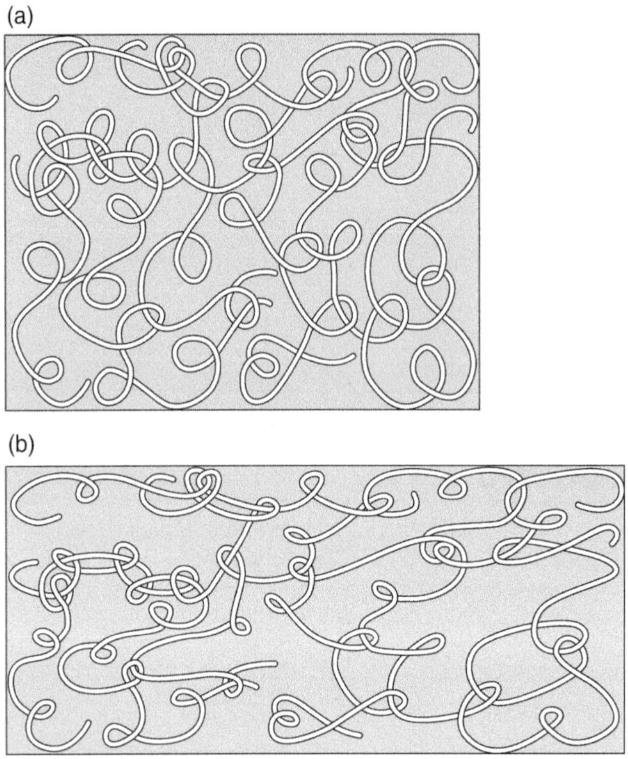

Figure 1.11 Schematic diagrams of (a) unoriented amorphous polymer and (b) oriented amorphous polymer.

Although such specimens are unoriented in the macroscopic sense, that is, they possess isotropic bulk mechanical properties, they are not homogeneous in the microscopic sense and often show a spherulitic structure under a polarizing microscope.

In summary, it may be said that for a polymer to crystallize the molecule must have a regular structure, the temperature must be below the crystal melting point and sufficient time must be available for the long molecules to become ordered in the solid state.

The structure of the crystalline regions of polymers can be deduced from wide-angle X-ray diffraction patterns of highly stretched specimens. When the stretching is uniaxial, the patterns are related to those obtained from fully oriented single crystals. The crystal structure of polyethylene is well established as determined by Bunn (1939) (Figure 1.12).

In addition to the discrete reflections from the crystallites, the diffraction pattern of a polymer shows diffuse scattering attributed to amorphous regions. Such polymers are said to be semi-crystalline, with the crystalline fraction being controlled by molecular regularity. By comparing the relative amounts of crystalline and amorphous scattering of X-rays, the crystallinity has been found to vary from more than 90% for linear polyethylene to about 30% for oriented polyethylene terephthalate.

The first model (Hill 1953) to describe the structure of a semi-crystalline polymer was the so-called fringed micelle model (Figure 1.13), which is a natural development of the

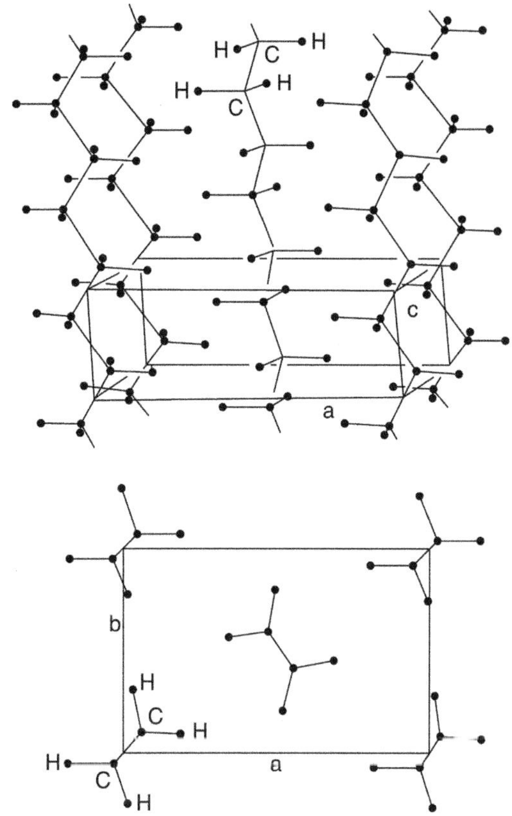

Figure 1.12 Arrangement of molecules in polyethylene crystallites. Adapted from Hill, 1953.

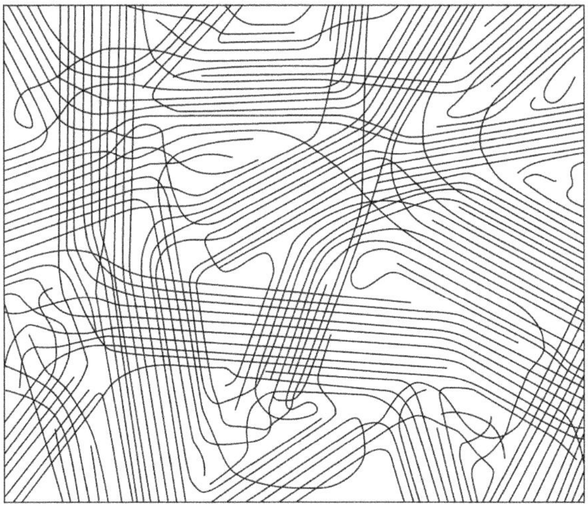

Figure 1.13 The fringed micelle representation of crystalline polymers. Adapted from Hill, 1953.

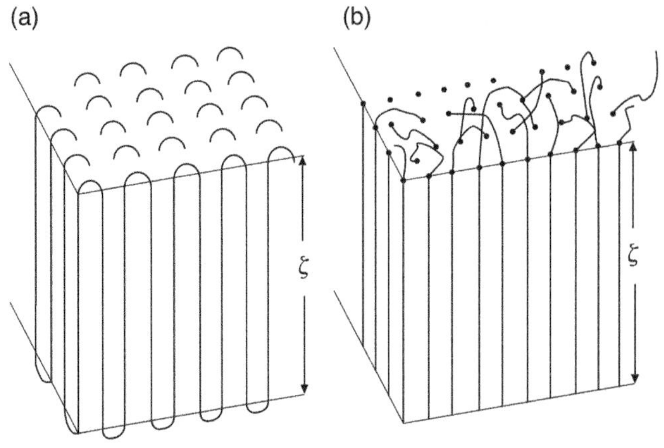

Figure 1.14 Crystallites with folded lamellar crystals of thickness ξ in the direction of the c axis for (a) regular folding and (b) irregular folding of the chain molecules.

imagined situation in an amorphous polymer. The molecular chains alternate between regions of order (the crystallites) and disorder (the amorphous regions).

The fringed micelle model was called into question by the discovery of polymer single crystals grown from solution (Fischer 1957; Keller 1957; Till 1957). Linear polyethylene, for example forms single-crystal lamellae with lateral dimensions of the order of 10–20 μm and thickness of 10 nm. Electron diffraction shows that the molecular chains are oriented approximately normal to the lamellar surface. As the molecules are typically about 1 μm in length, it follows that they must be folded back and forth within the crystals. The simplest geometric arrangement is that the folds are sharp and regular producing the adjacent re-entry model shown in Figure 1.14(a). This model provoked controversy, and an alternative switchboard model shown schematically in Figure 1.14(b) was proposed (Flory 1962).

The crystallization of polymers from the melt has proved even more controversial, as a single molecule is unlikely to be laid down on a crystalline substrate without interference from its neighbours, and it might be expected that the highly entangled topology of the chains that exists in the melt would be substantially retained in the crystalline state. These issues were explored to great effect by neutron-scattering measurements, notably by Fischer (1978), Stamm *et al.* (1979, 1981). The neutron-scattering measurements showed that the radii of gyration in the melt and in the semi-crystalline state for polyethylene quenched from the melt, for polyethylene oxide crystallized by slow cooling and for isotactic polypropylene, isothermally crystallized, were almost identical. On the basis of these results, Fischer (1978) proposed the solidification model shown in Figure 1.15, where straight sequences of the original melt are incorporated into the growing lamellae without long-range diffusional motion.

Although it is accepted that kinetic factors determine the growth rate of crystallization and the morphology, there is still debate in this area also. The theory proposed by Lauritzen and Hoffman (1960) led the field and predicted the growth rate as the function of the degree of supercooling, the temperature difference between the crystallization temperature and the melting point. It was assumed that the free-energy barrier associated with nucleation of the

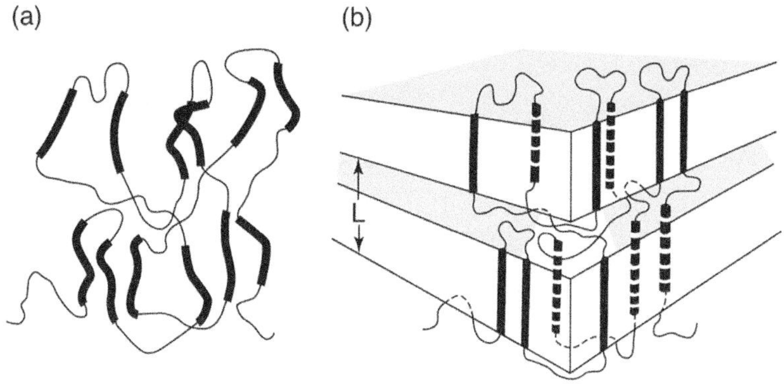

Figure 1.15 Chain conformation (a) in the melt and (b) in the crystal according to the solidification model. Stamm et al. (1979) / Royal Society of Chemistry.

crystallization was energetic in origin. An alternative model for chain folding in polymer crystals has been proposed by Sadler and Gilmer (1984), which assumes that the free-energy barrier for nucleation is predominantly entropic. For a comprehensive review of these theories and related issues, the reader is referred to an excellent review by Gedde (1995) and also to Strobl (1997).

There is, of course, much evidence to support the existence of a lamellar morphology in crystalline polymers. Typically, spherulites 1–10 μm in diameter are formed, which grow outwards until they impinge upon neighbouring spherulites (Figure 1.16). The spherulitic textures are formed by the growth of dominant lamellae from a central nucleus in all directions by a twisting of these lamellae along the fibrils, the intervening spaces being filled in by subsidiary lamellae, possibly due to low-molecular-weight material. This is shown schematically in Figure 1.17, where, for ease of illustration, regular chain folding is sketched (Takayanagi 1963). For a good review of polymer morphology, see the text by Bassett (1981) and also more recent work directed by the same author.

Figure 1.16 A photograph of typical spherulitic structure under a polarizing microscope.

Orientation through plastic deformation (drawing) destroys the spherulitic structure. What remains is determined to a large extent by the degree of crystallinity. Mechanical testing, described in the subsequent chapters, has helped to establish several models. At one extreme, some highly oriented, highly

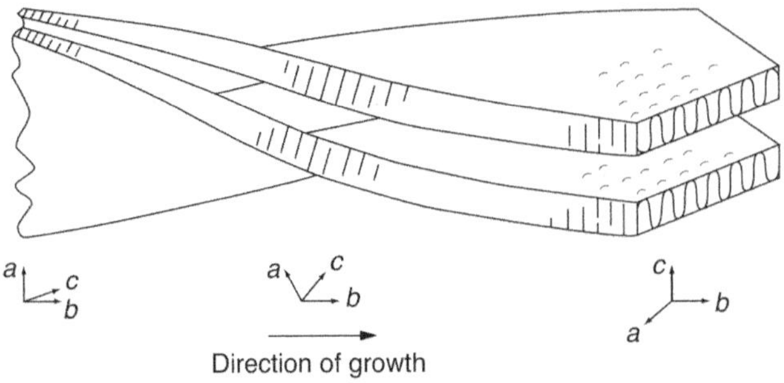

Figure 1.17 A model of the lamellar arrangement in a polyethylene spherulite. The small diagrams of the *a*, *b*, *c* axes show the orientation of the unit cell at various points. Adapted from Takayanagi, 1963.

crystalline specimens of linear polyethylene behave as blocks or lamellae of crystalline material, connected together by tie molecules or crystalline bridges and separated by the amorphous component. Such materials in some respects can be treated as microscopic composites. At the other extreme one has materials such as polyethylene terephthalate in which the crystalline and amorphous components are so intermixed that a single-phase model appears to be more appropriate.

The current state of knowledge suggests that both chain folding and the threading of molecules through the crystalline region occur in typical polymers.

A schematic attempt (Hosemann 1962) to illustrate this situation, and other types of irregularity, is given in Figure 1.18.

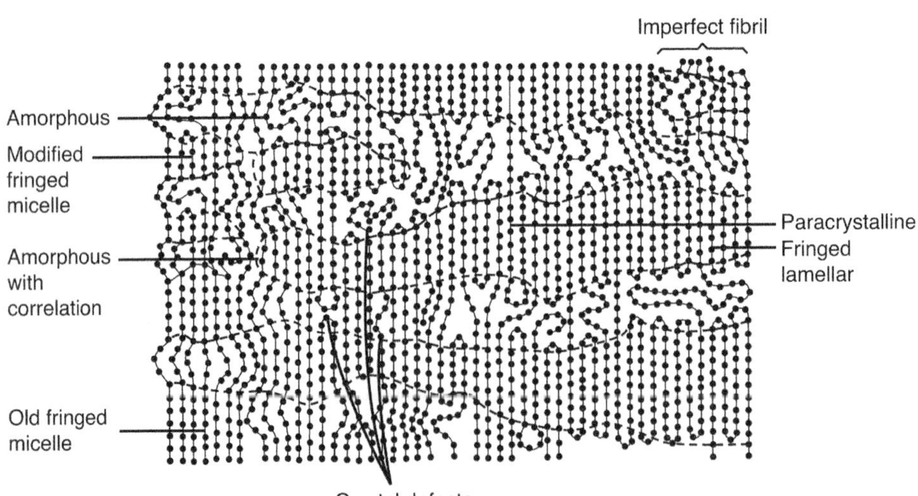

Figure 1.18 Schematic composite diagram of different types of order and disorder in oriented polymers. Adapted from Hosemann, 1962.

References

Bassett, D.C. (1981) *Principles of Polymer Morphology*, Cambridge University Press, Cambridge.

Bunn, C.W. (1939) The crystal structure of long-chain normal paraffin hydrocarbons: the "shape" of the >CH2 group. *Trans. Faraday Soc.*, **35**, 482.

Fischer, E.W. (1957) Stufen- und spiralförmiges Kristallwachstum bei Hochpolymeren. *Naturforschung*, **12a**, 753.

Fischer, E.W. (1978) Studies of structure and dynamics of solid polymers by elastic and inelastic neutron scattering. *Pure Appl. Chem.*, **50**, 1319.

Flory, P.J. (1962) On the morphology of the crystalline state in polymers. *J. Amer. Chem. Soc.*, **84**, 2857.

Gedde, U.W. (1995) *Polymer Physics*, Chapman and Hall, London.

Grime, D. and Ward, I.M. (1958) The assignment of infra-red absorptions and rotational isomerism in polyethylene terephthalate and related compounds. *Trans. Faraday Soc.*, **54**, 959.

Haider, T.P., Völker, C., Kramm, J., Landfester, K. and Wurm, F.R. (2019) Plastics of the future? Impact of biodegradable polymers on the environment and on society. *Angew. Chem. Int. Ed.*, **58**, 50–62.

Hill, R. (ed.). (1953) *Fibres from Synthetic Polymers*, Elsevier, Amsterdam.

Hosemann, R. (1962) Crystallinity in high polymers, especially fibres. *Polymer*, **3**, 393.

Keller, A. (1957) A note on single crystals in polymers: evidence for a folded chain configuration. *Philos. Mag.*, **2**, 1171.

Lauritzen, J.I. and Hoffman, J.D. (1960) Theory of formation of polymer crystals with folded chains in dilute solution. *J. Res. Nat. Bur. Std.*, **64A**, 73.

McCrum, N.G., Read, B.E. and Williams, G. (1991) *Anelastic and Dielectric Effects in Polymeric Solids*, Dover Publications, New York.

Mizushima, S.I. (1954) *Structure of Molecules and Internal Rotation*, Academic Press, New York.

Moore, J.C. (1964) Gel permeation chromatography. I. A new method for molecular weight distribution of high polymers. *J. Polym. Sci.: Part A*, **2**, 835.

Noël, C. and Navard, P. (1991) Liquid crystal polymers. *Prog. Polym. Sci.*, **16**, 55.

Robertson, R.E. (1965) Polymer order and polymer density. *J. Phys. Chem.*, **69**, 1575.

Sadler, D.M. and Gilmer, G.H. (1984) A model for chain folding in polymer crystals: rough growth faces are consistent with the observed growth rates. *Polymer*, **25**, 1446.

Stamm, M., Fischer, E.W., Dettenmaier, M. and Convert, P. (1979) Chain conformation in the crystalline state by means of neutron scattering methods. *Discuss. Chem. Soc. (London)*, **68**, 263.

Stamm, M., Schelten, J. and Ballard, D.G.H. (1981) Determination of the chain conformation of polypropylene in the crystalline state by neutron scattering. *Colloid Polym. Sci.*, **259**, 286.

Strobl, G. (1997) *The Physics of Polymers*, 2nd edn, Springer, Berlin.

Takayanagi, M. (1963) Viscoelastic properties of crystalline polymers. *Mem. Fac. Eng. Kyushu Univ.*, **23**, 1.

Till, P.H. (1957) The growth of single crystals of linear polyethylene. *J. Polym. Sci.*, **24**, 301.

Vaughan, M.F. (1960) Fractionation of polystyrene by gel filtration. *Nature*, **188**, 55.

Further Reading

Billmeyer, F.W. (1984) *Textbook of Polymer Science*, 3rd edn, John Wiley & Sons, New York.

Bower, D.I. (2002) *An Introduction to Polymer Physics*, Cambridge University Press, Cambridge.

Cowie, J.M.G. and Arrighi, V. (2008) *Polymers: Chemistry and Physics of Modern Materials*, 3rd edn, Taylor & Francis Group, Boca Raton, FL.

Hamley, I.W. (1998) *The Physics of Block Copolymers*, Oxford University Press, Oxford.

Mark, J., Ngai, K., Graessley, W. *et al.* (2004) *Physical Properties of Polymers*, 3rd edn, Cambridge University Press, Cambridge.

Odian, G.G. (2004) *Principles of Polymerization*, 4th edn, Wiley-Interscience, Hoboken, NJ.

Painter, P.C. and Coleman, M.M. (2008) *Essentials of Polymer Science and Engineering*, DEStech Publications, Lancaster Pennsylvania.

Rubinstein, M. and Colby, R.H. (2003) *Polymer Physics*, Oxford University Press, Oxford.

Sperling, L. (2006) *Introduction to Physical Polymer Science*, 4th edn, John Wiley & Sons, New York.

Tadokoro, H. (1979) *Structure of Crystalline Polymers*, John Wiley & Sons, Ltd, New York.

van Krevelen, D.W. and te Nijenhuis, K. (2009) *Properties of Polymers*, 4th edn, Elsevier, Oxford.

Wunderlich, B. *Macromolecular Physics, Vols 1 1973 Vol 2 1976, Vol 3 1980*, Academic Press, New York.

Young, R.J. and Lovell, P.A. (2011) *Introduction to Polymers*, 3rd edn, CRC Press/Taylor & Francis Group, Boca Raton, FL.

Sreekala, M.S., Ravindran, L., Goda, K. and Thomas, S. (2023) *Handbook of Natural Polymers, Volume 1: Sources, Synthesis, and Characterization*, Elsevier, Oxford.

Sreekala, M.S., Ravindran, L., Goda, K. and Thomas, S. (2024) *Handbook of Natural Polymers, Volume 2: Functionalization, Surface Modification, and Properties*, Elsevier, Oxford.

Olatunji, O. (2018) *Natural Polymers: Industry Techniques and Applications*, Springer International.

2

The Mechanical Properties of Polymers: General Considerations

2.1 Objectives

Discussions of the mechanical properties of solid polymers often contain two inter-related objectives. The first of these is to obtain an adequate macroscopic description of the particular facet of polymer behaviour under consideration. The second objective is to seek an explanation of this behaviour in molecular terms, which may include details of the chemical composition and physical structure. In this book, we will endeavour, where possible, to separate these two objectives and, in particular, to establish a satisfactory macroscopic or phenomenological description before discussing molecular interpretations.

This should make it clear that many of the established relationships are purely descriptive and do not necessarily have any implications with regard to an interpretation in structural terms. For engineering applications of polymers this is sufficient, because a description of the mechanical behaviour under conditions that simulate their end use is often all that is required, together with empirical information concerning their method of manufacture.

2.2 The Different Types of Mechanical Behaviour

It is difficult to classify polymers as particular types of materials such as a glassy solid or a viscous liquid, since their mechanical properties are so dependent on the conditions of testing, for example the rate of application of load, temperature and amount of strain.

A polymer can show all the features of a glassy, brittle solid or an elastic rubber or a viscous liquid depending on the temperature and time scale of measurement. Polymers are usually described as viscoelastic materials, a generic term which emphasizes their intermediate position between viscous liquids and elastic solids. At low temperatures, or high frequencies of measurement, a polymer may be glass-like with a Young's modulus of 1–10 GPa and will break or flow at strains greater than 5%. At high temperatures or low frequencies, the same polymer may be rubber-like with a modulus of 1–10 MPa, withstanding large extensions ($\sim$100%) without permanent deformation. At still higher temperatures, permanent deformation occurs under load, and the polymer behaves like a highly viscous liquid.

Mechanical Properties of Solid Polymers, Fourth Edition. John Sweeney and Peter Hine.

In an intermediate-temperature or -frequency range, commonly called the glass transition range, the polymer is neither glassy nor rubber-like. It shows an intermediate modulus, is viscoelastic and may dissipate a considerable amount of energy on being strained. The glass transition manifests itself in several ways, for example by a change in the volume coefficient of expansion, which can be used to define a glass transition temperature T_g. The glass transition is central to a great deal of the mechanical behaviour of polymers for two reasons. First there are the attempts to link the time–temperature equivalence of viscoelastic behaviour with the glass transition temperature T_g. Secondly, glass transitions can be studied at a molecular level by such techniques as nuclear magnetic resonance and dielectric relaxation. In this way, it is possible to gain an understanding of the molecular origins of the viscoelasticity.

The different features of polymer behaviour such as creep and recovery, brittle fracture, necking and cold drawing are usually considered separately, by comparative studies of different polymers. It is customary, for example, to compare the brittle fracture of poly-methyl methacrylate, polystyrene and other polymers, which show similar behaviour at room temperature. Similarly comparative studies have been made of the creep and recovery of polyethylene, polypropylene and other polyolefins. Such comparisons often obscure the very important point that the whole range of phenomena can be displayed by a single polymer as the temperature is changed. Figure 2.1 shows load–elongation curves for a polymer at four different temperatures. At temperatures well below the glass transition (curve A), where brittle fracture occurs, the load rises to the breaking point linearly with increasing elongation, and rupture occurs at low strains (−10%). At high temperatures (curve D), the polymer is rubber-like and the load rises to the breaking point with a sigmoidal relationship to the elongation, and rupture occurs at very high strains (∼30–1000%).

In an intermediate-temperature range below the glass transition (curve B), the load–deformation relationship resembles that of a ductile metal, showing a load maximum, i.e. a yield point before rupture occurs. At slightly higher temperatures (curve C), still below

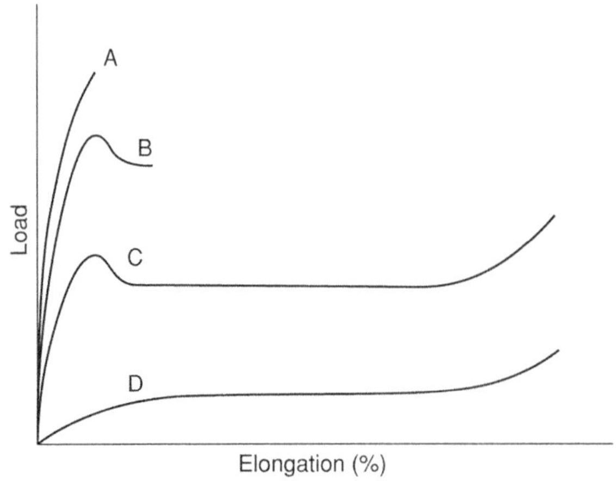

Figure 2.1 Load–elongation curves for a polymer at different temperatures. Curve A, brittle fracture; curve B, ductile failure; curve C, cold drawing; curve D, rubber-like behaviour.

the glass transition, the remarkable phenomenon of necking and cold drawing is observed. Here, the conventional load–elongation curve again shows a yield point and a subsequent decrease in conventional stress. However, with a further increase in the applied strain, the load falls to a constant level at which deformations of the order of 300–1000% are accomplished. At this stage, a neck has formed and the strain in the specimen is not uniform. (This is discussed in detail in Chapter 12.) Eventually, the load begins to rise again and finally fracture occurs.

It is usual to discuss the mechanical properties in the different temperature ranges separately, because different approaches and mathematical formalisms are adopted for the different features of mechanical behaviour. This conventional treatment will be followed here, although it is recognized that it somewhat arbitrarily isolates particular facets of the mechanical properties of polymers.

2.3 The Elastic Solid and the Behaviour of Polymers

Mechanical behaviour is, in most general terms, concerned with the deformations which occur under loading. In any specific case, the deformations depend on details such as the geometrical shape of the specimen or the way in which the load is applied. Such considerations are the province of the plastics engineer, who is concerned with predicting the performance of a polymer in a specified end use. In our discussion of the mechanical properties of polymers, we will ignore such questions as these, which relate to solving particular problems of behaviour in practice. We will concern ourselves only with the generalized equations termed *constitutive relations*, which relate stress and strain for a particular type of material. First it will be necessary to find constitutive relations that give an adequate description of the mechanical behaviour. Secondly, where possible, we will obtain a molecular understanding of this behaviour by a molecular model that predicts the constitutive relations.

One of the simplest constitutive relations is Hooke's law, which relates the stress σ to the strain e for the uniaxial deformation of an ideal elastic isotropic solid. Thus

$$\sigma = Ee,$$

where E is Young's modulus.

There are five important ways in which the mechanical behaviour of a polymer may deviate from that of an ideal elastic solid obeying Hooke's law. First, in an elastic solid the deformations induced by loading are independent of the history or rate of application of the loads, whereas in a polymer the deformations can be drastically affected by such considerations. This means that the simplest constitutive relation for a polymer should in general contain time or frequency as a variable in addition to stress and strain. Secondly, in an elastic solid all the situations pertaining to stress and strain can be reversed. Thus, if a stress is applied, a certain deformation will occur. On removal of the stress, this deformation will disappear exactly. This is not always true for polymers. Thirdly, in an elastic solid obeying Hooke's law, which in its more general implications is the basis of small-strain elasticity theory, the effects observed are *linearly* related to the influences applied. This is the essence of Hooke's law; stress is exactly proportional to strain. This is not generally true for polymers, but applies in many cases only as a good approximation for very small strains; in general, the constitutive relations are non-linear. It is important to note that *non-linearity* is not related to *recoverability*. In contrast to metals, polymers may recover from strains beyond the proportional limit without any permanent deformation.

Fourthly, the definitions of stress and strain in Hooke's law are only valid for small deformations. When we wish to consider larger deformations a new theory must be developed in which both stress and strain are defined more generally.

Finally, in many practical applications (such as films and synthetic fibres) polymers are used in an oriented or anisotropic form, which requires a considerable generalization of Hooke's law.

It will be convenient to discuss these various aspects separately as follows: (1) behaviour at large strains in Chapters 3 and 4 (finite elasticity and rubber-like behaviour, respectively); (2) time-dependent behaviour in Chapters 5–7 and 10 (viscoelastic behaviour); (3) the behaviour of oriented polymers in Chapters 8 and 9 (mechanical anisotropy); (4) non-linearity in Chapter 11 (non-linear viscoelastic behaviour); (5) the non-recoverable behaviour in Chapter 12 (plasticity and yield) and (6) fracture in Chapter 13 (breaking phenomena). However, it should be recognized that we cannot hold to an exact separation and that there are many places where these aspects overlap and can be brought together by the physical mechanisms, which underlie the phenomenological description.

2.4 Stress and Strain

It is desirable at this juncture to outline very briefly the concepts of stress and strain. For a more comprehensive discussion, the reader is referred to standard textbooks on the theory of elasticity (Love 1944; Bedford and Liechti 2000; Richards 2001; Solecki and Conant 2003; Hibbeler 2004; Timoshenko and Goodier 2008).

2.4.1 The State of Stress

The components of stress in a body can be defined by considering the forces acting on an infinitesimal cubical volume element (Figure 2.2) whose edges are parallel to the coordinate axes 1, 2 and 3. In equilibrium, the forces per unit area acting on the cube faces are:

P_x on the 23 plane,
P_y on the 31 plane,
P_z on the 12 plane.

These three forces are then resolved into their nine components in the 1, 2 and 3 directions as follows:

$$P_x : \sigma_{11}, \quad \sigma_{12}, \quad \sigma_{13};$$
$$P_y : \sigma_{21}, \quad \sigma_{22}, \quad \sigma_{23};$$
$$P_z : \sigma_{31}, \quad \sigma_{32}, \quad \sigma_{33}.$$

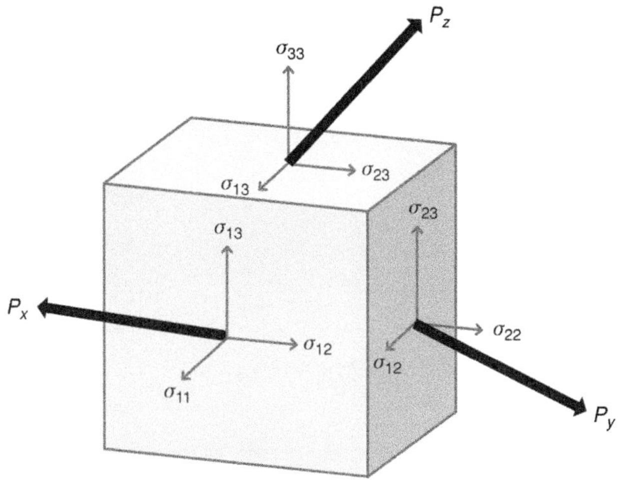

Figure 2.2 The stress components.

The first subscript refers to the direction of the normal to the plane on which the stress acts, and the second subscript to the direction of the stress. In the absence of body torques, the total torque acting on the cube must also be zero, and this implies three equalities:

$$\sigma_{12} = \sigma_{21}, \quad \sigma_{13} = \sigma_{31}, \quad \sigma_{23} = \sigma_{32}.$$

Therefore, the components of stress are defined by six independent quantities: σ_{11}, σ_{22} and σ_{33}, the normal stresses, and σ_{12}, σ_{23} and σ_{31}, the shear stresses.

These form the six independent components of the stress tensor Σ or σ_{ij}:

$$\Sigma = \sigma_{ij} = \begin{pmatrix} \sigma_{11} & \sigma_{12} & \sigma_{13} \\ \sigma_{12} & \sigma_{22} & \sigma_{23} \\ \sigma_{13} & \sigma_{23} & \sigma_{33} \end{pmatrix}.$$

The state of stress at a point in a body is determined when we can specify the normal components and the shear components of stress acting on a plane drawn in any direction through the point. If we know these six components of stress at a given point, the stresses acting on any plane through this point can be calculated. (See References Timoshenko and Goodier 2008, Section 67; and Love 1944, Section 47.)

2.4.2 The State of Strain – Engineering Components

The displacement of any point $\mathbf{X}$ (see Figure 2.3) in the body may be resolved into its components u_1, u_2 and u_3 parallel to 1, 2 and 3 (Cartesian coordinate axes chosen in the undeformed state) so that if the coordinates of the point in the undisplaced position were (X_1, X_2, X_3), they become $(X_1 + u_1, X_2 + u_2, X_3 + u_3)$ on deformation. In defining the strains, we are not interested in the *displacement* or rotation but in the *deformation*. The latter is the displacement of a point relative to adjacent points. Consider a point very close to $\mathbf{X}$,

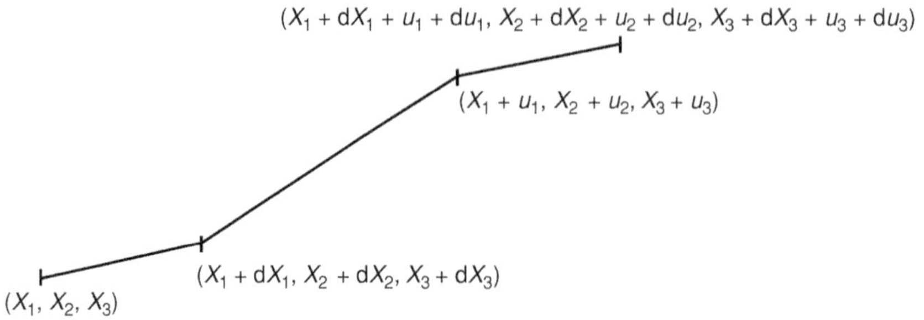

Figure 2.3 The displacements produced by deformation.

which in the undisplaced position had coordinates $(X_1 + dX_1, X_2 + dX_2, X_3 + dX_3)$ and let the displacement, which it has undergone, have components $(u_1 + du_1, u_2 + du_2, u_3 + du_3)$. The quantities required are then du_1, du_2 and du_3, the *relative* displacements.

If dX_1, dX_2 and dX_3 are sufficiently small, that is infinitesimal:

$$du_1 = \frac{\partial u_1}{\partial X_1}dX_1 + \frac{\partial u_1}{\partial X_2}dX_2 + \frac{\partial u_1}{\partial X_3}dX_3,$$

$$du_2 = \frac{\partial u_2}{\partial X_1}dX_1 + \frac{\partial u_2}{\partial X_2}dX_2 + \frac{\partial u_2}{\partial X_3}dX_3,$$

$$du_3 = \frac{\partial u_3}{\partial X_1}dX_1 + \frac{\partial u_3}{\partial X_2}dX_2 + \frac{\partial u_3}{\partial X_3}dX_3.$$

Thus, we require to define the nine quantities:

$$\frac{\partial u_1}{\partial X_1}, \frac{\partial u_1}{\partial X_2}, \ldots, \text{etc.}$$

For convenience, these nine quantities are regrouped and denoted as follows:

$$e_{11} = \frac{\partial u_1}{\partial X_1}, e_{22} = \frac{\partial u_2}{\partial X_2}, e_{33} = \frac{\partial u_3}{\partial X_3},$$

$$e_{23} = \frac{\partial u_3}{\partial X_2} + \frac{\partial u_2}{\partial X_3}, e_{31} = \frac{\partial u_1}{\partial X_3} + \frac{\partial u_3}{\partial X_1}, e_{12} = \frac{\partial u_2}{\partial X_1} + \frac{\partial u_1}{\partial X_2},$$

$$2\varpi_1 = \frac{\partial u_3}{\partial X_2} - \frac{\partial u_2}{\partial X_3}, 2\,\varpi_2 = \frac{\partial u_1}{\partial X_3} - \frac{\partial u_3}{\partial X_1}, 2\varpi_3 = \frac{\partial u_2}{\partial X_1} - \frac{\partial u_1}{\partial X_2}.$$

The first three quantities e_{11}, e_{22} and e_{33} correspond to the fractional expansions or contractions along the 1, 2 and 3 axes of an infinitesimal element at **X** – the normal strains. The second three quantities e_{23}, e_{31} and e_{12} correspond to the components of shear strain in the 23, 31 and 12 planes, respectively. The last three quantities ϖ_1, ϖ_2 and ϖ_3 do not correspond to a deformation of the element at **X**, but are the components of its rotation as a rigid body.

The concept of shear strain can be conveniently illustrated by a diagram showing the two-dimensional situation of shear in the 23 plane (see Figure 2.4). ABCD is an infinitesimal square that has been displaced and deformed into the rhombus A′B′C′D′, θ_1 and θ_2 being the angles that A′D′ and A′B′ make with the 2 and 3 axes, respectively.

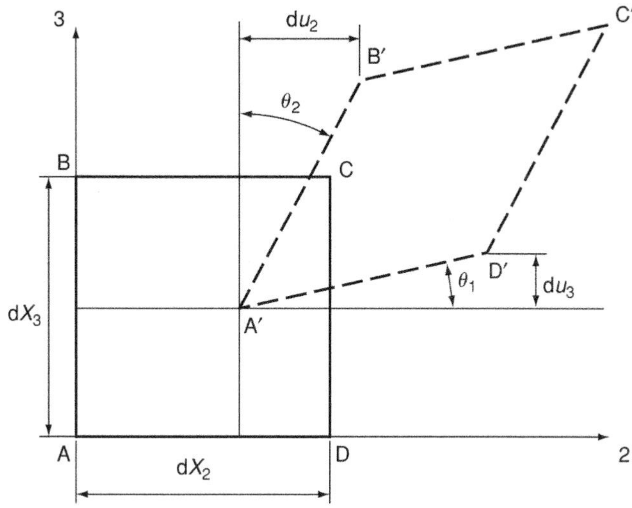

Figure 2.4 Shear strains.

Now,

$$\tan\theta_1 \approx \theta_1 = \frac{du_3}{dX_2} \rightarrow \frac{\partial u_3}{\partial X_2},$$

$$\tan\theta_2 \approx \theta_2 = \frac{du_2}{dX_3} \rightarrow \frac{\partial u_2}{\partial X_3}.$$

The shear strain in the 23 plane is given by

$$e_{23} = \frac{\partial u_3}{\partial X_2} + \frac{\partial u_2}{\partial X_3} = \theta_1 + \theta_2.$$

$2\varpi_1 = \theta_1 - \theta_2$ does not correspond to a deformation of ABCD but to twice the angle through which AC has been rotated.

Therefore, the deformation is defined by the first six quantities e_{11}, e_{22}, e_{33}, e_{23}, e_{31} and e_{12} that are called the components of strain. It is important to note that engineering strains have been defined. In Chapter 3, we take a more general approach and examine a number of strain-related tensor quantities in Section 3.1.5. For the purposes of this chapter, which concerns small strains, we define the strain tensor ε_{ij} as

$$\varepsilon_{ij} = \frac{1}{2}\left(\frac{\partial u_i}{\partial X_j} + \frac{\partial u_j}{\partial X_i}\right),$$

where i and j take values 1, 2 and 3, and we sum over all possible values. Then,

$$\varepsilon_{ij} = \begin{pmatrix} e_{11} & \frac{1}{2}e_{12} & \frac{1}{2}e_{13} \\ \frac{1}{2}e_{12} & e_{22} & \frac{1}{2}e_{23} \\ \frac{1}{2}e_{13} & \frac{1}{2}e_{23} & e_{33} \end{pmatrix}$$

in terms of the engineering components of strain.

2.5 The Generalized Hooke's Law

The most general *linear* relationship between stress and strain is obtained by assuming that each of the tensor components of stress is linearly related to all the tensor components of strain and vice versa. Thus

$$\sigma_{11} = a\varepsilon_{11} + b\varepsilon_{22} + c\varepsilon_{33} + d\varepsilon_{13} + \cdots \text{ etc.}$$

and

$$\varepsilon_{11} = a'\sigma_{11} + b'\sigma_{22} + c'\sigma_{33} + d'\sigma_{13} + \cdots \text{ etc.},$$

where a, b, ..., a', b', ... are constants. This is the generalized Hooke's law.

In tensor notation, we relate the second-rank tensor σ_{ij} to the second-rank strain tensor ε_{ij} by fourth-rank tensors $c_{ijk\ell}$ and $s_{ijk\ell}$. Thus,

$$\sigma_{ij} = c_{ijk\ell}\varepsilon_{k\ell}$$

or equivalently

$$\varepsilon_{ij} = s_{ijk\ell}\sigma_{k\ell},$$

where

$$\sigma_{ij} = \sigma_{11}, \sigma_{22}, ..., \text{etc.}$$

and

$$\varepsilon_{ij} = \varepsilon_{11}, \varepsilon_{22}, ..., \text{etc.}$$

The fourth-rank tensors $s_{ijk\ell}$ and $c_{ijk\ell}$ contain the compliance and stiffness constants, respectively, with i, j, k, ℓ taking values 1, 2 and 3.

It is customary to adopt an abbreviated nomenclature in which the generalized Hooke's law relates the six independent components of stress to the six independent components of the engineering strains.

In terms of finite strain, we have

$$\sigma_p = c_{pq}e_q \quad \text{and} \quad e_p = s_{pq}\sigma_q,$$

where σ_p represents σ_{11}, σ_{22}, σ_{33}, σ_{13}, σ_{23} or σ_{12} and e_p represents e_{11}, e_{22}, e_{33}, e_{13}, e_{23} or e_{12}. We form matrices c_{pq} and s_{pq} in which p and q take the values 1, 2, ..., 6. In the case of the stiffness constants, the values of p and q are obtained in terms of i, j, k, ℓ by substituting 1 for 11, 2 for 22, 3 for 33, 4 for 23, 5 for 13 and 6 for 12.

A typical relationship between stress and strain is now written as

$$\sigma_1 = c_{11}e_1 + c_{12}e_2 + c_{13}e_3 + c_{14}e_4 + c_{15}e_5 + c_{16}e_6.$$

The existence of a strain–energy function (see Love 1944, p. 149; Solecki and Conant 2003, p. 267) provides the relationships

$$c_{pq} = c_{qp}, \quad s_{pq} = s_{qp}$$

and reduces the number of independent constants from 36 to 21. We then have

$$c_{pq} = \begin{pmatrix} c_{11} & c_{12} & c_{13} & c_{14} & c_{15} & c_{16} \\ c_{12} & c_{22} & c_{23} & c_{24} & c_{25} & c_{26} \\ c_{13} & c_{23} & c_{33} & c_{34} & c_{35} & c_{36} \\ c_{14} & c_{24} & c_{34} & c_{44} & c_{36} & c_{46} \\ c_{15} & c_{25} & c_{35} & c_{36} & c_{55} & c_{56} \\ c_{16} & c_{26} & c_{36} & c_{46} & c_{56} & c_{66} \end{pmatrix}$$

and similarly

$$s_{pq} = \begin{pmatrix} s_{11} & s_{12} & s_{13} & s_{14} & s_{15} & s_{16} \\ s_{12} & s_{22} & s_{23} & s_{24} & s_{25} & s_{26} \\ s_{13} & s_{23} & s_{33} & s_{34} & s_{35} & s_{36} \\ s_{14} & s_{24} & s_{34} & s_{44} & s_{36} & s_{46} \\ s_{15} & s_{25} & s_{35} & s_{36} & s_{55} & s_{56} \\ s_{16} & s_{26} & s_{36} & s_{46} & s_{56} & s_{66} \end{pmatrix}.$$

These matrices define the relationships between stress and strain in a general elastic solid, whose properties vary with direction, that is an anisotropic elastic solid. In most of this book, we will be concerned with isotropic polymers; all discussion of anisotropic mechanical properties will be reserved for Chapter 8.

It is then most straightforward to use the compliance constants matrix, and note that measured quantities, such as Young's modulus E, Poisson's ratio ν and the torsional or shear modulus G, relate directly to the compliance constants.

For an isotropic solid, the matrix s_{pq} reduces to

$$s_{pq} = \begin{pmatrix} s_{11} & s_{12} & s_{12} & 0 & 0 & 0 \\ s_{12} & s_{11} & s_{12} & 0 & 0 & 0 \\ s_{12} & s_{12} & s_{11} & 0 & 0 & 0 \\ 0 & 0 & 0 & 2(s_{11}-s_{12}) & 0 & 0 \\ 0 & 0 & 0 & 0 & 2(s_{11}-s_{12}) & 0 \\ 0 & 0 & 0 & 0 & 0 & 2(s_{11}-s_{12}) \end{pmatrix}.$$

It can be shown that Young's modulus is given by

$$E = 1/s_{11},$$

Poisson's ratio by

$$\nu = -s_{12}/s_{11}$$

and the torsional modulus by

$$G = 1/2(s_{11} - s_{12}).$$

Thus, we obtain the stress–strain relationships, which are the starting point in many elementary textbooks of elasticity (see Timoshenko and Goodier 2008, pp. 7–9):

$$\begin{aligned}
e_{11} &= \frac{1}{E}(\sigma_{11} - \nu(\sigma_{22} + \sigma_{33})),\\
e_{22} &= \frac{1}{E}(\sigma_{22} - \nu(\sigma_{11} + \sigma_{33})),\\
e_{33} &= \frac{1}{E}(\sigma_{33} - \nu(\sigma_{11} + \sigma_{22})),\\
e_{13} &= \frac{1}{G}\sigma_{13},\\
e_{23} &= \frac{1}{G}\sigma_{23},\\
e_{12} &= \frac{1}{G}\sigma_{12},
\end{aligned} \tag{2.1}$$

where

$$G = \frac{E}{2(1 + \nu)}.$$

Another basic quantity is the bulk modulus K, which determines the dilatation $\Delta = e_{11} + e_{22} + e_{33}$, produced by a uniform hydrostatic pressure. Using the stress–strain relationships above, it can be shown that the strains produced by a uniform hydrostatic pressure p are given by

$$e_{11} = e_{22} = e_{33} = (s_{11} + 2s_{12})p.$$

Then,

$$K = \frac{p}{\Delta} = \frac{1}{3(s_{11} + 2s_{12})} = \frac{E}{3(1 - 2\nu)}.$$

Useful problems can be solved using a two-dimensional approach. Working in the 1–2 plane, two approaches are possible: plane stress with $\sigma_{33} = 0$; and plane strain, with $e_{33} = 0$. In the first case equations (2.1) reduce to

$$\begin{aligned}
e_{11} &= \frac{1}{E}(\sigma_{11} - \nu\sigma_{22}),\\
e_{22} &= \frac{1}{E}(\sigma_{22} - \nu\sigma_{11}),\\
e_{33} &= -\frac{\nu}{E}(\sigma_{11} + \sigma_{22}),\\
e_{12} &= \frac{1}{G}\sigma_{12},
\end{aligned} \tag{2.2}$$

For plane strain, $e_{33} = 0$ in the third of equations (2.1) produces

$$\sigma_{33} = \nu(\sigma_{11} + \sigma_{22})$$

which when substituted into the first two equations (2.1) results in

$$
\begin{aligned}
e_{11} &= \frac{1-\nu^2}{E}\left(\sigma_{11} - \frac{\nu}{1-\nu}\sigma_{22}\right), \\
e_{22} &= \frac{1-\nu^2}{E}\left(\sigma_{22} - \frac{\nu}{1-\nu}\sigma_{11}\right), \\
e_{12} &= \frac{1}{G}\sigma_{12}
\end{aligned}
\tag{2.3}
$$

The quantity $E/(1 - \nu^2)$ is sometimes referred to as the *plane strain modulus*. The values of stress and strain depend on the directions in space of the 1–2–3 axis set. The change in values resulting from a change in axis set is defined by the transformation equations, to be covered in the next chapter.

This completes our introduction to linear elastic behaviour at small strains. The extension to large strains will be considered in the next chapter on finite elasticity.

References

Bedford, A. and Liechti, K.M. (2000) *Mechanics of Materials*, Prentice Hall, Upper Saddle River, NJ.

Hibbeler, R.C. (2004) *Mechanics of Materials*, 6th edn, Prentice Hall, Upper Saddle River, NJ.

Love, A.E.H. (1944) *A Treatise on the Mathematical Theory of Elasticity*, 4th edn, Macmillan, New York.

Richards, R. (2001) *Principles of Solid Mechanics*, CRC Press LLC, Boca Raton, FL.

Solecki, R. and Conant, R.J. (2003) *Advanced Mechanics of Materials*, Oxford University Press, New York.

Timoshenko, S. and Goodier, J.N. (2008) *Theory of Elasticity*, 3rd edn, McGraw-Hill Higher Education, International Editions, New York.

3

Finite Strain Elasticity

In the rubber-like state, a polymer may be subjected to large deformations and still show complete recovery. The behaviour of a rubber band stretching to two or three times its original length and, when released, recovering essentially instantaneously to its original shape is a matter of common experience. To a good approximation, this is elastic behaviour at large strains. The first stage in developing an understanding of this behaviour is to seek a generalized definition of strain, which will not suffer the restriction of that derived in Chapter 2, i.e. that the strains are small. This is followed by a rigorous definition of stress for the situation where the deformations are not small. These considerations are the basis of finite elasticity theory. This subject has been considered in several notable texts (Green and Adkins 1970; Ogden 1997; Green and Zerna 2002), and it is not intended to duplicate the elegant treatments presented elsewhere. For the most part, the development of finite strain elasticity theory has been made using tensor calculus. In this book, the treatment is at a relatively elementary level, and it is hoped that in this way the exposition will be clear to those who only require an outline of the subject in order to understand the relevant mechanical properties of polymers.

3.1 The Generalized Definition of Strain

Figure 3.1 represents a deformed and an undeformed body. Points in a small area on the undeformed body are assigned co-ordinates $\mathbf{X} = (X_1, X_2, X_3)$. On deformation, this area is transformed so that in general both its shape and state of rotation are changed; the point $\mathbf{X} = (X_1, X_2, X_3)$ moves to the point $\mathbf{x} = (x_1, x_2, x_3)$. Locally this can be expressed as a linear transformation, so that the new co-ordinates are given in terms of the old co-ordinates as

$$\begin{aligned} x_1 &= f_{11}X_1 + f_{12}X_2 + f_{13}X_3 \\ x_2 &= f_{21}X_1 + f_{22}X_2 + f_{23}X_3 \, . \\ x_3 &= f_{31}X_1 + f_{32}X_2 + f_{33}X_3 \end{aligned} \tag{3.1}$$

This leaves one point, the origin, unchanged, and thus rigid-body translations have been automatically eliminated by this approach. We adopt the Lagrangian approach to strain, in which the reference state is the undeformed state of the material, in contrast to the

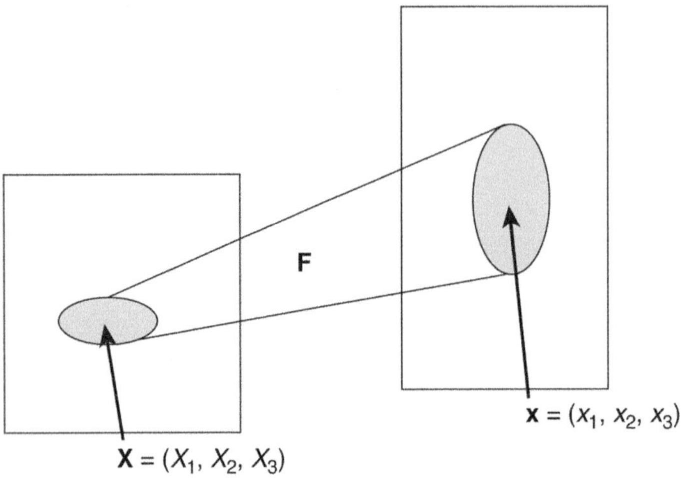

Figure 3.1 Undeformed and deformed material elements.

alternative Eulerian model, in which reference is made to the deformed state. We can rewrite Equation (3.1) in matrix form as

$$\mathbf{x} = \mathbf{F}\mathbf{X}, \tag{3.2}$$

where

$$\mathbf{F} = \begin{bmatrix} f_{11} & f_{12} & f_{13} \\ f_{21} & f_{22} & f_{23} \\ f_{31} & f_{32} & f_{33} \end{bmatrix}.$$

As a matter mainly of notation, Equation (3.1) can be differentiated to give

$$\frac{\partial x_i}{\partial X_j} = f_{ij} \quad (i,j = 1,2,3),$$

which we may express as

$$\mathbf{F} = \frac{\partial \mathbf{x}}{\partial \mathbf{X}}.$$

The matrix **F** and its components are referred to as *deformation gradients*; **F** is also termed the *deformation gradient tensor*. It defines a transformation of the undeformed state onto the deformed state.

3.1.1 The Cauchy–Green Strain Measure

The deformation gradient **F** contains geometrical information that is not related to strain (i.e. dimensional or shape change); namely, it includes rigid-body rotations. The matrix associated with rotation is familiar from its use in changing the representation of a vector when the axis set is rotated. It also defines the rigid-body rotation of an element of material, and so is in itself a deformation gradient. Suppose an element of material is specified within a 1–2–3 Cartesian axis system and is rotated so that lines of material points originally along the 1, 2 and 3 axes become aligned with new axes labelled 1′, 2′, 3′. Then the positions of the new axes can be defined in terms of their direction cosines within the old axis set. Let the 1′-axis have direction cosines (ℓ_{11}, ℓ_{12}, ℓ_{13}) with respect to the 1–2–3 axes, the 2′-axis direction cosines (ℓ_{21}, ℓ_{22}, ℓ_{23}) with respect to the 1–2–3 axes and the new 3-axis direction cosines (ℓ_{31}, ℓ_{32}, ℓ_{33}) with respect to the 1–2–3 axes. Then the matrix of rotation **R** is given by

$$R = \begin{bmatrix} \ell_{11} & \ell_{12} & \ell_{13} \\ \ell_{21} & \ell_{22} & \ell_{23} \\ \ell_{31} & \ell_{32} & \ell_{33} \end{bmatrix}. \tag{3.3}$$

This gives the new co-ordinate vector ($\mathbf{x}_1$, $\mathbf{x}_2$, $\mathbf{x}_3$) when operating on the old co-ordinates (X_1, X_2, X_3). Note that in two dimensions, it reduces to

$$\mathbf{R} = \begin{bmatrix} \cos\theta & \sin\theta \\ -\sin\theta & \cos\theta \end{bmatrix}, \tag{3.4}$$

where θ is the anticlockwise angle through which the new axis set has been rotated in the 1–2 plane.

Having defined the rotation matrix, the question we now address is how to remove rigid-body rotations from the deformation gradient **F**. In the following arguments, use will be made of the concept of the *transpose* $\mathbf{M}^T$ of the matrix **M**, in which the rows of elements of **M** are written as columns, such that an element a_{ij} of $\mathbf{M}^T$ is given by $a_{ij} = b_{ji}$, where b_{ji} is an element of **M**. In the case of a matrix of rotation **R**, $\mathbf{R}^T = \mathbf{R}^{-1}$, where $\mathbf{R}^{-1}$ is the inverse of **R**, corresponding to a rotation back to the original state, with $\mathbf{R}\mathbf{R}^{-1} = \mathbf{R}^{-1}\mathbf{R} = \mathbf{I}$, the identity matrix. We will also make use of the identity

$$(\mathbf{AB})^T = \mathbf{B}^T\mathbf{A}^T. \tag{3.5}$$

Suppose that the deformation gradient **F** consists of a 'pure' deformation **V** ('pure' in the sense that it does not include any rigid-body rotation) and a rotation **R**, then according to the *polar decomposition theorem* we may write

$$\mathbf{F} = \mathbf{RV}. \tag{3.6}$$

RV is known as the *polar decomposition* of **F**. Now let us form the matrix

$$\mathbf{C} = \mathbf{F}^T\mathbf{F}. \tag{3.7}$$

Using Equation (3.6), **F** becomes

$$\mathbf{C} = (\mathbf{RV})^T\mathbf{RV}.$$

and using the identity Equation (3.5)

$$\mathbf{C} = \mathbf{V}^T\mathbf{R}^T\mathbf{RV}.$$

As noted above, $\mathbf{R}^T = \mathbf{R}^{-1}$, so that $\mathbf{R}^T\mathbf{R} = \mathbf{I}$. Therefore,

$$\mathbf{C} = \mathbf{V}^T\mathbf{V}. \tag{3.8}$$

Notice now that the quantity **C** is not dependent on the rotation **R**; by performing the operation (Equation (3.7)), we have effectively stripped the rigid-body rotations off **F**, leaving a quantity that depends only on the pure deformation **V**. The diagonal entries of **V** are associated with changes in length or normal strains; and the off-diagonal entries with angle changes or shear strains. **C** is therefore a useful measure of strain and is known as the *right Cauchy–Green strain measure*. (The *left* Cauchy–Green strain measure is simply $\mathbf{FF}^T$.) Note that, if instead of Equation (3.6) it had been assumed that

$$\mathbf{F} = \mathbf{VR},$$

then the use of the left Cauchy–Green strain measure would have had the same effect of removing the rigid-body rotation.

C, unlike **F**, is always symmetric ($f_{ij} = f_{ji}$, $i = 1, 2, 3$). When a deformation gradient **F** is itself symmetric, it already corresponds to a deformation with no rigid-body motions, a pure deformation **V**. It is possible to derive such a quantity from the Cauchy–Green measure **C**. This requires knowledge of how to obtain principal values of **C** or **V**, and of how to transform **C** and **V** between different axis sets.

3.1.2 Principal Strains

Suppose we have a deformation gradient **V** that is symmetric and so does not include rigid-body rotations. As we have seen, **V** operates on a vector **X** corresponding to a point in space and moves it to **x**. By implication, **V** transforms a line of points to another line of points, with a straight line transformed to another straight line. For example, the line joining the origin to **X** will be transformed to the line joining the origin to **x**. The principal directions of the deformation **V** correspond to lines that are transformed parallel to themselves, so that perpendicular lines remain perpendicular and there is no angle change that would be identified with shear strain. In mathematical terms, the eigenvectors of **V** are the principal directions, and the corresponding eigenvalues are the stretch ratios along the principal directions, known as the *principal extension ratios*.

The condition for the transformed line being parallel to the initial line is

$$\mathbf{VX} = \lambda\mathbf{X},$$

where λ is a constant. As we will show, this defines a cubic equation in λ, with the three solutions of the principal extension ratios. In components, this equation becomes

$$\begin{aligned} v_{11}X_1 + v_{12}X_2 + v_{13}X_3 &= \lambda X_1 \\ v_{12}X_1 + v_{22}X_2 + v_{23}X_3 &= \lambda X_2 \\ v_{13}X_1 + v_{23}X_2 + v_{33}X_3 &= \lambda X_3 \end{aligned} \tag{3.9}$$

and it follows that

$$\begin{aligned} (v_{11} - \lambda)X_1 + v_{12}X_2 + v_{13}X_3 &= 0 \\ v_{12}X_1 + (v_{22} - \lambda)X_2 + v_{23}X_3 &= 0 \\ v_{13}X_1 + v_{23}X_2 + (v_{33} - \lambda)X_3 &= 0 \end{aligned} \tag{3.10}$$

or in matrix form

$$\begin{pmatrix} v_{11}-\lambda & v_{12} & v_{13} \\ v_{12} & v_{22}-\lambda & v_{23} \\ v_{13} & v_{23} & v_{33}-\lambda \end{pmatrix} \mathbf{X} = 0. \tag{3.11}$$

For solutions other than the trivial $\mathbf{X} = 0$, the matrix must be singular. This is true provided the determinant is zero:

$$\begin{vmatrix} v_{11}-\lambda & v_{12} & v_{13} \\ v_{12} & v_{22}-\lambda & v_{23} \\ v_{13} & v_{23} & v_{33}-\lambda \end{vmatrix} = 0. \tag{3.12}$$

This defines the cubic equation in λ. The three solutions for λ – the eigenvalues of $\mathbf{V}$ – are the principal extension ratios λ_{I}, λ_{II} and λ_{III}. We shall see that these quantities provide a simple way to the model of a molecular network, which is one of the key concepts in understanding the mechanical behaviour of polymers. The corresponding directions – the eigenvectors and principal directions – are obtained by substituting the λ values back into Equation (3.11) and solving for X_1, X_2 and X_3, subject to the condition that $\mathbf{X}$ is a unit vector, $X_1^2 + X_2^2 + X_3^2 = 1$, ensuring that its components are direction cosines.

Suppose $\mathbf{V}$ is represented by $\mathbf{V}'$ in principal axes, then, for principal extension ratios λ_{I}, λ_{II} and λ_{III},

$$\mathbf{V}' = \begin{pmatrix} \lambda_{\mathrm{I}} & 0 & 0 \\ 0 & \lambda_{\mathrm{II}} & 0 \\ 0 & 0 & \lambda_{\mathrm{III}} \end{pmatrix}, \tag{3.13}$$

where the directions of the I, II and III principal axes in the original 1–2–3 axes set are given by the direction cosines of the principal stretches λ_{I}, λ_{II} and λ_{III}.

Exactly the same technique can be used to analyze the Cauchy–Green tensor $\mathbf{C}$. When a deformation gradient $\mathbf{F}$ includes rigid-body rotation, it is necessary to first form the Cauchy–Green measure $\mathbf{C}$ and then find its principal components and directions using the methods outlined above for $\mathbf{V}$. The principal directions of $\mathbf{C}$ are the same as those of the pure deformation $\mathbf{V}$ that underlies $\mathbf{F}$ ($\mathbf{F} = \mathbf{VR}$). Writing the analogue of Equation (3.12) for $\mathbf{C}$

$$\begin{vmatrix} c_{11}-\mu & c_{12} & c_{13} \\ c_{12} & c_{22}-\mu & c_{23} \\ c_{13} & c_{23} & c_{33}-\mu \end{vmatrix} = 0 \tag{3.14}$$

gives three roots for μ: μ_{I}, μ_{II} and μ_{III}. The corresponding principal directions are obtained by substituting the values of μ into the analogue of Equation (3.11)

$$\begin{pmatrix} c_{11}-\mu & c_{12} & c_{13} \\ c_{12} & c_{22}-\mu & c_{23} \\ c_{13} & c_{23} & c_{33}-\mu \end{pmatrix} \mathbf{X} = \mathbf{0} \tag{3.15}$$

and solving for the three sets of direction cosines **X**. In principal axes, the Cauchy–Green strain is

$$\mathbf{C}' = \begin{pmatrix} \mu_{\mathrm{I}} & 0 & 0 \\ 0 & \mu_{\mathrm{II}} & 0 \\ 0 & 0 & \mu_{\mathrm{III}} \end{pmatrix}. \tag{3.16}$$

Making use of Equation (3.8) in principal axes

$$\mathbf{C}' = \mathbf{V}'^T\mathbf{V}'$$

and using Equations (3.13) and (3.16) gives the relationship between the principal values of **C** and the principal extension ratios:

$$\lambda_I = \sqrt{\mu_{\mathrm{I}}}; \quad \lambda_{\mathrm{II}} = \sqrt{\mu_{\mathrm{II}}}; \quad \lambda_{\mathrm{III}} = \sqrt{\mu_{\mathrm{III}}}. \tag{3.17}$$

To now find the deformation **V** in the original 1–2–3 axes, it is necessary to transform the deformation **V**′ given by Equation (3.13) using the appropriate rotation **R**, discussed later.

In Figure 3.2, we illustrate deformations both in arbitrary or global 1–2–3 axes and in principal I–II–III axes.

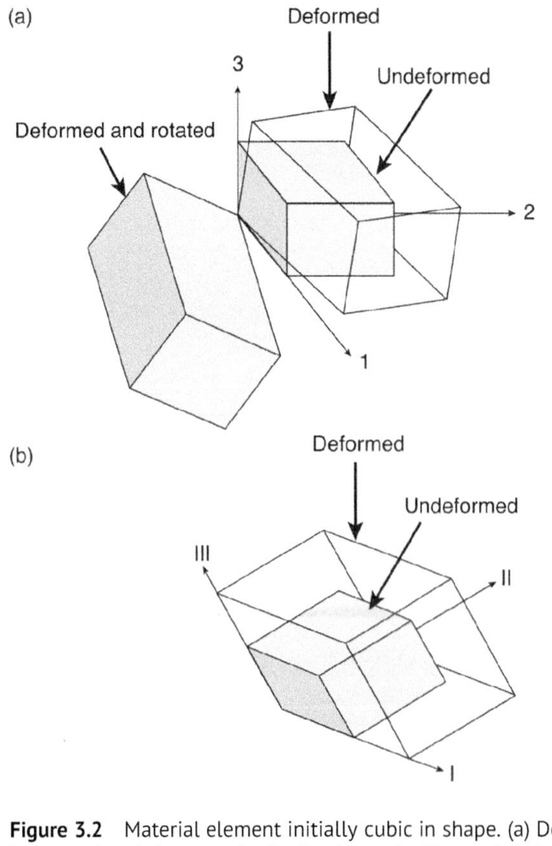

Figure 3.2 Material element initially cubic in shape. (a) Deformed in arbitrary global axes (1–2–3), by normal and shear strains both with and without rigid-body rotation. (b) Deformed in principal axes (I–II–III), in which there is no rotation or shear.

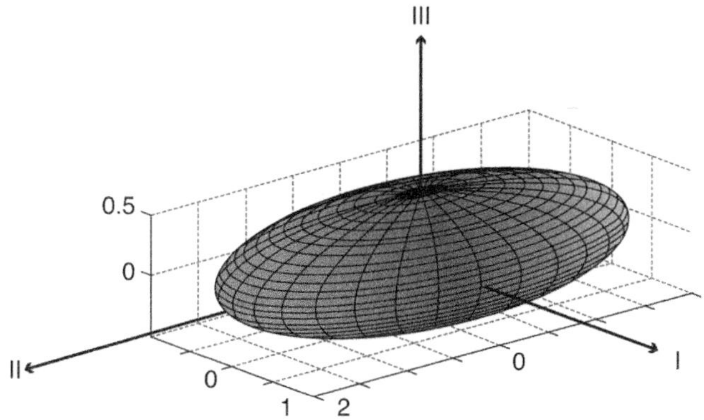

Figure 3.3 Strain ellipsoid for $\lambda_I = 1$, $\lambda_{II} = 2$ and $\lambda_{III} = 0.5$.

When viewed in principal directions, the deformation gradient has a simple physical interpretation. Consider a spherical surface of undeformed material of unit radius centred on the origin. A point on this surface is (ℓ_1, ℓ_2, ℓ_3) where

$$\ell_1^2 + \ell_2^2 + \ell_3^2 = 1 \tag{3.18}$$

and ℓ_1, ℓ_2 and ℓ_3 are in fact direction cosines. This point is transformed to a point (x_1, x_2, x_3) via the action of a deformation gradient in principal axes:

$$\begin{pmatrix} x_1 \\ x_2 \\ x_3 \end{pmatrix} = \begin{pmatrix} \lambda_\text{I} & 0 & 0 \\ 0 & \lambda_\text{II} & 0 \\ 0 & 0 & \lambda_\text{III} \end{pmatrix} \begin{pmatrix} \ell_1 \\ \ell_2 \\ \ell_3 \end{pmatrix}.$$

The three equations defined above can be used to substitute for the original co-ordinates (ℓ_1, ℓ_2, ℓ_3) in Equation (3.18). The result is

$$\frac{x_1^2}{\lambda_\text{I}^2} + \frac{x_2^2}{\lambda_\text{II}^2} + \frac{x_3^2}{\lambda_\text{III}^2} = 1.$$

This equation represents an ellipsoid with axes along the principal directions. The lengths of its semi-axes are λ_I, λ_II, λ_III along I, II and III principal directions, respectively, as shown in Figure 3.3.

3.1.3 Transformation of Strain

To obtain the components of strain in a rotated axis set, we start with the deformation gradient $\mathbf{F}$ and rewrite Equation (3.3):

$$\mathbf{x} = \mathbf{FX}.$$

In an axis set rotated according to the matrix $\mathbf{R}$ with respect to the original set, the new co-ordinates of points in the deformed body are given by

$$\mathbf{x}' = \mathbf{Rx} = \mathbf{RFX}. \tag{3.19}$$

In the rotated axes, the co-ordinates of the undeformed body are given by

$$\mathbf{X}' = \mathbf{RX}$$

so that

$$\mathbf{X} = \mathbf{R}^{-1}\mathbf{X}' = \mathbf{R}^T\mathbf{X}'$$

Substituting for **X** in Equation (3.19) now gives

$$\mathbf{x}' = \mathbf{RFR}^T\mathbf{X}'. \tag{3.20}$$

This resembles Equation (3.2). We may write

$$\mathbf{F}' = \mathbf{RFR}^T \tag{3.21}$$

to give Equation (3.20) as $\mathbf{x}' = \mathbf{F}'\mathbf{X}'$, and conclude that Equation (3.21) defines the deformation gradient in the rotated axis set.

It can be easily shown that the Cauchy–Green strain tensor also transforms like this.

$$\mathbf{C}' = \mathbf{RCR}^T \tag{3.22}$$

Both **C** and **F** are second-order tensors and so must transform in the same way.

3.1.4 Examples of Elementary Strain Fields

All uniform strain fields can, as we have seen, be expressed in terms of principal extension ratios. Some commonplace strain fields are naturally expressed in this way, such as the uniaxial field, which for incompressible stretching along the I axis is of the form $\lambda_\text{I} = \lambda$, $\lambda_\text{II} = \lambda_\text{III} = \lambda^{-1/2}$. Other strain states are customarily expressed in non-principal axes.

One such is that of *pure shear*. In the 1–2 plane, this can be expressed as the deformation gradient

$$F = \begin{pmatrix} 1 & \gamma \\ \gamma & 1 \end{pmatrix}.$$

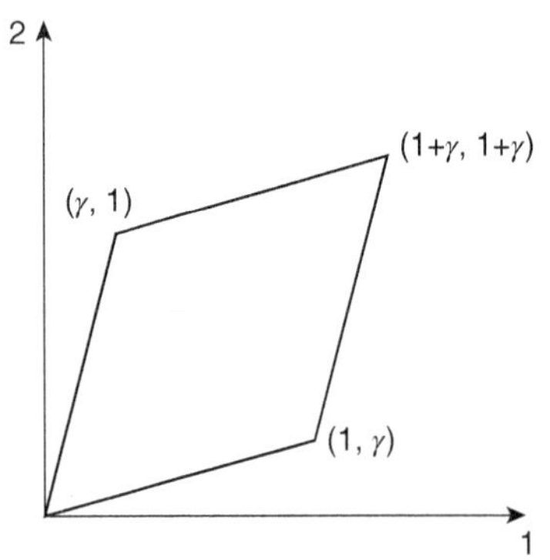

Figure 3.4 A pure shear deformation of a unit square.

Here **F** is symmetric and so no rigid-body rotation is included in the deformation. The effect of this deformation on a unit square of material is illustrated in Figure 3.4. Using the methods of Section 3.1.2, the principal extension ratios are the solutions of

$$\begin{vmatrix} 1-\lambda & \gamma \\ \gamma & 1-\lambda \end{vmatrix} = 0.$$

The principal values can be easily shown to be $\lambda_\text{I} = 1+\gamma$, $\lambda_\text{II} = 1-\gamma$. Using Equation (3.11), we can find the direction cosines (ℓ_1, ℓ_2) associated with λ_I as

$$\begin{pmatrix} 1-(1+\gamma) & \gamma \\ \gamma & 1-(1+\gamma) \end{pmatrix}\begin{pmatrix} \ell_1 \\ \ell_2 \end{pmatrix} = \begin{pmatrix} 0 \\ 0 \end{pmatrix}.$$

This gives $\ell_1 = \ell_2$. Hence, the principal I direction is at 45° to the 1 direction.

Simple shear is actually a more complex state than pure shear, in that it includes a rigid-body deformation. We shall see that it is a combination of tensile and compressive perpendicular principal strains of a rotated solid. Consider the deformation gradient

$$\mathbf{F} = \begin{pmatrix} 1 & \gamma \\ 0 & 1 \end{pmatrix}.$$

Its effect on a unit square of material is illustrated in Figure 3.5. The horizontal sides of the original square remain horizontal and the same length, whereas the initially vertical sides undergo an increase in length. Stress must be applied vertically to the horizontal sides to ensure this. The asymmetry of $\mathbf{F}$ indicates that it includes some rigid-body rotation. To find the principal extension ratios, we first eliminate this by finding the Cauchy–Green strain measure $\mathbf{C}$:

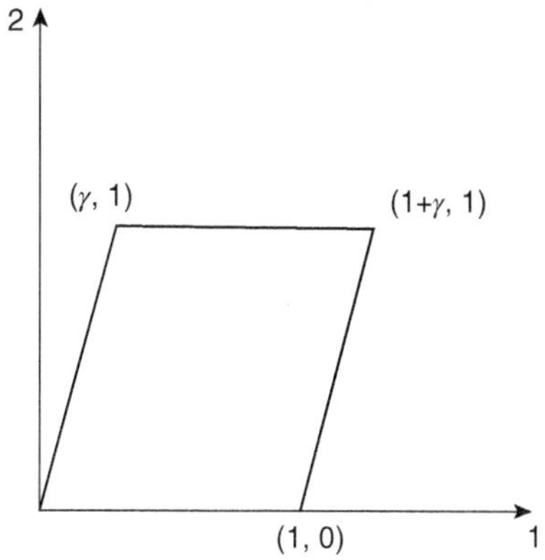

Figure 3.5 A simple shear deformation of a unit square.

$$\mathbf{C} = \mathbf{F}^T\mathbf{F} = \begin{pmatrix} 1 & 0 \\ \gamma & 1 \end{pmatrix}\begin{pmatrix} 1 & \gamma \\ 0 & 1 \end{pmatrix} = \begin{pmatrix} 1 & \gamma \\ \gamma & \gamma^2 + 1 \end{pmatrix}.$$

Principal values of $\mathbf{C}$ are given by the method of Equation (3.14). The principal extension ratios are given by the square roots of these quantities as specified in Equation (3.17). After some algebra these are found to be

$$\lambda_\mathrm{I} = \sqrt{1 + \gamma^2/2 + \sqrt{(1 + \gamma^2/2)^2 - 1}}$$

$$\lambda_\mathrm{II} = \sqrt{1 + \gamma^2/2 - \sqrt{(1 + \gamma^2/2)^2 - 1}}.$$

The deformation gradient in principal directions is thus given by

$$\mathbf{V}' = \begin{pmatrix} \lambda_\mathrm{I} & 0 \\ 0 & \lambda_\mathrm{II} \end{pmatrix}.$$

The eigenvectors of $\mathbf{C}$ define the direction cosines of the I and II directions with respect to the original 1–2 axes. They give the angle of transformation θ and the associated rotation matrix $\mathbf{R}_\theta$ required to transform $\mathbf{V}'$ back to the 1–2 axes set to give the deformation gradient $\mathbf{V}$:

$$\mathbf{V} = \mathbf{R}_\theta^T \mathbf{V}' \mathbf{R}_\theta.$$

Here, we have adapted Equation (3.21). $\mathbf{V}$, unlike $\mathbf{F}$, contains no rigid-body rotation, and the two are related (see Equation (3.6)) by

$$\mathbf{F} = \mathbf{R}_\phi \mathbf{V},$$

where $\mathbf{R}_\phi$ represents the rigid-body rotation inherent in $\mathbf{F}$ about the angle ϕ. This angle can be evaluated using the relation

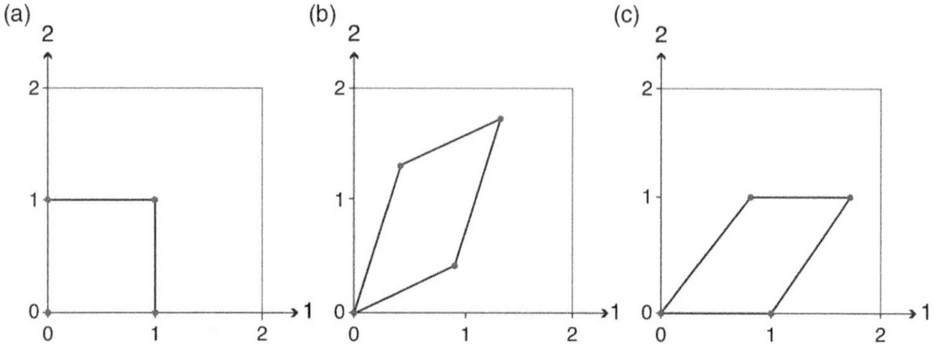

Figure 3.6 (a) The unit square, (b) the unit square operated on by **V** and (c) the unit square operated on by $R_\phi V$.

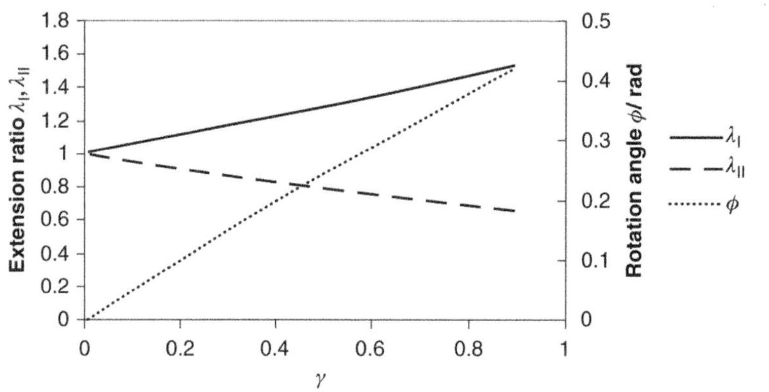

Figure 3.7 Development of principal stretches and clockwise rigid-body rotation angle in simple shear.

$$\mathbf{R}_\phi = \mathbf{F}\mathbf{V}^{-1}.$$

The application of the deformation $\mathbf{F} = \mathbf{R}_\phi \mathbf{V}$ is illustrated in Figure 3.6. A unit square in Figure 3.6(a) is operated on first by **V**, a combination of tension and shear, to give the shape in Figure 3.6(b). Then the shape is operated on further by $\mathbf{R}_\phi$ to give the simple shear of Figure 3.6(c). Figure 3.6 corresponds to a value $\gamma = 0.9$.

The algebraic expressions for **V** and ϕ are lengthy and will not be reproduced here. They can be readily evaluated numerically, and the principal extension ratios and the rigid body rotation are plotted as a function of shear strain in Figure 3.7.

3.1.5 Relationship of Engineering Strains to General Strains

Recall that, in developing small-strain theory, we made use of displacements u_1, u_2 and u_3 in the x_1, x_2 and x_3 directions The displacements are related to the initial and deformed positions X_i and x_i, respectively, by

$$\begin{aligned} u_1 &= x_1 - X_1 \\ u_2 &= x_2 - X_2 \, . \\ u_3 &= x_3 - X_3 \end{aligned} \tag{3.23}$$

Now introduce a deformation **V** that does not include rigid-body rotations. Then, $\mathbf{x} = \mathbf{VX}$ becomes

$$\begin{aligned} x_1 &= v_{11}X_1 + v_{12}X_2 + v_{13}X_3 \\ x_2 &= v_{12}X_1 + v_{22}X_2 + v_{23}X_3 \\ x_3 &= v_{13}X_1 + v_{23}X_2 + v_{33}X_3 \end{aligned} \tag{3.24}$$

and we can now make use of Equation (3.23) to rewrite Equation (3.24) in terms of displacements:

$$\begin{aligned} u_1 &= (v_{11} - 1)X_1 + v_{12}X_2 + v_{13}X_3 \\ u_2 &= v_{12}X_1 + (v_{22} - 1)X_2 + v_{23}X_3 \, . \\ u_3 &= v_{13}X_1 + v_{23}X_2 + (v_{33} - 1)X_3 \end{aligned} \tag{3.25}$$

We can now use the strain–displacement equations (Chapter 2) to derive the normal engineering strains e_{11}, e_{22} and e_{33}:

$$\begin{aligned} e_{11} &= \frac{\partial u_1}{\partial X_1} = v_{11} - 1 \\ e_{22} &= \frac{\partial u_2}{\partial X_2} = v_{22} - 1 \, . \\ e_{33} &= \frac{\partial u_3}{\partial X_3} = v_{33} - 1 \end{aligned} \tag{3.26}$$

and the shear strains:

$$\begin{aligned} e_{12} &= \frac{\partial u_1}{\partial X_2} + \frac{\partial u_2}{\partial X_1} = v_{12} + v_{12} = 2v_{12} \\ e_{23} &= \frac{\partial u_2}{\partial X_3} + \frac{\partial u_3}{\partial X_2} = v_{23} + v_{23} = 2v_{23} \, . \\ e_{13} &= \frac{\partial u_3}{\partial X_1} + \frac{\partial u_1}{\partial X_3} = v_{13} + v_{13} = 2v_{13} \end{aligned} \tag{3.27}$$

Putting the values of the v_{ij} derived from Equations (3.26) and (3.27) into the matrix **V** then gives

$$\mathbf{V} = \begin{pmatrix} 1 + e_{11} & \frac{1}{2}e_{12} & \frac{1}{2}e_{13} \\ \frac{1}{2}e_{12} & 1 + e_{22} & \frac{1}{2}e_{23} \\ \frac{1}{2}e_{13} & \frac{1}{2}e_{23} & 1 + e_{33} \end{pmatrix} . \tag{3.28}$$

Transformations as defined by Equation (3.1) and rotation operations such as axis transformations (3.21) can be performed on small strains provided that the components are first used to form the matrix **V**.

3.1.6 Logarithmic Strain

Principal true or logarithmic strains ε_{I}, $\varepsilon_{\mathrm{II}}$ and $\varepsilon_{\mathrm{III}}$ are defined in terms of the principal extension ratios by

$$\varepsilon_i = \ln(\lambda_i) \quad (i = \mathrm{I}, \mathrm{II}, \mathrm{III}) \tag{3.29}$$

These strains are exactly additive, in contrast with engineering strains that are only approximately so. To illustrate this, consider two pure deformations $\mathbf{V}^a$ and $\mathbf{V}^b$ that are applied successively to give a total deformation $\mathbf{V}$, so that

$$\mathbf{V} = \mathbf{V}^a\mathbf{V}^b.$$

$\mathbf{V}$ is said to be *multiplicatively decomposed* into $\mathbf{V}^a$ and $\mathbf{V}^b$. When operating with finite strains, deformation gradients must be used that, being tensors, are combined multiplicatively. Suppose that both $\mathbf{V}^a$ and $\mathbf{V}^b$ have the same principal directions and are expressed in diagonal form, with principal extension ratios λ_i^a and $\lambda_i^b (i = \mathbf{I}, \mathbf{II}, \mathbf{III})$, respectively, then $\mathbf{V}$ is diagonal with entries λ_i (i = I, II, III) such that

$$\lambda_i = \lambda_i^a \lambda_i^b \quad (i = \mathrm{I}, \mathrm{II}, \mathrm{III}).$$

It now follows that

$$\varepsilon_i = \ln(\lambda_i) = \ln\left(\lambda_i^a \lambda_i^b\right) = \ln\left(\lambda_i^a\right) + \ln\left(\lambda_i^b\right) = \varepsilon_i^a + \varepsilon_i^b (i = \mathrm{I}, \mathrm{II}, \mathrm{III}),$$

where the ε_i^a and ε_i^b are the logarithmic strains associated with $\mathbf{V}^a$ and $\mathbf{V}^b$, respectively.

Note also that the incompressibility (constant volume) condition

$$\lambda_{\mathrm{I}}\lambda_{\mathrm{II}}\lambda_{\mathrm{III}} = 1 \tag{3.30}$$

is exactly equivalent to

$$\varepsilon_{\mathrm{I}} + \varepsilon_{\mathrm{II}} + \varepsilon_{\mathrm{III}} = 0 \tag{3.31}$$

in contrast to the analogous sum for small strains, which is only approximately zero.

Physical justification for the logarithmic strain can be obtained by considering it as similar to the definition of small strain as extension/(original length), but taking account of the increasing length of the body upon which the increments in extension are applied as the deformation progresses. For stretching of a body from length L_1 to length L_2, the true strain ε is then given by the integral

$$\varepsilon = \int_{L_1}^{L_2} \frac{dl}{l} = \ln\left(\frac{L_2}{L_1}\right) = \ln\lambda.$$

The logarithmic strain has also been termed the *natural strain* by Nadai (1950), and is discussed by other authors, for example Rees (2006). In terms of the engineering strain e,

$$\varepsilon = \ln\lambda = \ln(1+e) = e - \frac{1}{2}e^2 + \frac{1}{3}e^3 - \cdots$$

so it is clear that e approximates to ε at small values.

3.2 The Stress Tensor

The stress tensor has been introduced in Chapter 2. In small-strain elasticity theory, the components of stress are defined by considering the equilibrium of an elemental cube within the body. When the strains are small, the dimensions of the body, and therefore the areas of the cube faces, are to a first approximation unaffected by the strain. It is then of no consequence whether the components of stress are defined with respect to the cube before deformation or the cube after deformation. For finite strains, however, this is not true and there are alternative definitions of stress depending on whether the deformed or undeformed state is chosen as a reference. We will choose to adopt the stress associated with the *deformed state* – the *true stress* or *Cauchy stress* – throughout this work. In our present axis notation, we can express this stress tensor $\boldsymbol{\Sigma}$ as

$$\sum = \begin{pmatrix} \sigma_{11} & \sigma_{12} & \sigma_{13} \\ \sigma_{12} & \sigma_{22} & \sigma_{23} \\ \sigma_{13} & \sigma_{23} & \sigma_{33} \end{pmatrix}. \tag{3.32}$$

To transform between different axis sets, we use the same method as for deformation gradients as given by Equation (3.21), since both stress and deformation gradients are second-order tensors. The equation for stress transformation is then

$$\boldsymbol{\Sigma}' = \mathbf{R}\boldsymbol{\Sigma}\mathbf{R}^T \tag{3.33}$$

Note that, unlike the deformation gradient, the stress tensor is always symmetric. This is the consequence of the equilibrium of torques applied to a material element, as pointed out in Chapter 2. As a second-order tensor, the stress is subject to the same axis transformation operations as the deformation gradient and Cauchy–Green measure (Equations (3.21) and (3.22)). The principal stresses are the eigenvalues of $\boldsymbol{\Sigma}$ and the associated principal directions are the eigenvectors, obtained in exactly the same way as for the deformation $\mathbf{V}$ and Cauchy–Green measure $\mathbf{C}$ as specified in Equations (3.12)–(3.16). Thus, in principal directions, the stress tensor is

$$\Sigma^{\text{principal}} = \begin{pmatrix} \sigma_{\text{I}} & 0 & 0 \\ 0 & \sigma_{\text{II}} & 0 \\ 0 & 0 & \sigma_{\text{III}} \end{pmatrix}$$

where the principal directions are the eigenvectors of $\boldsymbol{\Sigma}$.

Some equations familiar from two-dimensional can be recovered using relations (3.33) and (3.4). For a rotation θ in the 1–2 plane the stress becomes

$$\Sigma' = \begin{bmatrix} \cos\theta & \sin\theta \\ -\sin\theta & \cos\theta \end{bmatrix} \begin{bmatrix} \sigma_{11} & \sigma_{12} \\ \sigma_{12} & \sigma_{22} \end{bmatrix} \begin{bmatrix} \cos\theta & -\sin\theta \\ \sin\theta & \cos\theta \end{bmatrix}$$

$$= \begin{bmatrix} \frac{1}{2}[\sigma_{11}(1+\cos 2\theta) + \sigma_{22}(1-\cos 2\theta)] + \sigma_{12}\sin 2\theta & \frac{1}{2}(\sigma_{22}-\sigma_{11})\sin 2\theta + \sigma_{12}\cos 2\theta \\ \frac{1}{2}(\sigma_{22}-\sigma_{11})\sin 2\theta + \sigma_{12}\cos 2\theta & \frac{1}{2}[\sigma_{11}(1-\cos 2\theta) + \sigma_{22}(1+\cos 2\theta)] - \sigma_{12}\sin 2\theta \end{bmatrix}$$

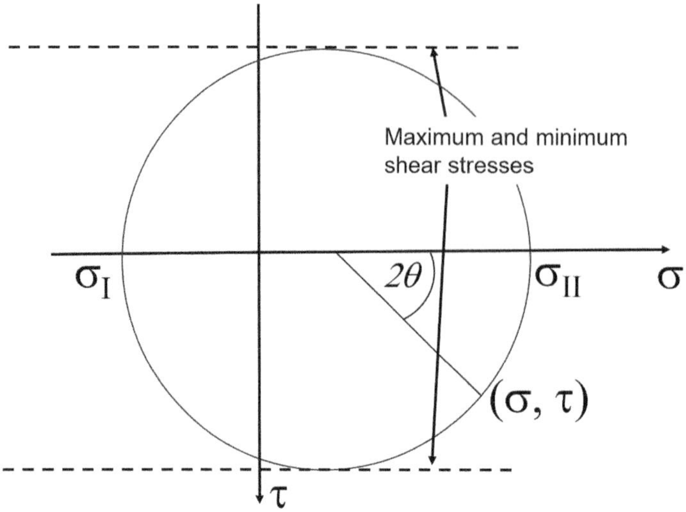

Figure 3.8 Mohr's circle for representation of stress transformation. The point (σ, τ) corresponds to a stress state in the direction at angle θ to the horizontal axis.

to give three transformation equations. These lead to the Mohr circle construction in the normal stress–shear stress $(\sigma - \tau)$ plane, shown in Figure 3.8.

The relationship of maximum and minimum shear stresses to the principal stresses can be obtained for the case $\sigma_{12} = 0$, $\sigma_{11} = \sigma_I$, $\sigma_{22} = \sigma_{II}$. The shear stress is given by the off-diagonal term in the matrix above, which has maximum and minimum values corresponding to $\theta = \pi/4$ and $\theta = -\pi/4$, respectively. Maximum shear stress is then

$$\sigma_{12}^{\max} = \frac{1}{2}(\sigma_{\mathrm{II}} - \sigma_{\mathrm{I}}).$$

This is clear from the Mohr circle. Similarly for the 2–3 and 1–3 planes we have the values

$$\sigma_{23}^{\max} = \frac{1}{2}(\sigma_{\mathrm{III}} - \sigma_{\mathrm{II}}), \sigma_{13}^{\max} = \frac{1}{2}(\sigma_{\mathrm{III}} - \sigma_{\mathrm{I}}).$$

These results are familiar from elementary treatments, e.g. Hibbeler (2004).

3.3 The Stress–Strain Relationships

Rather than developing a general theory of finite elasticity, we will introduce restrictions at an early stage that are appropriate for a representation of the behaviour of rubbers. The principal restrictions are driven by the simplifications that:

1) Rubber is isotropic in its undeformed state.
2) Volume changes associated with deformation are very small and may be neglected, i.e. a rubber is incompressible.

Restriction 2 will be relaxed at a later stage in the chapter.

First consider the impact of these assumptions on small-strain elasticity. From Chapter 2, using the expression for shear modulus in terms of modulus and Poisson's ratio, we may rewrite Hooke's law for isotropic material as

$$
\begin{aligned}
e_{11} &= \frac{1}{E}(\sigma_{11} - \nu(\sigma_{22} + \sigma_{33})) \\
e_{22} &= \frac{1}{E}(\sigma_{22} - \nu(\sigma_{33} + \sigma_{11})) \\
e_{33} &= \frac{1}{E}(\sigma_{33} - \nu(\sigma_{11} + \sigma_{22})) \\
e_{12} &= \frac{2}{E}(1 + \nu)\sigma_{12} \\
e_{23} &= \frac{2}{E}(1 + \nu)\sigma_{23} \\
e_{31} &= \frac{2}{E}(1 + \nu)\sigma_{31}
\end{aligned} \tag{3.34}
$$

Solving these expressions for the strain in terms of the stress,

$$
\begin{aligned}
e_{11} &= \frac{1+\nu}{E}\left(\sigma_{11} - \frac{\nu}{1+\nu}(\sigma_{11} + \sigma_{22} + \sigma_{33})\right) \\
e_{22} &= \frac{1+\nu}{E}\left(\sigma_{22} - \frac{\nu}{1+\nu}(\sigma_{11} + \sigma_{22} + \sigma_{33})\right) \\
e_{33} &= \frac{1+\nu}{E}\left(\sigma_{33} - \frac{\nu}{1+\nu}(\sigma_{11} + \sigma_{22} + \sigma_{33})\right)
\end{aligned} . \tag{3.35}
$$

Now we put

$$
p = \frac{\nu}{1+\nu}(\sigma_{11} + \sigma_{22} + \sigma_{33}). \tag{3.36}
$$

p is proportional to the hydrostatic pressure $(\sigma_{11} + \sigma_{22} + \sigma_{33})/3$, and equal to it when Poisson's ratio $\nu = 1/2$. Now Equation (3.35) becomes

$$
\begin{aligned}
e_{11} &= \frac{1+\nu}{E}(\sigma_{11} - p) \\
e_{22} &= \frac{1+\nu}{E}(\sigma_{22} - p) \\
e_{33} &= \frac{1+\nu}{E}(\sigma_{33} - p)
\end{aligned} . \tag{3.37}
$$

We now apply the incompressibility $e_{11} + e_{22} + e_{33} = 0$. Adding the above equations then gives

$$
0 = \frac{1+\nu}{E}(\sigma_{11} + \sigma_{22} + \sigma_{33} - 3p).
$$

Applying the definition for p from Equation (3.36), it is clear that the right-hand side is zero only when Poisson's ratio $\nu = 1/2$; this condition corresponds to incompressibility. Putting this condition into Equation (3.37) and appending the shear terms then gives

$$\begin{aligned} e_{11} &= \frac{3}{2E}(\sigma_{11}-p) \\ e_{22} &= \frac{3}{2E}(\sigma_{22}-p) \\ e_{33} &= \frac{3}{2E}(\sigma_{33}-p) \\ e_{12} &= \frac{3}{E}\sigma_{12} \\ e_{23} &= \frac{3}{E}\sigma_{23} \\ e_{31} &= \frac{3}{E}\sigma_{31} \end{aligned} \quad . \tag{3.38}$$

If the stresses are known, then p is known and the strains can be obtained from the above equations. If, however, the strains are known, then the normal stresses are only accessible via the quantities $\sigma_{11} - p$ and so on, and are indeterminate to the extent of what is now an arbitrary constant p. This reflects the fact that, when an incompressible material is subject to hydrostatic stress, it suffers no change in strain, so that for a general strain field the stress is uncertain within a hydrostatic pressure and in this sense indeterminate.

We will now propose a theory of finite deformations by generalizing Equation (3.38). Defining the behaviour in principal directions is sufficient to define the material behaviour, as the behaviour for arbitrary directions is obtainable by transformation. Rewriting Equation (3.38) in principal directions gives

$$\begin{aligned} 2e_{\mathrm{I}} &= \frac{3}{E}(\sigma_{\mathrm{I}}-p) \\ 2e_{\mathrm{II}} &= \frac{3}{E}(\sigma_{\mathrm{II}}-p) \\ 2e_{\mathrm{III}} &= \frac{3}{E}(\sigma_{\mathrm{III}}-p) \end{aligned} \quad . \tag{3.39}$$

We now create the theory of finite strain by replacing the engineering strains on the left-hand side of Equation (3.39) by measures of finite strain. We choose as a measure of finite strain the Cauchy–Green measure, which (see Equations (3.16) and (3.17)) has principal values $\lambda_{\mathrm{I}}^2, \lambda_{\mathrm{II}}^2$ and λ_{III}^2. Then,

$$\begin{aligned} \lambda_{\mathrm{I}}^2 &= \frac{3}{E}(\sigma_{\mathrm{I}}-p) \\ \lambda_{\mathrm{II}}^2 &= \frac{3}{E}(\sigma_{\mathrm{II}}-p) \\ \lambda_{\mathrm{III}}^2 &= \frac{3}{E}(\sigma_{\mathrm{III}}-p) \end{aligned} \quad . \tag{3.40}$$

This is essentially the scheme proposed by Rivlin (1949). Note that, unlike Equation (3.37), the sum of the left-hand sides of Equation (3.40) does not equal zero. Therefore, in these equations, the quantity p is no longer the prevailing hydrostatic pressure, but must be reinterpreted as an arbitrary pressure that is evaluated with reference to the conditions that apply in any particular case.

Take as an example the simple case of uniaxial stretching along the I direction with an applied stress $\sigma_{\mathrm{I}} = \sigma$, $\sigma_{\mathrm{II}} = \sigma_{\mathrm{III}} = 0$. The extension ratio along I is $\lambda_{\mathrm{I}} = \lambda$, with the symmetry so that the extension ratios in the other two principal directions are equal. Then, the incompressibility condition (3.30) becomes

$$\lambda\lambda_{\mathrm{II}}^2 = 1 \Rightarrow \lambda_{\mathrm{II}} = \lambda_{\mathrm{III}} = \lambda^{-1/2}.$$

Using the second or third of Equations (3.40), the values of stress and extension ratio yield

$$\lambda^{-1} = -\frac{3p}{E} \Rightarrow p = -\frac{E}{3\lambda}.$$

The first of Equation (3.40), when rearranged, gives for the stress

$$\sigma = \frac{E}{3}\lambda^2 + p,$$

which using the above value for p gives

$$\sigma = \frac{E}{3}\left(\lambda^2 - \frac{1}{\lambda}\right). \tag{3.41}$$

The value of p can easily be evaluated when one of the stresses is zero, such as in plane stress problems. For small strains, the use of $\lambda = 1 + e$ returns Hooke's law in one dimension:

$$\sigma = Ee.$$

provided third-order terms in e are neglected. It is a requirement that all finite strain theories reduce to Hooke's law at small strains.

In the above development, the stresses are understood to be *true* stresses, that is, the force per unit area of the deformed body. It is also of interest to find the nominal stress, that is, the stress obtained by dividing the force by the cross-sectional area of material before deformation. For a normal stress acting along the I direction, an original unit area is changed on deformation to the area $\lambda_{\mathrm{II}}\lambda_{\mathrm{III}}$. To regain the nominal stress σ^{nom}, we need to multiply the true stress by this area:

$$\sigma^{\mathrm{nom}} = \lambda_{\mathrm{II}}\lambda_{\mathrm{III}}\sigma. \tag{3.42}$$

Using the incompressibility condition (3.30) this becomes

$$\sigma^{\mathrm{nom}} = \frac{\sigma}{\lambda_{\mathrm{I}}} = \frac{\sigma}{\lambda}$$

and Equation (3.41) becomes

$$\sigma^{\mathrm{nom}} = \frac{E}{3}\left(\lambda - \frac{1}{\lambda^2}\right). \tag{3.43}$$

An expression of this form is often associated with the stretching of rubber elastic networks, having been derived on physical grounds (see Chapter 4). However, we can see here that it arises on purely phenomenological grounds once the assumptions of isotropy and incompressibility have been made, and does not imply that the material is rubber-like. These materials are sometimes referred to as neo-Hookean.

The finite deformation theory developed here is the result of a simple generalization of linear elasticity. There are much more sophisticated theories available, some of which are motivated by physical arguments. A more secure basis for the development of finite strain theories is the stored energy or strain-energy function, which will now be considered.

3.4 The Use of a Strain-Energy Function

3.4.1 Thermodynamic Considerations

The simplest (and 'weakest') definition of an elastic material is one for which the stress depends only on the current strain; these materials are termed *Cauchy elastic*. A subset of these materials is occupied by those for which the strain energy depends only on the current strain. These are termed *Green elastic* or *hyperelastic* and for these the strain energy is a function of the current strain only, and fully defines the material behaviour. For Cauchy elastic material, it is possible for the strain energy to depend on the strain path – the route or history via which the current strain level is attained – and thermodynamically non-physical behaviour may be permitted. Hyperelastic materials are not subject to this hazard and have become the standard way to define finite strain elasticity. The different definitions of elasticity have been discussed by Ogden (1997).

3.4.1.1 Development of Strain-Energy Functions

We shall develop the concept of strain energy using a small-strain approach. Because we will wish to follow the phenomenological treatment by one based on statistical mechanics, it is important at the outset to examine the different types of strain-energy function that can be defined, depending on experimental conditions. This introduces thermodynamic considerations.

Suppose we have a cylinder of material of axial length L and cross-sectional area A. A force f is acting along the axis such as to stretch the body. The body extends by a small length δx as this force f acts (the smallness of δx allows us to assume that f remains constant). The associated increment in energy is then $f\,\delta x$. We are generally more concerned with the energy per unit volume W. Dividing by the body's volume, the increment in this quantity is given by

$$\delta W = \frac{f\delta x}{AL},$$

which can be re-expressed in terms of nominal stress and strain as

$$\delta W = \frac{f}{A} \times \frac{\delta x}{L} = \sigma^{\text{nom}} \delta e.$$

The stress is nominal as the area A relates to the undeformed material. For small strains, the nominal and true stresses can be assumed equal. Then, when all six components of stress are acting, the energy is additive and, expressing the result in differential form, we have

$$dW = \sigma_{11}de_{11} + \sigma_{22}de_{22} + \sigma_{33}de_{33} + \sigma_{12}de_{12} + \sigma_{23}de_{23} + \sigma_{31}de_{31}. \tag{3.44}$$

Now consider a small-strain deformation of unit volume of an elastic solid occurring under adiabatic conditions. The first law of thermodynamics gives

$$dW = dU - dQ$$

relating the work done on the solid dW to the increase in internal energy dU and the mechanical value of the heat supplied dQ. For an adiabatic change of state, $dQ = 0$ and $dW = dU$. We can imagine that the deformation is produced by independent changes in each of the components of strain, i.e.

$$dW = dU = \frac{\partial U}{\partial e_{11}}de_{11} + \frac{\partial U}{\partial e_{22}}de_{22} + \frac{\partial U}{\partial e_{33}}de_{33} + \frac{\partial U}{\partial e_{12}}de_{12} + \frac{\partial U}{\partial e_{23}}de_{23} + \frac{\partial U}{\partial e_{31}}de_{31}. \tag{3.45}$$

Comparison of this equation with Equation (3.44) leads us to make the identifications

$$\sigma_{11} = \frac{\partial U}{\partial e_{11}}, \sigma_{22} = \frac{\partial U}{\partial e_{22}}, \sigma_{33} = \frac{\partial U}{\partial e_{33}}, \sigma_{12} = \frac{\partial U}{\partial e_{12}}, \sigma_{23} = \frac{\partial U}{\partial e_{23}}, \sigma_{31} = \frac{\partial U}{\partial e_{31}}. \tag{3.46}$$

We can therefore define a strain-energy function or stored energy function that defines the energy stored in the body as a result of the strain. Here we have performed an analysis involving energy per unit volume under the simplifying assumption that the conditions are adiabatic. For other conditions, other forms of energy are appropriate, as discussed by Baker (2005) and Houlsby and Puzrin (2006).

In the case of fluids, it is appropriate to define the enthalpy H as

$$H = U + pV, \tag{3.47}$$

where p is pressure and V is the volumetric strain. However, for solids an alternative formulation is used in which pV is replaced by another energy-like term, which involves the products of corresponding stress and strain components, in the form which we now define in compact notation:

$$\sigma_{ij}e_{ij} = \sigma_{11}e_{11} + \sigma_{22}e_{22} + \sigma_{33}e_{33} + \sigma_{12}e_{12} + \sigma_{23}e_{23} + \sigma_{31}e_{31}$$

Enthalpy is now defined as

$$H = U - \sigma_{ij}e_{ij}, \tag{3.48}$$

where the sign of the second term has changed since p denotes a negative stress. This form reduces to the more traditional form (Equation (3.47)) for gases or liquids with low viscosities, which can sustain only hydrostatic pressures. Given that U is a function of strain only, differentiation with respect to each stress component yields

$$e_{ij} = -\frac{\partial H}{\partial \sigma_{ij}} \quad (i,j = 1,2,3). \tag{3.49}$$

This gives an alternative set of stress–strain relationships parallel with Equation (3.46) but for the strain in terms of the stress.

To analyze non-adiabatic cases, it is useful to introduce free energies, which are essentially internal energies minus the heat exchanged. The Helmholtz free energy A is defined as

$$A = U - TS, \tag{3.50}$$

where S is the entropy. This can be used to analyze isothermal conditions. In this case the first law, with $\mathrm{d}Q \neq 0$, gives

$$\mathrm{d}W = \mathrm{d}U - \mathrm{d}Q = \mathrm{d}U - T\,\mathrm{d}S = \mathrm{d}A$$

and we can introduce a third strain-energy function that is identical to A and reproduces Equation (3.46) with A replacing U. The Helmholtz free energy is associated with constant volume conditions.

Finally, we introduce a second free energy, the Gibbs free energy G defined as

$$G = U - \sigma_{ij}e_{ij} - TS. \tag{3.51}$$

Differentiating with respect to the stress gives

$$e_{ij} = -\frac{\partial G}{\partial \sigma_{ij}} \quad (i,j = 1,2,3)$$

another set of stress–strain relationships resembling Equation (3.49). The Gibbs free energy is associated with constant pressure conditions.

Examining Equations (3.50) and (3.51), it is clear that

$$A - G = \sigma_{ij}e_{ij}.$$

A and $-G$ can be regarded as *complementary*. This can be interpreted for the one-dimensional case, with the areas under or to the right of a stress–strain curve being associated with A and $-G$, respectively (see Figure 3.9). Thus, A is the strain energy and $-G$ the complementary energy.

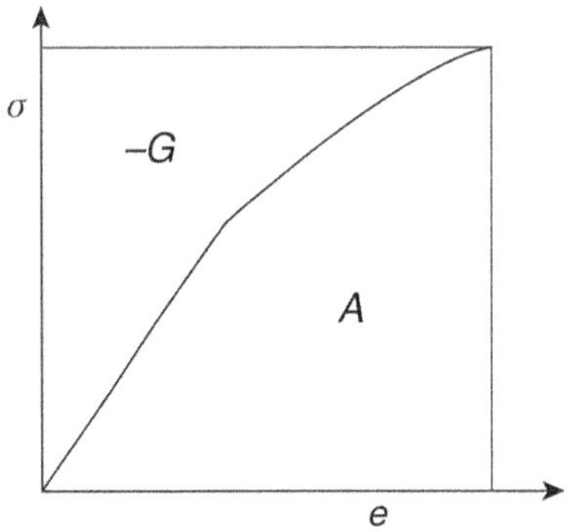

Figure 3.9 Areas defined by a stress–strain curve corresponding to Helmholtz free energy (A) and Gibbs free energy (G).

3.4.1.2 Finite Strains

For finite strains, we can adapt Equation (3.44) by recognizing that the stresses are nominal stresses. In this case, it is most effective to operate in principal directions. Choosing the principal extension ratios as strain measures, and noting that

$$\lambda_i = 1 + e_i \Rightarrow \mathrm{d}\lambda_i = \mathrm{d}e_i \; (i = \mathrm{I}, \mathrm{II}, \mathrm{III})$$

Equation (3.44) becomes

$$\mathrm{d}W = \mathrm{d}U = \sigma_{\mathrm{I}}^{\mathrm{nom}}\mathrm{d}\lambda_{\mathrm{I}} + \sigma_{\mathrm{II}}^{\mathrm{nom}}\mathrm{d}\lambda_{\mathrm{II}} + \sigma_{\mathrm{III}}^{\mathrm{nom}}\mathrm{d}\lambda_{\mathrm{III}}.$$

Introducing true stresses by correcting for the cross-sectional areas as in Equation (3.42), we have

$$\begin{aligned}\mathrm{d}W = \mathrm{d}U = {} & \lambda_{\mathrm{II}}\lambda_{\mathrm{III}}\sigma_{\mathrm{I}}\mathrm{d}\lambda_{\mathrm{I}} + \lambda_{\mathrm{III}}\lambda_{\mathrm{I}}\sigma_{\mathrm{II}}\mathrm{d}\lambda_{\mathrm{II}} \\ & + \lambda_{\mathrm{I}}\lambda_{\mathrm{II}}\sigma_{\mathrm{III}}\mathrm{d}\lambda_{\mathrm{III}}.\end{aligned} \tag{3.52}$$

Similarly, in principal directions Equation (3.45) becomes

$$dW = dU = \frac{\partial U}{\partial \lambda_{\mathrm{I}}} d\lambda_{\mathrm{I}} + \frac{\partial U}{\partial \lambda_{\mathrm{II}}} d\lambda_{\mathrm{II}} + \frac{\partial U}{\partial \lambda_{\mathrm{III}}} d\lambda_{\mathrm{III}}. \tag{3.53}$$

Comparison of Equations (3.52) and (3.53) leads to the relations for finite deformation

$$\sigma_{\mathrm{I}} = \frac{1}{\lambda_{\mathrm{II}}\lambda_{\mathrm{III}}} \frac{\partial U}{\partial \lambda_{\mathrm{I}}}, \quad \sigma_{\mathrm{II}} = \frac{1}{\lambda_{\mathrm{III}}\lambda_{\mathrm{I}}} \frac{\partial U}{\partial \lambda_{\mathrm{II}}}, \quad \sigma_{\mathrm{III}} = \frac{1}{\lambda_{\mathrm{I}}\lambda_{\mathrm{II}}} \frac{\partial U}{\partial \lambda_{\mathrm{III}}}. \tag{3.54}$$

For incompressible material, the expressions are slightly simpler, but involve the unknown hydrostatic term p as discussed in Section 3.3. Using the incompressibility relation (3.30) and incorporating the hydrostatic term results in

$$\sigma_{\mathrm{I}} = \lambda_{\mathrm{I}} \frac{\partial U}{\partial \lambda_{\mathrm{I}}} - p, \quad \sigma_{\mathrm{II}} = \lambda_{\mathrm{II}} \frac{\partial U}{\partial \lambda_{\mathrm{II}}} - p, \quad \sigma_{\mathrm{III}} = \lambda_{\mathrm{III}} \frac{\partial U}{\partial \lambda_{\mathrm{III}}} - p. \tag{3.55}$$

3.4.2 The Form of the Strain-Energy Function

The strain-energy function U is a function of the components of some measure of strain, such as the stretch **V** or the Cauchy–Green measure **C**. U is a physical quantity with a numerical value upon which all observers will agree – it is independent of the axis set. On the other hand, the components of **V** and **C** are entirely dependent on the axis set. Unless a function of these components is chosen with care, it will itself be dependent on the axis set and so will be inadmissible as a strain-energy function. Formally, it will violate the *principle of material indifference* (Noll 1958). This places restrictions on the form of U, which can be approached in two ways:

1) There are certain combinations of the strain components that are themselves independent of the axis set, known as *strain invariants*. Any function of these quantities is itself independent of the axis set, and in this sense will be a permissible form of U.
2) Alternatively, principal values of strain components are used, such as the extension ratios λ_{I}, λ_{II} and λ_{III}. Whilst the three values of these quantities will be agreed by all observers, the assignation of the labels I, II and III is arbitrary. The value of U defined in this way must be unchanged under the interchange of any pair of these labels; in other words, U must be a *symmetric function* of the principal strain values.

Both these approaches are discussed below. We give illustrative examples of the many strain-energy functions available. A useful bibliography (Vahapoğlu and Karadeniz 2006) has surveyed the field up to the year 2003.

3.4.3 The Strain Invariants

Any second-order tensor has a number of invariants associated with it. One such is the *trace* of the tensor, equal to the sum of its diagonal terms, applicable to any strain tensor. We define the *first invariant* I_1 as the trace of the Cauchy–Green strain measure tr(**C**):

$$I_1 = \mathrm{tr}(\mathbf{C}) = c_{11} + c_{22} + c_{33}.$$

Adopting principal directions, this can be expressed in terms of the extension ratios via Equation (3.17):

$$I_1 = \lambda_{\mathrm{I}}^2 + \lambda_{\mathrm{II}}^2 + \lambda_{\mathrm{III}}^2. \tag{3.56}$$

The second invariant I_2 can also defined in terms of the trace function and **C**:

$$I_2 = \frac{1}{2}\left[(\mathrm{tr}(\mathbf{C}))^2 - \mathrm{tr}(\mathbf{C}^2)\right],$$

which clearly as a function of invariants is also invariant, and can be readily re-expressed in terms of principal extension ratios as

$$I_2 = \lambda_{\mathrm{I}}^2\lambda_{\mathrm{II}}^2 + \lambda_{\mathrm{II}}^2\lambda_{\mathrm{III}}^2 + \lambda_{\mathrm{III}}^2\lambda_{\mathrm{I}}^2. \tag{3.57}$$

The third invariant is related to the volume of the material, which, like energy, is the same for all axis sets. A volume of material that is initially a unit cube is deformed to a volume equal to the product of the three principal extension ratios. The third invariant I_3 is defined directly as the square of this volume:

$$I_3 = \lambda_{\mathrm{I}}^2\lambda_{\mathrm{II}}^2\lambda_{\mathrm{III}}^2 \tag{3.58}$$

and is also equal to the determinant of **C**, det(**C**). For incompressible material $I_3 = 1$.

These three invariants are all that is necessary for our purposes in defining strain-energy functions.

3.4.4 Application of the Invariant Approach

In this section, we give examples of applications using the approach to the strain-energy function listed as (1) in Section 3.4.2. We shall restrict the discussion to incompressible materials, so that only the invariants I_1 and I_2 will be of concern.

The most elementary form is a function of I_1 only:

$$U = C(I_1 - 3).$$

Here C is a material constant and the 3 ensures that the strain energy is zero at zero strain. This form is known as the neo-Hookean or Gaussian model. It can be derived for rubbers using a physical argument based on thermodynamics (Treloar 1943, 2005), and will be discussed further in Chapter 4. In many cases, a more complex approach is required.

In a classic work of Rivlin and Saunders (1951) on vulcanized rubber, a strain-energy function of the form $U\,(I_1, I_2)$ was introduced. Experiments were performed on sheets of material in plane stress with the III direction normal to the plane. Homogeneous deformations were applied using a biaxial stretching machine so that the principal extension ratios λ_{I} and λ_{II} were under experimental control. Incompressibility applies and Equation (3.55) gives the principal stresses

$$\sigma_i = \lambda_i\frac{\partial U}{\partial \lambda_i} - p = \lambda_i\left(\frac{\partial U}{\partial I_1}\frac{\partial I_1}{\partial \lambda_i} + \frac{\partial U}{\partial I_2}\frac{\partial I_2}{\partial \lambda_i}\right) - p \quad (i = \mathrm{I}, \mathrm{II}, \mathrm{III}). \tag{3.59}$$

The incompressibility condition (3.30) leads to a simplified expression for I_2 derived from Equation (3.57):

$$I_2 = \frac{1}{\lambda_I^2} + \frac{1}{\lambda_{II}^2} + \frac{1}{\lambda_{III}^2}.$$

From this equation and Equation (3.56), expressions for the derivatives are obtained:

$$\frac{\partial I_1}{\partial \lambda_i} = 2\lambda_i; \quad \frac{\partial I_2}{\partial \lambda_i} = -\frac{2}{\lambda_i^3} \quad (i = \mathrm{I}, \mathrm{II}, \mathrm{III}).$$

Substitution of these expressions into Equation (3.59) now yields the stresses

$$\begin{aligned}\sigma_I &= 2\left(\lambda_I^2 \frac{\partial U}{\partial I_1} - \frac{1}{\lambda_I^2}\frac{\partial U}{\partial I_2}\right) - p \\ \sigma_{II} &= 2\left(\lambda_{II}^2 \frac{\partial U}{\partial I_1} - \frac{1}{\lambda_{II}^2}\frac{\partial U}{\partial I_2}\right) - p \\ \sigma_{III} &= 2\left(\lambda_{III}^2 \frac{\partial U}{\partial I_1} - \frac{1}{\lambda_{III}^2}\frac{\partial U}{\partial I_2}\right) - p = 0\end{aligned}.$$

The last component is equated to zero according to the plane stress condition, giving an expression for p. On substituting p back into the first two equations and making use of the incompressibility condition (3.30) to eliminate λ_{III}, the in-plane stresses are

$$\begin{aligned}\sigma_I &= 2\left(\lambda_I^2 - \frac{1}{\lambda_I^2\lambda_{II}^2}\right)\left(\frac{\partial U}{\partial I_1} + \lambda_{II}^2\frac{\partial U}{\partial I_2}\right) \\ \sigma_{II} &= 2\left(\lambda_{II}^2 - \frac{1}{\lambda_I^2\lambda_{II}^2}\right)\left(\frac{\partial U}{\partial I_1} + \lambda_I^2\frac{\partial U}{\partial I_2}\right)\end{aligned}.$$

Both the stresses and the extension ratios are available from experimental data, so the above equations provide sufficient information to evaluate the derivatives $\frac{\partial U}{\partial I_1}$ and $\frac{\partial U}{\partial I_2}$. They can be rearranged to give

$$\begin{aligned}\frac{\partial U}{\partial I_1} &= \left(\frac{\lambda_I^2\sigma_I}{\lambda_I^2 - 1/\lambda_I^2\lambda_{II}^2} - \frac{\lambda_{II}^2\sigma_{II}}{\lambda_{II}^2 - 1/\lambda_I^2\lambda_{II}^2}\right)/2(\lambda_I^2 - \lambda_{II}^2) \\ \frac{\partial U}{\partial I_2} &= \left(\frac{\sigma_I}{\lambda_I^2 - 1/\lambda_I^2\lambda_{II}^2} - \frac{\sigma_{II}}{\lambda_{II}^2 - 1/\lambda_I^2\lambda_{II}^2}\right)/2(\lambda_I^2 - \lambda_{II}^2)\end{aligned}. \tag{3.60}$$

Rivlin and Saunders used these equations to plot the graphs shown in Figure 3.10. Pairs of values of principal extension ratio were chosen such as to provide a range of values of I_1 for a constant I_2 and a range of values of I_1 for a constant I_2. It can be seen from the figure that $\frac{\partial U}{\partial I_1}$ is approximately constant and independent of both invariants. $\frac{\partial U}{\partial I_2}$, however, is independent of I_1 but decreases with increasing I_2. This is suggestive of a form of strain-energy function

$$U = C_1(I_1 - 3) + f(I_2 - 3) \tag{3.61}$$

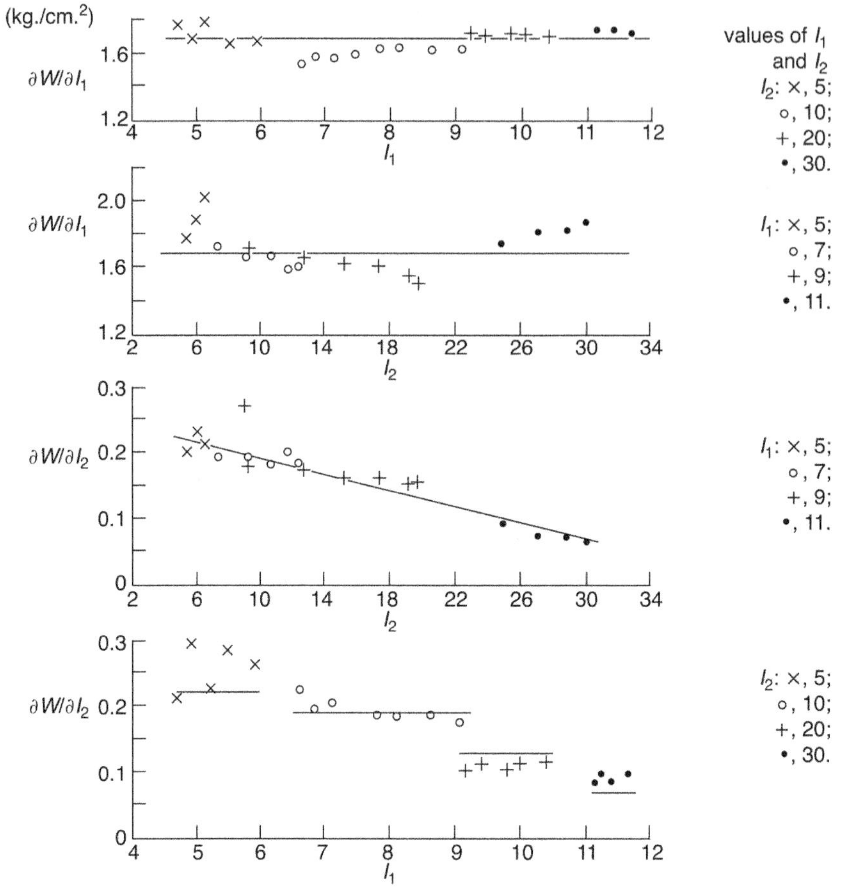

Figure 3.10 Derivatives of strain energy plotted against invariants. Rivlin and Saunders, 1951 / The Royal Society.

a generalization of the neo-Hookean model discussed above, where f is a function with $f(0) = 0$. The second term is generally smaller than the first and decreases as I_2 increases. In this case,

$$\frac{\partial U}{\partial I_1} \sim 170\,\text{kPa},$$

whereas

$$\frac{\partial U}{\partial I_2} \sim 15\,\text{kPa}.$$

This suggests that the use of a constant in place of the function f could be a useful model. This form is then

$$U = C_1(I_1 - 3) + C_2(I_2 - 3). \tag{3.62}$$

This is known as the Mooney–Rivlin model (Mooney 1940).

Rivlin and Saunders carried out experiments in simple tension, torsion, pure shear and pure shear superposed on simple extension to extend the range of combinations of principal extension ratio. This showed behaviour generally consistent with Equation (3.61). Their work illustrates the importance of exploring a wide range of combinations of extension ratio to establish the function U. The traditional form of materials test – the uniaxial stretch – involves a very specific mode of deformation. Finding materials parameters by least squares fitting to data generated in this way, or in any other way involving a single combination of extension ratios, is likely to produce suspect results that give rise to poor predictions in general.

Kawabata et al. (1981) have applied a similar approach to an isoprene rubber vulcanization. They showed that the strain-energy function of Equation (3.61) was applicable to this material. They modelled temperature dependence satisfactorily by modifying the first term to be proportional to the absolute temperature.

Yeoh (1993) created a power law model using only the first invariant, taking the form

$$U = C_{10}(I_1 - 3) + C_{20}(I_1 - 3)^2 + C_{30}(I_1 - 3)^3 \tag{3.63}$$

Also using the first invariant only, Gent (1996) created a material model that includes an upper limit on the level of strain. This property is referred to as 'finite strain extensibility' when applied to polymers. The strain-energy density U takes the form

$$U = -\frac{E}{6} J_m \ln(1 - J_1/J_m) \tag{3.64}$$

where J_1 is simply related to the first invariant:

$$J_1 = I_1 - 3.$$

E is the small-strain modulus and J_m is the value of J_1 that defines the limit of extensibility. The limitation on strain arises since, as J_1 approaches J_m, U approaches infinity. The model of Equation (3.64) has been successfully applied in many contexts, as documented by Horgan (2015) and Puglisi and Saccomandi (2015).

Pucci and Saccomandi (2002) showed that the Gent model, whilst effective at high strain, could not give good predictions of stress at low-to-moderate strains, and proposed a modification, of the form

$$U = -\frac{E}{6} J_m \ln(1 - J_1/J_m) + C_2 \ln\left(\frac{I_2}{3}\right) \tag{3.65}$$

This model, sometimes named the Gent-Gent or GG model, was shown to give a superior model of rubber in uniaxial tensile stretching at all strains up to extension ratios of 7. It has since been applied to the dynamics of elastomer balloons by Alibakhshi and Heidari (2020).

3.4.5 Application of the Principal Stretch Approach

Rivlin adopted a formulation for the strain-energy function U that involved the squares of the extension ratios because he envisaged that negative values of these quantities were possible, whilst U must be always positive. However, as demonstrated in Section 3.1.2, any strain field can be expressed in terms of principal extension ratios λ_{I}, λ_{II}, λ_{III} that are essentially positive quantities and can be interpreted as the dimensions of the principal axes of the strain ellipsoid.

The use of the invariant approach, which is closely linked to Rivlin's use of strain in the form of even powers of the extension ratios as exemplified in Equation (3.40), now appears to be unnecessarily restrictive. In this section, we give examples of applications using the approach to the strain-energy function listed as (2) in Section 3.4.2. We shall again restrict the discussion to incompressible materials.

Valanis and Landel (1967) have introduced a strain-energy function of the form

$$U = u(\lambda_{\mathrm{I}}) + u(\lambda_{\mathrm{II}}) + u(\lambda_{\mathrm{III}}). \tag{3.66}$$

This form has the essential property specified in (2) in Section 3.4.2, that U is a symmetric function of the principal extension ratios. Stresses are derived using Equation (3.55) that gives

$$\sigma_i = \lambda_i u'(\lambda_i) - p \quad (i = \mathrm{I},\ \mathrm{II}, \mathrm{III}).$$

Valanis and Landel used the expression for the stress difference in the I–II plane

$$\sigma_{\mathrm{I}} - \sigma_{\mathrm{II}} = \lambda_{\mathrm{I}} u'(\lambda_{\mathrm{I}}) - \lambda_{\mathrm{II}} u'(\lambda_{\mathrm{II}}) \tag{3.67}$$

to analyze the validity of their approach. They chose, for the range $1 < \lambda < 2.5$, the function

$$u'(\lambda) = 2\mu \ln(\lambda)$$

so that Equation (3.67) becomes

$$\sigma_{\mathrm{I}} - \sigma_{\mathrm{II}} = 2\mu(\lambda_{\mathrm{I}} \ln(\lambda_{\mathrm{I}}) - \lambda_{\mathrm{II}} \ln(\lambda_{\mathrm{II}})).$$

In the small-strain limit, the logarithmic strain approaches the engineering strain (see Section 3.1.6) and comparison of this expression with Hooke's law then reveals that μ is the shear modulus. In order to validate the model over a range of material data, a plot was made of

$$\frac{\sigma_{\mathrm{I}} - \sigma_{\mathrm{II}}}{2\mu} \text{ versus } \{\lambda_{\mathrm{I}} u'(\lambda_{\mathrm{I}}) - \lambda_{\mathrm{II}} u'(\lambda_{\mathrm{II}})\}/2\mu$$

to give a common plot of slope unity. This is shown in Figure 3.11, which combines data on natural rubber from three sources. The analysis gives excellent support to the Valanis–Landel proposal.

The form (3.66) of U has also been used by Ogden (1972), who proposed an n-term energy function

$$U = \sum_n \frac{\mu_n}{\alpha_n} (\lambda_{\mathrm{I}}^{\alpha_n} + \lambda_{\mathrm{II}}^{\alpha_n} + \lambda_{\mathrm{III}}^{\alpha_n} - 3). \tag{3.68}$$

For $n = 1$, u in Equation (3.66) is a power law function

$$u(\lambda) = \frac{\mu_1}{\alpha_1} \lambda^{\alpha_1} - 1$$

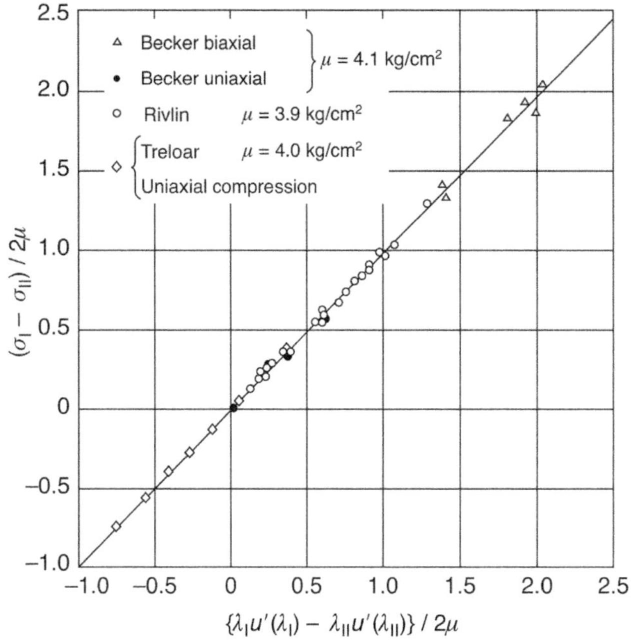

Figure 3.11 Various data sets plotted according to the proposal of Valanis and Landel. Valanis, K.C. and Landel, R.F. (1967) / AIP Publishing.

and for higher values of n, u is a summation. Principal stress components are given via Equation (3.55) as

$$\sigma_i = \sum_n \mu_n \lambda_i^{\alpha_n} - p \quad (i = \mathrm{I},\ \mathrm{II}, \mathrm{III}).$$

Ogden showed that the data of Treloar (1944) for the stretching modes simple tension, equibiaxial tension and pure shear could be fitted well with a four-term model, in contrast to the poor fits obtained for the neo-Hookean model and the single-term Ogden model with exponent $\alpha = 1$. The results, in Figure 3.12, show a highly effective representation. There have been numerous applications of the Ogden model both to rubbers and in polymers in general, owing to its effectiveness and ease of use.

Other forms of the Landel–Valanis functions u are of course possible. Recently, Darijani et al. (2010) explored a number of possibilities, involving polynomial, logarithmic and exponential functions. In this scheme, strain-energy functions are constructed using a set of basic functions of the principal stretches. These are:

Power law $(\lambda^m - 1)$
Polynomial $(\lambda - 1)^m$
Logarithmic $(\ln \lambda)^m$
Exponential $\exp(m(\lambda - 1)) - 1$

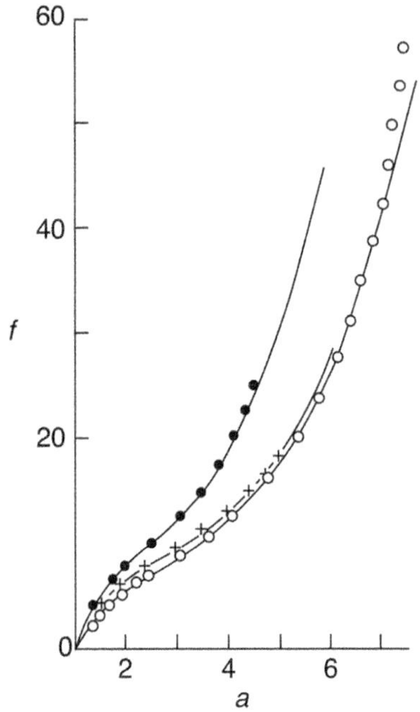

Figure 3.12 Ogden's four-term model applied to Treloar's results, plotted as nominal stress against extension ratio. ∘ – simple tension; + – pure shear; • equibiaxial tension. Ogden, 1972 / The Royal Society.

The strain-energy functions are combinations of these basic forms, set up to ensure physically realistic values of stress at zero, infinitely large or infinitely small strain. Darijani, Naghdabadi and Kargarnovin list eight possible forms of strain-energy function. For instance, the 'exponential–exponential' form is

$$u(\lambda) = \sum_{k=1}^{M} A_k(\exp(m_k(\lambda - 1)) - 1) + \sum_{k=1}^{N} B_k\left(\exp\left(n_k\left(\lambda^{-1} - 1\right)\right) - 1\right) \tag{3.69}$$

and the 'power law-logarithmic' form is

$$u(\lambda) = \sum_{k=1}^{M} A_k(\lambda^{m_k} - 1) + \sum_{k=1}^{N} B_k(\ln \lambda)^k.$$

The strain energy U is created from these functions using Equation (3.64). Darjini, Naghdabadi and Kargarnovin applied these methods to a number of published results on rubbers, including those of Treloar that are the subject of Figure 3.12. The performance of the exponential–exponential (3.69) with this set of data is examined in Figure 3.13, for two versions of the model: three-term ($M = 2$, $N = 1$) and two-term ($M = N = 1$). The three-term model gives an excellent fit to the data. Other published data, due to Kawabata et al. (1981), Lambert-Diani and Rey (1999) and Alexander (1968), were also subject to analysis using a range of the available strain-energy functions. It was concluded that strain-energy functions involving exponential terms gave the best performance.

A major conclusion of this work is that a programme of multiaxial strain experiments at a range of combinations of principal stretches is essential for the evaluation of the strain-energy function. It is additionally clear that the principal stretch approach has produced a greater number of advances in recent years than the invariant approach. It also seems more amenable to being adapted for physically based models, as it is more obviously related to the stretching of molecular chains. As pointed out by Hopkins (1976), the use of extension ratios seems wholly natural in any physical theory, and this is especially true when we are translating the extension ratios to changes in the molecular conformation, i.e. directly to events at a microscopic level. These ideas will be pursued further in the next chapter.

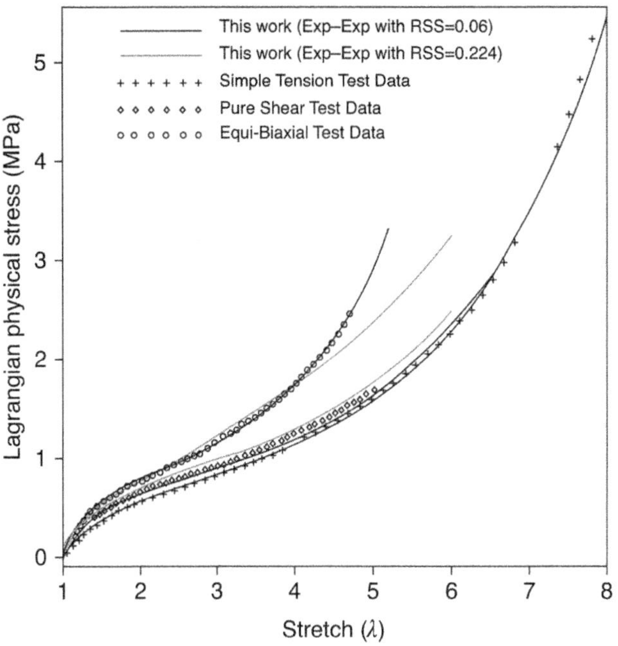

Figure 3.13 Predictions of Treloar's data using Equations (3.69) and (3.64). The RSS (residual sum of square error) = 0.06 corresponds to a three-term model, and the RSS = 0.224 to a two-term model. Darijani et al., 2010 / John Wiley & Sons.

3.4.6 Compressible Material

Strain-energy functions can be defined for which the incompressibility condition $I_3 = 1$ or equivalently $\lambda_I\lambda_{II}\lambda_{III} = 1$ does not apply. Simply copying a strain-energy function for incompressible material and relaxing the incompressibility condition will not generally result in a viable model. To derive stresses for compressible material, Equation (3.54) rather than Equation (3.55) must be applied. Suppose we take single-term Ogden model of Equation (3.68) so that with $n = 1$

$$U = \frac{\mu}{\alpha}\left(\lambda_I^\alpha + \lambda_{II}^\alpha + \lambda_{III}^\alpha - 3\right) \tag{3.70}$$

Applying Equation (3.54) to derive σ_I then gives

$$\sigma_I = \frac{1}{\lambda_{II}\lambda_{III}}\mu\lambda_I^{\alpha-1}.$$

This form has the undesirable property that, in the undeformed state $\lambda_I = \lambda_{II} = \lambda_{III} = 1$,

$$\sigma_I = \mu \neq 0.$$

In the incompressible case, the process of eliminating p produces a zero value of stress for the undeformed state independently of the form of U. For the compressible case U has to be of the correct form to give zero initial stress, and also to contain a term that represents the energy due to volume change. Introducing the invariant J

$$J = \sqrt{I_3} = \lambda_\mathrm{I}\lambda_\mathrm{II}\lambda_\mathrm{III} \tag{3.71}$$

we may generalize Equation (3.70) to, for instance,

$$U = \frac{\mu}{\alpha}\left(\lambda_\mathrm{I}^\alpha + \lambda_\mathrm{II}^\alpha + \lambda_\mathrm{III}^\alpha - 3 - \alpha\ln(J)\right) + \frac{B}{2}(J-1)^2 \tag{3.72}$$

The second term represents energy associated with volume change, parameterized by the bulk modulus type quantity B; other forms are allowable provided they give a positive contribution for all J. Applying Equation (3.54) gives for the stress along I:

$$\sigma_\mathrm{I} = \frac{\mu}{\alpha\lambda_\mathrm{II}\lambda_\mathrm{III}}\left(\alpha\lambda_I^{\alpha-1} - \alpha\frac{\lambda_\mathrm{II}\lambda_\mathrm{III}}{J}\right) + B(J-1).$$

When $\lambda_\mathrm{I} = \lambda_\mathrm{II} = \lambda_\mathrm{III} = 1$ both terms are zero and the stress along I is zero. Equation (3.72) is a valid generalization of the single-term Ogden model (3.70) for compressible material. Other models of the Valanis–Landel form can be adapted for compressibility using a similar method.

Another approach to compressibility is to begin by separating the volumetric component of deformation from the total, to give two components: volumetric or hydrostatic strain, and the remaining deviatoric strain (Kaliske and Rothert 1997). Using the principal extension ratios, we generate the deviatoric strain measures λ_i^*:

$$\lambda_i^* = \lambda_i J^{-1/3} \quad i = \mathrm{I, II, III} \tag{3.73}$$

Then,

$$\lambda_\mathrm{I}^*\lambda_\mathrm{II}^*\lambda_\mathrm{III}^* = 1.$$

We proceed by creating a strain-energy function that is the sum of separate functions of deviatoric and volumetric strain, S and B, respectively:

$$U = S\left(\lambda_\mathrm{I}^*, \lambda_\mathrm{II}^*, \lambda_\mathrm{III}^*\right) + V(J) \tag{3.74}$$

We now generate principal stresses using Equation (3.54). Adopting the notation

$$S_i = \frac{\partial S}{\partial\lambda_i^*} \quad i = \mathrm{I, II, III}$$

this results in

$$\sigma_\mathrm{I} = \frac{1}{\lambda_\mathrm{II}\lambda_\mathrm{III}}\left[S_\mathrm{I}\frac{\partial\lambda_\mathrm{I}^*}{\partial\lambda_\mathrm{I}} + S_\mathrm{II}\frac{\partial\lambda_\mathrm{II}^*}{\partial\lambda_\mathrm{I}} + S_\mathrm{III}\frac{\partial\lambda_\mathrm{III}^*}{\partial\lambda_\mathrm{I}}\right] + \frac{\partial V}{\partial J}$$

$$\sigma_\mathrm{II} = \frac{1}{\lambda_\mathrm{III}\lambda_\mathrm{I}}\left[S_\mathrm{I}\frac{\partial\lambda_\mathrm{I}^*}{\partial\lambda_\mathrm{II}} + S_\mathrm{II}\frac{\partial\lambda_\mathrm{II}^*}{\partial\lambda_\mathrm{II}} + S_\mathrm{III}\frac{\partial\lambda_\mathrm{III}^*}{\partial\lambda_\mathrm{II}}\right] + \frac{\partial V}{\partial J}$$

$$\sigma_\mathrm{III} = \frac{1}{\lambda_\mathrm{I}\lambda_\mathrm{II}}\left[S_\mathrm{I}\frac{\partial\lambda_\mathrm{I}^*}{\partial\lambda_\mathrm{III}} + S_\mathrm{II}\frac{\partial\lambda_\mathrm{II}^*}{\partial\lambda_\mathrm{III}} + S_\mathrm{III}\frac{\partial\lambda_\mathrm{III}^*}{\partial\lambda_\mathrm{III}}\right] + \frac{\partial V}{\partial J}$$

Using Equation (3.73) to create the derivatives $\frac{\partial \lambda_I^*}{\partial \lambda_I}$, etc. we find

$$\begin{aligned}
\sigma_I &= \frac{2}{3}\lambda_I^{-1/3}\lambda_{II}^{-4/3}\lambda_{III}^{-4/3}S_I - \frac{1}{3}\lambda_I^{-4/3}\lambda_{II}^{-1/3}\lambda_{III}^{-4/3}S_{II} - \frac{1}{3}\lambda_I^{-4/3}\lambda_{II}^{-4/3}\lambda_{III}^{-1/3}S_{III} + \frac{\partial V}{\partial J} \\
\sigma_{II} &= -\frac{1}{3}\lambda_I^{-1/3}\lambda_{II}^{-4/3}\lambda_{III}^{-4/3}S_I + \frac{2}{3}\lambda_I^{-4/3}\lambda_{II}^{-1/3}\lambda_{III}^{-4/3}S_{II} - \frac{1}{3}\lambda_I^{-4/3}\lambda_{II}^{-4/3}\lambda_{III}^{-1/3}S_{III} + \frac{\partial V}{\partial J} \\
\sigma_{III} &= -\frac{1}{3}\lambda_I^{-1/3}\lambda_{II}^{-4/3}\lambda_{III}^{-4/3}S_I - \frac{1}{3}\lambda_I^{-4/3}\lambda_{II}^{-1/3}\lambda_{III}^{-4/3}S_{II} + \frac{2}{3}\lambda_I^{-4/3}\lambda_{II}^{-4/3}\lambda_{III}^{-1/3}S_{III} + \frac{\partial V}{\partial J}
\end{aligned} \tag{3.75}$$

Now let us investigate the mean or hydrostatic stress $\overline{\sigma} = \frac{1}{3}(\sigma_I + \sigma_{II} + \sigma_{III})$ (see Chapter 2) by adding Equation (3.75). The coefficients of S_I, S_{II} and S_{III} are then all zero, and we have

$$\overline{\sigma} = \frac{\partial V}{\partial J}.$$

This means that the hydrostatic stress depends only on V, the volumetric strain-energy density. Likewise, according to Equation (3.75) the deviatoric stresses ($\sigma_i^{dev} = \sigma_i - \overline{\sigma}$; $i =$ I, II, III; see Chapter 2) now depend only on the deviatoric strain-energy density S. This method of creating a compressible material model is generally applicable to any incompressible strain-energy function and is used widely (e.g. Chen et al. 2017; Felder et al. 2020; Popa et al. 2020; Sweeney et al. 2021) and in the finite element package ABAQUS (Dassault Systems 2022).

The S and V functions of Equation (3.74) are sometimes referred to as energies of shape change and energies of volume change, respectively. When values of V are large compared with values of S, the material shows little resistance to shape change and is approximately incompressible; this resembles an elastic liquid. Conversely, a material with low values of V relative to S will resemble a highly compressible solid, with a Poisson's ratio near to zero when stretched in tension.

Numerical examples of both these types of material are shown in Figure 3.14 using Ogden models. A single-term compressible form is used:

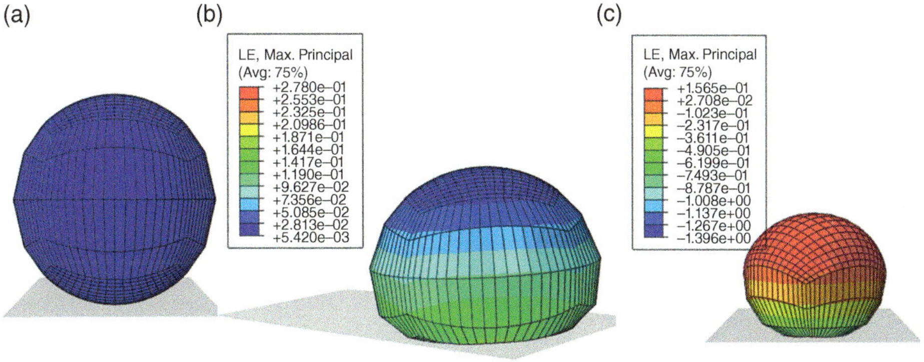

Figure 3.14 Effects for different V/S value ratios for a sphere under gravitational loading. (a) shows the undeformed case. (b) High V/S with B = 100 MPa. The material behaves like a liquid drop, spreading over the surface and keeping volume almost constant. (c) Low V/S with B = 0.01 MPa. Spherical shape is largely retained, but volume decreases markedly.

$$U = \frac{\mu}{\alpha}\left(\lambda_{\mathrm{I}}^{*} + \lambda_{\mathrm{II}}^{*} + \lambda_{\mathrm{III}}^{*} - 3\right) + B(J-1)^2.$$

In both cases $\boldsymbol{\mu} = \mathbf{0.1}$ **MPa**, $\boldsymbol{\alpha} = \mathbf{2.5}$. For the high V, low S material, $B = 100$ MPa; and for the high S, low V, material $B = 0.01$ MPa. Results are for an initially spherical body of material resting on a rigid surface subject to gravitational loading. Figure 3.14(a) shows the undeformed state, with Figure 3.14(b) showing the case $B = 100$ MPa and 3.14(c) shows the case $B = 0.01$ MPa. Densities differ with values 2.5×10^{-5} kg/mm^3 for case (b) and 5.0×10^{-6} kg/mm^3 for case (c). The analysis was performed using ABAQUS Standard (Dassault Systems 2022).

References

Alexander, H. (1968) A constitutive relation for rubber-like materials. *Int. J. Eng. Sci.*, **6**, 549.

Alibakhshi, A. and Heidari, H. (2020) Nonlinear dynamics of dielectric elastomer balloons based on the Gent-Gent hyperelastic model. *Eur. J. Mech. A Solids*, **82**, 103986.

Baker, G. (2005) Thermodynamics in solid mechanics: a commentary. *Phil. Trans. R. Soc. A*, **363**, 2465.

Chen, L., Chang, Z. and Qin, T. (2017) Elastic wave propagation in simple-sheared hyperelastic materials with different constitutive models. *Int. J. Solids Struct.*, **126–127**, 1.

Darijani, H., Naghdabadi, R. and Kargarnovin, M.H. (2010) Constitutive modeling of rubber-like materials based on consistent strain energy density functions. *Polym. Eng. Sci.*, **50**, 1058.

Dassault Systèms (2022), Johnston, RI, USA.

Felder, S., Holthusen, H., Hesseler, S., Pohlkemper, F., Gries, T., Simon, J.W. and Reesae, S. (2020) Incorporating crystallinity distributions into a thermos-mechanically coupled constitutive model for semi-crystalline polymers. *Int. J. Plast.*, **135**, 102751.

Gent, A.N. (1996) A new constitutive relation for rubber. *Rubber Chem. Technol.*, **69**, 59–61.

Green, A.E. and Adkins, J.E. (1970) *Large Elastic Deformations and Non-Linear Continuum Mechanics*, Clarendon Press, Oxford.

Green, A.E. and Zerna, W. (2002) *Theoretical Elasticity*, Courier Dover Publications, Mineola, NY.

Hibbeler, R.C. (2004) *Mechanics of Materials*, 6th edn, Prentice Hall, Upper Saddle River, NJ.

Hopkins, H.G. (1976) The mechanics of rubber elasticity [and discussions]. *Proc. R. Soc. A*, **351**, 322.

Horgan, C.O. (2015) The remarkable Gent constitutive model for hyperelastic materials. *Int. J. Non-Linear Mech.*, **68**, 9–16.

Houlsby, G.T. and Puzrin, A.M. (2006) *Principles of Hyperplasticity*, Springer, London.

Kaliske, M. and Rothert, H. (1997) On the finite element implementation of rubber-like materials at finite strains. *Eng. Comput.*, **14**, 216–232.

Kawabata, S., Matsuda, M., Tei, K. and Kawai, H. (1981) Experimental survey of the strain energy density function of isoprene rubber vulcanizate. *Macromolecules*, **14**, 154.

Lambert-Diani, J. and Rey, C. (1999) New phenomenological behavior laws for rubbers and thermoplastic elastomers. *Eur. J. Mech. A Solids*, **18**, 1027.

Mooney, M. (1940) A theory of large elastic deformation. *J. Appl. Phys.*, **2**, 582.

Nadai, A. (1950) *Theory of Flow and Fracture of Solids*, Vol. **1**, 2nd edn, McGraw-Hill, New York.

Noll, W. (1958) A mathematical theory of the mechanical behavior of continuous media. *Arch. Ration. Mech. Anal.*, **2**, 197.

Ogden, R.W. (1972) Large deformation isotropic elasticity – on the correlation of theory and experiment for incompressible rubber-like solids. *Proc. R. Soc. A*, **326**, 565.

Ogden, R.W. (1997) *Non-linear Elastic Deformations*, Courier Dover Publications, Mineola, NY.

Popa, C.M., Gebhardt, C., Raje, N., Steenwyk, B. and Kaliske, M. (2020) Investigation of cord-rubber composite durability by the material force method. *Eng. Fract. Mech.*, **229**, 106909.

Pucci, E. and Saccomandi, G. (2002) A note on the Gent model for rubber-like materials. *Rubber Chem. Technol.*, **75**, 839–852.

Puglisi, G. and Saccomandi, G. (2015) The Gent model for rubber-like materials: an appraisal for an ingenious and simple idea. *Int. J. Non-Linear Mech.*, **68**, 17–24.

Rees, D.W.A. (2006) *Basic Engineering Plasticity*, Butterworth-Heinemann, Oxford.

Rivlin, R.S. (1949) Large elastic deformations of isotropic materials. I. Fundamental concepts. *Phil. Trans. R. Soc. A*, **240**, 459.

Rivlin, R.S. and Saunders, D.W. (1951) Large elastic deformations of isotropic materials. VII. Experiments on the deformation of rubber. *Phil. Trans. R. Soc. A*, **243**, 251.

Sweeney, J., Spencer, P.E., Thompson, G.P., Barker, D.J. and Coates, P.D. (2021) Constitutive modelling of polylactic acid at large deformation using multiaxial strains. *Polymers*, **13**, 2967.

Treloar, L.R.G. (1943) The elasticity of a network of long-chain molecules. II. *Trans. Faraday Soc.*, **39**, 241.

Treloar, L.R.G. (1944) Stress-strain data for vulcanised rubber under various types of deformation. *Trans. Faraday Soc.*, **40**, 59.

Treloar, L.R.G. (2005) *The Physics of Rubber Elasticity*, Oxford University Press, Oxford.

Vahapoğlu, V. and Karadeniz, S. (2006) Constitutive equations for isotropic rubber-like materials using phenomenological approach: a bibliography (1930–2003). *Rubber Chem. Technol.*, **79**, 489–499.

Valanis, K.C. and Landel, R.F. (1967) The strain-energy function of a hyperelastic material in terms of the extension ratios. *J. Appl. Phys.*, **38**, 2997.

Yeoh, O.H. (1993) Some forms of the strain energy function for rubber. *Rubber Chem. Technol.*, **66**, 754–771.

Further Reading

Holzapfel, G.A. (2000) *Nonlinear Solid Mechanics: A Continuum Approach for Engineering*. John Wiley and Sons Ltd., Chichester.

4

Rubber-Like Elasticity

4.1 General Features of Rubber-Like Behaviour

The most noticeable feature of natural rubber and other elastomers is the ability to undergo large and reversible elastic deformation. It is not unexpected that stress can cause polymeric molecules to adopt an extended configuration, but at first sight it may seem surprising that on removal of the stress the molecules retract, on average, to their initial coiled form. Simple theories of rubber-like elasticity assume, as an approximation, that both extension and retraction occur instantaneously, and neglect any permanent deformation. Natural rubber (*cis*-polyisoprene) in its native state does not satisfy this last criterion, as molecules in extended configurations tend to slide past one another and do not recover completely. Molecules need to be chemically cross-linked by sulphur bonds (vulcanization) to prevent any permanent flow, and we shall show that the degree of cross-linking determines the extensibility of the rubber for a given stress.

The application of stress is considered to cause molecules to change from a coiled to an extended configuration instantaneously. For this reason, it is possible to apply equilibrium thermodynamics to determine how the stress is related to changes in both internal energy and entropy. The general nature of thermodynamics implies that this type of approach can give no direct information on molecular rearrangements but, when augmented by molecular theories of a statistical nature, it is possible to derive an equation of state that relates the force causing extension to molecular parameters. We will show that this equation of state is identical to the neo-Hookean form of Section 3.4.4, which was derived as the equivalent of Hooke's law for finite deformations. It will be shown that the reason for this direct link between the behaviour at a molecular level and the mechanics of finite elasticity arises because of the applicability of the so-called affine deformation assumption. Affine deformation in the molecular theory of rubber elasticity means that we can assume that the changes in the length and orientation of lines joining adjacent cross-links in the molecular network are identical to the changes in lines marked on the macroscopic rubber. It is also assumed that there is no change in volume on deformation. This assumption can be justified as a good approximation because the bulk modulus (K) is some 10^4 times greater than the shear modulus (G): typical values are 10^{10} Pa and 10^6 Pa. As a consequence, at low strains, Poisson's ratio, given by

Mechanical Properties of Solid Polymers, Fourth Edition. John Sweeney and Peter Hine.

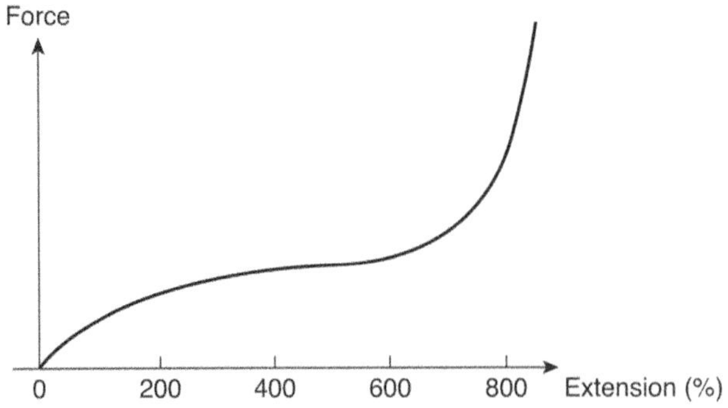

Figure 4.1 The force developed in uniaxial extension of a typical lightly cross-linked rubber.

$$\nu = \frac{3K - 2G}{2(3K + 2G)}$$

is effectively ½ and deformation occurs essentially at constant volume.

A schematic force–extension curve for a typical rubber is shown in Figure 4.1, with the maximum extensibility varying between 500 and 1000%, depending on the extent of cross-linking. The behaviour is Hookean, with a linear relationship between stress and strain only at strains of the order of 1% of so. At larger strains the force–extension relation is non-linear, and we will show that its form is determined essentially by changes in configurational entropy rather than internal energy.

High extension results in a greatly reduced entropy, so that retraction is a consequence of the necessity for entropy to be maximized. A fully extended chain is a state of zero entropy because there is only one possible conformation of bonds through which it can occur. In contrast, there are a very large number of ways of obtaining a given end-to-end distance for a contracted configuration of the chain. As all configurations have approximately the same internal energy, in the absence of external stress an extended chain will return to a more probable state. For this reason, rubber is sometimes referred to as a 'probability spring' or 'entropy spring', in contrast to the 'energy spring' characteristics of the elasticity of materials of low molecular mass, where extension causes an increase in internal energy. For a fuller discussion, see Treloar (2005).

Useful reviews of the subject area are available such as Boyce and Arruda (2000) and Mihai and Goriely (2017).

4.2 The Thermodynamics of Deformation

The change in internal energy during deformation $\mathrm{d}U$ is given by

$$\mathrm{d}U = \mathrm{d}Q + \mathrm{d}W, \tag{4.1}$$

where $\mathrm{d}Q$ and $\mathrm{d}W$ are the heat absorbed and the work due to external forces, respectively. For a reversible process $\mathrm{d}Q$ relates to the entropy change $\mathrm{d}S$ by the relationship

$$\mathrm{d}Q = T\,\mathrm{d}S. \tag{4.2}$$

Hence, for a reversible process we have

$$dU = T\,dS + dW. \tag{4.3}$$

When an elastic solid of initial length L is extended uniaxially under a tensile force f, the work done on the solid in an infinitesimal displacement is

$$dW = f\,dL. \tag{4.4}$$

By combining Equations (4.3) and (4.4) we have

$$dU = f\,dL + T\,dS. \tag{4.5}$$

It is convenient to introduce the Helmholtz free energy A, which relates to changes that occur at constant volume (see Section 3.4.1)

$$A = U - TS. \tag{4.6}$$

Thus, for a change that occurs at constant temperature and constant volume we have

$$dA = dU - T\,dS. \tag{4.7}$$

By combining Equations (4.3) and (4.7) we see that $dA = dW$ for isothermal changes at constant volume. Hence the tension f is given by

$$f = \left(\frac{\partial W}{\partial L}\right)_{T,V} = \left(\frac{\partial A}{\partial L}\right)_T = \left(\frac{\partial U}{\partial L}\right)_T - T\left(\frac{\partial S}{\partial L}\right)_T. \tag{4.8}$$

For any change in the Helmholtz free energy A (not necessarily an isothermal change)

$$dA = dU - T\,dS - S\,dT. \tag{4.9}$$

But $dU = f\,dL + T\,dS$ from Equation (4.3); hence,

$$dA = f\,dL - S dT. \tag{4.10}$$

Then

$$\left(\frac{\partial A}{\partial L}\right)_T = f \quad \text{and} \quad \left(\frac{\partial A}{\partial T}\right)_L = -S \tag{4.11}$$

But

$$\frac{\partial}{\partial l}\left(\frac{\partial A}{\partial T}\right)_L = \frac{\partial}{\partial T}\left(\frac{\partial A}{\partial L}\right)_T.$$

Substituting

$$\left(\frac{\partial S}{\partial L}\right)_T = -\left(\frac{\partial f}{\partial T}\right)_L. \tag{4.12}$$

Hence, Equation (4.8) becomes

$$\left(\frac{\partial U}{\partial L}\right)_T = f - T\left(\frac{\partial f}{\partial T}\right)_L. \tag{4.13}$$

'freely jointed' chain, a mathematical abstraction in which the atoms are reduced to mere points, joined by one-dimensional equal links, with no restriction on the angle between adjacent links. It is assumed in this simple model that there is no difference in internal energy between the different molecular conformations along the chain.[1]

4.3.2 Average Length of a Molecule between Cross-Links

Consider a freely jointed chain with n links, each of vector length ℓ_i. The vector end-to-end length of a single chain is then given by

$$\mathbf{r} = \sum_{i=1}^{n} \ell_i. \tag{4.14}$$

For a large number of chains (q), or one chain considered at many different times, the mean length

$$\bar{\mathbf{r}}(q) = \frac{1}{q}\sum_{i=1}^{q} \mathbf{r}_i = 0 \tag{4.15}$$

as the vector length is equally likely to be positive or negative.

We follow the procedure used, for example, with sinusoidally varying quantities such as alternating current and voltage, where the mean value is zero, and calculate the mean square chain length

$$\overline{r^2} = \frac{1}{q}\sum_{j=1}^{q} \mathbf{r}_j^2 = \frac{1}{q}\sum_{j=1}^{q}\left(\sum_{i=1}^{n} \ell_i\right)_j^2. \tag{4.16}$$

Expand, giving

$$\overline{r^2} = \frac{1}{q}\sum_{1}^{q}\left(\ell_1^2 + \ell_2^2 + \cdots \ell_n^2 + \underline{\ell}_1 \bullet \underline{\ell}_2 + \underline{\ell}_1 \bullet \underline{\ell}_3 + \cdots \underline{\ell}_{n-1} \bullet \underline{\ell}_n\right)$$

but $\ell_1^2 = \ell_2^2 = \cdots \ell_n^2$ and $\underline{\ell}_m \bullet \underline{\ell}_n = \ell_m \ell_n \cos\theta$. In a freely jointed chain, θ can have any value with equal probability; hence, $\sum \underline{\ell}_m \bullet \underline{\ell}_n = 0,\ m \neq n$. Therefore

$$\overline{r^2} = \frac{1}{q}\sum_{1}^{q} n\ell^2 = n\ell^2.$$

The root mean square chain length is then

$$\sqrt{\overline{r^2}} = \ell\sqrt{n} = r_{\text{rms}} \tag{4.17}$$

compared with the fully extended chain length of ln; that is, for a chain of 100 bonds between cross-links the maximum extensibility is 10.

---

1 Conformation is used to denote differences in the immediate situation of a bond, for example, *trans* or *gauche* conformations. Configuration is retained to refer to the arrangement of the whole molecular chain.

4.3.3 The Entropy of a Single Chain

The expression just derived indicates the reason for the high extensibility of lightly cross-linked rubbers and serves to introduce the important concept of a mean square length, but yields no information on the probability of a chain having a particular end-to-end length. This latter problem was first analysed mathematically by Kuhn (1934, 1936) and by Guth and Mark (1934).

Consider a chain of n links each of length l, which has a configuration such that one end P is at the origin (Figure 4.4). The probability distribution for the position of the end Q is derived using approximations that are valid provided that the distance between the chain ends P and Q is much less than the extended chain length nl. The probability that Q lies within the elemental volume $\mathrm{d}X_1\ \mathrm{d}X_2\ \mathrm{d}X_3$ at the point (X_1, X_2, X_3) can be shown to be

$$p(X_1, X_2, X_3)\mathrm{d}X_1\mathrm{d}X_2\mathrm{d}X_3 = \frac{b^3}{\pi^{3/2}} \exp(-b^2r^2)\mathrm{d}X_1\mathrm{d}X_2\mathrm{d}X_3, \quad (4.18)$$

where $b^2 = 3/2nl^2$.

This distribution has the form of the Gaussian error function and is spherically symmetrical about the origin, where the value is a maximum (Figure 4.5). The most probable

Figure 4.4 The freely jointed chain.

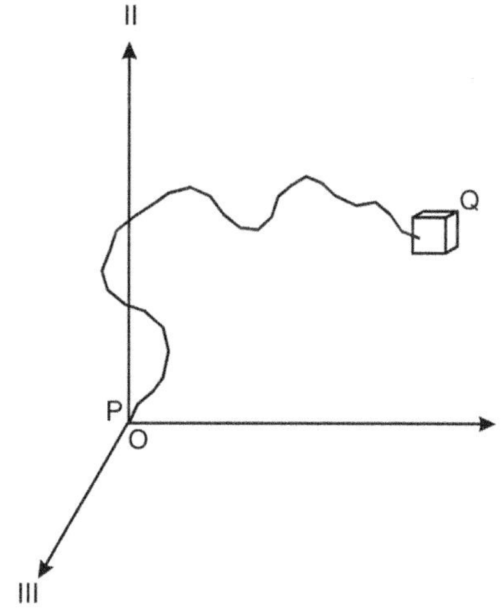

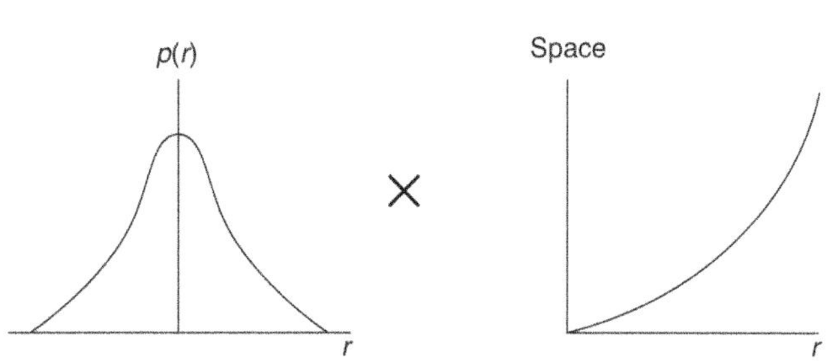

Figure 4.5 The Gaussian probability distribution for the free end of a chain must be multiplied by the volume in which that end can reside: $P(r) = p(r) \times 4\pi r^2\mathrm{d}r$.

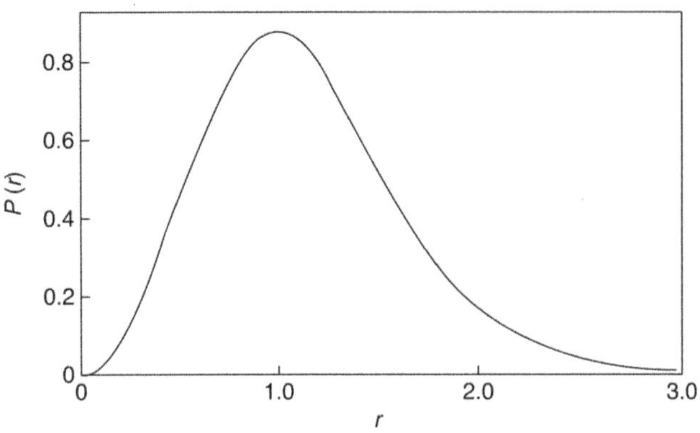

Figure 4.6 The distribution function $P(r)$ = constant × r^2 exp $(-b^2r^2)$.

end-to-end length is not, however, zero, as the probability that Q falls within an elemental volume situated between r and $(r + \mathrm{d}r)$ from the origin, irrespective of direction, is the product of the probability distribution $p(r)$ and the volume of the concentric shell, $4\pi r^2\mathrm{d}r$. The overall probability is then

$$\begin{aligned} P(r)\mathrm{d}r &= p(r)4\pi r^2\mathrm{d}r = \left(b^3/\pi^{3/2}\right)\exp\left(-b^2r^2\right)4\pi r^2\mathrm{d}r \\ &= \left(4b^3/\pi^{1/2}\right)r^2\exp\left(-b^2r^2\right)\mathrm{d}r, \end{aligned} \tag{4.19}$$

which is illustrated in Figure 4.6.

It is seen that the most probable end-to-end distance, irrespective of direction, is not zero, but it is a function of b, that is, of the length l of the links and the number n of links in the chain, as shown in Section 4.3.2.

Another important quantity is the root mean square chain length $\left(\overline{r^2}\right)^{1/2}$

$$\overline{r^2} = \int_0^\infty r^2 P(r)\mathrm{d}r.$$

Substitution of the above expression for $P(r)$ gives

$$\overline{r^2} = 3/2b^2 = n\ell^2, \tag{4.20}$$

so that the root mean square length $\left(\overline{r^2}\right)^{1/2} = \ell\sqrt{n}$, that is, it is proportional to the square root of the number of links in the chain, as shown in Section 4.3.2.

The entropy of the freely jointed chain s is proportional to the logarithm of the number of configurations Q so that

$$s = k \ln \Omega,$$

where k is Boltzmann's constant. If $\mathrm{d}x_1\, \mathrm{d}x_2\, \mathrm{d}x_3$ is constant, the number of configurations available to the chain is proportional to the probability per unit volume $p(x_1, x_2, x_3.)$. The entropy of the chain is thus, according to Equation (4.18), given by

$$s = c - kb^2r^2 = c - kb^2\left(x_1^2 + x_2^2 + x_3^2\right), \tag{4.21}$$

where c is an arbitrary constant.

4.3.4 The Elasticity of a Molecular Network

We wish to calculate the strain energy function for a molecular network, assuming that this is given by the change in entropy of a network of chains as a function of strain.

The actual network is replaced by an ideal network in which each segment of a molecule between successive points of cross-linkage is considered to be a Gaussian chain.

Three additional assumptions are introduced:

1) In either the strained or unstrained state, each junction point may be regarded as fixed at its mean position.
2) The effect of the deformation is to change the components of the vector length of each chain in the same ratio as the corresponding dimensions of the bulk material (the 'affine' deformation assumption).
3) The mean square end-to-end distance for the whole assembly of chains in the unstrained state is the same as for a corresponding set of free chains and is therefore given by Equation (4.20).

In effect, it is necessary to calculate the difference in probability between a spherical distribution of chain end-to-end vectors in the unstrained state and an ellipsoidal distribution for uniaxial extension (Figure 4.7). This difference is related to changes in entropy, and so to tensile force.

As discussed in Section 2.2, we can restrict our discussion to the case of normal strain without loss of generality. We choose principal extension ratios λ_{I}, λ_{II} and λ_{III}, parallel to the three principal coordinate axes I, II and III. The affine deformation assumption implies that the relative displacement of the chain ends is defined by the macroscopic deformation. Thus, in Figure 4.8 we take a system of coordinates I, II and III in the undeformed body.

In this coordinate system, a representative chain PQ has one end P at the origin. We refer any point in the deformed body to this system of coordinates. Thus, the origin, i.e., the end of

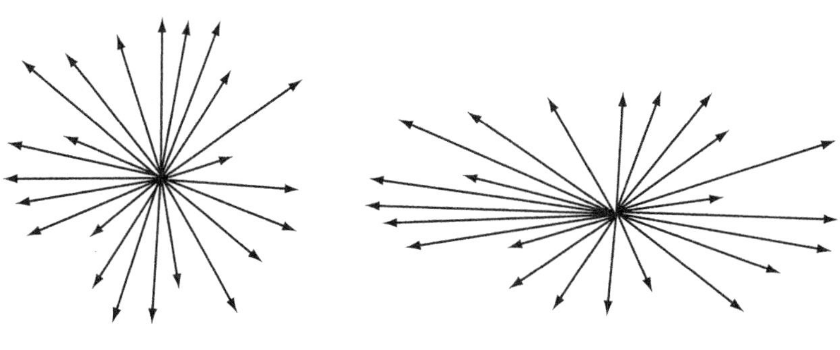

Figure 4.7 Schematic representation of chain end-to-end vectors in the initial and final states.

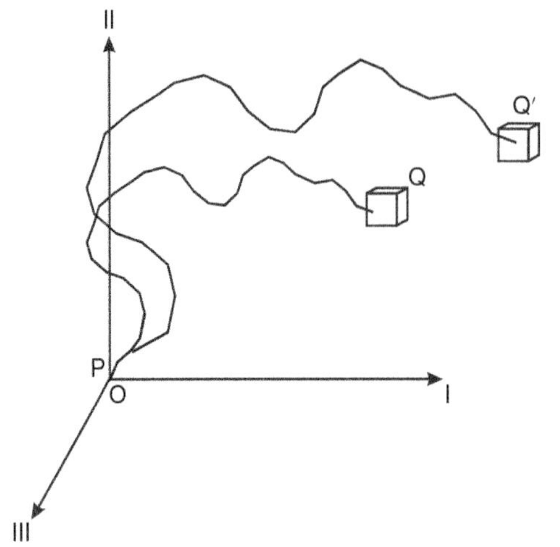

Figure 4.8 The end of the chain $Q(X_I, X_{II}, X_{III})$ is displaced to the point $Q'(x_I, x_{II}, x_{III})$.

the chain P, could be moved bodily during the deformation. The other end $Q(X_I, X_{II}, X_{III})$ is displaced to the point $Q'(x_I, x_{II}, x_{III})$, and from the affine deformation assumption we have (see Section 3.1.2)

$$x_I = \lambda_I X_I, x_{II} = \lambda_{II} X_{II}, x_{III} = \lambda_{III} X_{III}.$$

The entropy of the chain in the undeformed state is given by

$$s = c - kb^2\left(X_I^2 + X_{II}^2 + X_{III}^2\right).$$

After deformation the entropy becomes

$$s' = c - kb^2\left(\lambda_I^2 X_I^2 + \lambda_{II}^2 X_{II}^2 + \lambda_{III}^2 X_{III}^2\right), \tag{4.22}$$

giving the change in entropy

$$\Delta s = s' - s = -kb^2\left\{\left(\lambda_I^2 - 1\right)X_I^2 + \left(\lambda_{II}^2 - 1\right)X_{II}^2 + \left(\lambda_{III}^2 - 1\right)X_{III}^2\right\}. \tag{4.23}$$

Let there be N chains per unit volume in the network, with m of these having a given value of b (say b_p). The total entropy change for this particular group of chains is

$$\Delta s_b = \sum_1^m \Delta s = kb_p^2\left\{\left(\lambda_I^2 - 1\right)\sum_1^m X_I^2 + \left(\lambda_{II}^2 - 1\right)\sum_1^m X_{II}^2 + \left(\lambda_{III}^2 - 1\right)\sum_1^m X_{III}^2\right\}, \tag{4.24}$$

where $\sum_1^m X_I^2$ is the sum of the squares of the I components for these m chains in the undeformed network. As there is no preferred direction for the chain vectors in the undeformed (isotropic) state, there is no preference for the I, II or III directions, so that

$$\sum_1^m X_I^2 = \sum_1^m X_{II}^2 = \sum_1^m X_{III}^2,$$

but

$$\sum_1^m X_{\rm I}^2 + \sum_1^m X_{\rm II}^2 + \sum_1^m X_{\rm III}^2 = \frac{1}{3}\sum_1^m r^2,$$

giving

$$\sum_1^m X_{\rm I}^2 = \sum_1^m X_{\rm II}^2 = \sum_1^m X_{\rm III}^2 = \frac{1}{3}\sum_1^m r^2. \tag{4.25}$$

From Equation (4.20)

$$\sum_1^m r^2 = mr^2 = m\left(\frac{3}{2b_p^2}\right). \tag{4.26}$$

Combining Equations (4.25), (4.26) and (4.24)

$$\Delta s_b = -\frac{1}{2}mk\{\lambda_{\rm I}^2 + \lambda_{\rm II}^2 + \lambda_{\rm III}^2 - 3\}. \tag{4.27}$$

We can now add the contribution of all the chains in the network (N per unit volume) and obtain the entropy change of the network S where

$$\Delta S = \sum_1^N \Delta s = -\frac{1}{2}Nk\left(\lambda_{\rm I}^2 + \lambda_{\rm II}^2 + \lambda_{\rm III}^2 - 3\right). \tag{4.28}$$

Assuming no change in internal energy on deformation, this gives the change in the Helmholtz free energy

$$\Delta A = -T\Delta S = \frac{1}{2}NkT\left(\lambda_{\rm I}^2 + \lambda_{\rm II}^2 + \lambda_{\rm III}^2 - 3\right).$$

If we assume that the strain energy function U is zero in the undeformed state, this gives

$$U = \Delta A = \frac{1}{2}NkT\left(\lambda_{\rm I}^2 + \lambda_{\rm II}^2 + \lambda_{\rm III}^2 - 3\right). \tag{4.29}$$

The above equation reproduces the neo-Hookean strain energy form of Section 3.4.4. Consider a simple uniaxial elongation $\lambda_{\rm I} = \lambda$. The incompressibility condition gives $\lambda_{\rm I}\lambda_{\rm II}\lambda_{\rm III} = 1$. Hence, by symmetry $\lambda_{\rm II} = \lambda_{\rm III} = \lambda^{-1/2}$ and

$$U = \frac{1}{2}NkT\left(\lambda^2 + \frac{2}{\lambda} - 3\right). \tag{4.30}$$

From Equation (3.43), the nominal stress along I denoted here denoted by f is

$$f = \frac{\partial U}{\partial \lambda} = NkT\left(\lambda - \frac{1}{\lambda^2}\right). \tag{4.31}$$

Comparison of Equation (4.31) with Equation (3.43) shows that $NkT = E/3$, where E is Young's modulus. For small strain, we can put $\lambda = 1 + e_{\rm I}$ and it follows from Equation (4.31) that

$$f = 3NkTe_{\rm I} = Ee_{\rm I}.$$

Since for an incompressible material $E \equiv 3G$, we see that the quantity NkT in Equation (4.31) is equivalent to the shear modulus of rubber, G. This term is sometimes written in terms of the mean molecular mass M_c of the chains, that is, between successive points of cross-linkage. Then

$$G = NkT = \rho RT/M_c,$$

where ρ is the density of the rubber and R is the gas constant.

4.4 Modifications of Simple Molecular Theory

The theory of the deformation of a molecular network described above formed the starting point for substantial research activity in polymer science. It does, however, rest on many simplifying assumptions, most notably freely jointed chains that can deform independently with a Gaussian distribution of end-to-end lengths that give rise to no changes in internal energy on deformation. There have been a considerable number of different approaches to developing a better theoretical basis, and because these often address different issues it is convenient to consider them in turn. At this stage, it is important to recognize that in the discussion of this chapter, molecular theories are being considered to refine and modify the theory of rubber elasticity as distinct from phenomenological equations based on solid mechanics that have been described in Chapter 3. Because both approaches often address similar issues relating to the deviation of experimental results from the simple equation of state (4.30), some authors such as Meissner (2000) have combined the two approaches in an ad hoc manner to achieve satisfactory empirical fits to experimental data. Boyce and Arruda (2000) have reviewed both approaches.

4.4.1 The Phantom Network Model

A key assumption of the single molecular theory is that the junction points in the network move affinely with the macroscopic deformation; that is, they remain fixed in the macroscopic body. It was soon proposed by James and Guth (1947) that this assumption is unnecessarily restrictive. It was considered adequate to assume that the network junction points fluctuate around their most probable positions (James and Guth 1947; Mark and Erman 1988) and the chains are portrayed as being able to transect each other. This has been termed the phantom network model. The vector $\mathbf{r}$ joining the two junction points is considered as the sum of a time average mean $\bar{\mathbf{r}}$ and the instantaneous fluctuation $\Delta\mathbf{r}$ from the mean so that

$$\mathbf{r} = \bar{\mathbf{r}} + \Delta\mathbf{r},$$

where the fluctuations $\Delta\mathbf{r}$ are independent of deformation and the mean vectors $\bar{\mathbf{r}}$ deform affinely following the macroscopic strain. This changes the $\frac{1}{2}NkT$ term in Equation (4.29) by a factor $\xi = 1 - 2/\varphi$, where φ is the number of chains emanating from a network junction point, so that $G = \frac{1}{2}(1 - 2/\phi)NkT$.

4.4.2 The Constrained Junction Model

Later network models of rubber elasticity by Ronca and Allegra (1975) and Flory and Erman (1982) are based on the phantom network model but assume that the junction point fluctuations are restricted due to the presence of entanglements. The strength of the constraints is defined by a parameter

$$\kappa = \frac{< (\Delta R)^2 >}{< (\Delta s)^2 >},$$

where $< (\Delta R)^2 >^{1/2}$ is the radius of the average region in which the junction points would fluctuate in the absence of constraints and $< (\Delta s)^2 >$ is the average radius of the region in which the junction points would fluctuate under the action of constraints only. The total Helmholtz free energy is then

$$A = \frac{1}{2}\xi kT\left(\lambda_{\mathrm{I}}^2 + \lambda_{\mathrm{II}}^2 + \lambda_{\mathrm{III}}^2 - 3\right) + \frac{1}{2}NkT\sum_{i=I}^{\mathrm{III}}[B_i + D_i - \ln(B_i + 1) - \ln(D_i + 1)], \tag{4.32}$$

where $B_i = \kappa^2\left(\lambda_i^2 - 1\right)\left(\lambda_i^2 + \kappa\right)^{-2}$ and $D_i = \lambda_i^2\kappa^{-1}B_i$ for principal directions i = I, II and III.

Later refinements of the constrained junction model place the effects of the constraints on the centres of mass of the network chains (Erman and Monnerie 1989) or consider distributing the constraints continuously along the chains (Kloczkowski *et al.* 1995).

4.4.3 The Slip-Link Model

Edwards and co-workers have proposed a different molecular model that subsumes the considerations discussed above, which are based on phantom, affine and constraint junctions to describe the deformation of a molecular network. Edwards and his colleagues have advanced purely geometrical ideas to explain both the strain softening at low strains and the rapid strain hardening at high strains. Ball *et al.* (1981) considered that in a rubber network there will be two types of junction points: the permanent cross-links where the chains are joined by chemical bonds and entanglements that are temporary junction points caused by chains becoming entangled. In an ingenious intuitive step, these temporary junction points are replaced by slip-links (Figure 4.9), which allow the chains to slide over one another. The *change* in free energy (note that in these theories the energy function is not zero for the undeformed state) associated with the slip-links that results from the deformation is given by

$$\frac{F_s}{kT} = \frac{1}{2}N_s\sum_{i=I}^{\mathrm{III}}\left[\frac{(1+\eta)\lambda_i^2}{1+\eta\lambda_i^2} + \log\left(1+\eta\lambda_i^2\right)\right], \tag{4.33}$$

where N_s is the number of slip-links per unit volume, the λ_i are the principal extension ratios and η is the slipperiness factor that in principle can vary between infinity for perfect sliding and zero for no sliding, i.e., a cross-link.

The total free energy change is then

$$F = F_c + F_s, \tag{4.34}$$

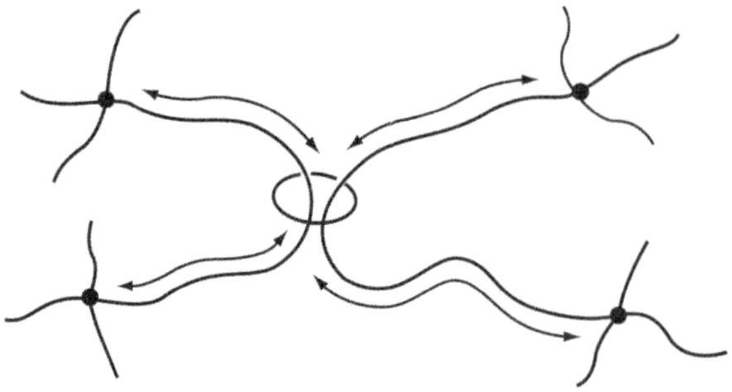

Figure 4.9 Slip-link model. Ball et al. (1981) / with permission of Elsevier.

where

$$\frac{F_c}{kT} = \frac{1}{2} N_c \sum_{i=I}^{III} \lambda_i^2 \tag{4.35}$$

and N_c is the number of permanent cross-links per unit volume.

It can be seen that the first term in Equation (4.34) is the free energy of a phantom network, equivalent to the first term in Equation (4.32) of the constrained junction model. The second term can be thought of as equivalent to the second term in Equation (4.32), but because there is no fixed limit to the ratio of slip-links N_s to permanent cross-links N_c, the maximum value of the free energy can be greater than that of the affine network that would replace $\frac{1}{2}N_c$ in the first term by N_c.

Edwards and Vilgis (1986) subsequently extended this theory to include the limitation of finite extensibility for the network that they regarded as arising when the network chains are fully extended, that is, $\lambda_{\max} \propto n^{1/2}$, where n is the number of links in the polymer chain between junction points (see Equation (4.17)).

This introduces into Equations (4.33) and (4.35) another constant α, simply related to the first invariant I_1, (see Chapter 3) given by $\alpha = 1/\sqrt{(I_1)_{\max}} = \left[\left(\lambda_I^2 + \lambda_{II}^2 + \lambda_{III}^3\right)_{\max}\right]^{-1/2}$. The terms involving α impose a limit on strain by inducing a singularity in the strain energy. The cross-link and slip-link energy changes are then

$$\frac{F_c}{kT} = \frac{1}{2} N_c \left[\frac{(1-\alpha^2)I_1}{1-\alpha^2 I_1} + \ln\left(1-\alpha^2 I_1\right)\right] \tag{4.36}$$

and

$$\frac{F_s}{kT} = \frac{1}{2} N_s \left[\sum_{i=I}^{III} \left\{\frac{\lambda_i^2(1+\eta)(1-\alpha^2)}{\left(1+\eta\lambda_i^2\right)(1-\alpha^2 I_1)} + \ln\left(1+\eta\lambda_i^2\right)\right\} + \ln\left(1-\alpha^2 I_1\right)\right] \tag{4.37}$$

Edwards and Vilgis showed that their theory could produce an excellent fit to data for natural rubber by assuming reasonable values of N_c, N_s, η and α (Figure 4.10).

Following the ideas of Edwards and co-workers, Heinrich and Kaliske (1997) proposed a tube model, and later Kaliske and Heinrich (1999) a constrained tube model. The Gaussian

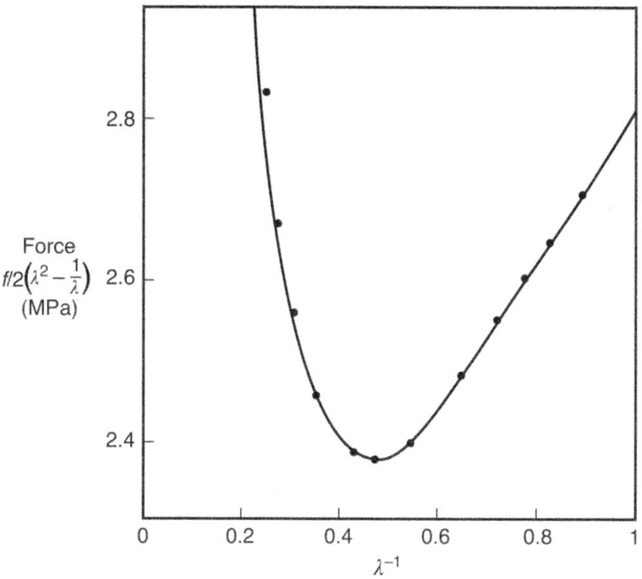

Figure 4.10 Stress–strain data for natural rubber according to Mullins. The solid line is calculated by the Edwards and Vilgis theory using the parameters $N_ckT = 1.2$ MPa, $N_skT = 2.1$ MPa, $\eta = 0.2$, $\alpha^{-1} = \lambda_{max} = 7.5$. Edwards et al. (1986) / with permission of Elsevier.

distribution of the Edwards papers is replaced by the non-Gaussian Langevin distribution (see Section 4.4.4), and an inextensibility parameter identical to that of Edwards and Vilgis is introduced. Marckmann and Verron (2006) have reviewed these models and an attempt by Miehe *et al.* (2004) to incorporate these ingredients into a comprehensive model.

4.4.4 The Inverse Langevin Approximation

An early development of the simple molecular network theory is the so-called 'inverse Langevin approximation' for the probability distribution. The Gaussian approximation is only valid for end-to-end distances that are much less than the extended chain length. It was shown by Kuhn and Grün (1942) and James and Guth (1947) that removing this restriction (but still maintaining the other assumptions of freely jointed chains) gives a probability distribution $p(r)$ as

$$\ln p(r) = \mathrm{c} - n\left[\frac{r}{n\ell}\beta + \ln\frac{\beta}{\sinh\beta}\right]. \tag{4.36}$$

In this equation β is defined by

$$\frac{r}{n\ell} = \coth\beta - \frac{1}{\beta} = \mathcal{L}(\beta),$$

where $\mathcal{L}$ is the Langevin function and $\beta = \mathcal{L}^{-1}(r/n\ell)$ is the inverse Langevin function. This expression may be expanded to give

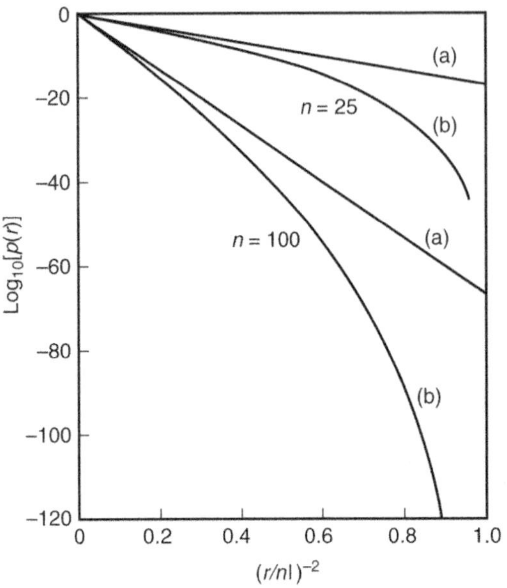

Figure 4.11 Distribution functions for 25- and 100-link random chains: (a) Gaussian approximation and (b) inverse Langevin approximation. (Reproduced from Treloar (1975) The Physics of Rubber Elasticity, 3rd edn, Oxford University Press, Oxford. Copyright (1975) Oxford University Press.)

$$\ln p(r) = \mathrm{c} - n\left[\frac{3}{2}\left(\frac{r}{n\ell}\right)^2 + \frac{9}{20}\left(\frac{r}{n\ell}\right)^4 + \frac{99}{350}\left(\frac{r}{n\ell}\right)^6 + \cdots\right], \tag{4.37}$$

from which it can be seen that the Gaussian distribution is the first term of the series, an adequate approximation for $r = nl$.

The distribution functions for 25- and 100-link random chains obtained from the Gaussian and inverse Langevin approximations, respectively, are compared in Figure 4.11.

For the inverse Langevin approximation, we have from Equation (4.36)

$$S = k\ln p(r) = \mathrm{c} - kn\left(\frac{r}{n\ell}\beta + \ln\frac{\beta}{\sinh\beta}\right), \tag{4.38}$$

and the tension in the chain

$$f = T\frac{\partial S}{\partial r} = \frac{kT}{\ell}\mathcal{L}^{-1}\left(\frac{r}{n\ell}\right), \tag{4.39}$$

which, following an expansion similar to Equation (4.37), gives

$$f = \frac{kT}{l}\left\{3\left(\frac{r}{n\ell}\right) + \frac{9}{5}\left(\frac{r}{n\ell}\right)^3 + \frac{297}{175}\left(\frac{r}{n\ell}\right)^5 + \frac{1539}{875}\left(\frac{r}{n\ell}\right)^7 + \cdots\right\}. \tag{4.40}$$

The final part of the exercise is to reconsider the stress–strain relations using the inverse Langevin distribution function. This was done by James and Guth using an analogous development to that for the Gaussian distribution function.

The tensile force per unit unstrained area is

$$f = \frac{NkT}{3}n^{1/2}\left\{\mathcal{L}^{-1}\left(\frac{\lambda}{n^{1/2}}\right) - \lambda^{-3/2}\mathcal{L}^{-1}\left(\frac{1}{\lambda^{1/2}n^{1/2}}\right)\right\}. \tag{4.41}$$

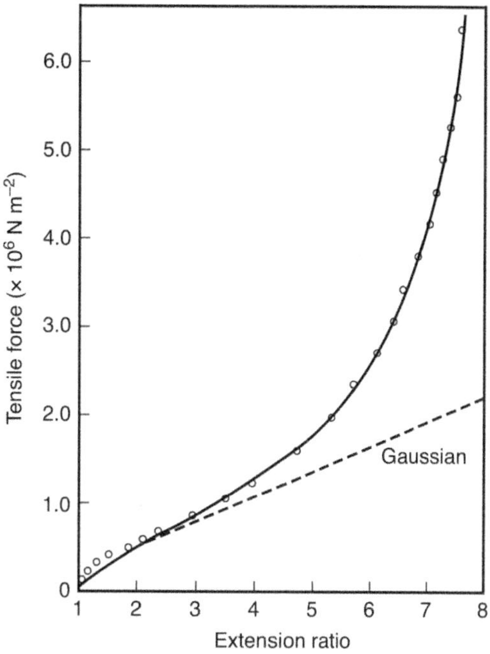

Figure 4.12 Theoretical non-Gaussian force–extension curve obtained by fitting experimental data to the James and Guth theory, with $NkT = 0.273$ MPa, $n = 75$. (Reproduced from Treloar (1975) The Physics of Rubber Elasticity, 3rd edn, Oxford University Press, Oxford. Copyright (1975) Oxford University Press.)

Figure 4.12 shows Treloar's fit to the experimental data for natural rubber, using this relationship and a suitable choice of the parameters N and n. The maximum extension of the network is primarily determined by the number of random links n. This result is relevant to the cold-drawing (Chapter 12 Section 12.9) and crazing behaviour of polymers (Chapter 13 Section 13.6), where the basic deformation also involves the extension of a molecular network.

It can be seen from Figure 4.12 that there is broad agreement between the form of the force–extension curve predicted by the inverse Langevin approximation and that observed in practice.

Amore recent approach along similar lines, pioneered by Arruda and Boyce (1993), involves further development of the non-Gaussian model of James and Guth (1947). The James and Guth model was essentially one of three mutually perpendicular molecular chains, and this is generalized by increasing the number of chains. Thus, Arruda and Boyce introduced the eight-chain model, in which the chains are envisaged as being attached at one end to the corners of a cube of material, and at the other end to the cube centre. The cube is aligned with the principal strain directions. On deformation of the cube into a rectangular element with extension ratios along its sides λ_{I}, λ_{II} and λ_{III}, each of the eight chains develops the same extension ratio λ_{chain} defined by

$$\lambda_{\mathrm{chain}} = \frac{1}{\sqrt{3}}\left(\lambda_{\mathrm{I}}^2 + \lambda_{\mathrm{II}}^2 + \lambda_{\mathrm{III}}^2\right)^{1/2}.$$

The analogue of the three-chain Equation (4.41) for uniaxial stretching becomes

$$f = \frac{NkT}{3} n^{1/2} \mathcal{L}^{-1}\left[\frac{\lambda_{\mathrm{chain}}}{n^{1/2}}\right] \frac{\lambda - 1/\lambda^2}{\lambda_{\mathrm{chain}}}. \tag{4.42}$$

Arruda and Boyce (1993) showed that the eight-chain model performed much better than three- and four-chain models in predicting the behaviour of vulcanized rubber in uniaxial and biaxial tension and shear, and also performed excellently in modelling uniaxial and plane strain compression of gum and neoprene rubber.

Wu and van der Giessen (1993) produced a further generalization to an infinitely large number of chains – the full network model. Their expression for force is more complex than Equations (4.41) or (4.42), but they showed that it could be approximated very well with a linear combination of the three- and eight-chain expressions. Contrary to expectations, the full network model performed less well than the eight-chain model for uniaxial and biaxial testing of natural gum-rubber and silicone rubber. Wu and van der Giessen pointed out that both models were approximations, in that entanglement (slip-link) effects and other intermolecular interactions are ignored, as is the existence of non-affine deformation. They raised the possibility that the eight-chain model was fortuitously compensating for these effects.

Sweeney (1999) has compared the eight-chain, full network and Edwards–Vilgis models and shown that the first two can be approximated very well by the Edwards–Vilgis model, provided the chain extensibility limit is not approached too closely. This latter proviso is a result of the different forms of the singularity in strain energy that controls the approach to the extensibility limit. The additional feature of slip-links in the Edwards–Vilgis model means that is the most general of the three, and this is reflected in the greater number of fitting parameters. Rickaby and Scott (2015) have produced a comprehensive comparison of various models that include finite strain extensibility.

4.4.5 The Conformational Exhaustion Model

Another approach to rubber elasticity has been developed by Stepto and Taylor (1995a, 1995b), which also accounts for the Mooney–Rivlin softening. Their approach is not phenomenological, but is based on structural considerations that give an accurate description of the molecular conformational states of the units in the polymer chains as the network is stretched. They have proposed a method for calculating the free energy of a stretched molecular network based on the rotational isomeric state of the network chains with conformational energies determined from observations of conformational properties.

Using a series of Monte Carlo calculations, the elastic properties of the network are derived from the network chain end-to-end distance distribution, and are assumed to arise solely as a result of allowed conformational changes in individual network chains. Figure 4.13 shows the calculation for the probability density functions $p(r)$ calculated from the simulated radial end-to-end distance distribution functions $P(r)$, where

$$p(r) = \frac{P(r)}{4\pi r^2}$$

as explained in Section 4.3.3.

Comparison of the Monte Carlo $p(r)$ with those predicted on the basis of the Gaussian distribution function (Equation (4.18)) and the Langevin function (Equation (4.36)) shows clear differences. In particular, the molecular-structure-based Monte Carlo $p(r)$ reflects clearly the limited extensibility of chains in the true network. The Gaussian curve grows

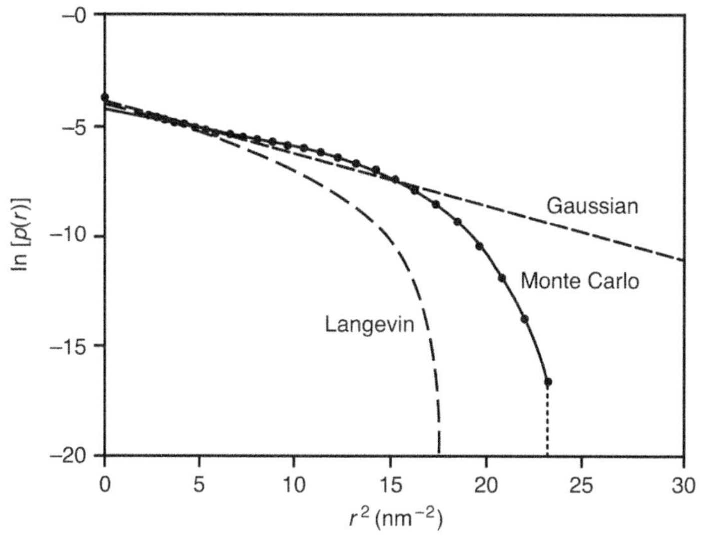

Figure 4.13 Plot of ln[$p(r)$] versus r^2 at 298 K for polydimethyl siloxane chains of 40 bonds compared with Gaussian and Langevin treatments of the freely jointed chain. Stepto and Taylor, 1995 / Royal Society of Chemistry.

without limit, while the Langevin curve approaches a limiting stretch at extremely low probability. By contrast, the Monte Carlo method genuinely reaches a finite limit. At this stage the material may continue to deform, ensuring that the deformation is non-affine and that the stretched chain can no longer behave entropically.

In the Stepto theory, the free energy of the network in uniaxial tension is given by

$$F = \frac{RT\rho}{M_c} s\left(\lambda^2 + \frac{2}{\lambda} - 3\right),$$

where R, T, ρ and M_c have the same meaning as in Section 4.3.4, but there is a new factor s that is a function of λ and quantitatively accounts for the Mooney–Rivlin softening. In this theory, softening arises because some chains in the deformed network reach their full extensibility so that on further deformation these chains do not give a contribution to the reduction in entropy and hence to the network stress, which is correspondingly reduced. Stepto and Taylor have shown that their theory gives an accurate quantitative fit to the stress–strain behaviour of polydimethyl siloxane networks (Stepto and Taylor 1995a, 1995b), based only on the values for the rotational isomeric states obtained from the Flory–Crescenzi–Mark rotation isomeric state model for polydimethyl siloxane (Flory *et al.* 1964). In addition, the optical birefringence, that is, the stress-optical coefficient, for cross-linked polyethylenes, was also quantitatively predicted (Taylor *et al.* 1999; Cail *et al.* 2000).

In a series of investigations on PET (Cail and Stepto 2003; Saunders *et al.* 2007; Cail and Stepto 2012) the method was used to simulate elastic behaviour and the deformations of the

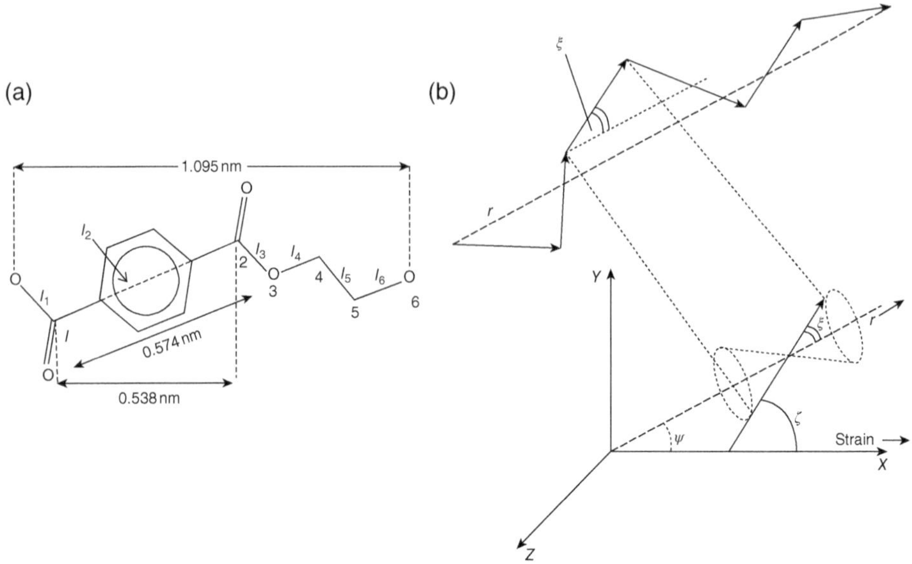

Figure 4.14 (a) PET molecule, with glycol l_3, l_4, l_5, l_6. (b) Orientation directions between segment and chain axis are generated randomly on the cone defined by ξ. Saunders, L.S. and Ward, I.M. (2007) / with permission of Elsevier. Experimental studies and molecular modelling of the stress-optical and stress-strain behaviour of poly(ethylene terephthalate). Part I: Infra-red spectroscopic investigation and modelling of chain conformation and orientation changes on drawing. Polymer **49**, 1360–1366.)

molecules. The proportion of trans and gauche transformations in the glycol residue was predicted and confirmed using infra-red spectroscopy (Saunders *et al.* 2007). In Figure 4.14 we show a diagram of the PET molecule and illustrate how the model molecules are generated by varying randomly the conformation angles in the glycol.

4.4.6 The Effect of Strain-Induced Crystallization

Although we have seen how the extension to non-Gaussian statistics gives rise to a very large increase in tensile stress at large extensions, in the case of natural rubber it has been proposed that the observed increase in tensile stress occurs primarily because of strain-induced crystallization (Mark 1979). The basic physical idea is that the melting point T_c of the rubber is increased due to extension. $T_c = \Delta H/\Delta S$, where ΔH and ΔS are the enthalpy and entropy of fusion, respectively. Because the entropy of the extended rubber is low, the change in entropy on crystallization is reduced and T_c correspondingly increased. A higher degree of supercooling then gives rise to crystallization and the crystallites act to increase the modulus by forming additional physical cross-links.

The precise effect of crystallization is, however, difficult to determine. Experiments on butadiene rubbers (Doherty *et al.* 1980), which do not crystallize on extension, when directly compared with natural rubber, suggested that the influence of crystallization was relatively unimportant.

4.5 The Internal Energy Contribution to Rubber Elasticity

The simple treatment of rubber elasticity given above makes two assumptions, which require further consideration. First, it has been assumed that the internal energy contribution is negligible, which implies that different molecular conformations of the chains have identical internal energies. Secondly, the thermodynamic formulae that have been derived are, strictly, only applicable to measurements at constant volume, whereas most experimental results are obtained at a known pressure. These two assumptions are interrelated in the sense that the experimental work of Gee (see Section 4.2.1) based on the approximation

$$\left(\frac{\partial f}{\partial T}\right)_{P,\lambda} = \left(\frac{\partial f}{\partial T}\right)_{V,L} \tag{4.43}$$

(where λ is the extension ratio relative to natural length) leads to the conclusion that the internal energy contribution was zero. Here a specimen is stretched uniaxially with a nominal tensile stress f, in the presence of a superposed hydrostatic pressure P – hence the hydrostatic pressure of the system is $P - f/3$. Although Gee's approximation is much better than the assumption that

$$\left(\frac{\partial f}{\partial T}\right)_{P,L} = \left(\frac{\partial f}{\partial T}\right)_{V,L}, \tag{4.44}$$

it is based on the premise that the material is isotropically compressible under hydrostatic pressure even when strained, that is,

$$\frac{\mathrm{d}V}{V} = 3\frac{\mathrm{d}L}{L} \quad or \quad \left(\frac{\partial(\ln L)}{\partial(\ln V)}\right)_{f,T} = \frac{1}{3}.$$

More rigorous consideration suggests that this is not correct and that there is an anisotropic compressibility in the strained state. Gee's experimental result that the internal energy contribution is negligible, based on measurements of the change in stress with temperature at constant length, is therefore capable of another interpretation, which leads to the conclusion that the internal energy contribution may not be zero.

The internal energy component of the tensile force f is given by

$$f_e = \left(\frac{\partial U}{\partial L}\right)_{T,V} \tag{4.45}$$

(see Equation (4.8)). Volkenltein and Ptitsyn (1953, 1955) showed that, if the unperturbed dimensions of an isolated chain are temperature dependent, f_e is given by

$$\frac{f_e}{f} = T\frac{\partial\left(\ln \overline{r_0^2}\right)}{\partial T}, \tag{4.46}$$

where $\overline{r_0^2}$ is the root mean square length (where the subscript 0 indicates a free chain unconstrained by cross-linkages).

Experimental data on dilute solutions of polymers using light scattering and viscosity measurements show that $\overline{r_0^2}$ depends in general on temperature. This implies that the energy of a chain depends on its conformation and that for a rubber, f_e will in general differ from zero. Flory and his collaborators (Ciferri *et al.*1961) were particularly prominent in performing stress–temperature measurements on polymer networks, together with physicochemical measurements to confirm these points and to obtain the energy difference between different conformations, for example, the *trans* and *gauche* conformations of the polyethylene chain.

The value f_e can be expressed in terms of the tensile force–temperature relationship by the equation

$$\frac{f_e}{f} = -T\left[\frac{\partial\{\ln(f/T)\}}{\partial T}\right]_{V,L}, \tag{4.47}$$

which follows by manipulating Equation (4.13) while acknowledging the validity of Equation (4.45). However, this expression for constant volume is not easily related to the interpretation of most laboratory experiments, where an equation at constant pressure P is more appropriate.

In addressing this problem, Flory *et al.* (1960) adopted a procedure based on the theory of the Gaussian network. They showed that if the rubber network obeys Gaussian statistics, the expression for measurement of simple extension at constant pressure is

$$\left[\frac{\partial\{\ln(f/T)\}}{\partial T}\right]_{P,L} + \frac{\alpha}{\lambda^3 - 1} = -\frac{\mathrm{d}\left(\ln \overline{r_0^2}\right)}{\mathrm{d}T}, \tag{4.48}$$

where α is the coefficient of volume expansion of the rubber at constant pressure, that is,

$$\alpha = \frac{1}{V}\left(\frac{\partial V}{\partial T}\right)_P. \tag{4.49}$$

Using Equations (4.46) and (4.48), they then went on to derive an expression for f_e/f

$$\frac{f_e}{f} = -T\left[\frac{\partial\{\ln(f/T)\}}{\partial T}\right]_{P,L} - \frac{\alpha T}{(\lambda^3 - 1)}. \tag{4.50}$$

Equating Equations (4.47) and (4.50) shows that the derivatives involving f are unequal, and that Equation (4.44) is false. A correction factor can be derived that enables the constant pressure derivative to be derived from the constant volume derivative. Graessley and Fetters (2001) introduced such a factor $\Delta_{P,L}$ equivalent to

$$\Delta_{P,L} = \left[\frac{\partial(f/T)}{\partial(1/T)}\right]_{V,L} - \left[\frac{\partial(f/T)}{\partial(1/T)}\right]_{P,L},$$

which can itself be shown to be equivalent to

$$\Delta_{P,L} = fT\left(\left[\frac{\partial\{\ln(f/T)\}}{\partial T}\right]_{P,L} - \left[\frac{\partial\{\ln(f/T)\}}{\partial T}\right]_{V,L}\right). \tag{4.51}$$

By equating Equations (4.47) and (4.50) we can derive an expression for the right-hand side of (4.51), showing that for this analysis

$$\Delta_{P,L} = \frac{\alpha f T}{\lambda^3 - 1}. \tag{4.52}$$

Treloar (1969a, 1969b) considered the behaviour in torsion and showed that the relationship corresponding to (4.48) is

$$\left[\frac{\partial\{\ln(M/T)\}}{\partial T}\right]_{P,L,\psi} = \frac{-\mathrm{d}\left(\ln \overline{r_0^2}\right)}{\mathrm{d}T} + \alpha,$$

where M is the torsional moment and ψ is the twist (expressed as radians per unit length of the strained axis). M_e, the internal energy contribution to the couple at constant volume, is given by an equation similar to (4.47):

$$\frac{M_e}{M} = -T\left[\frac{\partial\{\ln(M/T)\}}{\partial T}\right]_{V,L,\psi}.$$

Allen *et al.* (1963) and Allen *et al.* (1971) adopted the alternative approach of constructing apparatus to undertake measurements at constant volume, so that Equation (4.47) could be used directly. They were also able to compare $\left[\frac{\partial f}{\partial T}\right]_{V,L}$, $\left[\frac{\partial f}{\partial T}\right]_{P,L}$, and $\left[\frac{\partial f}{\partial P}\right]_{T,L}$.

Although it was shown that the Flory equation was not correct, the errors involved tended to compensate, so the values of f_e/f obtained by both sets of workers were similar.

In a review of this research, Price (1976) concluded that the Gaussian theory is able to account satisfactorily for the temperature dependence of the force, but it does not give an adequate representation of the pressure dependence of the force and the strain-induced dilatation. In a later review, Graessley and Fetters (2001) followed the approximation and the Gaussian network model to deduce their improved correction factor $\Delta_{P,L}$ that was in much better agreement with the dilation data:

$$\Delta_{P,L} = \frac{\alpha f T}{3}\frac{\lambda^3 + 2}{\lambda^3 - 1}.$$

This correction will clearly begin to deviate from that of Equation (4.52) at sufficiently high λ.

For a comprehensive account of early work in this area, the reader is referred to Chapter 13 in Treloar (2005).

4.6 Applications Using Finite Element Modelling

One the first studies in this area is that of Liu *et al.* (1994), who developed large deformation three-dimensional finite element code. They implemented the Mooney–Rivlin strain energy function in problems involving rubber rectangular blocks, cylinders and vehicle tyres. Böl and Reese (2006) have incorporated the inverse Langevin chain model, expanded to ten terms, directly into a finite element scheme. This method has the potential for including non-affine deformations. They compared their method's capability of modelling experimental data in uniaxial and equibiaxial tension and pure shear with that of the Neo-Hookean, Mooney–Rivlin, Arruda–Boyce, Ogden and Yeoh models. They found it gave results comparable with the Arruda–Boyce model, which generally gave the best performance. They applied it to the modelling of a rubber boot.

However, the majority of finite element applications now make use of commercially available packages, some of which incorporate better known material models as options for the user. The work discussed here covers material models of both this chapter and Chapter 3. Thus, Saidou *et al.* (2021) have used the ADINA finite element package to model bridge rubber bearings in a study in which they compared the performances of neo-Hookean, Mooney–Rivlin, Arruda–Boyce and three-term Ogden models. They concluded that the Ogden model gave the most accurate results. Yildiz (2014) has worked on rubber fender devices for the marine industry, using the ANSYS package to compare the performances of the Mooney–Rivlin, Yeoh and single-term Ogden models. All the material models were adapted to be compressible using the method of Kaliske and Rothert in Chapter 3, Section 3.4.6. Other finite element studies in which material models are compared are Sujendra *et al.* (2018), Seibert and Schöche (2000) and Huri and Mankovits (2018). Commercial packages also offer the possibility of incorporating a material model via a user-defined subroutine, so that innovative constitutive models can be used, as in the case of Ghaemi *et al.* (2005) using the MSC Marc software.

4.7 Conclusions

The phenomenological theories following the approach of Rivlin and others were based on empirically developed strain energy functions. These have been overtaken by molecular network models notably by Edwards and colleagues, Boyce and colleagues and Stepto and colleagues.

It can be concluded that these latest developments offer the way forward, and that it is less rewarding to attempt to ascribe physical significance to higher-order terms in invariant-based strain energy functions; rather we should accept more complex formulations in terms of the principal extension ratios directly. An important issue is to incorporate the internal energy contribution into the purely geometrical considerations concerned with the deformation of the molecular network.

In finite element applications the approach is often pragmatic, based on how well the material model fits the available data. This can result in a poor choice of model parameters if uniaxial testing is the sole source of the data, as pointed out by Yeoh (1997); the addition of biaxial and shear results will give better discrimination. The strains and strain states generated in finite element models should be compatible with the experimental data that defines the material parameters.

References

Allen, G., Bianchi, U. and Price, C. (1963) Thermodynamics of elasticity of natural rubber. *Trans. Faraday Soc.*, **59**, 2493.

Allen, G., Kirkham, M.C., Padget, J. *et al.* (1971) Thermodynamics of rubber elasticity at constant volume. *Trans. Faraday Soc.*, **67**, 1278.

Arruda, E.M. and Boyce, M.C. (1993) A 3-dimensional constitutive model for the large stretch behavior of rubber elastic-materials. *J. Mech. Phys. Solids*, **41**, 389.

Ball, R.C., Doi, M., Edwards, S.F. *et al.* (1981) Elasticity of entangled networks. *Polymer*, **22**, 1010.

Böl, M. and Reese, S. (2006) Finite element modelling of rubber-like polymers based on chain statistics. *Int. J. Solids Struct.*, **43**, 2–26.

Boyce, M.C. and Arruda, E. (2000) Constitutive models of rubber elasticity: A review. *Rubber Chem. Technol.*, **73**, 504.

Cail, J.I. and Stepto, R.F.T. (2003) Molecular modelling of the elastic behaviour of poly(ethyleneterephthalate) network chains. *Polymer*, **44**, 6077–6087.

Cail, J.I. and Stepto, R.F.T. (2012) Molecular modelling of the conformational flexibility and elastic behaviour of short poly(ethylene terephthalate) chains. *J. Inorg. Organomet. Polym. Mater.*, **22**, 646–652.

Cail, J.I., Taylor, D.J.R. and Stepto, R.F.T. (2000) Computer simulation studies of molecular orientation in polyethylene networks: orientation functions and the legendre addition theorem. *Macromolecules*, **33**, 4966.

Ciferri, A., Hoeve, C.A.J. and Flory, P.J. (1961) Stress-temperature coefficients of polymer networks and conformational energy of polymer chains. *J. Am. Chem. Soc.*, **83**, 1015.

Doherty, W.O.S., Lee, K.L. and Treloar, L.G.R. (1980) Non-Gaussian effects in styrene-butadiene rubber. *Br. Polym. J.*, **12**, 19.

Edwards, S.F. and Vilgis, T. (1986) The effect of entanglements in rubber elasticity. *Polymer*, **27**, 483.

Erman, B. and Monnerie, L. (1989) Theory of elasticity of amorphous networks – effect of constraints along chains. *Macromolecules*, **22**, 3342.

Flory, P.J. and Erman, B. (1982) Theory of elasticity of polymer networks, 3. *Macromolecules*, **15**, 800.

Flory, P.J., Ciferri, A. and Hoeve, C.A.J. (1960) The thermodynamic analysis of thermoelastic measurements on high elastic materials. *J. Polym. Sci.*, **45**, 235.

Flory, P.J., Crescenzi, V. and Mark, J.E. (1964) Configuration of poly- (dimethylsiloxane) chain. 3. Correlation of theory and experiment. *J. Am. Chem. Soc.*, **86**, 146.

Gee, G. (1946) The interaction between rubber and liquids. 9. The elastic behaviour of dry and swollen rubbers. *Trans. Faraday Soc.*, **42**, 585.

Ghaemi, H., Spence, A. and Behdinan, K. (2005) On the development of compressible pseudo-strain energy density function for hyperelastic material: Experiment, theory and FEM. *Trans. Can. Soc. Mech. Eng.*, **29**, 459–475.

Graessley, W.W. and Fetters, L.J. (2001) Thermoelasticity of polymer networks. *Macromolecules*, **34**, 7147.

Guth, E. and Mark, H. (1934) Zur Innermolekularen, Statistik, Insbesondere Bei Kettenmolekiilen I. *Monatsh. Chem.*, **65**, 93.

Heinrich, G. and Kaliske, M. (1997) Theoretical and numerical formulation of a molecular based constitutive tube-model of rubber elasticity. *Comput. Theo. Polym. Sci.*, **7**, 227.

Huri, D. and Mankovits, T.(2018). Comparison of the material models in rubber finite element analysis. *IOP Conf. Series: Materials Science and Engineering* 393 012018.

James, H.M. and Guth, E. (1947) Theory of the increase in rigidity of rubber during cure. *J. Chem. Phys.*, **15**, 669.

Kaliske, M. and Heinrich, G. (1999) An extended tube-model for rubber elasticity: statistical-mechanical theory and finite element implementation. *Rubber Chem. Technol.*, **72**, 602.

Kloczkowski, A., Mark, J.E. and Erman, B. (1995) A diffused-constraint theory for the elasticity of amorphous polymer networks. 1. Fundamentals and stress-strain isotherms in elongation. *Macromolecules*, **28**, 5089.

Kuhn, W. (1934) Concerning the shape of thread shapes molecules in solution. *Kolloidzeitschrift*, **68**, 2.

Kuhn, W. (1936) Relationship between molecular size, static molecular shape and elastic characteristics of high polymer materials. *Kolloidzeitschrift*, **76**, 258.

Kuhn, W. and Grün, F. (1942) Relations between elastic constants and the strain birefringence of high-elastic substances. *Kolloidzitschrift*, **101**, 248.

Liu, C.H., Hofstetter, G. and Mang, H.A. (1994) 3D finite element analysis of rubber-like materials at finite strains. *Eng. Comput.*, **11**, 111–128.

Marckmann, G. and Verron, E. (2006) Comparison of hyperelastic models for rubber-like materials. *Rubber Chem. Technol.*, **79**, 835.

Mark, J.E. (1979) Effect of strain-induced crystallization on the ultimate properties of an elastomeric polymer network. *Polym. Eng. Sci.*, **19**(6), 409.

Mark, J.E. and Erman, B. (1988) *Rubberlike Elasticity. A Molecular Primer*, John Wiley & Sons, Chichester.

Meissner, B. (2000) Tensile stress-strain behaviour of rubberlike networks up to break. Theory and experimental comparison. *Polymer*, **41**, 7827.

Meyer, K.H. and Ferri, C. (1935) On elasticity of rubber. *Helv. Chim. Acta*, **18**, 570.

Meyer, K.H., Von Susich, G. and Valko, E. (1932) The elastic properties of organic high polymers and their kinetic explanation. *Kolloidzeitschrift*, **59**, 208.

Miehe, C., Goktepe, S. and Lulei, F. (2004) A micro-macro approach to rubber-like materials – part 1: the non-affine micro-sphere model of rubber elasticity. *J. Mech. Phys. Solids*, **52**, 2617.

Mihai, L.A. and Goriely, A. (2017) How to characterize a nonlinear elastic material? A review on nonlinear constitutive parameters in isotropic finite elasticity. *Proc. R. Soc. A*, **473**, 20170607.

Price, C. (1976) Thermodynamics of rubber elasticity. *Proc. R. Soc. Lond.*, **A351**, 331.

Rickaby, S.R. and Scott, N.H. (2015) A comparison of limited-stretch models of rubber elasticity. *Int. J. Non-Linear Mech.*, **68**, 71–86.

Ronca, G. and Allegra, G.J. (1975) Approach to rubber elasticity with internal constraints. *Chem. Phys.*, **63**, 4990.

Saidou, A., Gauron, O., Busson, A.Q. and Paultre, P. (2021) High-order finite element model of bridge rubber bearings for the prediction of buckling and shear failure. *Eng. Struct.*, **240**, 112314.

Saunders, L.S., Ward, I.M., Cail, J.I. and Stepto, R.F.T. (2007) Experimental studies and molecular modelling of the stress-optical and stress-strain behaviour of poly(ethylene terephthalate). Part I: infra-red spectroscopic investigation and modelling of chain conformation and orientation changes on drawing. *Polymer*, **49**, 1360–1366.

Seibert, D.J. and Schöche, N. (2000) Direct comparison of some recent elasticity models. *Rubber Chem. Technol.*, **73**, 366–384.

Stepto, R.F.T. and Taylor, D.J.R. (1995a) Modeling the elastic behavior of real chains in polymer networks. *Macromol. Symp.*, **93**, 261.

Stepto, R.F.T. and Taylor, D.J.R. (1995b) Molecular modeling of the elastic behavior of polymer-chains in networks – comparison of polymethylene and poly (dimethylsiloxane). *J. Chem. Soc. Faraday Trans.*, **91**, 2639.

Sujendra, H.M., Jayaram, A.S., Chandan, R. and Hattaraki, G. (2018) Finite element modelling and simulation of rubber component in predicting hyperelastic material model. *Int. Res. J. Eng. Technol.*, **5**, 751–754.

Sweeney, J. (1999) A comparison of three polymer network models in current use. *Comput. Theo. Polym. Sci.*, **9**, 27.

Taylor, D.J.R., Stepto, R.F.T. and Jones, R.A. (1999) Computer simulation studies of molecular orientation and the stress-optical properties of polyethylene networks. *Macromolecules*, **32**, 1978.

Treloar, L.R.G. (1969a) Volume changes and mechanical anisotropy of strained rubbers. *Polymer*, **10**, 279.

Treloar, L.R.G. (1969b) Theoretical stress/temperature relations for rubber in torsion. *Polymer*, **10**, 291.

Treloar, L.R.G. (2005) *The Physics of Rubber Elasticity*, Oxford University Press, Oxford.

Volkenshtein, M.V. and Ptitsyn, O.B. (1953) O Rastyazhenii Polimernykh Tsepochek. *Dokl Akad SSR*, **91**, 1313.

Volkenshtein, M.V. and Ptitsyn, O.B. (1955) Vnutrennee Vrashchenie V Polimernykh Tsepyakh I Ikh Fizicheskie Svoistva. 2. Rastyazhenie Polimernoi Tsepi. *Zhur Tech Fiz*, **25**(649), 662.

Wu, P.D. and van der Giessen, E. (1993) On improved network models for rubber elasticity and their applications to orientation hardening in glassy-polymers. *Mech. Phys. Solids*, **41**, 427.

Yeoh, O.H. (1997) On the Ogden strain energy function. *Rubber Chem. Technol.*, **70**, 175–182.

Yildiz, F. (2014) The effect of different strain energy functions on rubber fender: Experiment and finite element simulation. *J. Elastomers Plast.*, **46**, 722–736.

Further Reading

Gent, A.N. (2012) *Engineering with Rubber: How to Design Rubber Components*, 3rd edn, Carl Hanser Verlag, Munich.

Mark, J.E. and Erman, B. (2007) *Rubberlike Elasticity: A Molecular Primer*, 2nd edn, Cambridge University Press, Cambridge.

Erman, B., Mark, J.E. and Roland, C.M. (eds). (2013) *The Science and Technology of Rubber*, 4th edn, Elsevier Academic Press, Waltham, MA.

5

Linear Viscoelastic Behaviour

In this chapter, we describe the common forms of viscoelastic behaviour and discuss the phenomena in terms of the deformation characteristics of elastic solids and viscous fluids. The discussion is confined to linear viscoelasticity, for which the Boltzmann superposition principle enables the response to multistep loading processes to be determined from simpler creep and relaxation experiments. Phenomenological mechanical models are considered and used to derive retardation and relaxation spectra, which describe the timescale of the response to an applied deformation. Finally, we show that in alternating strain experiments the presence of the viscous component leads to a phase difference between stress and strain.

5.1 Viscoelasticity as a Phenomenon

The behaviour of materials of low relative molecular mass is usually discussed in terms of two particular types of ideal material: the elastic solid and the viscous liquid. The former has a definite shape and is deformed by external forces into a new equilibrium shape; on removal of these forces it reverts instantaneously to its original form. The solid stores all the energy that it obtains from the external forces during the deformation, and this energy is available to restore the original shape when the forces are removed. By contrast, a viscous liquid has no definite shape and flows irreversibly under the action of external forces.

One of the most interesting features of polymers is that a given polymer can display all the intermediate range of properties between an elastic solid and a viscous liquid depending on the temperature and the experimentally chosen timescale. Bouncing putty, a silicone product, flows over a period of hours, fractures like a ductile solid when deformed rapidly and bounces like an elastomer when dropped. Of greater commercial importance are the rubber-like, and in extreme cases brittle, characteristics exhibited by molten polymers at high processing rates. This form of response, which combines both liquid-like and solid-like features, is termed viscoelasticity.

Mechanical Properties of Solid Polymers, Fourth Edition. John Sweeney and Peter Hine.

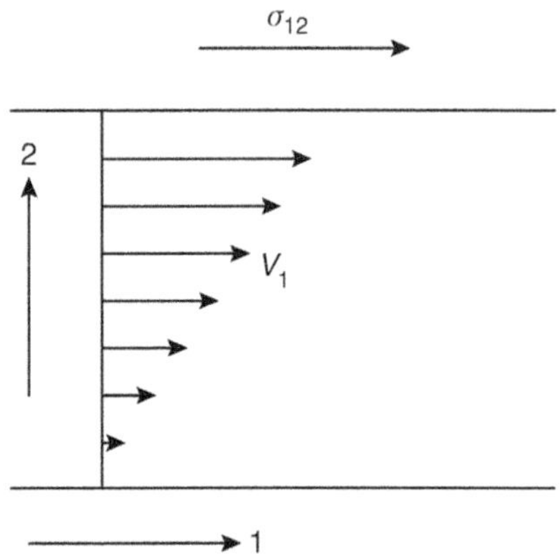

Figure 5.1 The velocity gradient.

5.1.1 Linear Viscoelastic Behaviour

Newton's law of viscosity defines viscosity η by stating that stress σ is proportional to the velocity gradient in the liquid:

$$\sigma = \eta \frac{\partial V}{\partial y},$$

where V is the velocity and y is the direction of the velocity gradient. For a velocity gradient in the 12 plane,

$$\sigma_{12} = \eta \left(\frac{\partial V_1}{\partial X_2} + \frac{\partial V_2}{\partial X_1} \right),$$

where $\frac{\partial V_1}{\partial X_2}$ and $\frac{\partial V_2}{\partial X_1}$ are the velocity gradients in the 2 and 1 directions, respectively (see Figure 5.1 for the case where the velocity gradient is in the 2 directions).

Since $V_1 = \partial u_1/\partial t$ and $V_2 = \partial u_2/\partial t$, where u_1 and u_2 are the displacements in the 1 and 2 directions, respectively, we have

$$\begin{aligned} \sigma_{12} &= \eta \left[\frac{\partial}{\partial X_2} \left(\frac{\partial u_1}{\partial t} \right) + \frac{\partial}{\partial X_1} \left(\frac{\partial u_2}{\partial t} \right) \right] \\ &= \eta \frac{\partial}{\partial t} \left(\frac{\partial u_1}{\partial X_2} + \frac{\partial u_2}{\partial X_1} \right) \\ &= \frac{\partial e_{12}}{\partial t} \end{aligned}$$

It can be seen that the shear stress σ_{12} is directly proportional to the rate of change of shear strain with time. This formulation brings out the analogy between Hooke's law for elastic solids and Newton's law for viscous liquids. In the former, the stress is linearly related to the strain, in the latter the stress is linearly related to the *rate of change of strain* or *strain rate*.

Hooke's law describes the behaviour of a linear *elastic* solid and Newton's law that of a linear *viscous* liquid. A simple constitutive relation for the behaviour of a linear viscoelastic solid is obtained by combining these two laws.

For elastic behaviour $(\sigma_{12})_E = Ge_{12}$, where G is the shear modulus.
For viscous behaviour $(\sigma_{12})_V = \eta(\partial e_{12}/\partial t)$.

A possible formulation of linear viscoelastic behaviour combines these equations; thus,

$$\sigma_{12} = (\sigma_{12})_E + (\sigma_{12})_V = Ge_{12} + \eta \frac{\partial e_{12}}{\partial t}$$

This makes the simplest possible assumption that the shear stresses related to strain and strain rate are additive. The equation represents one of the simple models for linear viscoelastic behaviour (the Voigt or Kelvin model) and will be discussed in detail in Section 5.2.5.

Most of the experimental work on linear viscoelastic behaviour is confined to a single mode of deformation, usually corresponding to a measurement of Young's modulus or the shear modulus. Our initial discussion of linear viscoelasticity will therefore be confined to the one-dimensional situation, recognizing that greater complexity will be required to describe the viscoelastic behaviour fully. For the simplest case of an isotropic polymer at least two of the modes of deformation defining two of the quantities E, G and K for an elastic solid must be examined, if the behaviour is to be completely specified.

In defining the constitutive relations for an elastic solid, we have assumed that the strains are small and that there are linear relationships between stress and strain. We now ask how the principle of linearity can be extended to materials where the deformations are time dependent. The basis of the discussion is the Boltzmann superposition principle. This states that in linear viscoelasticity effects are simply additive, as in classical elasticity, the difference being that in linear viscoelasticity it matters at which instant an effect is created. Although the application of stress may now cause a time-dependent deformation, it can still be assumed that each increment of stress makes an independent contribution. From the present discussion, it can be seen that the linear viscoelastic theory must also contain the additional assumption that the strains are small. In Chapter 11, we will deal with attempts to extend linear viscoelastic theory either to take into account non-linear effects at small strains or to deal with the situation at large strains.

5.1.2 Creep

It is convenient to introduce the discussion of linear viscoelastic behaviour with the one-dimensional situation of creep under a fixed load. For an elastic solid, the following is observed at the two levels of stress σ_0 and $2\sigma_0$ (Figure 5.2(a)).

The strain follows the pattern of the loading programme exactly and in exact proportionality to the magnitude of the loads applied.

The effect of applying a similar loading programme to a linear viscoelastic solid has several similarities (Figure 5.2(b)). In the most general case, the total strain e is the sum of three separate parts e_1, e_2 and e_3. e_1 and e_2 are often termed the *immediate* elastic deformation and the *delayed* elastic deformation, respectively. e_3 is the Newtonian flow, that is that part of the deformation, which is identical with the deformation of a viscous liquid obeying Newton's law of viscosity.

Because the material shows linear behaviour, the magnitude of the strains e_1, e_2 and e_3 is exactly proportional to the magnitude of the applied stress. Thus the simple loading experiment defines a creep compliance $J(t)$ that is only a function of time:

$$\frac{e(t)}{\sigma} = J(t) = J_1 + J_2 + J_3 \tag{5.1}$$

where J_1, J_2 and J_3 correspond to e_1, e_2 and e_3, respectively.

The term J_3, which defines the Newtonian flow, can be neglected for rigid polymers at ordinary temperatures, because their flow viscosities are very large. Linear amorphous polymers do show a finite J_3 at temperatures above their glass transitions, but at lower temperatures their behaviour is dominated by J_1 and J_2. Cross-linked polymers do not show a J_3 term, and this is true to a very good approximation for highly crystalline polymers as well.

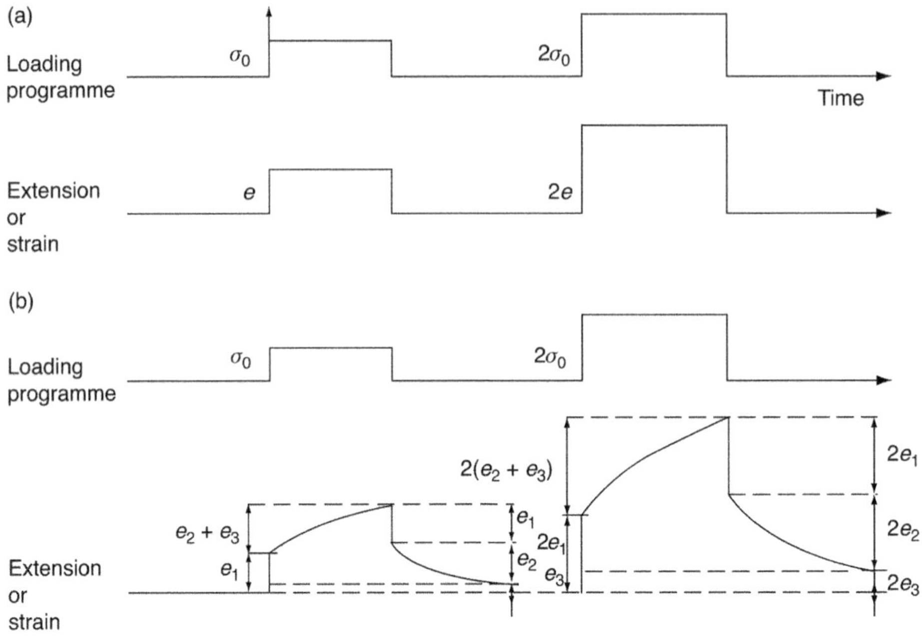

Figure 5.2 (a) Deformation of an elastic solid and viscoelastic solid and (b) deformation of a linear viscoelastic solid.

This leaves J_1 and J_2. At any given temperature, the separation of the compliance into terms J_1 and J_2 may involve an arbitrary division, which expresses the fact that at the shortest experimentally accessible times we will observe a limiting compliance J_1. We will assume, however, that there is a real distinction between the elastic and delayed responses. In some texts, the immediate elastic response in a creep experiment is called the 'unrelaxed' response to distinguish it from the 'relaxed' response that is observed at times sufficiently long for the various relaxation mechanisms to have occurred. To emphasize that the values of such terms as J_1 are sometimes arbitrary, we will enclose them in brackets.

We have already discussed in the introductory chapter how polymers can behave as glassy solids, viscoelastic solids, rubbers or viscous liquids depending on the timescale or on the temperature of the experiment. How does this fit in with our present discussion? Figure 5.3 shows the variation in compliance with time at constant temperature over a very wide timescale for an idealized amorphous polymer with only one relaxation transition. This diagram shows that for short-time experiments the observed compliance is 1 GPa^{-1}, that for a glassy solid. It is also time independent. At very long times the observed compliance is 10 MPa^{-1}, that for a rubbery solid, and it is again time independent. At intermediate times, the compliance lies between these values and is time dependent; this is the general situation of viscoelastic behaviour.

These considerations suggest that the observed behaviour will depend on the timescale of the experiment relative to some basic time parameter of the polymer. For creep, this parameter is called the retardation time τ' and falls in the middle range of our timescale as shown in the diagram. The distinction between a rubber and a glassy plastic can then be seen as

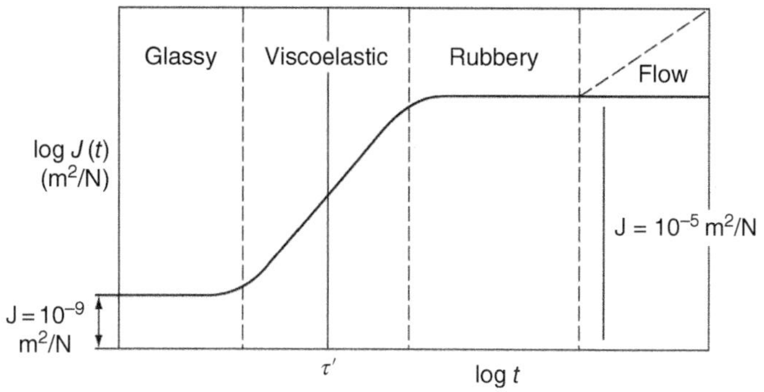

Figure 5.3 The creep compliance $J(t)$ as a function of time t. τ' is the characteristic time (the retardation time).

somewhat artificial in that it depends only on the value of τ' at room temperature for each polymer. Thus, for a rubber, τ' is very small at room temperature compared with normal experimental times that are greater than say 1 s, whereas the opposite is true for a glassy plastic. The value of this parameter τ' for a given polymer relates to its molecular constitution, as will be discussed later.

These considerations lead immediately to a qualitative understanding of the influence of temperature on polymer properties. With increasing temperature, the frequency of molecular rearrangements is increased, reducing the value of τ'. Thus, at very low temperatures, a rubber will behave like a glassy solid, as is well known, and equally a glassy plastic will soften at high temperatures to become rubber-like.

In the diagram illustrating creep under constant load recovery curves are also displayed. We will presently show that the recovery behaviour is basically similar to the creep behaviour if we neglect the quantity e_3, the Newtonian flow. This is a direct consequence of linear viscoelastic behaviour.

5.1.3 Stress Relaxation

The counterpart of creep is stress relaxation, where the sample is subjected to constant strain e, and the decay of stress $\sigma(t)$ is observed. This is illustrated in Figure 5.4.

The assumption of linear behaviour enables us to define the stress relaxation modulus $G(t) = \sigma(t)/e$. In the case of stress relaxation, the presence of viscous flow will affect the limiting value of the stress. Where viscous flow occurs, the stress can decay to zero at sufficiently long times, but where there is no viscous flow the stress decays to a finite value, and we obtain an equilibrium or relaxed modulus G_r, at infinite time. Figure 5.5 is a schematic graph of the stress relaxation modulus as a function of time. This is to be compared with the corresponding graph for creep (Figure 5.3).

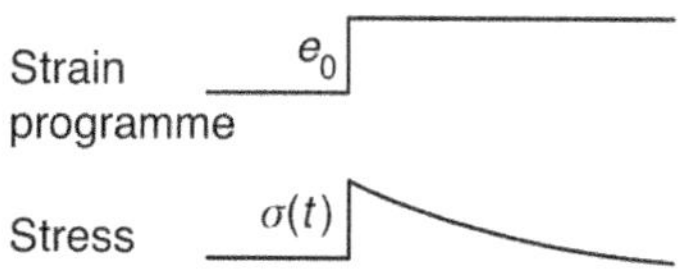

Figure 5.4 Stress relaxation.

The same regions of behaviour, viz. glassy, viscoelastic, rubbery and flow, can be identified, and a transition time τ is defined that characterizes the timescale of the viscoelastic behaviour. We will shortly discuss

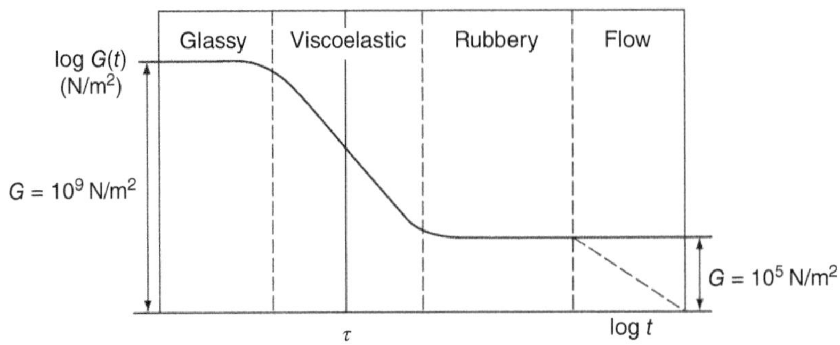

Figure 5.5 The stress relaxation characteristic modulus $G(t)$ as a function of time. τ is the characteristic time (the retardation time).

stress relaxation and creep in detail and show that as customarily defined, the characteristic times τ' and τ', although of the same order of magnitude, are not identical. Similar considerations to those discussed for creep apply to the effect of changing temperature on stress relaxation; i.e., changing temperature is equivalent to changing the timescale. Time–temperature equivalence is applicable to all linear viscoelastic behaviour in polymers and is considered fully in Chapter 7. The measurement of G_r may present difficulties as in the case of the elastic response. We will assume that there is a relaxed response to which it relates, but encloses the term involving it in brackets as for those involving the elastic response.

5.2 Mathematical Representation of Linear Viscoelasticity

The models discussed here, which are phenomenological and have no direct relation with chemical composition or molecular structure, in principle enable the response to a complicated loading pattern to be deduced from a single creep (or stress–relaxation) plot extending over a long time interval. Interpretation depends on the assumption in linear viscoelasticity that the total deformation can be considered as the sum of independent elastic (Hookean) and viscous (Newtonian) components. In essence, the simple behaviour is modelled by a set of either integral or differential equations, which are then applicable in other situations.

5.2.1 The Boltzmann Superposition Principle

Boltzmann (1876) proposed that

1) The creep is a function of the entire past loading history of the specimen.
2) Each loading step makes an independent contribution to the final deformation, so that the total deformation can be obtained by the addition of all the contributions.

Figure 5.6 illustrates the creep response to a multistep loading programme, in which incremental stresses $\Delta\sigma_1$, $\Delta\sigma_2$, ..., are added at times τ_1, τ_2, ..., respectively. The total creep at time t is then given by

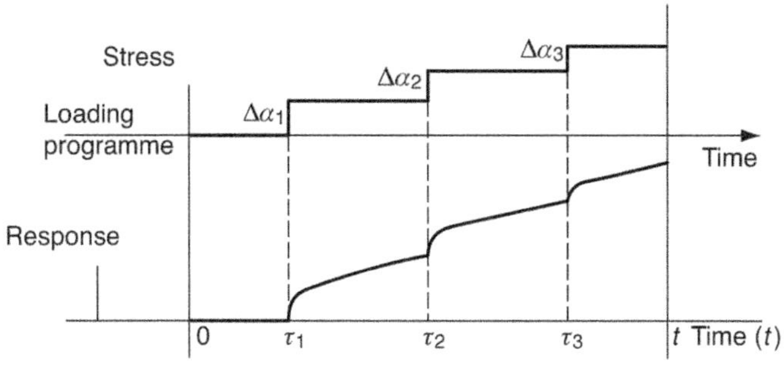

Figure 5.6 The creep behaviour of a linear viscoelastic solid.

$$e(t) = \Delta\sigma_1 J(t-\tau_1) + \Delta\sigma_2 J(t-\tau_2) + \Delta\sigma_3 J(t-\tau_3) + \cdots, \tag{5.2}$$

where $J(t - \tau)$ is the creep compliance function. For a particular loading step the relevant form of the function is that for the time interval between the present instant and that at which the load increment was applied.

The summation of Equation (5.2) can be generalized in integral form as

$$e(t) = \int_{-\infty}^{t} J(t-\tau)\mathrm{d}\sigma(\tau). \tag{5.3}$$

It is usual to separate out the instantaneous elastic response in terms of the unrelaxed modulus G_u, giving

$$e(t) = \frac{\sigma}{G_u} + \int_{-\infty}^{t} J(t-\tau)\frac{\mathrm{d}\sigma(\tau)}{\mathrm{d}\tau}\mathrm{d}\tau, \tag{5.4}$$

where σ represents the total stress at the end of the experiment. Note that the integral extends from $-\infty$ to t, which implies that all previous elements of loading history must be taken into account and, in principle, the user must know the history of each specimen since its manufacture. In fact, when creep levels are low enough for linearity to apply, the deformation effectively levels off at sufficiently long times, so that only comparatively recent history is relevant, and this can be standardized by a conditioning treatment (see Section 6.1.1). For this reason, viscoelastic solids are sometimes said to be materials with 'fading memory'.

The integral in Equation (5.4) is called a Duhamel integral, and it is a useful illustration of the consequences of the Boltzmann superposition principle to evaluate the response for a number of simple loading programmes. Recalling the development that leads to Equation (5.3), it can be seen that the Duhamel integral is most simply evaluated by treating it as the summation of a number of response terms. Consider two specific cases:

1) *Single-step loading of a stress σ_0 at time $\tau = 0$* (Figure 5.7). For this case

$$J(t-\tau) = J(t) \text{ and } e(t) = \sigma_0 J(t)$$

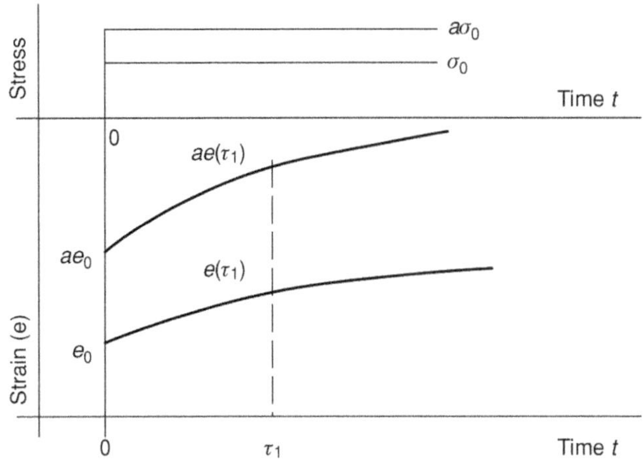

Figure 5.7 Creep for single step loading. For stress a σ_0 the strain at any time ($\tau \geq 0$) is a times greater than that for stress σ_0.

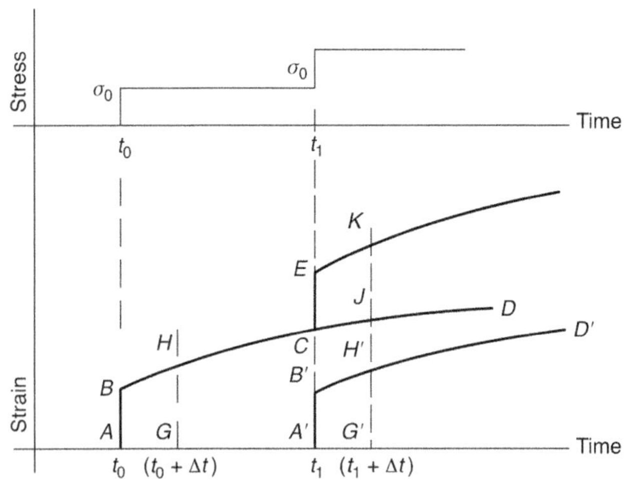

Figure 5.8 Creep for two equal loading steps. Additional instantaneous strain CE = AB. Additional total strain at ($t_1 + \Delta t$): JK = GH.

2) *Two-step loading of a stress σ_0 at time $\tau = 0$ followed by an additional stress σ_0 at time $\tau = t_1$* (Figure 5.8). For this case the creep deformation produced by the two loading steps are

$$e_1 = \sigma_0 J(t) \quad \text{and} \quad e_2 = \sigma_0 J(t - t_1)$$

so that

$$e(t) = e_1 + e_2 = \sigma_0 J(t) + \sigma_0 J(t - t_1)$$

The 'additional creep' $e'_c(t - t_1)$ produced by the second loading step is given by

$$e'_c(t - t_1) = \sigma_0 J(t) + \sigma_0 J(t - t_1) - \sigma_0 J(t) = \sigma_0 J(t - t_1)$$

This illustrates one consequence of the Boltzmann principle, viz. that the additional creep $e'_c\,(t - t_1)$ produced by adding the stress σ_0 is identical with the creep that would have occurred had this stress σ_0 been applied without any previous loading at the same instant in time t_1.

The principle is illustrated in Figure 5.8, where *ABD* represents creep under σ_0 alone. The response to an additional stress σ_0 at t_1 is found by sliding *ABD* along the time axis by t_1 to give curve $A'B'D'$, and at any time adding together the individual strains due to *ABD* and $A'B'D'$; e.g., at Δt after the initial loading step the deformation is *GH*; at Δt after t_1 the deformation due to the initial σ_0 is $G'J$. The total deformation at $(t_1 + \Delta t)$ is found by adding to $G'J$ the strain $JK = G'H' = GH$.

If the second load had been $a\sigma_0$, where a is a constant, then $CE = A'B' = aAB$; $JK = aG'H' = aGH$, etc.

3) *Creep and recovery.* In this case (Figure 5.9), the stress σ_0 is applied at time $\tau = 0$ and removed at time $\tau = t_1$. The deformation $e(t)$ at a time $t > t_1$ is given by the addition of two terms $e_1 = \sigma_0 J(t)$ and $e_2 = -\sigma_0 J(t - t_1)$, which express the application and removal of the stress σ_0, respectively. Thus, $e(t) = \sigma_0 J(t) - \sigma_0 J(t - t_1)$.

The recovery $e_r(t - t_1)$ will be defined as the difference between the anticipated creep under the initial stress and the actual measured response.

Thus

$$e_r(t - t_1) = \sigma_0 J(t) - [\sigma_0 J(t) - \sigma_0 J(t - t_1)] = \sigma_0 J(t - t_1),$$

which is identical to the creep response to a stress σ_0 applied at a time t_1. This procedure demonstrates a second consequence of the Boltzmann superposition principle, that the creep and recovery responses are identical in magnitude.

The initial creep curve, *ABD* in Figure 5.9, is again moved along the time axis by t_1 to give $A'B'D'$. At any subsequent time, the overall deformation is given by the difference between

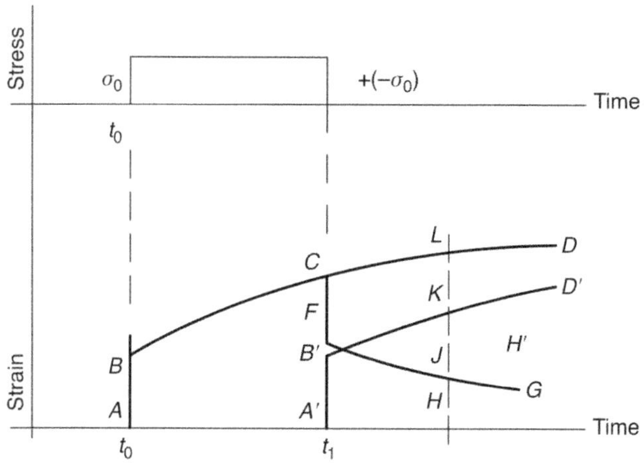

Figure 5.9 Recovery considered as the addition of a negative stress increment that is subtract B′D′ from CD.

the two curves (*CFG*). It is important to realize that at a time $t_2 > t_1$, the residual deformation is obtained by subtracting $A'B'D'$ from the deformation that would have occurred had unloading not taken place; for example, residual deformation $HJ = HL - HK$, and not $A'C - HK$, where $A'C$ is the maximum strain attained before unloading.

5.2.2 The Stress Relaxation Modulus

Stress relaxation behaviour can be represented in an exactly complementary fashion using the Boltzmann superposition principle. Consider a stress relaxation programme in which incremental strains Δe_1, Δe_2, Δe_3 and so on are added at times τ_1, τ_2, τ_3, etc., respectively. The total stress at time t is then given by

$$\sigma(t) = \Delta e_1 G(t - \tau_1) + \Delta e_2 G(t - \tau_2) + \Delta e_3 G(t - \tau_3) + \cdots, \tag{5.5}$$

where $G(t - \tau)$ is the stress relaxation modulus. Equation (5.5) may be generalized in an identical manner in which Equation (5.2) leads to Equations (5.3) and (5.4) to give

$$\sigma(t) = [G_r e] + \int_{-\infty}^{t} G(t - \tau) \frac{\mathrm{d}e(\tau)}{\mathrm{d}\tau} \mathrm{d}\tau, \tag{5.6}$$

where G_r is the equilibrium or relaxed modulus.

5.2.3 The Formal Relationship Between Creep and Stress Relaxation

We have already seen that stress relaxation is the converse of creep in a general sense. It is therefore to be expected that they will be formally related through a simple mathematical relationship.

For simplicity consider only the time-dependent terms in Equation (5.4). Then

$$e(t) = \int_{-\infty}^{t} J(t - \tau) \frac{\mathrm{d}\sigma(\tau)}{\mathrm{d}\tau} \mathrm{d}\tau.$$

Consider a stress programme starting at time $\tau = 0$ in which the stress decreases exactly as the relaxation function $G(t)$. In this case, the corresponding strain must remain constant as in a typical stress relaxation experiment. Thus, if

$$\frac{\mathrm{d}\sigma(\tau)}{\mathrm{d}\tau} = \frac{\mathrm{d}G(\tau)}{\mathrm{d}\tau}$$

then

$$\int_{0}^{t} \frac{\mathrm{d}G(\tau)}{\mathrm{d}\tau} J(t - \tau) \mathrm{d}\tau = \text{constant}. \tag{5.7}$$

For simplicity we can normalize the definitions of $G(\tau)$ and $J(\tau)$ so that the constant is unity. We then have

$$\int_0^t \frac{\mathrm{d}G(\tau)}{\mathrm{d}\tau} J(t-\tau)\mathrm{d}\tau = 1. \tag{5.8}$$

This expression is sometimes integrated to give

$$\int_0^t G(\tau)J(t-\tau) = t. \tag{5.9}$$

These equations provide a formal connection between the creep and stress relaxation functions. Using Equation (5.9), J can be calculated if G is known and *vice versa*. This can be achieved using numerical methods, as originally proposed by Hopkins and Hamming (1957, 1958) and detailed more recently by Lin (2005). The problem of interchangeability of creep and stress relaxation data can also be dealt with via relaxation or retardation spectra, and by approximate methods.

5.2.4 Mechanical Models, Relaxation and Retardation Time Spectra

The Boltzmann superposition principle is one starting point for a theory of linear viscoelastic behaviour, and is sometimes called the '*integral representation* of linear viscoelasticity', because it defines an integral equation. An equally valid starting point is to relate the stress to the strain by a linear differential equation, which leads to a differential representation of linear viscoelasticity. In its most general form, the equation is expressed as

$$P\sigma = Qe,$$

where P and Q are linear differential operators with respect to time. This representation has been found of particular value in obtaining solutions to specific problems in the deformation of viscoelastic solids (Lee 1960; Christensen 2003).

Most generally the differential equation is

$$a_0\sigma + a_1\frac{\mathrm{d}\sigma}{\mathrm{d}t} + a_2\frac{\mathrm{d}^2\sigma}{\mathrm{d}t^2} + \cdots = b_0\sigma + b_1\frac{\mathrm{d}e}{\mathrm{d}t} + b_2\frac{\mathrm{d}^2 e}{\mathrm{d}t^2} + \cdots. \tag{5.10}$$

It is often adequate to represent the experimental data obtained over a limited timescale by including only one or two terms on each side of this equation. We will now show that this is equivalent to describing the viscoelastic behaviour by mechanical models constructed of elastic springs that obey Hooke's law and viscous dashpots that obey Newton's law of viscosity.

The simplest models consist of a single spring and a single dashpot either in series or in parallel, and these are known as the Maxwell model and the Kelvin or Voigt models, respectively.

5.2.5 The Kelvin or Voigt Model

This model (Figure 5.10(a)) consists of a spring of modulus E_K, in parallel with a dashpot of viscosity η_K. If a constant stress σ is applied at time $t = 0$, there can be no instantaneous extension of the spring, as it is retarded by the dashpot. Deformation then occurs at a varying rate, with the stress shared between the two components until, after a time dependent on the dashpot viscosity, the spring approaches a finite maximum extension. When the stress is removed the reverse process occurs: there is no instantaneous retraction, but the initial unstretched length is eventually recovered (Figure 5.10(b)). The model does represent the time-dependent component of creep to a first approximation.

The stress–strain relations are, for the spring, $\sigma_1 = E_K\, e_1$ and for the dashpot

$$\sigma_2 = \eta_K \frac{\mathrm{d}e_2}{\mathrm{d}t}$$

The total stress σ is shared between spring and dashpot: $\sigma = \sigma_1 + \sigma_2$; but the strain in each component is the total strain: $e = e_1 = e_2$.

$$\therefore \sigma = E_K e + \eta_K \frac{\mathrm{d}e}{\mathrm{d}t}. \tag{5.11}$$

Solving for $0 < t < t_1$, when the stress is σ

$$\frac{E_K}{\eta_K}\int_0^t \mathrm{d}t = \int_0^e \frac{\mathrm{d}e}{\sigma/E_K - 1},$$

where η_K/E_K has the dimensions of time and represents the rate at which the deformation occurs: it is the retardation time τ'. Hence by integration

$$\frac{t}{\tau'} = \ln\left(\frac{\sigma/E_K}{\sigma/E_K - e}\right)$$

giving

$$\frac{\sigma}{E_K} = \left(\frac{\sigma}{E_K} - e\right)\exp(t/\tau')$$

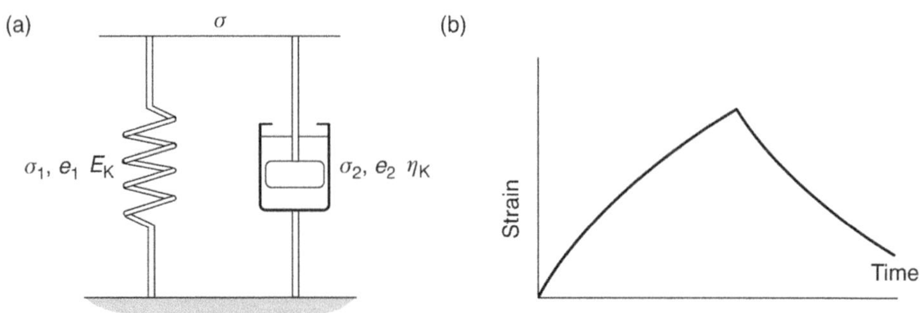

Figure 5.10 (a) The Kelvin, or Voigt, unit; (b) creep and recovery behaviour.

Rearranging, we obtain

$$e = \frac{\sigma}{E_K}[1 - \exp(-t/\tau')]. \tag{5.12}$$

It is convenient in creep experiments to replace E_K by $1/J$, where J is the spring compliance, to give

$$e = J\sigma[1 - \exp(-t/\tau')]. \tag{5.13}$$

For $t > t_1$ after unloading, the solution becomes

$$e = e_{t_1} \exp\left(\frac{t_1 - t}{\tau'}\right), \quad \text{where } e_{t_1} = J\sigma[1 - \exp(-t_1/\tau')]. \tag{5.14}$$

The retardation time τ' is the time after loading for the strain to reach

$$\left(1 - \frac{1}{\exp(1)}\right)$$

of its equilibrium value; after stress removal the strain decays to (1/exp(0)) of its maximum value in time τ'.

The Kelvin model is unable to describe stress relaxation, as at constant strain the dashpot cannot relax. In mathematical terms $(de/dt) = 0$, giving $\sigma = E_K e$.

5.2.6 The Maxwell Model

The Maxwell model consists of a spring and dashpot in series as shown in Figure 5.11(a).

The equations for the stress–strain relations are

$$\sigma_1 = E_m e_1, \tag{5.15a}$$

relating the stress σ_1 and the strain e_1 in the spring, and

$$\sigma_2 = \eta_m \frac{de_2}{dt} \tag{5.15b}$$

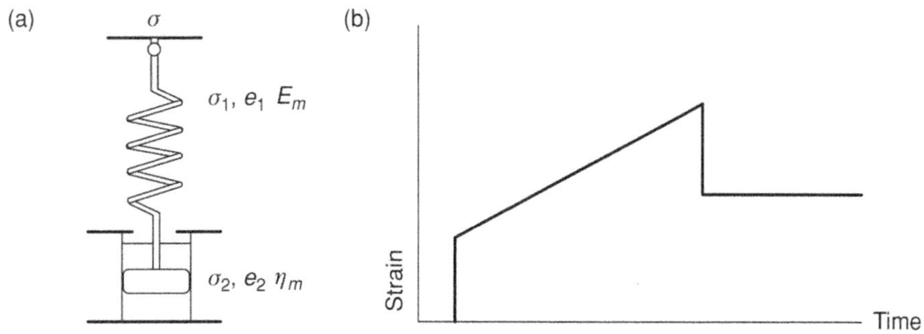

Figure 5.11 (a) The Maxwell unit and (b) creep and recovery behaviour.

relating the stress σ_2 and the strain e_2 in the dashpot. Because the stress is identical for the spring and the dashpot, the total stress $\sigma = \sigma_1 = \sigma_2$. The total strain e is the sum of the strain in the spring and the dashpot, that is $e = e_1 + e_2$.

To find the relationship between total stress and total strain, Equation (5.15a) can be written as

$$\frac{d\sigma_1}{dt} = E_m \frac{de_1}{dt}$$

and added to Equation (5.15b), giving

$$\frac{de}{dt} = \frac{1}{E_m}\frac{d\sigma}{dt} + \frac{\sigma}{\eta_m}. \tag{5.16}$$

The Maxwell model is of particular value in considering a stress relaxation experiment. In this case,

$$\frac{de}{dt} = 0 \text{ and } \frac{1}{E_m}\frac{d\sigma}{dt} + \frac{\sigma}{\eta_m} = 0.$$

Thus,

$$\frac{d\sigma}{\sigma} = -\frac{E_m}{\eta_m}dt.$$

At time $t = 0$, $\sigma = \sigma_0$, the initial stress, and integrating we have

$$\sigma = \sigma_0 \exp\left(\frac{-E_m}{\eta_m}\right)t. \tag{5.17}$$

This equation shows that the stress decays exponentially with a characteristic time constant $\tau = \eta_m/E_m$:

$$\sigma = \sigma_0 \exp\left(\frac{-t}{\tau}\right),$$

where τ is called the 'relaxation time'. There are two inadequacies of this simple model that can be understood immediately.

First, under conditions of constant stress, i.e.,

$$\frac{d\sigma}{d\tau} = 0, \quad \frac{de}{dt} = \frac{\sigma}{\eta_m}$$

and Newtonian flow is observed. This is clearly not generally true for viscoelastic materials, where the creep behaviour is more complex.

Secondly, the stress relaxation behaviour cannot usually be represented by a single exponential decay term, nor does it necessarily decay to zero at infinite time.

5.2.7 The Standard Linear Solid

We have seen that the Maxwell model describes the stress relaxation of a viscoelastic solid to a first approximation, and the Kelvin model describes the creep behaviour, but that neither model is adequate for the general behaviour of a viscoelastic solid where it is necessary to describe both stress relaxation and creep.

Consider again the general linear differential equation, which represents linear viscoelastic behaviour. From the present discussion, it follows that to obtain even an approximate description of both stress relaxation and creep, at least the first two terms on each side of Equation (5.10) must be retained, that is the simplest equation will be of the form

$$a_0\sigma + a_1\frac{\mathrm{d}\sigma}{\mathrm{d}t} = b_0 + b_1\frac{\mathrm{d}e}{\mathrm{d}t}. \tag{5.18}$$

This will be adequate to a first approximation for creep (when $\mathrm{d}\sigma/\mathrm{d}t = 0$) and for stress relaxation (when $\mathrm{d}e/\mathrm{d}t = 0$), giving an exponential response in both cases.

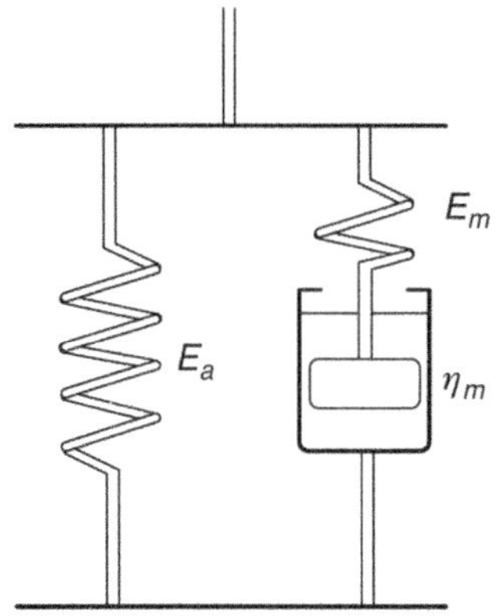

Figure 5.12 The standard linear solid.

It is very easy to show that the model shown in Figure 5.12 has this form. The stress–strain relationship is

$$\sigma + \tau\frac{\mathrm{d}\sigma}{\mathrm{d}t} = E_a e + (E_m + E_a)\tau\frac{\mathrm{d}e}{\mathrm{d}t}, \quad \text{where } \tau = \frac{\eta_m}{E_m}. \tag{5.19}$$

This model is known as the 'standard linear solid' and is usually attributed to Zener (1948). It provides an approximate representation to the observed behaviour of polymers in their main viscoelastic range. As has been discussed, it predicts an exponential response only. To describe the observed viscoelastic behaviour quantitatively would require the inclusion of many terms in the linear differential Equation (5.10). These more complicated equations are equivalent to either a large number of Maxwell elements in parallel or a large number of Voigt elements in series (Figure 5.13(a) and (b)).

5.2.8 Relaxation Time Spectra and Retardation Time Spectra

It is next required to obtain a quantitative description of stress relaxation and creep that will help to form a link with the original mathematical description in terms of the Boltzmann integrals. It is simple and instructive to do this by development of the Maxwell and Kelvin models.

Equation (5.17) shows that the Maxwell model gives

$$\sigma(t) = E_m e \exp\frac{-t}{\tau}$$

and the stress relaxation modulus $G(t) = E_m \exp(-t/\tau)$. For a series of Maxwell elements joined in parallel, again at constant strain e, the stress is given by

$$\sigma(t) = e\sum_n \exp\frac{-t}{\tau_n},$$

where E_n, τ_n are the spring constant and relaxation time, respectively, of the nth Maxwell element.

The summation can be written as an integral, giving

$$\sigma(t) = [G_r e] + e\int_0^{\infty} f(\tau)\exp\frac{-t}{\tau}\mathrm{d}\tau, \tag{5.20}$$

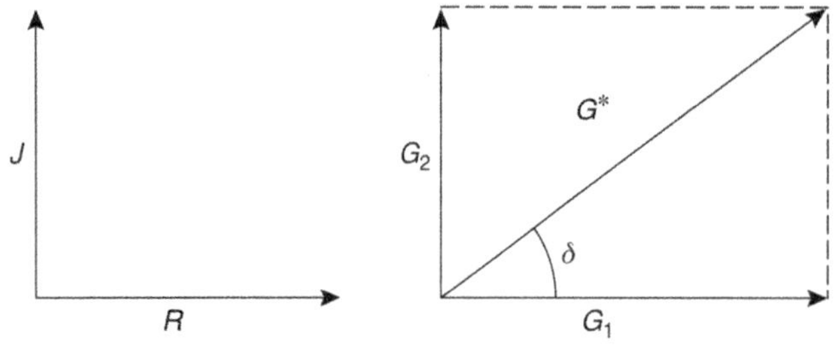

Figure 5.14 Phasor diagram for complex modulus $G^* = G_1 + iG_2$ and phase lag $\tan\delta = G_2/G_1$.

A phasor diagram (Figure 5.14) then indicates that G_1 and G_2 define a complex modulus G^*. If $e = e_0 \exp(i\omega t)$, $\sigma = \sigma_0 \exp\,[i\,(\omega t + \delta)]$, so that

$$\begin{aligned} G^* &= \frac{\sigma}{e} = \frac{\sigma_0}{e_0}\exp(i\delta) \\ &= \frac{\sigma_0}{e_0}(\cos\delta + i\sin\delta) \\ &= G_1 + G_2, \end{aligned} \tag{5.25}$$

where G_1, which is in phase with the strain, is called the storage modulus because it defines the energy stored in the specimen due to the applied strain, and G_2, which is $\pi/2$ out of phase with the strain, defines the dissipation of energy and is called the loss modulus, for a reason that becomes evident through calculating the energy (ΔE) dissipated per cycle:

$$\Delta E = \oint \sigma\,\mathrm{de} = \int_0^{\frac{2\pi}{\omega}} \sigma \frac{\mathrm{de}}{\mathrm{dt}}\mathrm{dt}.$$

Substituting for σ and e,

$$\Delta E = \omega e_0^2 \int_0^{2\pi/\omega} \left(G_1 \sin\omega t \cos\omega t + G_2 \cos^2\omega t\right)\mathrm{d}t. \tag{5.26}$$

The integral is solved by using $\mathbf{\sin\omega t \cos\omega t = \frac{1}{2}\sin 2\omega t}$ and $\mathbf{\cos^2\omega t = \frac{1}{2}(1 + \cos 2\omega t)}$, to give

$$\Delta E = \pi G_2 e_0^2. \tag{5.27}$$

If the integral for ΔE is evaluated over a quarter-cycle rather than over the complete period, the first term

$$G_1 \omega e_0^2 \int_0^{\pi/2\omega} \sin\omega t \cos\omega t \ \mathrm{d}t \tag{5.28}$$

gives the maximum stored elastic energy (E).

Evaluating as before, we obtain

$$E = \frac{1}{2} G_1 e_0^2, \tag{5.29}$$

which, as expected, is independent of frequency. Equations (5.27) and (5.29) can be rewritten as

$$G_1 = \frac{2E}{e_0^2} G_2 = \frac{\Delta E}{\pi e_0^2}.$$

Hence

$$\frac{G_2}{G_1} = \tan\,\delta = \frac{\Delta E}{2\pi E}. \tag{5.30}$$

The ratio $\Delta E/E$ is called the specific loss

$$\frac{\Delta E}{E} = 2\pi \tan\,\delta. \tag{5.31}$$

Typical values of G_1, G_2 and $\tan\delta$ for a polymer are 1 GPa, 10 MPa and 0.01, respectively. In such cases, $|G^*|$ is approximately equal to G_1, and it is customary to define the dynamic mechanical behaviour in terms of the 'modulus' $G \approx G_1$ and the phase angle δ or $\tan\delta = G_2/G_1$.

A complementary treatment can be developed to define a complex compliance $J^* = J_1 - iJ_2$, which is directly related to the complex modulus, as $G^* = 1/J^*$.

5.3.1 Experimental Patterns for G_1, G_2 and so on as a Function of Frequency

Now consider the complex moduli and compliances as a function of frequency for a typical viscoelastic solid, in a similar manner to the creep and stress relaxation as a function of time.

Figure 5.15 shows the variation in G_1, G_2 and $\tan\delta$ with frequency for a polymer that shows no flow. At low frequencies, the polymer is rubber-like and has a low modulus G_1 of ~0.1 MPa, which is independent of frequency. At high frequencies, the polymer is glassy with a modulus of ~1 GPa, which is again independent of frequency. At intermediate frequencies, the polymer behaves as a viscoelastic solid, and its modulus G_1 increases with increasing frequency.

The complementary pattern of behaviour is shown by the loss modulus G_2. At low and high frequencies G_2 is zero, the stress and strain being exactly in phase for the rubbery and glassy states. At intermediate frequencies, where the polymer is viscoelastic, G_2 rises to a maximum value, this occurring at a frequency close to that for which the storage modulus is changing most rapidly with frequency. The viscoelastic region is also characterized by a maximum in the loss factor $\tan\delta$, but this occurs at a slightly lower frequency than that in G_2, since $\tan\delta = G_2/G_1$ and G_1 is also changing rapidly in this frequency range.

An analogous diagram (Figure 5.16) shows the variation in the compliances J_1 and J_2 with frequency.

The next development is to obtain a mathematical representation for the dynamic mechanical behaviour as a function of frequency. As in the case of stress relaxation and creep, a very easy starting point for the argument is based on the Maxwell and Voigt models.

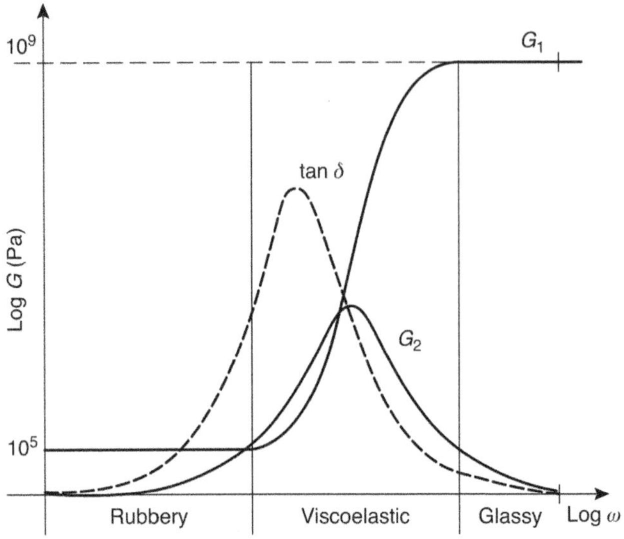

Figure 5.15 The complex modulus $G_1 + iG_2$ as a function of frequency ω.

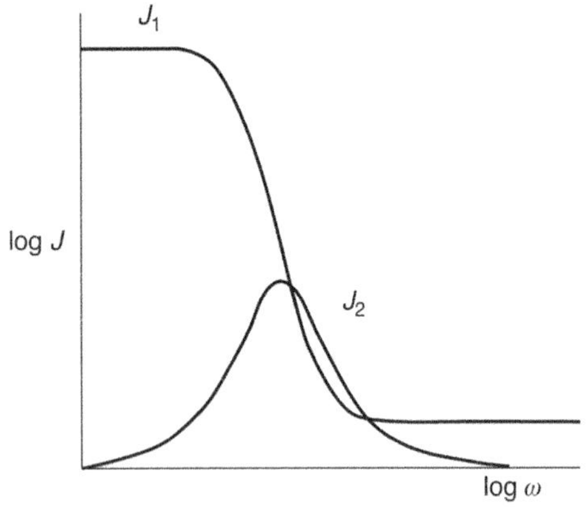

Figure 5.16 The complex compliance $J^* = J_1 - iJ_2$ as a function of frequency.

Using Equation (5.12) for the Maxwell model,

$$\frac{de}{dt} = \frac{1}{E_m}\frac{d\sigma}{dt} + \frac{\sigma}{\eta_m}\eta$$

and the definition of the relaxation time as $\tau = \eta_m/E_m$, we can write

$$\sigma + \tau\frac{d\sigma}{dt} = E_m\tau\frac{de}{dt}.$$

Put

$$\sigma = \sigma_0 e^{i\omega t} = (G_1 + G_2)e.$$

This gives

$$\sigma_0 e^{i\omega t} + i\omega\tau\sigma_0 e^{i\omega t} = \frac{E_m \tau i\omega\sigma_0 e^{i\omega t}}{G_1 + iG_2}$$

from which it follows that

$$G_1 + iG_2 = \frac{E_m i\omega\tau}{1 + i\omega\tau},$$

i.e.,

$$G_1 = \frac{E_m \omega^2 \tau^2}{1 + \omega^2 \tau^2}, \quad G_2 = \frac{E_m \omega\tau}{1 + \omega^2 \tau^2}, \quad \text{and} \quad \tan\,\delta = \frac{1}{\omega t}. \tag{5.32}$$

This result gives the pattern shown in Figure 5.17 for G_1, G_2 and $\tan\delta$ as a function of frequency (or $\omega\tau$, which is more convenient). It is seen that the qualitative features are correct in the case of G_1, G_2, but not for $\tan\delta$.

A similar manipulation of the equation representing the Voigt model, introducing the complex compliances, leads to a comparable qualitative picture of J_1, J_2 and $\tan\delta$ as a function of frequency. Again the qualitative features are correct for J_1 and J_2 but not for $\tan\delta$, which in this case is equal to $\omega\tau'$.

The Maxwell and Voigt models are therefore inadequate to describe the dynamic mechanical behaviour of a polymer, as they do not provide an adequate representation of both the creep and stress relaxation behaviour. A good measure of qualitative improvement could be gained, as in the previous discussion of creep and stress relaxation, by using a three-parameter model, e.g., the standard linear solid, and it is an interesting exercise to show that this model gives a more realistic variation in G_1, G_2 and $\tan\delta$ with frequency.

It is, however, desirable to move directly to the general representation, using the relaxation time spectrum.

The general representation for the stress relaxation modulus (Equation 5.21),

$$G(t) = \frac{\sigma(t)}{e} = [G_r] + \int_{-\infty}^{\infty} H(\tau) \exp \frac{-t}{\tau} \mathrm{d}(\ln\tau),$$

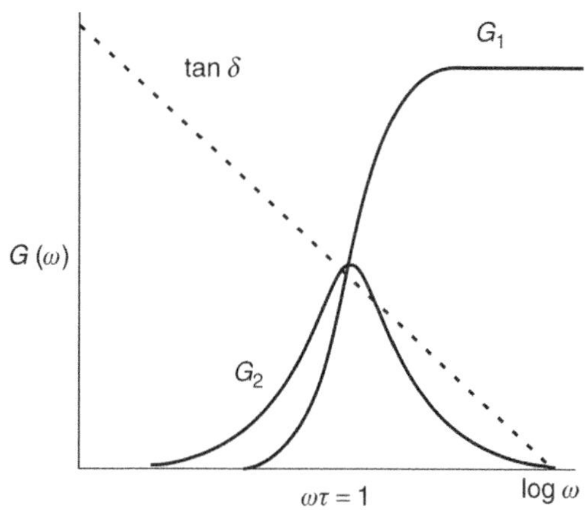

Figure 5.17 The complex modulus $G^* = G_1 + iG_2$ as a function of frequency ω.

follows by generalizing the stress relaxation response from a single Maxwell element, where $G(t) = E_m \exp(-t/\tau)$.

The response of a Maxwell element to an alternating strain is defined by the relationships

$$G_1 = \frac{E_m\omega^2\tau^2}{1+\omega^2\tau^2} \text{ and } G_2 = \frac{E_m\omega\tau}{1+\omega^2\tau^2}.$$

An identical generalization to the previous one then gives

$$G_1(\omega) = [G_r] + \int_{-\infty}^{\infty} \frac{H(\tau)\omega^2\tau^2}{1+\omega^2\tau^2} \mathrm{d}(\ln\tau) \tag{5.33}$$

and

$$G_2(\omega) = \int_{-\infty}^{\infty} \frac{H(\tau)\omega\tau}{1+\omega^2\tau^2} \mathrm{d}(\ln\tau). \tag{5.34}$$

As previously, the spring constant E_m is replaced by the weighting function $H(\tau)\mathrm{d}(\ln\tau)$ that defines the contribution to the response of elements whose relaxation time is between $\ln\tau$ and $\ln\tau + \mathrm{d}(\ln\tau)$. It is seen that the stress relaxation modulus $G(t)$ and the real and imaginary parts of the complex compliance G_1 and G_2 can all be directly related to the same relaxation time spectrum $H(\tau)$.

Similar relationships hold between the creep compliance $J(t)$, the real and imaginary parts of the complex compliance J_1 and J_2 and the retardation time spectrum $L(t)$. These relationships can be readily derived by consideration of the response of a Voigt element to an alternating stress. They will not be derived here, but the results are quoted for completeness:

$$J_1(\omega) = [J_u] + \int_{-\infty}^{\infty} \frac{L(\tau)}{1+\omega^2\tau^2} \mathrm{d}(\ln\tau) \tag{5.35}$$

$$J_2(\omega) = \int_{-\infty}^{\infty} \frac{J(\tau)\omega\tau}{1+\omega^2\tau^2} \mathrm{d}(\ln\tau). \tag{5.36}$$

5.4 The Relationships Between the Complex Moduli and the Stress Relaxation Modulus

The exact formal relationships between the various viscoelastic functions are conveniently expressed using Fourier or Laplace transform methods (cf. Section 5.4.2). However, it is often adequate to use simple approximations due to Alfrey in which the exponential term for a single Kelvin or Maxwell unit is replaced by a step function, as shown schematically in Figure 5.18.

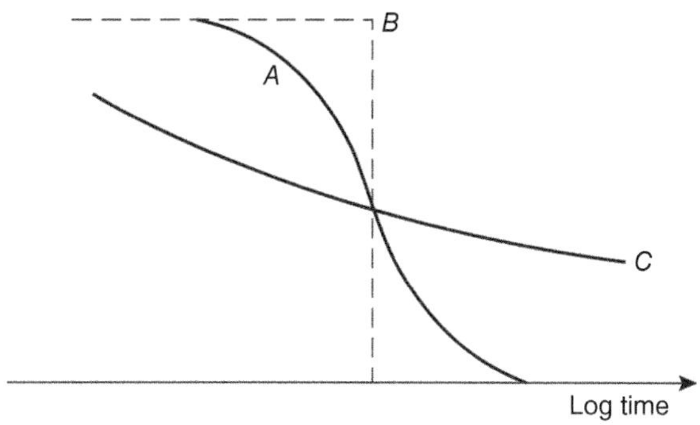

Figure 5.18 The Alfrey approximation: the stress relaxation of a Maxwell unit A is replaced by a step function B. The curve C represents relaxation of a typical viscoelastic polymer.

Consider

$$G(t) = [G_r] + \int_{-\infty}^{\infty} H(\tau) \exp \frac{-t}{\tau} \mathrm{d}(\ln \tau). \tag{5.37}$$

If we assume that $e^{-t/\tau} = 0$ up to the time $\tau = t$, and $e^{-t/\tau} = 1$ for $\tau > t$, we can write

$$G(t) = [G_r] + \int_{\ln t}^{\infty} H(\tau)\mathrm{d}(\ln \tau).$$

This gives the relaxation time spectrum

$$H(\tau) = -\left[\frac{\mathrm{d}G(t)}{\mathrm{d}\ln t}\right]_{t=\tau}, \tag{5.38}$$

which is known as the 'Alfrey approximation' (Alfrey and Doty 1945; Alfrey 1948).

The relaxation time spectrum can be expressed to a similar degree of approximation in terms of the real and imaginary parts of the complex modulus:

$$H(\tau) = \left[\frac{\mathrm{d}G_1(\omega)}{\mathrm{d}\ln\omega}\right]_{\frac{1}{\omega}=\tau} = \frac{2}{\pi}[G_2(\omega)]_{\frac{1}{\omega}=\tau} \tag{5.39}$$

These relationships are illustrated diagrammatically for the case of a single relaxation transition in Figure 5.19. To obtain the complete relaxation time spectrum, the longer time part of $H(\tau)$ will be found from the stress relaxation modulus data of Figure 5.19(a) and the shorter time part from the dynamic mechanical data of Figure 5.19(b).

Figure 5.19 shows that the relaxation time spectrum can be determined directly from the gradient of plots of the relaxation modulus or the dynamic modulus G_1 versus logarithm of time, or from G_2 even more directly.

Complementary relationships can be used to obtain the retardation time spectrum in terms of the complex compliances and the creep compliance.

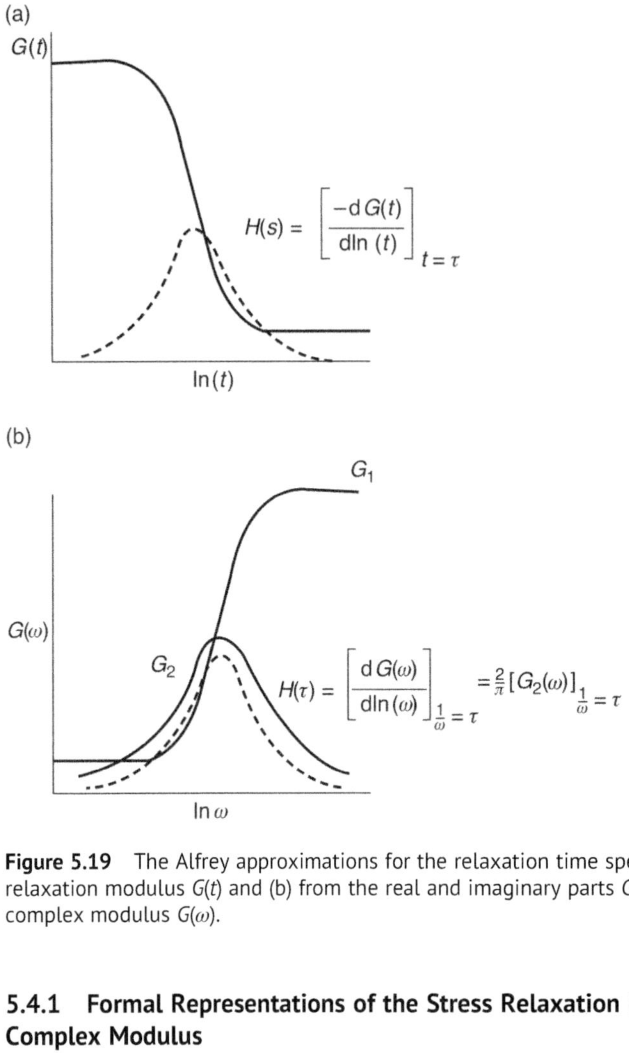

Figure 5.19 The Alfrey approximations for the relaxation time spectrum $H(\tau)$: (a) from the stress relaxation modulus $G(t)$ and (b) from the real and imaginary parts G_1 and G_2, respectively, of the complex modulus $G(\omega)$.

5.4.1 Formal Representations of the Stress Relaxation Modulus and the Complex Modulus

A complete exposition of the mathematical structure of linear viscoelasticity has been given by Gross (1953). Here we will only summarize certain parts of his argument to illustrate the use of Laplace and Fourier transforms in establishing the formal connections between various viscoelastic functions:

1) *The stress relaxation modulus.* This modulus is a continuous, decreasing function that goes to zero at infinite time. In Gross' nomenclature, it is represented in integral form as

$$G(t) = [G_r] + \int_0^{\infty} \overline{\beta F}(\tau) \exp \frac{-t}{\tau} d\tau, \qquad (5.40)$$

where $\overline{F}(\tau)\mathrm{d}\tau$ is the relaxation spectrum and $\overline{\beta}$ is a normalization factor such that

$$\int_0^\infty \overline{F}(\tau)\mathrm{d}\tau = 1, \quad \overline{\beta} = G(0).$$

This equation, in terms of our models, represents an infinite series of Maxwell elements and it is formally identical to Equation (5.21).

It can be transformed into a Laplace integral or Laplace transform by putting $1/\tau = s$, the relaxation frequency, and introducing a frequency function $\overline{\mathbf{N}}(\boldsymbol{s})\mathbf{d}\boldsymbol{s}$ defined as

$$\overline{N}(s) = \frac{\overline{\beta}\overline{F}(1/s)}{s^2}.$$

Thus

$$G(t) = [G_r] + \int_0^\infty \overline{N}(s)e^{-ts}\mathrm{d}s. \tag{5.41}$$

The importance of this representation is that when $G(t)$ has been determined, the relaxation time spectrum can be found, in principle, by standard methods for the inversion of the Laplace integral. In practice this requires computation methods, as it is not usually possible to find an analytical expression to fit the stress relaxation modulus.

The Alfrey approximation is now given by putting

$$e^{-ts} - 1 \quad \text{for} \quad s \leq 1/t$$

and

$$e^{-ts} = 0 \quad \text{for} \quad s > 1/t$$

and

$$G(t) = [G_r] + \int_0^{1/t} \overline{N}(s)\mathrm{d}s. \tag{5.42}$$

2) *The complex modulus.* The Boltzmann Superposition Principle gives us

$$\sigma(t) = [G_r e(t)] + \int_{-\infty}^{\infty} G(t-\tau)\frac{\mathrm{d}e(\tau)}{\mathrm{d}\tau}\mathrm{d}\tau.$$

Put $e(\tau) = e_0 e^{i\omega t}$. Then

$$\sigma(t) = [G_r e(t)] + i\omega \int_{-\infty}^{t} G(t-\tau)e_0 e^{i\omega\tau}\mathrm{d}\tau. \tag{5.43}$$

Put $t - \tau = T$. Then

$$\sigma(t) = [G_r e(t)] + i\omega \int_0^\infty G(T)e^{-i\omega t}\mathrm{d}T e_0 e^{i\omega t}. \tag{5.44}$$

Now $e(t) = e_0 e^{i\omega t}$. Thus

$$\frac{\sigma(t)}{e(t)} = [G_r] + i\omega \int_0^\infty G(\tau) e^{-i\omega\tau} d\tau = G^*(\omega),$$

the complex modulus, where we have changed the dummy variable from T back to τ.

Therefore

$$G_1(\omega) = \omega \int_0^\infty G(\tau) \sin \omega\tau d\tau \tag{5.45}$$

and

$$G_2(\omega) = \omega \int_0^\infty G(\tau) \cos \omega\tau d\tau, \tag{5.46}$$

where $G^*(\omega) = (G_r + G_1) + iG_2$. Equations (5.45) and (5.46) are one-sided Fourier transforms.

Inversion gives the stress relaxation modulus

$$G(t) = \frac{2}{\pi} \int_0^\infty \frac{G_1(\omega)}{\omega} \sin \omega t d\omega \tag{5.47}$$

and

$$G(t) = \frac{2}{\pi} \int_0^\infty \frac{G_2(\omega)}{\omega} \cos \omega t d\omega. \tag{5.48}$$

These equations imply a relationship between $G_1(\omega)$ and $G_2(\omega)$, the dispersion or compatibility relations, which are the viscoelastic analogue of the Kramers–Krönig relations for optical dispersion and magnetic relaxation.

5.4.2 Formal Representations of the Creep Compliance and the Complex Compliance

Similar relationships hold for the creep compliance and complex compliance to those derived for the stress relaxation modulus and the complex modulus. The details of the derivations will not be given, but the results are quoted for completeness.

1) *Creep compliance*. In this case the *rate of change* of creep compliance is expressed as a Laplace integral. Thus

$$\frac{dJ(t)}{dt} = \int_0^\infty sN(s) e^{-ts} ds, \tag{5.49}$$

where

$$N(s) = \frac{F(1/s)}{s^2}, \quad s = \frac{1}{\tau}$$

and $F(\tau)\mathrm{d}\tau$ is the distribution of retardation times. Note that $N(s) \neq \overline{N}(s)$, and that $F(\tau) \neq \overline{F}(\tau)$, i.e., the retardation time spectrum is not identical to the relaxation time spectrum.

2) *Complex compliance.* Here it is found that

$$J_1(\omega) = \int_0^{\infty} \frac{\mathrm{d}J(\tau)}{\mathrm{d}\tau} \cos\omega\tau\mathrm{d}\tau \tag{5.50}$$

and

$$J_2(\omega) = -\int_0^{\infty} \frac{\mathrm{d}J(\tau)}{\mathrm{d}\tau} \sin\omega\tau\mathrm{d}\tau. \tag{5.51}$$

Again both $J_1(\omega)$ and $J_2(\omega)$ are Fourier transforms, which may be inverted to give the creep compliance in terms of the components of the complex compliance. Both the inversion formulae give the creep compliance, implying a relationship between the real and imaginary parts of the complex compliance, as in the case of the complex modulus.

5.4.3 The Formal Structure of Linear Viscoelasticity

Gross (1953) has discussed the formal structure of the theory of linear viscoelasticity. A summary of his treatment will be presented here, as a suitable conclusion to our discussion.

There are two groups of experiments:

Group 1: Experiments that take place under a given stress, either fixed or alternating. These define the creep compliance or the complex compliance.

Group 2: Experiments that take place under a given strain, either fixed or alternating. These define the stress relaxation modulus or the complex modulus.

Within each group, the viscoelastic functions exist in three levels:

(a) Top level	Complex compliance	(Group 1)
	Complex modulus	(Group 2)
(b) Medium level	Creep function	(Group 1)
	Relaxation function	(Group 2)
(c) Bottom level	Retardation spectrum	(Group 1)
	Relaxation spectrum	(Group 2)

To go *up* a level, one applies either a Laplace transform or a one-sided complex Fourier transform.

To go *down* a level, one applies either an inverse Laplace transform or an inverse Fourier transform.

The relationships between the groups vary in complexity. At the top level, the complex compliance is merely the inverse of the complex modulus. The relationships between the creep function and the relaxation function and between the retardation spectrum and the relaxation spectrum involve integral equations and integral transforms, respectively.

5.5 The Relaxation Strength

A concept that is of value in considering the relationship of viscoelastic behaviour to physical and chemical structure is that of 'relaxation strength'. In a stress relaxation experiment, the modulus relaxes from a value G_u at very short times to G_r, at very long times (Figure 5.20(b)). Similarly in a dynamic mechanical experiment, the modulus changes from G_r at low frequencies to G_u at very high frequencies. G_u is the unrelaxed modulus and G_r is the relaxed modulus (Figure 5.20(c)).

This behaviour is shown by the standard linear solid of Figure 5.20(a). Consider in turn the behaviour in the initial unrelaxed state and in the final relaxed state.

1) A total applied stress σ is given in terms of the initial unrelaxed strain e_1 by adding the stresses in both springs. Thus

$$\sigma = G_r e_1 + (G_u - G_r)e_1 = G_u e_1$$

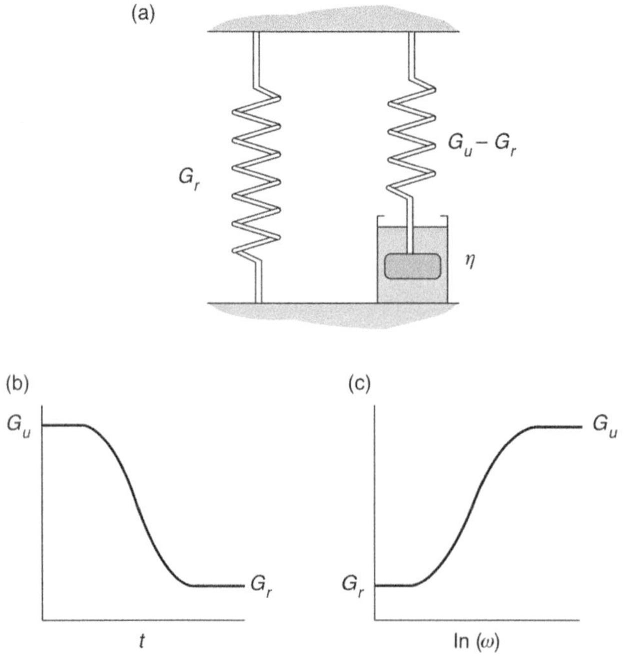

Figure 5.20 The standard linear solid (a) gives a response in a stress relaxation test shown in (b) and in a dynamic mechanical test in (c). G_u is the unrelaxed modulus and G_r, the relaxed modulus.

and the initial unrelaxed strain

$$e_1 = \frac{\sigma}{G_u} = \frac{\sigma}{\text{Unrelaxed modulus}}$$

2) The final relaxed strain e_2 is given in terms of an applied stress σ by

$$e_2 = \frac{\sigma}{G_r} = \frac{\sigma}{\text{Relaxed modulus}}$$

because the spring $(G_u - G_r)$ is now ineffective. The relaxation strength is conventionally defined as

$$\frac{\text{Final strain} - \text{Initial strain}}{\text{Initial strain}}$$

This is

$$\frac{e_2 - e_1}{e_1} = \left\{\frac{1}{G_r} - \frac{1}{G_u}\right\} G_u = \frac{G_u - G_r}{G_r} \tag{5.52}$$

The equation for the standard linear solid in Figure 5.20(a) is

$$\sigma + \tau_1 \frac{d\sigma}{dt} = G_r \left\{ e + \tau_2 \frac{de}{dt} \right\}, \tag{5.53}$$

where

$$\tau_1 = \frac{\eta}{G_u - G_r} \text{ and } \tau_2 = \frac{\tau_1}{G_r}.$$

Then for dynamic mechanical measurements it may be shown that

$$G_1(\omega) = \frac{G_r(1 + \omega^2\tau_1\tau_2)}{1 + \omega^2\tau_1^2} = \frac{G_r + \omega^2\tau^2 G_u}{1 + \omega^2\tau^2} \tag{5.54}$$

$$G_2(\omega) \frac{G_r(\tau_2 - \tau_1)\omega}{1 + \omega^2\tau_1^2} = \frac{G_u + \omega^2\tau^2 G_u}{1 + \omega^2\tau^2} \tag{5.55}$$

and

$$\tan\,\delta = \frac{(\tau_2 - \tau_1)\omega}{1 + \omega^2\tau_1\tau_2} = \frac{(G_u - G_r)\omega\tau}{G_r + \omega^2\tau^2 G_u}, \tag{5.56}$$

where we have put

$$\tau = \tau_1 = \frac{\eta}{G_u - G_r}.$$

The following relationships then hold:

$$\tan\,\delta_{\max}\left(\omega^2\tau^2 = G_r/G_u\right) = \frac{G_u - G_r}{2\sqrt{G_u G_r}} \tag{5.57}$$

$$G_{2_{\max}}\left(\omega^2\tau^2 = 1\right) = \frac{G_u - G_r}{2} \tag{5.58}$$

$$\int_{-\infty}^{\infty} G_2(\omega)\mathrm{d}(\ln\omega) = \frac{\pi}{2}(G_u - G_r) \tag{5.59}$$

$$\int_{-\infty}^{\infty} \tan\,\mathrm{d}(\ln\omega) = \frac{\pi}{2}\frac{(G_u - G_r)}{\sqrt{G_u G_r}}. \tag{5.60}$$

All the relationships defined above are proportional to $G_u - G_r$, and hence to the relaxation strength. $\tan\delta_{\max}$ and $\int_{-\infty}^{\infty}$ **tan** $\boldsymbol{\delta}$**d**$(\ln\omega)$ are closest to our original definition in being normalized to a dimensionless quantity. This provides some formal justification for the use of $\tan\delta$ rather than G_2 for estimating the relaxation strength to correlate with structural parameters (Gray and McCrum 1968).

References

Alfrey, T. (1948) *Mechanical Behaviour of High Polymers*, John Wiley & Sons, New York.

Alfrey, T. and Doty, P. (1945) The methods of specifying the properties of viscoelastic materials. *J. Appl. Phys.*, 16, 700.

Boltzmann, L. (1876) Zur Theorie der elastischen Nachwirkung. *Pogg. Ann. Phys. Chem.*, 7, 624.

Christensen, R.M. (2003) *Theory of Viscoelasticity*, 2nd edn, Dover Publications, Inc., Mineola, NY.

Gray, R.W. and McCrum, N.G. (1968) On the γ relaxations in linear polyethylene and polytetrafluoroethylene. *J. Polym. Sci. B. Polym. Lett.*, 6, 691.

Gross, B. (1953) *Mathematical Structure of the Theories of Viscoelasticity*, Hermann, Paris.

Hopkins, I.L. and Hamming, R.W. (1957) On creep and relaxation. *J. Appl. Phys.*, 28, 906–909.

Hopkins, I.L. and Hamming, R.W. (1958) Note on paper "On creep and relaxation". *J. Appl. Phys.*, 29, 742.

Lee, E.H. (1960) Viscoelastic stress analysis, in *Proceedings of the First Symposium on Naval Structural Mechanics*, (eds J.N. Goodier and J.F. Hoff), Pergamon Press, Oxford, p. 456.

Lin, Y.-H. (2005) Whole range of chain dynamics in entangled polystyrene melts revealed from creep compliance: thermorheological complexity between glassy-relaxation region and rubber-to-fluid region. 1. *J. Phys. Chem. B*, 109, 17654–17669.

Zener, C. (1948) *Elasticity and Anelasticity of Metals*, Chicago University Press, Chicago.

Further Reading

Drozdov, A.D. (1998) *Mechanics of Viscoelastic Solids*, John Wiley & Sons, Chichester.

Ferry, J.D. (1980) *Viscoelastic Properties of Polymers*, John Wiley & Sons, New York.

Lin, Y.-H. (2010) *Polymer Viscoelasticity: Basics, Molecular Theories, Experiments and Simulations*, 2nd edn, Scientific Publishing Company, Singapore.

Shaw, M.T. and MacKnight, W.J. (2005) *Introduction to Polymer Viscoelasticity*, 3rd edn, John Wiley & Sons, Hoboken.

Wineman, A.S. and Rajagopal, K.R. (2000) *Mechanical Response of Polymers*, Cambridge University Press, New York.

6

The Measurement of Viscoelastic Behaviour

For a satisfactory understanding of the viscoelastic behaviour of polymers, data are required over a wide range of frequency (or time) and temperature. The number of experiments required can sometimes be reduced by using either the equivalence of creep, stress relaxation and dynamic mechanical data (described in Chapter 5) or the equivalence of time and temperature as variables (to be discussed in Chapter 7). Nevertheless a variety of techniques need to be combined to cover a wide range of both time and temperature.

There are three main classes of experiment, which will be discussed in turn:

1) Static measurements: creep and stress relaxation
2) Vibrational methods
3) Wave propagation methods.

The approximate frequency scale for each class of technique is indicated in Figure 6.1.

6.1 Creep and Stress Relaxation

These forms of testing are essentially static. They are the simplest tests, in that one of the principal quantities – stress or strain – is held constant, whilst the other is allowed to evolve with time.

6.1.1 Creep

In the case of creep, to conform with viscoelastic theory the true stress should be held constant. However, traditionally creep testing has been conveniently carried out dead loading so that the force, and therefore the engineering stress, is held constant. The difference becomes significant as the specimen cross-sectional area varies at high axial strains. Leaderman (1962) introduced a cam mechanism to generate a load that decreased with time to correct for this effect. More recently, Nitta and Maeda (2010) developed apparatus for creep at constant true stress that functioned by monitoring the transverse strain and using this data to continuously control the applied force. Sheet specimens of high-density polyethylene (HDPE) were subject to tensile loading whilst their thicknesses, which were in the range 200–500 μm, were measured using laser technology to an accuracy of ±10 μm. Axial

Mechanical Properties of Solid Polymers, Fourth Edition. John Sweeney and Peter Hine.

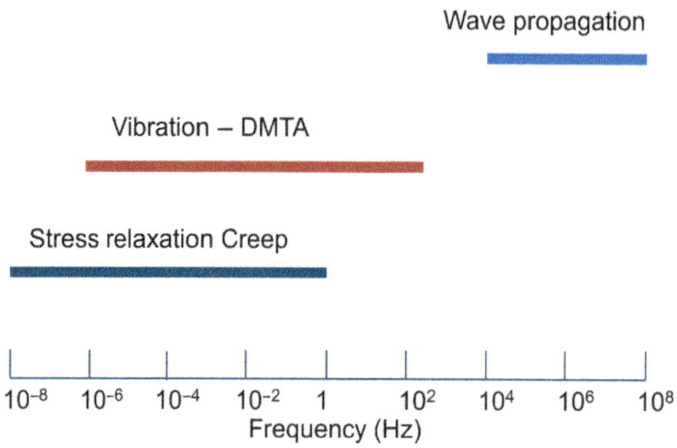

Figure 6.1 Approximate frequency scales for different experimental techniques.

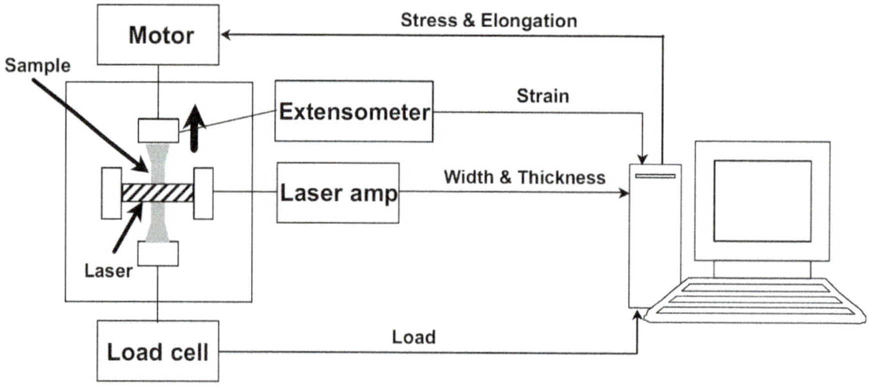

Figure 6.2 Servo-controlled creep apparatus. Nitta, K-h. and Maeda, H. (2010) / with permission of Elsevier.

strains were as high as 0.25, corresponding to a significant reduction in cross-section. The apparatus is shown in diagrammatic form in Figure 6.2.

Results for an HDPE with 54% crystallinity, shown in Figure 6.3, are clearly nonlinear viscoelastic. This is reinforced in Figure 6.4, where the creep compliance function at time 1000 s is plotted against stress and shown to be nonconstant. However, it achieves constancy at a critical stress value. The use of constant true stress creep data adds confidence in the precision of this finding.

When creep testing at constant load is considered acceptable, there are possibilities other than dedicated creep machines. Dynamic mechanical thermal analysis (DMTA) apparatus (see Section 6.2) can be used in a static mode and programmed to maintain a constant load whilst monitoring tensile extension. Tensile testing machines often have a 'load control' setting and thus the same capability.

6.1.2 Stress Relaxation

Tensile testing machines with controlled displacement cross-head motion are the traditional apparatus for stress relaxation. However, as with creep, DMTA sets can be used in

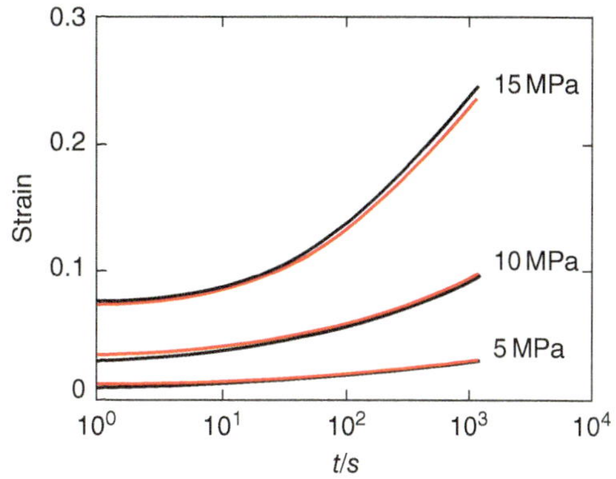

Figure 6.3 Creep at 22°C of HDPE with 54% crystallinity. Different colour curves denote different specimen geometries. Nitta, K-h. and Maeda, H. (2010) / with permission of Elsevier.

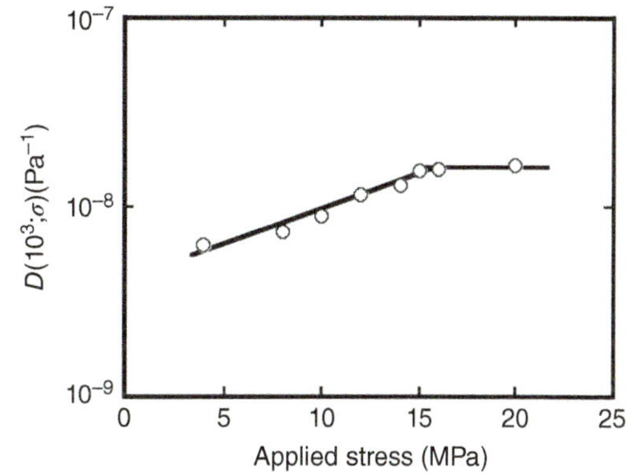

Figure 6.4 Creep compliance function D at 22°C at 1000 s of HDPE with 54% crystallinity. Nonlinearity is clear up to around 15 MPa. Nitta, K-h. and Maeda, H. (2010) / with permission of Elsevier.

a static mode and in this case used to maintain a constant extension whilst the decaying load is monitored. An example of stress relaxation by DMTA for a range of temperatures is shown in Figure 6.5. The use at elevated temperatures of DMTA rather than a conventional tensile testing machine has the advantage that tests can be performed in rapid succession as the DMTA's small environmental chamber cools down quickly.

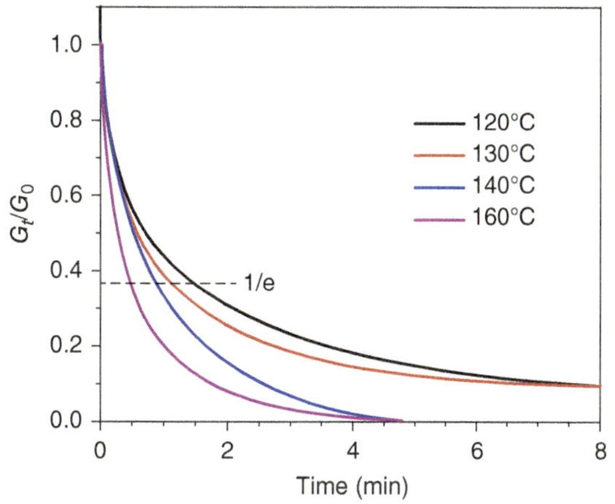

Figure 6.5 Relaxation modulus G normalized by its initial value for cross-linked sulphur-based polymer. (Yan, P. et al. (2020) Inverse vulcanized polymers with shape memory, enhanced mechanical properties, and vitrimer behavior. Angew. Chem. **132**, 13473–13480.)

6.2 Dynamic Mechanical Thermal Analysis (DMTA)

There are a number of commercial machines available for dynamic testing under varying temperature conditions. An actuator imposes an oscillatory (linear or angular) displacement, and typically a strain-gauge load cell measures force. Testing is frequently in bending mode, producing results that may be difficult to interpret in terms of viscoelasticity owing to the non-uniform nature of the stress field. However, such testing is useful in locating the temperatures of relaxation transitions (see Chapter 10). Temperature ranges are typically −150 to 600°C, and frequency ranges are 10^{-6} to 200 Hz. In some cases, a single machine can operate in a range of testing modes – bend, tension, compression, shear and torsion – by the use of different loading fixtures. Modulus and tan δ data (see Chapter 5) are routinely derived using the manufacturer's proprietary software. Examples are due to Damman and Buijs (1994), investigating liquid–crystal polymer in tension; Beaudoin et al. (1998), examining the properties of a polybutylene terephthalate composite in compression–tension; and Zhou et al. (1995), comparing drawn and undrawn polycarbonate in torsion. A commercial DMTA set-up is shown in Figure 6.6.

6.3 Wave-Propagation Methods

Wave-propagation methods are in three broad categories:

1) In the kilohertz frequency range.
2) In the megahertz frequency range: ultrasonic methods.
3) In the gigahertz frequency range: Brillouin spectroscopy.

6.3.1 The Kilohertz Frequency Range

At frequencies in the kilohertz range, the wavelength of the stress waves is of the order of the length of a viscoelastic specimen. Typically (Hillier and Kolsky 1949), a thin monofilament is stretched longitudinally. Piezoelectric devices are used to transmit and receive the signals (Tanaka et al. 1989). As shown in Figure 6.7 for low-density polyethylene, a plot of the phase (θ) against distance (l) takes the form of damped oscillations about the line $\theta = kl$, where k is the propagation constant.

Where the attenuation coefficient α is small, it has been shown (Hillier 1961) that $V_{\max}/V_{\min} = (\alpha l + \beta)$, where V is the signal amplitude. A plot of

$$\tanh^{-1}\left(\frac{V_{\max}}{V_{\min}}\right)$$

against l then gives a line of slope α.

It is possible to relate α and k to the storage and loss moduli E_1 and E_2, and to tan δ. For a filament of density ρ, in which c is the longitudinal wave velocity

$$E_1 = \frac{\omega^2}{k^2}\rho, \quad E_2 = \frac{2\alpha c^3 \rho}{\omega}$$

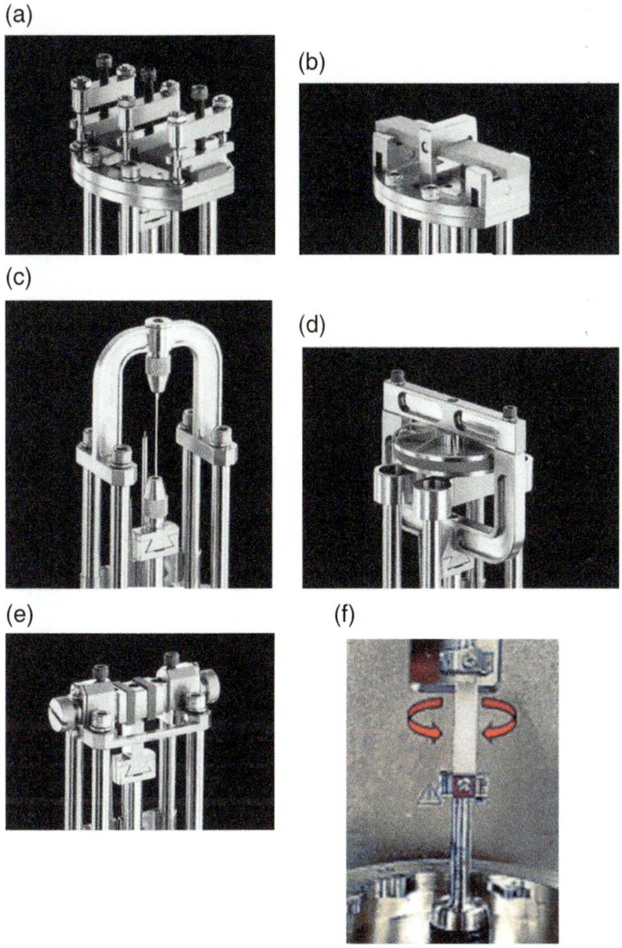

Figure 6.6 DMTA testing heads for various modes: (a) dual cantilever, (b) three-point bend, (c) tension, (d) compression, (e) shear sandwich (showing two specimens) and (f) torsion. (Images: (a)–(e) TA Instruments, Delaware, United States. Mehdipour et al. (2024) / with permission of Elsevier.

and

$$\tan\delta = \frac{2\alpha c}{\omega}.$$

For further information the reader is referred to Kolsky (2003).

6.3.2 The Megahertz Frequency Range: Ultrasonic Methods

Measurements of the velocity and attenuation of elastic waves at ultrasonic frequencies are important, especially for oriented polymers and composites. Compact solid specimens with dimensions of the order of 10 mm are required.

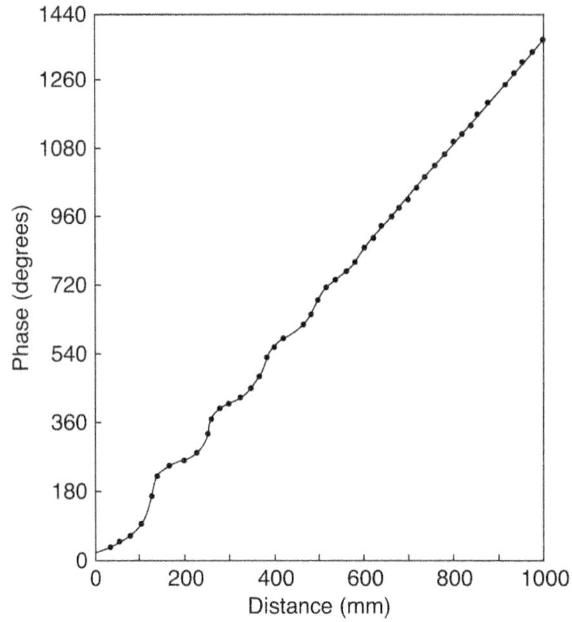

Figure 6.7 The variation in phase angle with distance along a polyethylene monofilament for transmission of sound waves at 3000 Hz. Adapted from Chan et al., 1978.

In a typical application of this technique, Chan et al. (1978) measured the elastic constants of a uniaxially oriented rod 12 mm in diameter, by cutting discs of thickness 4–8 mm parallel, perpendicular and at 45° to the axis of the rod (Figure 6.8). Quartz transducers were bonded to the discs, so that longitudinal and transverse waves were propagated along the geometrical axes of each disc. In principle, nine different velocities ν_{ab} can be measured, where a refers to the direction of polarization and b to the direction of wave propagation. For a specimen of density ρ, we can then define $Q_{ab} = \rho\nu_{ab}^2$, where Q_{ab} is either an elastic stiffness constant or a linear combination of such constants. Velocities were measured using the pulse echo-overlap technique (Papadakis 1964; Kwan et al. 1975), and $\tan\delta$ was obtained by making attenuation measurements (Roderick and Truell 1952).

An alternative approach (Markham 1970; Rawson and Rider 1974; Darras et al. 1995) is to immerse a specimen, thickness d, in a water-filled tank fitted with both a transmitter and a receiver of ultrasonic waves, and measure the change (τ) in transit time with and without the specimen in the beam. If V is the velocity in the polymer and V_w that in the water, we have

$$\frac{1}{V} = \frac{1}{V_w} - \frac{\tau}{d}.$$

The various wave velocities, which can be derived from measurements made over a range of incident angles, are related to the elastic stiffness constants.

In a variation in the above-mentioned method, Wright et al. (1971) detected the component of the incident beam that was reflected from the immersed specimen and hence measured the critical angle of incidence.

This technique has also been developed to measure the anisotropy of uniaxial composites (Dyer et al. 1992). A specimen of uniform thickness, placed between the transmitting and receiving heads in a water-filled container, could be rotated about a vertical axis to change the angle of incidence and hence the direction of the beam in the sample (Figure 6.9). The velocity V and angle of refraction r of the wave are then calculated following the method of Markham (1970). Let X_1 and X_2 axes define the isotropic plane perpendicular to the fibre axis (the 3 axis). It can then be shown (Musgrave 1954) that for quasi-tensile waves propagating in the X_1–X_3 plane at an angle r to the X_1 axis, with the specimen axis X_2 vertical,

$$V_t^2 = \frac{B_{11} + B_{33} + \left[(B_{33} - B_{11})^2 + 4B_{13}^2\right]^{1/2}}{2\rho}$$

and for quasi-shear waves

$$V_s^2 = \frac{B_{11} + B_{33} - \left[(B_{33} + B_{11})^2 + 4B_{13}^2\right]^{1/2}}{2\rho}.$$

where ρ is the specific gravity of the specimen.

The elastic stiffness constants C_{ij} are obtained from

$$B_{11} = C_{11}\cos^2 r + C_{44}\sin^2 r$$
$$B_{33} = C_{33}\sin^2 r + C_{44}\cos^2 r.$$
$$B_{13} = (C_{44} + C_{13})\sin r \cos r$$

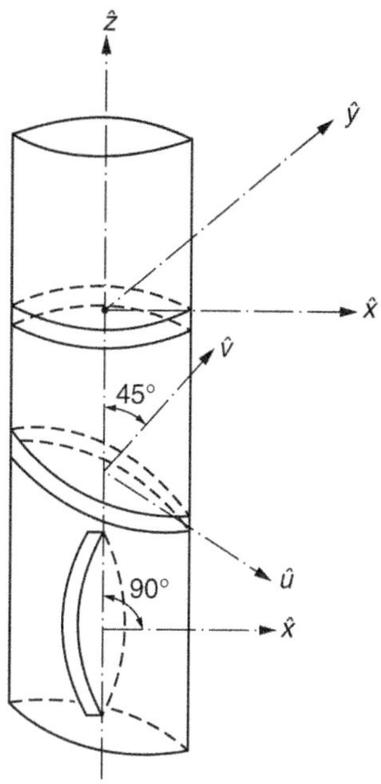

Figure 6.8 Schematic diagram illustrating the sample discs employed in ultrasonic measurements. Dyer et al., 1992 / IOP Publishing.

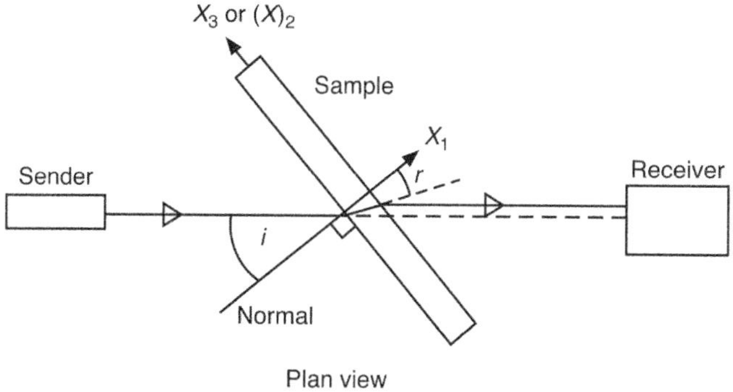

Figure 6.9 Schematic diagram showing layout of ultrasonic apparatus used for measurement of elastic constants.

6.3.3 The Hypersonic Frequency Range: Brillouin Spectroscopy

In the ultrasonic method described in the previous section 6.3.2, a fixed frequency is set for the propagating sound wave (in the MHz range) and the elastic properties are measured as a result of the propagation of this wave. The result is a bulk measurement in the direction of the propagating sound wave. However, this technique cannot detect sound waves produced by thermal vibrations of atoms (called acoustic phonons), as these can have a much weaker strength. A technique that can investigate this is Brillouin scattering, which examines the inelastic scattering of light (usually laser light) from these phonon vibrations. It has the advantage of being a non-contact and non-destructive method, and it can provide results for the full elastic constant tensor of a material.

Brillouin spectroscopy enables the elastic constants of polymeric materials to be determined at frequencies of several gigahertz, i.e., three orders of magnitude higher than those pertaining to ultrasonic measurements, which are known as hypersonic frequencies. The principle of the method is to use Fabry–Perot spectroscopy to measure the frequency shift in laser light scattered after passage through a parallel sample from thermal motion of atoms in the material. A range of geometries can be employed to investigate different scattering modes. The review article of Philipp et al. (2012) has a useful discussion of these different scattering geometries and many other aspects of using Brillouin spectroscopy.

The hypersonic velocity V_s is obtained from the equation

$$V_s = \frac{f_B \lambda_i}{\sqrt{2}},$$

where λ_i is the laser wavelength and f_B is the measured Brillouin shift.

For an isotropic polymer, the hypersonic sound velocity is determined as a function of direction, and the elastic constants are obtained by fitting the data to a set of equations known as the Christoffel equations, which essentially relate the values of V_s to the stiffness constants C_s through $C_s = \rho V_s^2$, where ρ is the density (Auld 1973). For further detailed discussion of this technique, the reader is referred to further papers by Kruger, Pietralla and coworkers (Krbecek et al. 1993; Krbecek et al. 1996; Kruger et al. 1986). More recently, the development of non-scanning Fabry—Perot interferometry has made possible faster acquisition of data to enable study of liquid–glass transitions (Ike and Kojima 2006). Further discussion of the background and impact of this technique can be found in the review paper of Kojima (2022).

References

Auld, B.A. (1973) *Acoustic Fields and Waves in Solids*, Wiley, New York, p. 211.

Beaudoin, O., Bergeret, A., Quantin, J.C. *et al.* (1998) Viscoelastic behaviour of glass beads reinforced poly(butylene terephthalate): experimental and theoretical approaches. *Compos. Interfaces*, **5**, 543.

Chan, O.K., Chen, F.C., Choy, C.L. *et al.* (1978) The elastic constants of extruded polypropylene and polyethylene terephthalate. *J. Phys. D*, **11**, 617.

Damman, S.B. and Buijs, J.A.H.M. (1994) Liquid-crystalline main-chain polymers with a poly (*p*-phenylene terephthalate) backbone. 5. Dynamic-mechanical behavior of the polyester with dodecyloxy side-chains. *Polymer*, **35**, 2359.

Darras, O., Duckett, R.A., Hine, P.J. and Ward, I.M. (1995) Anisotropic elasticity of oriented polyethylene materials. *Compos. Sci. Technol.*, **55**, 131–138.

Dyer, S.R.A., Lord, D., Hutchinson, I.J. *et al.* (1992) Elastic anisotropy in unidirectional fiber reinforced composites. *J. Phys. D*, **25**, 66.

Hillier, K.W. (1961) Measurement of dynamic elastic properties, in *Progress in Solid Mechanics*, (eds I.N. Sneddon and R. Hill), North-Holland, Amsterdam, pp. 199–243.

Hillier, K.W. and Kolsky, H. (1949) An investigation of the dynamic elastic properties of some high polymers. *Proc. Phys. Soc. B*, **62**, 111.

Ike, Y. and Kojima, S. (2006) Brillouin scattering study of polymer dynamics. *Mater. Sci. Eng.*, **A442**, 383–386.

Kojima, S. (2022) 100th Anniversary of Brillouin Scattering: Impact on Materials Science. *Materials*, **15**, 3517.

Kolsky, H. (2003) *Stress Waves in Solids*, Courier Dover Publications, Mineola, New York.

Krbecek, H., Kruger, J.K. and Pietralla, M. (1993) Poisson ratios and upper bounds of intrinsic birefringence from brillouin scattering of oriented polymers. *J. Polym. Sci. Phys. Ed.*, **31**, 1477.

Krbecek, H.H., Kupisch, W. and Pietralla, M. (1996) A new Brillouin scattering analysis of high frequency relaxations in liquids demonstrated at the hypersound relaxation of PPG. *Polymer*, **37**, 3483–3491.

Kruger, J.K., Mars, A. and Peets, L. (1986) Simultaneous determination of elastic and optical properties of polymers by high performance Brillouin spectroscopy using different scattering geometries. *Colloid. Polym. Sci.*, **264**, 403.

Kwan, S.F., Chen, F.C. and Choy, C.L. (1975) Ultrasonic studies of 3 fluoropolymers. *Polymer*, **16**, 481.

Leaderman, H. (1962) Large longitudinal retarded elastic deformation of rubberlike network polymers. *Trans. Soc. Rheol.*, **6**, 361.

Markham, M.F. (1970) Measurement of the elastic constants of fibre composites by ultrasonics. *Composites*, **1**, 145.

Musgrave, M.J.P. (1954) On the propagation of elastic waves in aeolotropic media. I. General principles. *Proc. R. Soc.*, **A226**, 339.

Nitta, K.-H. and Maeda, H. (2010) Creep behavior of high density polyethylene under a constant true stress. *Polym. Test.*, **29**, 60–65.

Papadakis, E.P. (1964) Ultrasonic attenuation and velocity in 3 transformation products in steel. *J. Appl. Phys.*, **35**, 1474.

Philipp, M., Müller, U., Sanctuary, R., Seck, P. and Krüger, J.K. (2012) Scanning brillouin microscopy: Acoustic microscopy at gigahertz frequencies. *Archives des Sciences Naturelles, Physiques et Mathématiques*, **NS46**.

Rawson, F.F. and Rider, J.G. (1974) Elastic constants of oriented polyvinyl-chloride. *J. Phys. D*, **7**, 41.

Roderick, R.L. and Truell, R. (1952) The measurement of ultrasonic attenuation in solids by the pulse technique and some results in steel. *J. Appl. Phys.*, **23**, 267.

Tanaka, A., Nitta, K. and Onogi, S. (1989) Ultrasonic velocity and attenuation of polymeric solids under oscillatory deformation: apparatus and preliminary results. *Polym. Eng. Sci.*, **29**, 1124–1130.

Wright, H., Faraday, C.S.N., White, E.F.T. *et al.* (1971) Elastic constants of oriented glassy polymers. *J. Phys. D*, **4**, 2002.

Zhou, Z., Chudnovsky, A. and Bodnyak, C.P. (1995) Cold-drawing (necking) behavior of polycarbonate as a double glass-transition. *Polym. Eng. Sci.*, **35**, 304.

7

Experimental Studies of Linear Viscoelastic Behaviour as a Function of Frequency and Temperature: Time–Temperature Equivalence

7.1 General Introduction

An introduction to the extensive experimental studies of linear viscoelastic behaviour in polymers falls conveniently into three parts, in which amorphous polymers, crystalline polymers and temperature dependence are discussed in turn.

7.1.1 Amorphous Polymers

Many of the earlier investigations of linear viscoelastic behaviour in polymers were confined to amorphous polymers. During the early 1950s Marvin (1954) and later Marvin and Oser (1962) of the National Bureau of Standards in the United States coordinated the assembly of data from many laboratories who had measured the complex shear modulus and complex shear compliance of a specimen of polyisobutylene $(CH_2{-}CCH_3CH_3)_n$ of high relative molecular mass over a wide range of frequencies. The results, redrawn in Figure 7.1, show clearly the four regions, that is the glassy, the viscoelastic, rubbery and flow regions that are characteristic for amorphous high polymers. At high frequencies, the complex modulus has a value around 10^9 Pa. For material of high-molecular mass, a plateau in modulus occurs, and appreciable molecular flow occurs only at frequencies below 10^{-5} Hz (i.e., period >1 day).

The reader may use the *Alfrey* approximation (see Section 5.4) to derive relaxation and retardation time spectra from the data of Figure 7.1. These spectra can be approximated by a 'wedge and box' distribution (Manero et al. 1997), shown by the dotted lines in Figure 7.2.

The observed plateau in the rubbery region is a consequence of high-molecular mass, as the long molecules tend to entangle, with the formation of physical cross-links that restrict molecular flow through the formation of temporary networks. At long times such physical entanglements are usually labile and lead to some irreversible flow, in contrast with the situation for permanent chemical cross-links, such as those introduced when rubber is vulcanized. It follows directly from the theory of rubber elasticity (Chapter 4) that the value of the modulus in the rubber-like plateau region is directly related to the number of effective cross-links per unit volume.

The influence of molecular entanglements is illustrated by Figure 7.3, which shows the stress relaxation behaviour for two samples of polymethyl methacrylate. It is seen that the

Mechanical Properties of Solid Polymers, Fourth Edition. John Sweeney and Peter Hine.

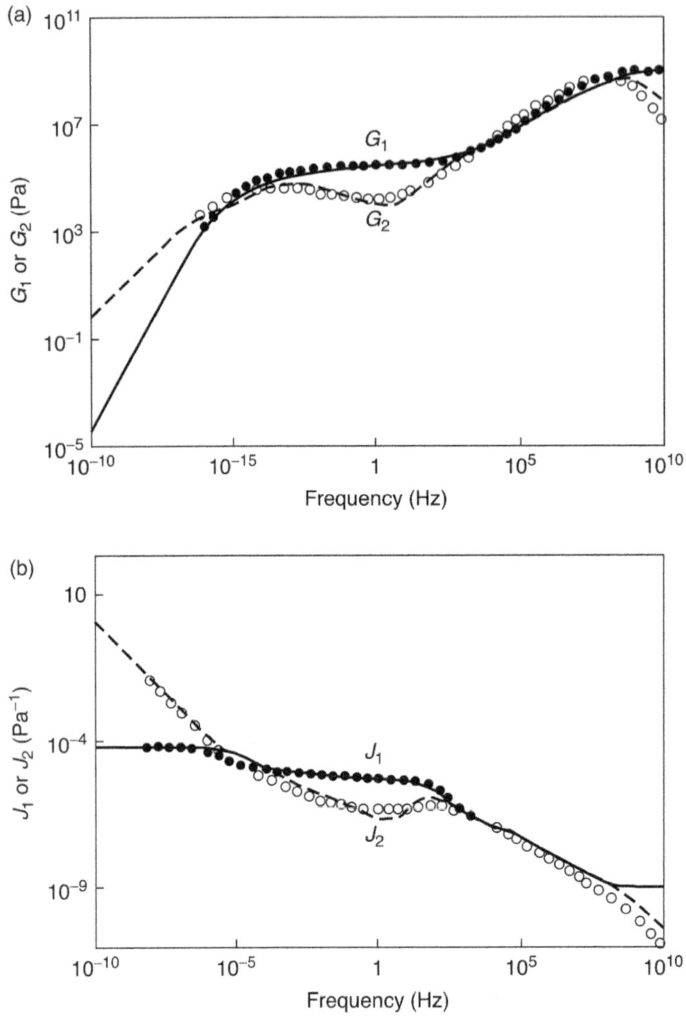

Figure 7.1 Complex shear modulus (a) and complex shear compliance (b) for 'standard' polyisobutylene reduced to 25°C. Points from averaged experimental measurements; curves from a theoretical model for viscoelastic behaviour. Adapted from Marvin and Oser, 1962.

lower-molecular mass sample does not show a rubbery plateau region of modulus but passes directly from the viscoelastic region to the region of permanent flow.

7.1.2 Temperature Dependence of Viscoelastic Behaviour

Previously we have referred only indirectly to the effect of temperature on viscoelastic behaviour. From a practical viewpoint, however, the temperature dependence of polymer properties is of paramount importance because plastics and rubbers show very large changes in properties with changing temperature.

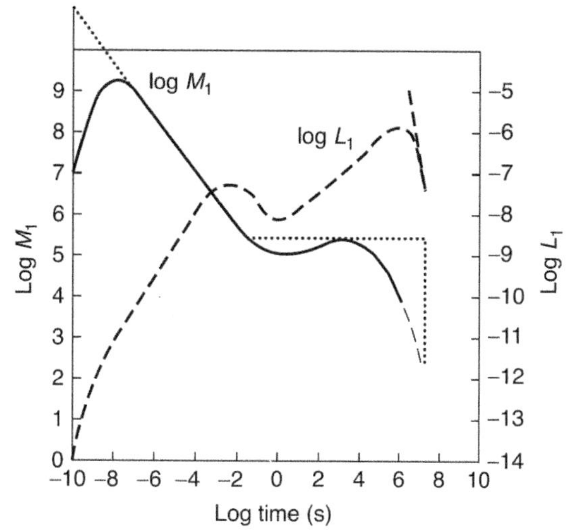

Figure 7.2 Approximate distribution functions of relaxation (M_1) and retardation (L_1) times for polyisobutylene. Adapted from Marvin, 1954.

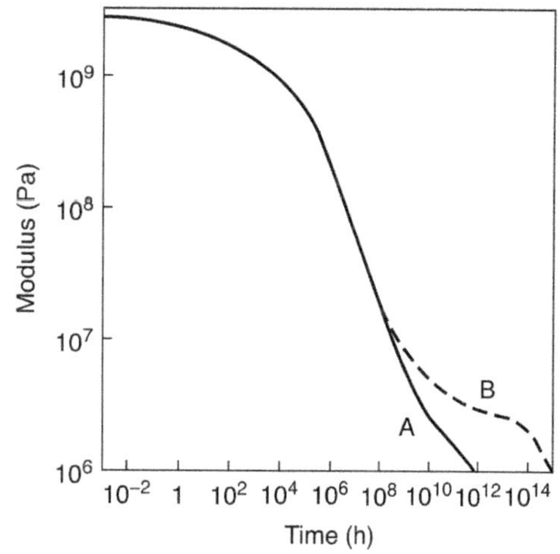

Figure 7.3 Master stress-relaxation curves for low-molecular mass (molecular mass 1.5×10^5 daltons, curve A) and high-molecular mass (molecular mass 3.6×10^6 daltons, curve B) polymethyl methacrylate. (Reproduced with permission from McLoughlin and Tobolsky, J. Colloid Sci., 7, 555. Copyright (1952) Elsevier Ltd.)

In purely scientific terms, the temperature dependence has two primary points of interest. In the first place, as we have seen in Chapter 6, it is not possible to obtain from a single experimental technique a complete range of measuring frequencies to evaluate the relaxation spectrum at a single temperature. It is therefore a matter of considerable experimental convenience to change the temperature of the experiment, and so bring the relaxation processes of interest within a timescale that is readily available. This procedure, of course, assumes that a simple interrelation exists between timescale and temperature, and we will discuss shortly the extent to which this assumption is justified.

Secondly, there is the question of obtaining a molecular interpretation of the viscoelastic behaviour. In most general terms, polymers change from glass-like to rubber-like behaviour as either the temperature is raised or the timescale of the experiment is increased. In the glassy state at low temperatures, we would expect the stiffness to relate to changes in the stored elastic energy on deformation that are associated with small displacements of the molecules from their equilibrium positions. In the rubbery state at high temperatures, on the other hand, the molecular chains have considerable flexibility, so that in the undeformed state they can adopt conformations that lead to maximum entropy (or, more strictly, minimum free energy). The rubber-like elastic deformations are then associated with changes in the molecular conformations.

The molecular physicist is interested in understanding how this conformational freedom is achieved in terms of molecular motions, for example to establish which bonds in the structure become able to rotate as the temperature is raised. One approach, which has proved successful to some degree, has been to compare the viscoelastic behaviour with dielectric relaxation behaviour and more particularly with nuclear magnetic resonance behaviour.

We have tacitly assumed that there is only one viscoelastic transition, corresponding to the change from the glassy low-temperature state to the rubbery state. In practice there are several relaxation transitions. For a typical amorphous polymer, the situation is summarized in Figure 7.4. At low temperatures, there are usually several secondary transitions involving comparatively small changes in modulus. These transitions are attributable to such features as side-group motions, for example methyl ($-CH_3$) groups in polypropylene

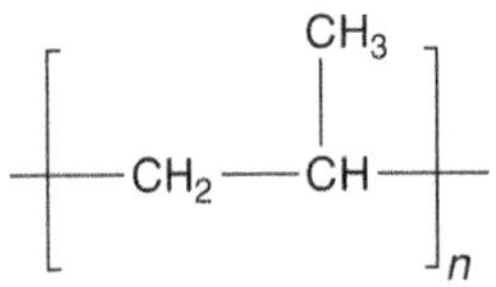

In addition, there is one primary transition, usually called the 'glass transition' that involves a large change in modulus. The temperature at which it occurs is commonly denoted by T_g.

7.1.3 Crystallinity and Inclusions

Although the viscoelastic behaviour of semi-crystalline polymers gives some indication of the four characteristic regions that can be identified for amorphous polymers, they are

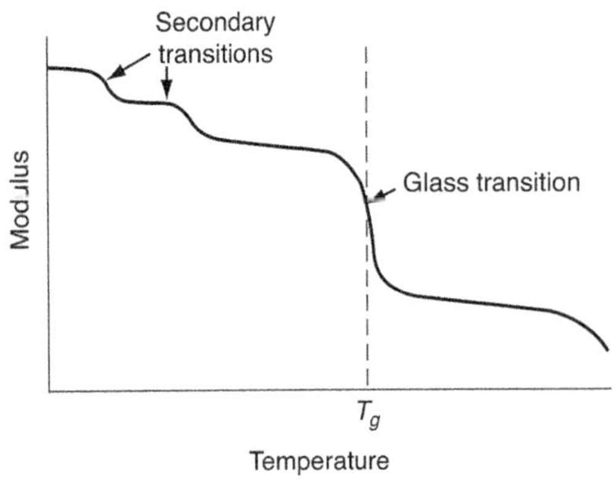

Figure 7.4 Temperature dependence of modulus in a typical polymer.

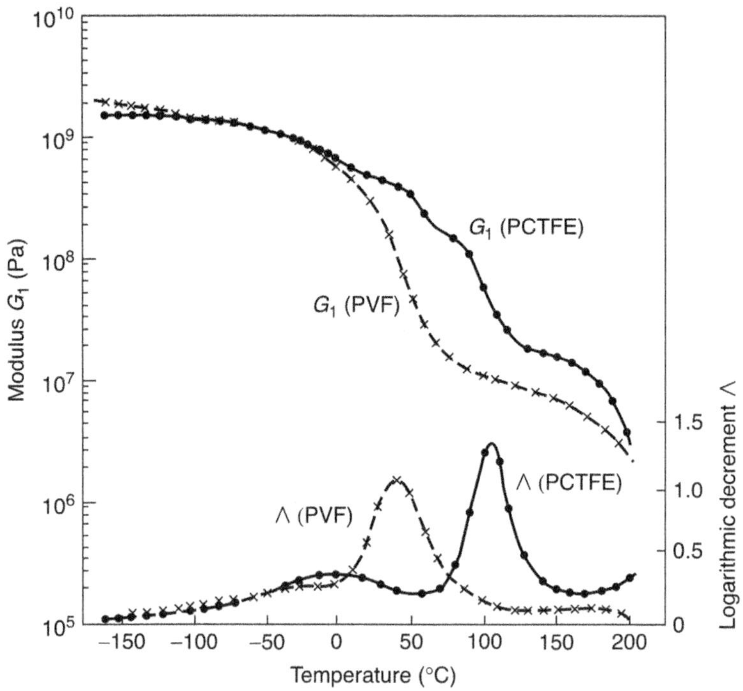

Figure 7.5 Shear modulus G_1 and logarithmic decrement Λ of polychlorotrifluoroethylene (PCTFE) and polyvinyl fluoride (PVF) as a function of temperature at ~3 Hz. Schmieder and Wolf, 1953 / Springer Nature.

much less clearly defined, as is illustrated in Figure 7.5, which shows data for polychlorotrifluoroethylene $(CClF{-}CF_2)_n$ and polyvinyl fluoride $(CH_2CHF)_n$ obtained by Schmieder and Wolf (1953). The fall in modulus over the glass transition region for semi-crystalline materials is, at between one and two orders of magnitude, much less than for amorphous polymers, and the change in modulus or loss factor with temperature or frequency is much more gradual, indicating a broader relaxation time spectrum. At high temperatures (or low frequencies), molecular mobility is severely curtailed by the crystalline regions, so it is no longer correct to regard the polymer as rubber-like. These differences are clearly illustrated by the data of Thompson and Woods (1956) for polyethylene terephthalate, a material that is amorphous when quenched rapidly from the melt (Figure 7.6(a)), but semi-crystalline when slowly cooled or subsequently heat treated (Figure 7.6(b)).

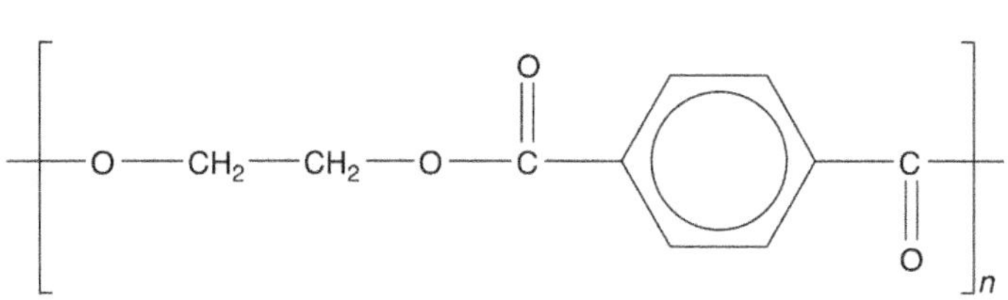

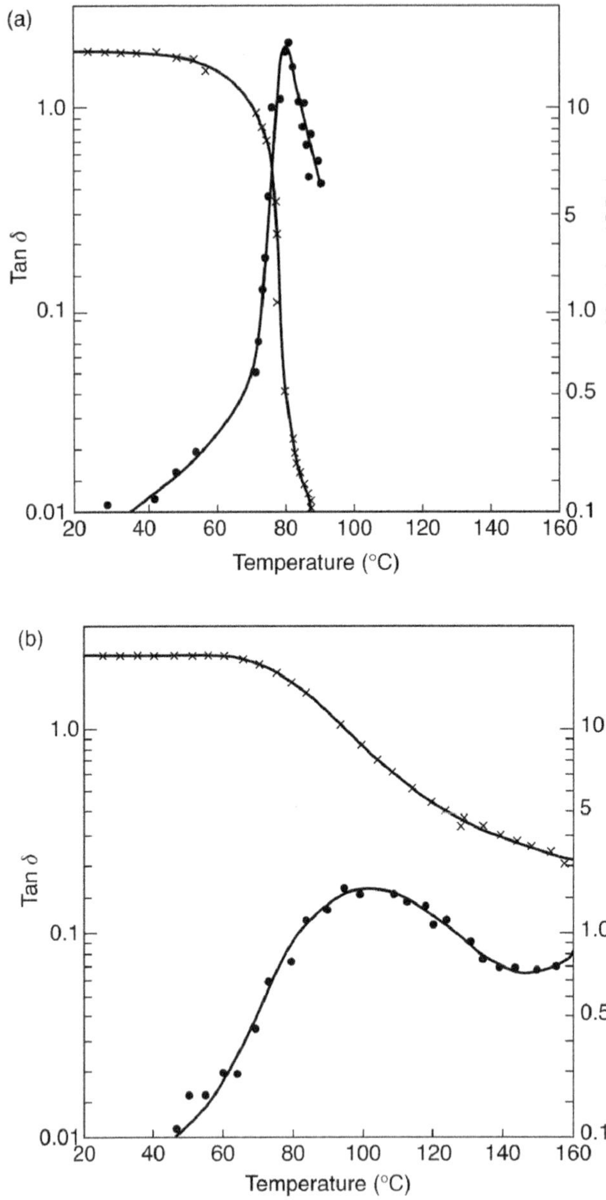

Figure 7.6 Tensile modulus and loss factor tan δ for unoriented amorphous polyethylene terephthalate (a) and unoriented crystalline polyethylene terephthalate (b) as a function of temperature at ~1.2 Hz: (x) modulus; (●) tan δ. Adapted from Thompson and Woods, 1956.

Molecular mobility may be restricted by other factors, such as the addition of nanometre-scale inclusions. This and other factors have been reviewed by Aharoni (1998), with particular reference to changes in the glass transition temperature T_g.

7.2 Time–Temperature Equivalence and Superposition

Time–temperature equivalence in its simplest form implies that the viscoelastic behaviour at one temperature can be related to that at another temperature by a change in the time-scale only. Consider the idealized double logarithmic plots of creep compliance versus time shown in Figure 7.7(a). The compliances at temperatures T_1 and T_2 can be superimposed exactly by a horizontal displacement log a_t, where a_t is called the shift factor. Similarly (Figure 7.7(b)), in dynamic mechanical experiments, double logarithmic plots of tan δ versus frequency show an equivalent shift with temperature.

The experimental procedure is illustrated in Figures 7.8 and 7.9. A series of creep compliance curves each typically extending over 2 h, so that individual tests can be performed on successive days, is plotted using a specimen that has been mechanically conditioned at the highest temperature used. The individual plots are then transposed along the

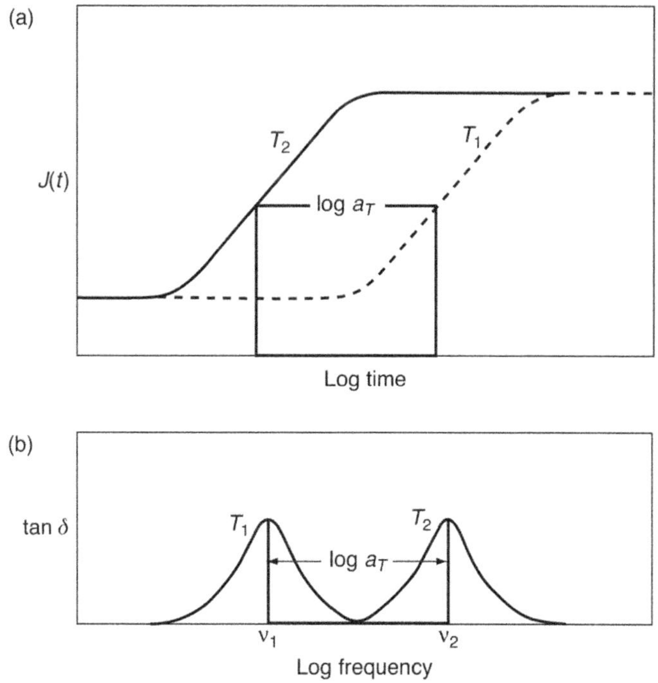

Figure 7.7 Schematic diagrams illustrating the simplest form of time–temperature equivalence for (a) compliance, $J(t)$ and (b) loss factor, tan δ.

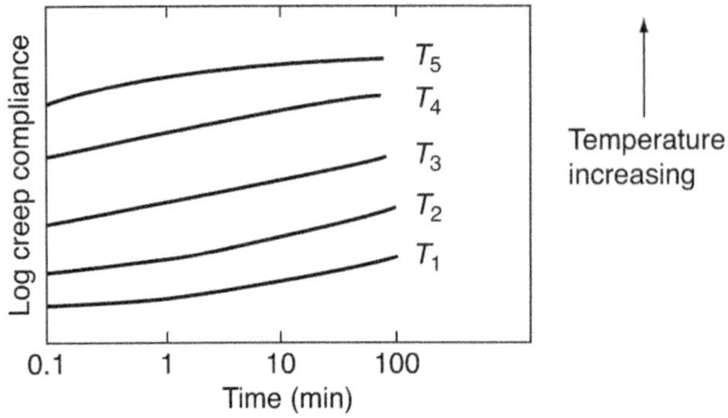

Figure 7.8 Creep plots at different temperatures (schematic).

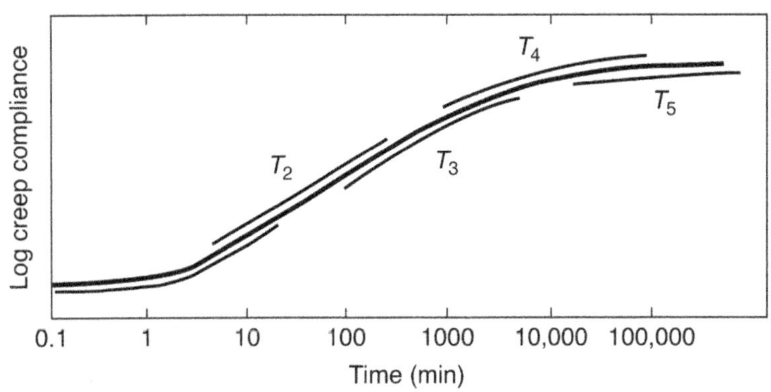

Figure 7.9 Master curve of creep from superposing plots of Figure 7.8.

logarithmic time axis until they coincide, using any required temperature within the experimental range as the reference value. The variation in shift factor with temperature should also be recorded, for comparison with the predictions of theoretical interpretations to be discussed shortly.

Williams et al. (1955), on the basis of the molecular theory of viscoelasticity, proposed that superposition should incorporate a small vertical shift factor $T_0\rho_0/T\rho$, where ρ is the density at the experimental temperature T and ρ_0 relates to the reference temperature T_0. Further corrections have been suggested by McCrum and Morris (1964) to deal with the changes in unrelaxed and relaxed compliances with temperature.

The situation is illustrated schematically in Figure 7.10. When we compare the creep compliance curves at the two temperatures T_1 and T_2, we see that both the relaxed and unrelaxed compliances are changing with temperature. McCrum and Morris (1964) propose a scaling procedure for obtaining a modified or 'reduced' compliance curve at the temperature T_1, to give the dashed curve $J_\rho^T(t)$ in Figure 7.10. The shift factor is now obtained by a horizontal shift of $J_\rho^{T_1}(t)$ to superimpose $J^{T_2}(t)$.

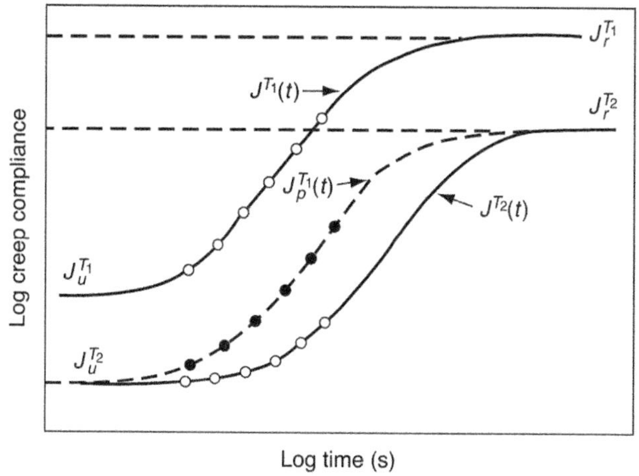

Figure 7.10 Schematic diagram illustrating McCrum's reduction procedure for superposition of creep data: $J_r^{T_1}$ and $J_u^{T_1}$ are the relaxed and unrelaxed compliances, respectively, at the temperature T_1; $J_r^{T_2}$ and $J_u^{T_2}$ are the corresponding quantities at the temperature T_2.

7.3 Transition-State Theories

The simplest theories that attempt to deal with the temperature dependence of viscoelastic behaviour are the transition-state or barrier theories. The transition-state theory of time-dependent processes stems from the theory of chemical reactions and is associated with the names of Eyring, Glasstone et al. (1941). The basic idea is that for two molecules to react they must first form an *activated complex* or *transition state*, which then decomposes to give the final products of the reaction.

The potential energy diagram for two reacting molecules is closely analogous to that for the internal rotation of molecules discussed in Section 1.2.1. Figure 7.11 shows the change

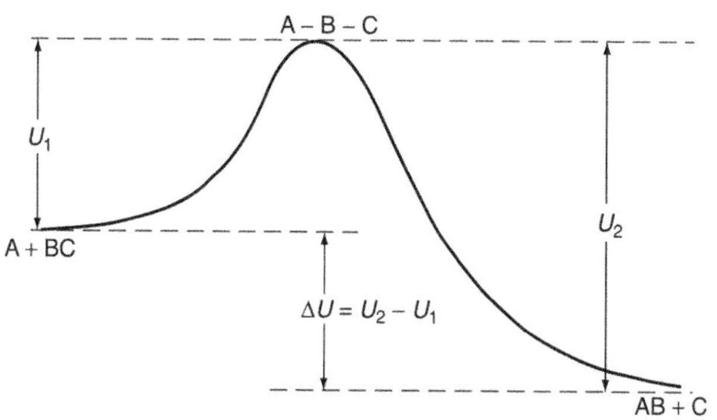

Figure 7.11 Change of potential energy in a chemical reaction. (Adapted from Glasstone (1953) Textbook of Physical Chemistry, 2nd edn, Macmillan, London. Copyright (1953) Macmillan Publishing.)

in potential energy for the reaction between an atom A and a diatomic molecule BC that results in the formation of a diatomic molecule AB and an atom C. The intermediate step is the formation of the activated complex A–B–C.

The theory of absolute reaction rates now argues as follows. The activated complex can be treated by statistical methods as a normal molecule, except that in addition to having three translational degrees of freedom, it has a fourth degree of freedom of movement along what is termed the reaction coordinate. The reaction coordinate is the direction leading to the lower potential energy of the final reactants. The theory shows that the rate of reaction is the product of two quantities, the probability of forming the activated complex and the effective rate of crossing the energy barrier by the activated complexes. It can be shown that the effective rate of crossing the energy barrier, which is the low-frequency vibration of the activated complex in the direction of the reaction coordinate, is equal to kT/h. This is a universal frequency whose value is dependent only on the temperature and is independent of the nature of the reactants and the type of reaction (k is Boltzmann's constant and h is Planck's constant). Because there is equilibrium between the initial reacting species and the transition state, the probability of forming the activated complex is determined in absolute terms by the Boltzmann factor $e^{-\Delta G/RT}$, where ΔG is the free-energy difference per mole between the system when the reactants are relatively far from each other and when they form the activated complex. For reactions occurring under constant pressure conditions, ΔG is the Gibbs free-energy difference per mole.

We now argue that, by analogy, the frequency of molecular jumps between two rotational isomeric states of a molecule (Section 1.2.1) is given by

$$\nu = \frac{kT}{h} e^{-\Delta G/RT}, \tag{7.1}$$

where ΔG is the Gibbs free-energy barrier height per mole.

This equation states that the frequency of molecular conformational changes depends on the *barrier height* and not on the free-energy difference between the equilibrium sites. Equation (7.1) may be written as

$$\nu = \frac{kT}{h} e^{\Delta S/R} e^{-\Delta H/RT} = \nu_0 e^{-\Delta H/RT}. \tag{7.2}$$

This form of Equation (7.1) emphasizes the way in which temperature affects ν primarily through the activation energy ΔH. To a good approximation the activation energy for the process (actually an enthalpy) is thus given by

$$\Delta H = -R \left[\frac{\partial(\ln \nu)}{\partial(1/T)} \right]_P \tag{7.3}$$

Equation (7.2) is known as the 'Arrhenius equation', because it was first shown by Arrhenius (1889) that it describes the influence of temperature on the velocity of chemical reactions.

We now take the intuitive step (to be justified below by the site model theory) that the viscoelastic behaviour can be directly related to a controlling molecular rate process with a constant activation energy.

Consider the tan δ curves of Figure 7.7(b). At the temperatures T_1 and T_2 the peak value of tan δ occurs at frequencies ν_1 and ν_2, respectively. The assumption is that ν_1 and ν_2 are related by the equation

$$\frac{\nu_1}{\nu_2} = \frac{e^{-\Delta H/RT_1}}{e^{-\Delta H/RT_2}},$$

that is

$$\log \frac{\nu_1}{\nu_2} = \log a_T = \frac{\Delta H}{R}\left\{\frac{1}{T_2} - \frac{1}{T_1}\right\}. \tag{7.4}$$

The activation energy for the process can therefore be obtained from a plot of log a_T against the reciprocal of the absolute temperature. For large values of ΔH, changes in temperature give very large changes in frequency. Dynamic mechanical data on polymers are often dealt with in terms of the Arrhenius equation and a constant activation energy. In some cases, this can be regarded as only an approximate treatment due to the limited range of experimental frequencies available. In general, it has been found that the temperature dependence of the glass transition relaxation behaviour of amorphous and crystalline polymers does not fit a constant activation energy, in contrast to more localized molecular relaxations.

7.3.1 The Site Model Theory

The site model theory is based on transition-state theory, and although first developed to explain the dielectric behaviour of crystalline solids (Debye 1945; Fröhlich 1968) has also been applied to mechanical relaxations in polymers (Hoffman et al. 1966).

In its simplest form there are two sites, separated by an equilibrium free-energy difference $\Delta G_1 - \Delta G_2$, the barrier heights being ΔG_1 and ΔG_2 per mole, respectively (Figure 7.12).

The transition probability for a jump from site 1 to site 2 is given by

$$\omega_{12}^0 = A' e^{-\Delta G_1/RT} \tag{7.5}$$

and for a jump from site 2 to site 1 by

$$\omega_{21}^0 = A' e^{-\Delta G_2/RT}, \tag{7.6}$$

where A' is a constant.

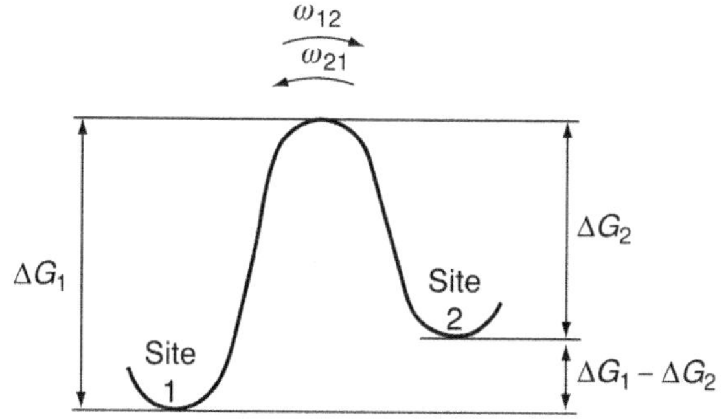

Figure 7.12 The two-site model.

(In some treatments, the change in molecular conformation is imagined to be a simple rotation of 180° around one bond. It is then considered that the transition probability is $2\omega_{12}^0$, where ω_{12}^0 is the probability for a jump in either a clockwise or anticlockwise direction.)

To give rise to a mechanical relaxation process, the energy difference between the two sites must be changed by the application of the applied stress. There is then a change in the populations of sites 1 and 2, and it is assumed that this relates directly to the strain. It is not difficult to imagine how this might arise at a molecular level if, for example, the uncoiling of a molecular chain involved internal rotations. Locally, the chain conformations could be changing from crumpled *gauche* conformations to extended *trans* conformations (see Section 1.2.1).

Assume that the applied stress σ causes a small linear shift in the free energies of the sites such that

$$\delta G_1' = \lambda_1 \sigma \tag{7.7}$$

and

$$\delta G_2' = \lambda_2 \sigma \tag{7.8}$$

for sites 1 and 2, respectively, where λ_1 and λ_2 are constants with the dimensions of volume. The transition probabilities ω_{12} and ω_{21} in the presence of the applied stress are then given by

$$\omega_{12} \simeq \omega_{12}^0 \left[1 - \frac{\delta G_1'}{RT}\right] = \omega_{12}^0 \left[1 - \frac{\lambda_1 \sigma}{RT}\right], \tag{7.9}$$

where ω_{12}^0 is the transition probability in the absence of the stress. Similarly

$$\omega_{21} \simeq \omega_{21}^0 \left[1 - \frac{\lambda_2 \sigma}{RT}\right]. \tag{7.10}$$

The rate equations for sites 1 and 2 are then

$$\frac{dN_1}{dt} = -N_1\omega_{12} + N_2\omega_{21}, \tag{7.11}$$

$$\frac{dN_2}{dt} = -N_2\omega_{21} + N_1\omega_{12}, \tag{7.12}$$

where we can write the occupation number N_1 of state 1 as $N_1 = N_1^0 + n$ and similarly $N_2 = N_2^0 - n$, where N_1^0 and N_2^0 are the occupation numbers at zero stress, $N_1^0 + N_2^0 = N_1 + N_2 = N$.

Combining these equations and making suitable approximations gives a rate equation

$$\frac{dn}{dt} + n\left(\omega_{12}^0 + \omega_{21}^0\right) = N_1^0\omega_{12}^0 \left[\frac{\lambda_1 - \lambda_2}{RT}\right]\sigma, \tag{7.13}$$

which describes the change in the site population n as a function of time. Assuming that this change in site population is directly related to the observed strain e, e is given by

$$e = e_u + n\bar{e}. \tag{7.14}$$

In this equation e_u is the instantaneous or unrelaxed elastic deformation and it is considered that each change in site population produces a proportionate change in strain by an amount $\bar{e}$.

Equation (7.13) can then be seen to have the form

$$\frac{\mathrm{d}e}{\mathrm{d}t} + Be = C,$$

where B and C are constants. This is formally identical to the equation of a Voigt element, with a characteristic retardation time given by $\tau' = 1/B$ which is

$$\tau' = \frac{1}{(\omega_{12}^0 + \omega_{21}^0)} = \frac{e^{\Delta G_2/RT}}{A'[\exp\{-(\Delta G_1 - \Delta G_2)/RT\} + 1]}. \tag{7.15}$$

Since RT is usually small compared with the equilibrium free-energy difference we may approximate to

$$\tau' = \frac{1}{A'} e^{\Delta G_2/RT}. \tag{7.16}$$

Equation (7.16) is formally equivalent to Equation (7.1).

It follows that the time–temperature behaviour of the relaxation process is governed by the unperturbed transition probabilities, and to a good approximation by ΔG_2, a free energy of activation. The *magnitude* of the relaxation (Wachtman 1963; Hoffman et al. 1966), on the other hand, is proportional to

$$p\left[\frac{\exp[-(\Delta G_1 - \Delta G_2)/RT]}{(1 + \exp[-(\Delta G_1 - \Delta G_2)/RT])^2}\right]\frac{(\lambda_1 - \lambda_2)^2}{RT},$$

where p is the number of species per unit volume. Thus, the intensity of the relaxation on this model is low at both high and low temperatures and passes through a maximum when the free-energy difference $(\Delta G_1 - \Delta G_2)$ and RT are of the same order of magnitude.

It must be emphasized that the site model is applicable only to relaxation processes showing a constant activation energy, examples being those associated with localized motions in the crystalline regions of semi-crystalline polymers. The temperature dependence of the glass transition relaxation behaviour of polymers does not fit a constant activation energy model, and where this has appeared to be true it is probably a consequence of the limited range of experimental frequencies that were available.

7.4 The Time–Temperature Equivalence of the Glass Transition Viscoelastic Behaviour in Amorphous Polymers and the Williams, Landel and Ferry (WLF) Equation

In considering time–temperature equivalence of the glass transition behaviour in amorphous polymers, we will follow a treatment very close to that given by Ferry (1980). To fix our ideas, consider the storage compliance J_1 of an amorphous polymer (poly-*n*-octyl methacrylate) as a function of temperature and frequency (Figure 7.13). It can be seen that

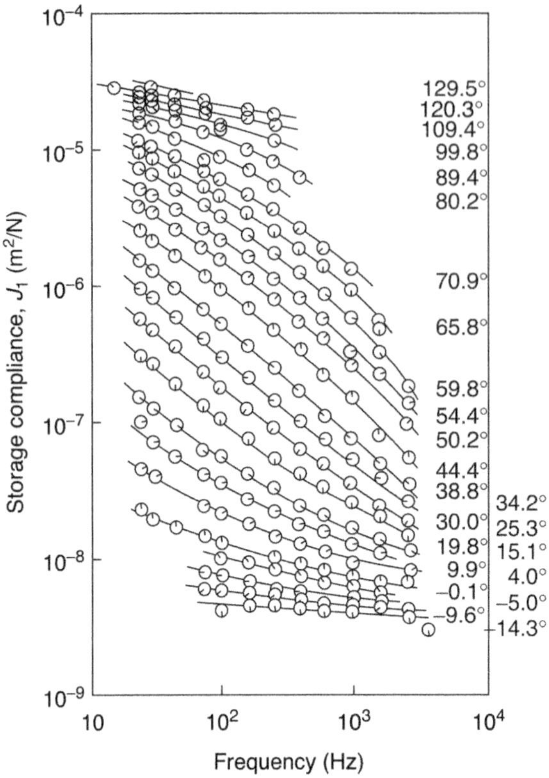

Figure 7.13 Storage compliance of poly-n-octyl methacrylate in the glass transition region plotted against frequency at 24 temperatures as indicated. Adapted from Ferry, 1980.

there is an overall change in the shape of the compliance–frequency curve as the temperature changes. At high temperatures, there is an approximately constant high compliance, the rubbery compliance. At low temperatures, the compliance is again approximately constant but at a low value, the glassy compliance. At intermediate temperatures, there is the frequency-dependent viscoelastic compliance.

The simplest way of applying time–temperature equivalence is to produce a 'master compliance curve' by choosing one particular temperature and applying only a horizontal shift on a logarithmic timescale to make the compliance curves for other temperatures join as smoothly as possible to the curve at this particular temperature. This simple procedure is very nearly, but not quite, the procedure adopted by Ferry and his co-workers. The molecular theories of viscoelasticity suggest that there should be an additional small vertical shift factor $T_0\rho_0/T\rho$ in changing from the actual temperature T in kelvins (at a density ρ) to the reference temperature T_0 in kelvins (at a density ρ_0). The physical meaning of this vertical correction factor is that the molecular theories suggest that the equilibrium modulus changes with temperature in the transition range in a manner according to the theory of rubber elasticity (see Chapter 4). This is quite distinct from changes in the molecular

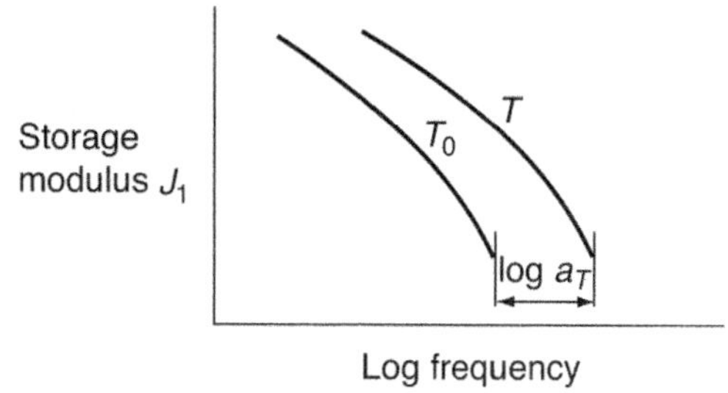

Figure 7.14 Diagram illustrating shift factor log a_T for change in temperature T to T_0.

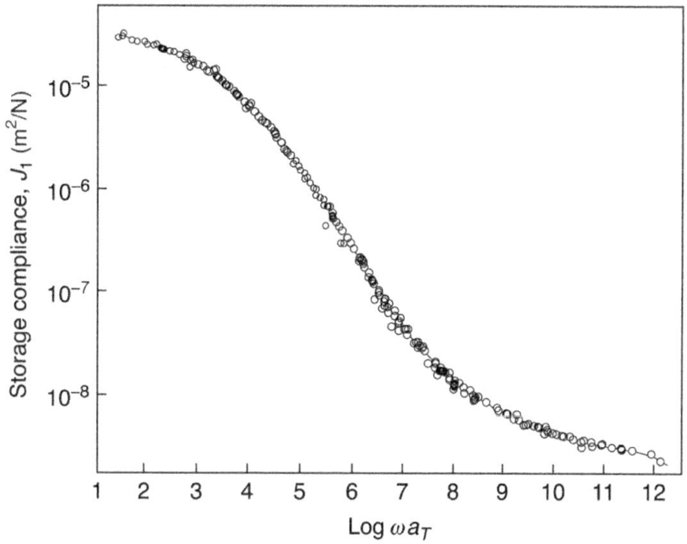

Figure 7.15 Composite curve obtained by plotting the data of Figure 7.13 with suitable shift factors, giving the behaviours over an extended frequency scale at temperature T_0. Adapted from Ferry, 1980.

relaxation times, which affect the measured modulus at a given time or frequency due to affecting the viscoelastic behaviour. In practice, the correction factor has a very small effect in the viscoelastic range of temperatures compared with the large changes in the viscoelastic behaviour. Thus, it is usually adequate to apply a simple horizontal shift on the timescale only (see Figure 7.14).

This procedure gives the storage compliance as a function of frequency over a very wide range of frequencies, as shown in Figure 7.15. Thus, it is now possible to calculate the retardation time spectrum and compare this with any theoretical models that may be proposed.

We may also consider the significance of the horizontal shift on the logarithmic timescale, as shown in Figure 7.16.

The remarkable observation, which was established largely by the work of Williams, Landel and Ferry, is that for all amorphous polymers this shift factor–temperature relationship is *approximately* identical.

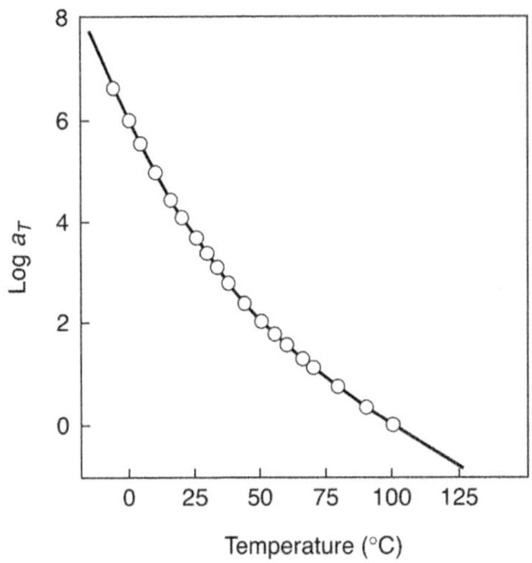

Figure 7.16 Temperature dependence of the shift factor a_T used in plotting Figure 7.15 Points, chosen empirically; curve is WLF equations with a suitable choice of T_g (or T_s). Adapted from Ferry, 1980.

It was found that the relationship

$$\log a_T = \frac{C_1(T - T_s)}{C_2 + (T - T_s)},$$

where C_1 and C_2 are constants, and T_s is a reference temperature peculiar to a particular polymer, holds extremely well over the temperature range $T = T_s \pm 50°\text{C}$ for all amorphous polymers. This equation, known as the 'WLF equation' (Williams et al. 1955) (we shall see that there are other forms for the WLF equation), was originally considered to be only an empirical equation, and the constants C_1 and C_2 were originally determined by arbitrarily choosing $T_s = 243$ K for polyisobutylene.

Following this empirical discovery, there was naturally some speculation as to whether the WLF equation has a more fundamental interpretation. This brings us to considerations of the dilatometric glass transition and to discussion of the use of the concept of free volume.

The glass transition can be defined on the basis of dilatometric measurements. As shown in Figure 7.17 if the specific volume of the polymer is measured against temperature, a change of slope is observed at a characteristic temperature, which we may call T_g. In the first place, this change in slope may be somewhat less sharp than this diagram suggests. Secondly, it is known that if dilatometric measurements are carried out at very slow rates of temperature change, one approaches a roughly constant value for the glass transition temperature T_g. The value of T_g will vary by only 2–3 K when the heating rate is decreased from 1 K/min to 1 K/day. Thus, it appears possible to define a rate-independent value of T_g to at least a very good approximation.

It has subsequently been shown that the original WLF equation can be rewritten in terms of this dilatometric transition temperature such that

$$\log a_T = \frac{C_1^g(T - T_g)}{C_2^g + (T - T_g)},$$

where C_1^g and C_2^g are new constants and $T_g = T_s - 50°\text{C}$.

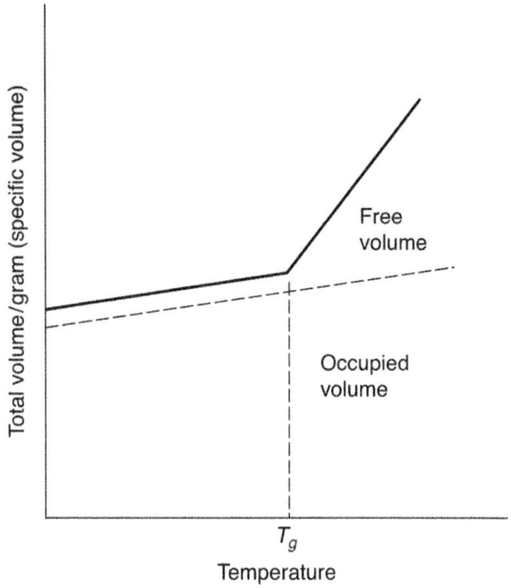

Figure 7.17 The volume–temperature relationship for a typical amorphous polymer.

Moreover, it is now possible to give a plausible theoretical basis to the WLF equation in terms of the concept of free volume (Williams et al. 1955).

In liquids, the concept of free volume has proved useful in discussing transport properties such as viscosity and diffusion. These properties are considered to relate to the difference $\nu_f = \nu - \nu_0$, where ν is the total macroscopic volume, ν_0 is the actual molecular volume of the liquid molecules, the 'occupied volume' and ν_f is the proportion of holes or voids, the 'free volume'. Figure 7.17 shows the schematic division of the total volume of the polymer into both occupied and free volumes. It is argued that the occupied volume increases uniformly with temperature. The discontinuity in the expansion coefficient at T_g then corresponds to a sudden onset of expansion in the free volume. This suggests that certain molecular processes that control the viscoelastic behaviour commence at T_g, and not merely that T_g is the temperature when their timescale becomes comparable with that of the measuring timescale. This would seem to imply that T_g is a genuine thermodynamic temperature. This point is not, however, completely resolved, and it has been shown by Kovacs (1958) that the T_g measured dilatometrically is still sensibly dependent on the timescale, that is the rate of heating. However, as already mentioned, this time dependence is small. Thus, to a good approximation it can be assumed that the free volume is constant up to T_g and then increases linearly with increasing temperature.

The fractional free volume $f = v_f/v$ can therefore be written as

$$f = f_g + \alpha_f (T - T_g), \tag{7.17}$$

where f_g is the fractional free volume at the glass transition T_g and α_f is the coefficient of expansion of the free volume.

The WLF equation can now be obtained in a simple manner. The model representations of linear viscoelastic behaviour all show that the relaxation times are given by expressions of the form $\tau = \eta/E$ (see the Maxwell model in Section 5.2.6), where η is the viscosity of a dashpot and E the modulus of a spring.

If we ignore the changes in the modulus E with temperature compared with changes in the viscosity η, this suggests that the shift factor a_T for changing temperature from T_g to T will be given by

$$a_T = \frac{\eta_T}{\eta_{T_g}}. \tag{7.18}$$

At this juncture, we introduce Doolittle's viscosity equation (Doolittle 1951), which relates the viscosity to the free volume. This equation is based on experimental data for monomeric liquids and gives

$$\eta = a\exp\left(bv/v_f\right), \tag{7.19}$$

where a and b are constants. Using Equations (7.18) and (7.19), it can be shown that the Doolittle equation becomes

$$\ln a_T = b\left\{\frac{1}{f} - \frac{1}{f_g}\right\}. \tag{7.20}$$

Substituting $f = f_g + \alpha_f(T - T_g)$, we have

$$\log a_T = -\frac{\left(b/2.303f_g\right)\left(T - T_g\right)}{f_g/\alpha_f + T - T_g}, \tag{7.21}$$

which is the WLF equation.

Ferry and his co-workers have given further consideration to the exact form of the WLF equation. It can be shown that a better fit to data for different polymers can be obtained by changing the constants C_1^g and C_2^g; and that the actual values obtained for C_1^g and C_2^g yield values for f_g and α_f, which are plausible on physical grounds. The reader is referred to Ferry's book (Ferry 1980) for detailed discussion of these points. We will, however, note here that the fractional free volume at the glass transition temperature f_g is 0.025 ± 0.003 for most amorphous polymers. The thermal coefficient of expansion of free volume α_f is a more variable quantity, but has the physically reasonable 'universal' average value of $4.8 \times 10^{-4}\ \mathrm{K}^{-1}$.

It is of some interest to complete our discussion of the WLF equation by indicating the lines of its derivation by Bueche (1953) using a transition-state model.

It is possible to develop the transition-state theory on the basis of free volume by expressing the frequency ν of the controlling molecular process by the equation

$$\nu = A\int_{f_c}^{\infty} \phi(f)\mathrm{d}f.$$

It is assumed that the required unit of structure can move when the local fractional free volume f exceeds some critical value f_c.

Bueche evaluated $\phi(f)$ and showed that with some approximations

$$\nu = \nu_g \exp\left\{-Nf_c\left[\frac{1}{f} - \frac{1}{f_g}\right]\right\}, \tag{7.22}$$

where ν_g is the frequency at T_g. If $f = f_g + \alpha_f(T - T_g)$ it may be shown that

$$\ln \frac{\nu}{\nu_g} = \frac{\left(Nf_c/f_g\right)\left(T - T_g\right)}{T - T_g + f_g/f}. \tag{7.23}$$

Assuming that there is a direct link between the shift factor a_T and the ratio of the frequencies of the controlling molecular process, Equation (7.23) is identical in form to Equation (7.21). Note also that Equation (7.22) is Bueche's analogy to the Doolittle equation (7.20).

In conclusion, we observe that for time–temperature equivalence to be exact, a necessary simplicity is implied. At a molecular level, the individual relaxation times for molecular processes must shift uniformly with temperature. In phenomenological terms, the spectrum of relaxation times must shift as a unit on a logarithmic timescale to shorter times with increasing temperature.

Staverman and Schwarzl (1956) called these materials thermorheologically simple, and Lee (1960) has worked out the theoretical consequences of this assumption, so that complex problems concerning the deformation of viscoelastic solids in variable temperature situations can be solved.

7.4.1 The Williams, Landel and Ferry Equation, the Free Volume Theory and Other Related Theories

The WLF equation gives the shift factor for time–temperature superposition as

$$\log a_T = \frac{C_1^g\left(T - T_g\right)}{C_2^g + \left(T - T_g\right)}.$$

We have seen that this relationship can be regarded as describing the change in the internal viscosity of the polymer as we change the temperature from the glass transition temperature T_g to the test temperature T (Equation (7.18)).

We may therefore write the WLF equation as

$$\log \eta_T = \log \eta_{T_g} + \frac{C_1^g\left(T - T_g\right)}{C_2^g + \left(T - T_g\right)},$$

where η_T, η_{T_g} is the viscosity of the polymer at temperatures T, T_g, respectively. In this form, the equation implies that at a temperature $T = T_g - C_2^g$ (i.e., $T = T_g - 51.6$ for the WLF equation in its universal form) the viscosity of the polymer is infinite.

This has led to the view that the WLF equation should be related at molecular level to the temperature $T = T_g - 51.6$, which we will call T_2, rather than to the dilatometric glass transition T_g.

There have been two basic approaches along these lines:

1) The free volume theory is modified so that the changes in free volume with temperature relate to a discontinuity that occurs at T_2 rather than T_g. This is discussed in Section 7.4.2.
2) It is considered that T_2 represents a true thermodynamic transition temperature. A modified transition-state theory is developed in which the frequency of molecular

jumps relates to the cooperative movement of a group of segments of the chain. The number of segments acting cooperatively is then calculated from statistical thermodynamic considerations. This is the theory of Adam and Gibbs (1965), which is described in Section 7.4.3.

7.4.2 The Free Volume Theory of Cohen and Turnbull

Cohen and Turnbull (1959) have proposed that the free volume ν_f corresponds to that part of the excess volume $\nu - \nu_0$ (ν = total measured specific volume, ν_0 = occupied volume as in Section 7.4), which can be redistributed without a change in energy. It is then assumed on the basis of arguments concerning the nature of the 'cage' formed round a molecule by its neighbours that the redistribution can take place without a change in energy at temperatures above a critical temperature, which is to be identified with T_2, where the cage reaches a critical size.

Thus

$$v_f = 0 \quad \text{for } T < T_2$$

and

$$v_f = \alpha \bar{v}_m (T - T_2) \quad \text{for } T \geq T_2,$$

where α is the average expansion coefficient and $\bar{v}_m$ the average value of the molecular volume ν_0 in the temperature range T_2 to T.

Manipulating the viscosity Equation (7.19) gives

$$\eta = a \exp \frac{B'}{T - T_2},$$

where B' is a constant, and correspondingly for the average relaxation time

$$\tau = \tau_0 \exp \frac{B'}{T - T_2}.$$

There is much experimental evidence from dielectric relaxation for the validity of this equation for amorphous polymers. As we have discussed, putting $T_2 = T_g - 51.6$ gives us the WLF equation and the relaxation time will become infinitely long as we approach T_2 due to the disappearance of free volume.

7.4.3 The Statistical Thermodynamic Theory of Adam and Gibbs

Gibbs and di Marzio (1958a, 1958b) proposed that the dilatometric T_g is a manifestation of a true equilibrium second-order transition at the temperature T_2. In a further development, Adam and Gibbs (1965) have shown how the WLF equation can then be derived. On their theory, the frequency of molecular jumps is given by

$$v_c = A \exp - \frac{n\Delta G^*}{kT}, \tag{7.24}$$

where A is a constant ($A = kT/h$ on the transition-state theory), ΔG^* is the free-energy difference hindering rearrangement *per segment* (the barrier height) and n is the number of segments acting cooperatively as a unit to make a configurational rearrangement.

The essence of the Adam and Gibbs theory is that n can be calculated on thermodynamic equilibrium grounds as follows:

If S is the configurational entropy of the system, that is the entropy for a mole of segments,

$$S = \frac{N_A}{n} s_n, \tag{7.25}$$

where N_A is Avogadro's number and s_n is the entropy of a unit of n segments. Thus

$$n = \frac{N_A s_n}{S} \quad \text{and} \quad \upsilon_c = A \exp \frac{-N_A s_n \Delta G^*}{SkT}. \tag{7.26}$$

It is assumed that s_n is independent of temperature and that S, the configurational entropy of the system, can be calculated directly for any temperature from the specific heat at constant pressure.

A further assumption is that $S = 0$ at the thermodynamic transition temperature T_2. In molecular terms, n becomes infinite and there are no configurations available into which the system may rearrange. We may note that although the entropy S is assumed to be zero at T_2, this is not necessarily (or ever, in practice) a state of complete order.

This gives the entropy $S(T)$ at a temperature T as

$$S(T) = \Delta C_p \ln \frac{T}{T_2}, \tag{7.27}$$

where ΔC_p is the difference in specific heat between the supercooled liquid and the glass at T_g and is assumed to be constant over the temperature range considered.

Substituting for $S(T)$ and approximating somewhat we find that

$$\upsilon_c = A \exp - \frac{N_A \Delta G^* s_n}{k \Delta C_p (T - T_2)}. \tag{7.28}$$

This gives a relaxation time equation of the form

$$\tau = \tau_0 \exp\{B/(T - T_2)\},$$

which as we have seen reduces to the WLF equation if we put

$$T_2 = T_g - 51.6.$$

7.4.4 An Objection to Free Volume Theories

Hoffman et al. (1966) raised a serious objection to the free volume ideas. Williams (1964) showed that the β-relaxations of polymethyl acrylate and polypropylene oxide behave somewhat similarly under constant pressure and constant volume conditions. It would be expected, however, on the free volume concept, that because the occupied volume ν_0 would increase with temperature, the results for these two conditions would be very different. Williams concluded that the dielectric relaxation time was not a unique function of volume. He suggested that this implied that the free volume did not remain constant for constant *total* volume, whilst temperature and pressure are varied. An alternative view is that relationships of the form

$$\eta = \eta_0 \exp U_\eta / R(T - T_2) \quad \text{and} \quad \tau = C \exp U_\tau / R(T - T_2)$$

should indeed be based on concepts that are more fundamental than those of free volume.

7.5 Normal-Mode Theories Based on Motion of Isolated Flexible Chains

We have so far discussed two types of theories, those based on the site model, and those based on the WLF equation and its ramifications, which deal with time–temperature equivalence. The site model theories predict constant activation energies and are more applicable to relaxation transitions originating from localized chain motions, whereas the WLF equation theories deal with the glass transition behaviour in amorphous polymers.

In the introductory section on amorphous polymers (Section 7.1.1), we considered the relaxation spectrum of amorphous polymers and noted that it was quite complex. The normal-mode theories, now to be discussed, attempt to predict the relaxation spectrum for amorphous polymers, as well as the time–temperature equivalence.

These theories are associated with the names of Rouse (1953), Bueche (1954) and Zimm (1956) and are based on the idea of representing the motion of polymer chains in a viscous liquid by a series of linear differential equations. The three theories have been discussed by Ferry (1991). They are essentially *dilute solution* theories, but we shall see that, rather unexpectedly perhaps, they can be extended to predict the behaviour of the pure polymer. Because of its simplicity, we will give an account of the theory due to Rouse (1953).

Each polymer chain is considered to consist of a number of submolecules. This is similar to the composition of a rubber network where molecular chains join the cross-link points (see Figure 7.18(a)). We can then represent the polymer molecules as a system of beads

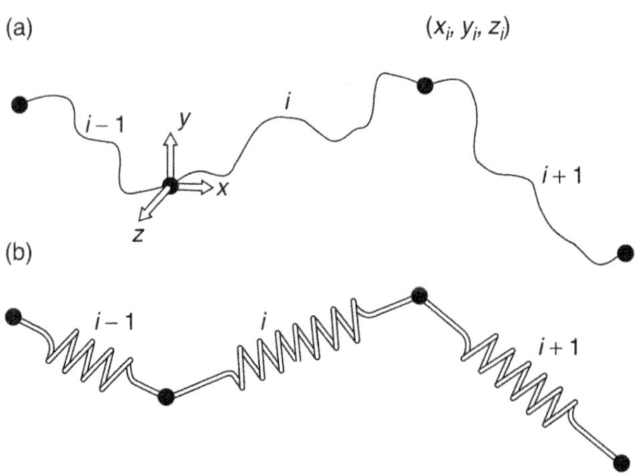

Figure 7.18 The Rouse model: (a) the network of chains and (b) the representation of the network as a combination of springs and beads.

connected by springs whose behaviour is that of a freely jointed chain on the Gaussian theory of rubber elasticity (Figure 7.18(b)). The molecular chains between the beads are all of equal length, this portion of the polymer chain being long enough for the separation of its ends to approximate to a Gaussian probability distribution. It is assumed that only the beads interact directly with the solvent molecules. If a bead is displaced from its equilibrium position, there are two types of forces acting on it; first, the forces due to this viscous interaction with the solvent molecules and secondly, the forces due to the tendency of the molecular chains to return to a state of maximum entropy by Brownian diffusional movements.

Consider the motion of the bead situated at the point $(x_iy_iz_i)$ between the ith and $(i+1)$th submolecules. The origin of coordinates is the bead between the $(i-1)$th and ith submolecules, that is at the other end of the ith submolecule.

For a Gaussian distribution of links in the submolecule, the probability that this bead will lie at the point $x_iy_iz_i$ in the volume element $dx_i\, dy_i,\, dz_i$ is

$$p_i(x_iy_iz_i)dx_i\, dy_i\, dz_i = \frac{b^3}{\pi^{3/2}} \exp\{-b^2(x_i^2 + y_i^2 + z_i^2)\}\, dx_i\, dy_i\, dz_i,$$

where

$$b^2 = 3/2zl^2$$

with l the length of each link, n the total number of links in the molecular chain, m the number of submolecules, giving $z = n/m =$ number of links in a submolecule.

The conformational probability of the entire chain can be represented by a point in $3m$-dimensional space. The probability that this point lies at the point $x_1y_1, \ldots, z_m$ in the volume element $dx_1\, dy_1, \ldots, dz_m$ is given by

$$\begin{aligned} P_m dx_1 \ldots dz_m &= \prod_{i=1}^{m} p_i(x_i\, y_i\, z_i)dx_i\, dy_i\, dz_i \\ &= \left(\frac{b^3}{\pi^{3/2}}\right)^m \exp\left\{-b^2\left[\sum_{i=1}^{m} x_i^2 + y_i^2 + z_i^2\right]\right\} dx_1 \ldots dz_m. \end{aligned}$$

At equilibrium, the most probable values of the x_i, y_i and z_i coordinates are zero, that is each submolecule is in a coiled-up configuration. Any change from the equilibrium position will result in a decrease in entropy ΔS, or an increase in Helmholtz free energy $\Delta A = -T\,\Delta S$. (All conformations are assumed to have the same internal energy.)

Consider the change of x_i due to displacements of the ith submolecule from the equilibrium situation (0, 0, 0), that is the change of x_i is referred to a private coordinate system with its origin at the bead between the $(i-1)$th and ith submolecules.

There will be a restoring force

$$T\left(\frac{\partial S_m}{\partial x_i} - \frac{\partial S_m}{\partial x_{i-1}}\right)$$

due to displacements of the bead between the submolecules $i-1$ and i and a restoring force

$$T\left(\frac{\partial S_m}{\partial x_{i+1}} - \frac{\partial S_m}{\partial x_i}\right)$$

due to displacements of the bead between the submolecules i and $i+1$.

The total equation of motion is

$$\eta\dot{x}_i = T\left(2\frac{\partial S_m}{\partial x_i} - \frac{\partial S_m}{\partial x_{i-1}} - \frac{\partial S_m}{\partial x_{i+1}}\right) \tag{7.29}$$

η is the coefficient of friction defining the viscous interaction between the beads and the solvent. S_m is the entropy of a molecule of conformation $x_1y_1, ..., z_m$ and is given by

$$S_m = k\ln P_m. \tag{7.30}$$

Combining Equations (7.29) and (7.30) we have

$$\eta\dot{x}_i + \frac{3kT}{zl^2}(2x_i - x_{i-1} - x_{i+1}) = 0, \tag{7.31}$$

with $3m$ equations for coordinates $x_1y_1, ..., z_m$. If we make the intuitive connection between displacement and strain, we can see that these equations of motion for the chain molecules are directly equivalent to the equation of a Voigt element that has the form $\eta\dot{e} + Ee = 0$.

It therefore follows that these equations can be regarded as defining a set of creep compliances and stress relaxation moduli, or complex compliances and moduli.

The mathematical problem is to uncouple the $3m$ equations using a normal coordinate transformation. This involves obtaining the eigenfunctions that are linear combinations of the positions of the submolecules. Each eigenfunction then describes a configuration that decays with a time constant given by an associated eigenvalue, that is a single viscoelastic element with characteristic time-dependent properties.

For stress relaxation and dynamic mechanical experiments it can be shown that the stress relaxation modulus $G(t)$ and the real part of the complex modulus $G_1(\omega)$, respectively, are given by

$$G(t) = NkT\sum_{p=1}^{m} e^{-t/\tau_p} \tag{7.32}$$

and

$$G_1(\omega) = NkT\sum_{p=1}^{m}\frac{\omega^2\tau_p^2}{1+\omega^2\tau_p^2}, \tag{7.33}$$

where N is the number of molecules per cubic centimetre and τ_p, the relaxation time of the pth mode, is given by

$$\tau_p = zl^2\eta\left[24kT\sin^2\{p\pi/2(m+1)\}\right]^{-1}, \quad p = 1, 2, ..., m. \tag{7.34}$$

These equations predict that $G(t)$ and $G_1(\omega)$ are determined by a discrete spectrum of relaxation times, each of which characterizes a given normal mode of motion. These normal modes are shown schematically in Figure 7.19. In the first mode, corresponding to $p = 1$, the ends of the molecule move whilst the centre of the molecule remains stationary. In the second mode, there are two nodes in the molecule. The general case of the pth mode has p nodes, with motion of the molecule occurring in $p + 1$ segments.

On this model the submolecule is the shortest length of chain that can undergo relaxation and the motion of segments *within* the submolecules are ignored. But such motions

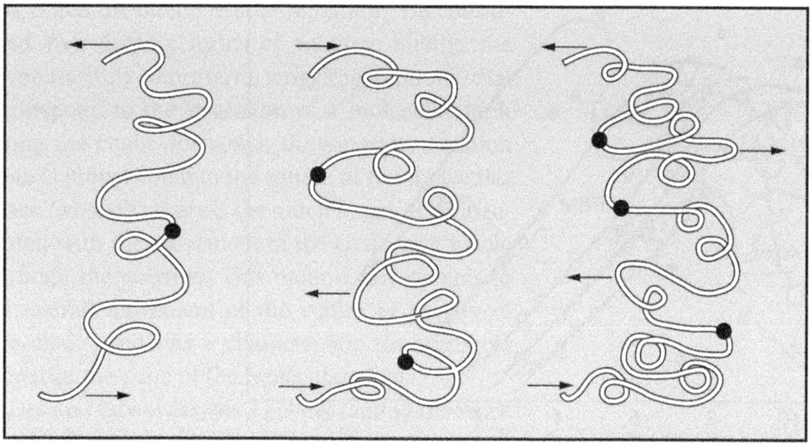

Figure 7.19 Illustration of the first three normal modes of a chain molecule.

contribute to the relaxation spectrum for values of $m \gtrsim 5$. Thus, we would only expect the Rouse theory to be applicable for $m \gg 1$ where the equation for τ_p reduces to

$$\tau_p = \frac{m^2 z l^2 \eta}{6\pi^2 p^2 kT} = \frac{n^2 l^2 \eta_0}{6\pi^2 p^2 kT}, \tag{7.35}$$

where $\eta_0 = \eta/z$ is the friction coefficient per random link. The relaxation times depend on temperature directly through the factor $1/T$, through the quantity nl^2 that defines the equilibrium mean square separation of the chain ends and may change due to differences in the energy of different chain conformations, and through changes in the friction coefficient η_0. η_0 changes rapidly with temperature and is primarily responsible for changes in τ_p. The fact that each τ_p has the same temperature dependence on this molecular theory shows that it satisfies the requirements of thermorheological simplicity and gives theoretical justification for time–temperature equivalence.

Rouse's theory is the simplest molecular theory of polymer relaxation. A later theory of Zimm (1956) does not assume that the velocity of the liquid solvent is unaffected by the movement of the polymer molecules (the 'free draining approximation'). The hydrodynamic interaction between the moving submolecules is taken into account and this gives a modified relaxation spectrum.

The Rouse, Zimm and Bueche theories are satisfactory for the longer relaxation times, which involve movement of submolecules. This has been confirmed for dilute polymer solutions, where the theory would be expected to be most appropriate (Tschoegl and Ferry 1963; Lamb and Matheson 1964). The theories have also been extended to the amorphous solid state in a theory known as the Enhanced Friction Model (Hansel et al. 1976; Matsuoka et al. 1978).

Ferry has shown that if the three longest relaxation times are ignored, the distribution of relaxation times $H(\ln \tau)$ is given by

$$H(\ln \tau)\mathrm{d}(\ln \tau) = NkT\left(\frac{\mathrm{d}p}{\mathrm{d}\tau}\right)\mathrm{d}\tau \tag{7.36}$$

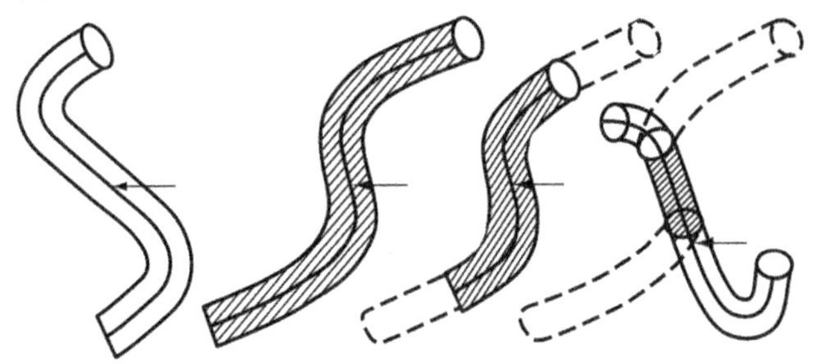

Figure 7.22 Explanation of the stress relaxation after a large step strain. (a) Before deformation the conformation of the primitive chain is in equilibrium ($t = -0$). (b) Immediately after deformation, the primitive chain is in the affinely deformed conformation ($t = +0$). (c) After time τ_R, the primitive chain contracts along the tube and recovers the equilibrium contour length ($t \approx \tau_R$). (d) After the time τ_d, the primitive chain leaves the deformed tube by reptation ($t \approx \tau_d$). The oblique lines indicate the deformed part of the tube. Adapted from Doi and Edwards, 1986.

becomes large, so that the viscosity increases faster than the asymptotic N^3 relationship: hence the apparent 3.4 relationship with molecular weight. Doi (1983) identified this faster process with what he called contour-length fluctuations. These fluctuations involve the chain contracting within the tube and then stretching out again, so that orientation of the ends of the tube is forgotten and part of the stress is relaxed. Milner and McLeish (1998) propose a quantitative theory for the stress relaxation in monodisperse linear polymer melts that incorporates reptation and contour-length fluctuations. The basis for the theory is that linear chains can be regarded as two-armed stars. In star polymers, reptation cannot occur because the arms do not have a single tube. The stress relaxes by arm retraction, where the star contracts by fluctuations down its tube towards the branch point, i.e., similar to the contour-length fluctuations of Doi.

The stress relaxation modulus $G(t)$ then has three components:

1) High-frequency modes due to contour-length fluctuations at the ends of the tubes.
2) Lower-frequency modes associated with reptation.
3) Very-high-frequency (short-time) Rouse modes where the tube constraints are not felt.

It was shown that the Milner–McLeish theory predicted the loss modulus $G_2(\omega)$ for monodisperse polystyrene melts, in addition to achieving the 3.4 molecular weight relationship for the viscosity that converges to 3 for high-molecular weight. More recently, Likhtman and McLeish (2002) have produced a more satisfactory expression for the stress relaxation modulus, which incorporates contour-length fluctuations, longitudinal stress relaxation along the tube and constraint release, where the effect of tube breakdown is taken into account in a less *ad hoc* fashion. The previous predictions of Milner and McLeish are modified to some extent, but the main advance is that the Likhtman–McLeish theory is more self-consistent and avoids many of the approximations in the original theory.

The Doi–Edwards model also has some shortcomings for describing non-linear behaviour at high shear rates where the viscosity is not linearly proportional to the shear rate.

The Doi–Edwards theory predicts that the viscosity $\eta(\dot{\gamma})$ reduces with $\dot{\gamma}$ (shear thinning) as $\dot{\gamma}^{-3/2}$, which would imply that there is a maximum in the shear stress as a function of $\dot{\gamma}$, which is not observed. Shear thinning is due to the orientation of the entangled chain network under flow. Marrucci (1996) proposed that the severity of the orientation is reduced due to what he termed convective constraint release, which relates to the relaxation (i.e., contraction) of chains extended affinely by the flow, by retraction to their equilibrium length in their tubes. Marrucci and others have developed theories for the non-linear behaviour based on this idea (Ianniruberto and Marrucci 1996; Mead et al. 1998).

An alternative approach is that convective constraint release can be regarded as due to a hopping motion of the tube, most simply following Rouse dynamics (Viovy et al. 1991). This has been developed analytically by Milner et al. (2001), whose theory predicts a monotonically increasing shear stress with increasing shear strain rate, with no stress maximum for polymer melts.

For an account of the application of Doi–Edwards theory to the prediction of dynamic moduli G_1 and G_2 as functions of frequency, the reader is referred to papers by Janeschitz-Kriegl and co-workers on polystyrenes (Janeschitz-Kriegl 1982; Schausberger et al. 1985; Schindlauer et al. 1985).

The application of tube theory to the dynamics of entangled flexible polymers in the melt has been comprehensively reviewed by McLeish (2002). This review addresses the current understanding of reptation, contour-length fluctuations and constraint release and gives attention to polymers of complex topology including long-chain branching.

References

Adam, G. and Gibbs, J.H. (1965) On temperature dependence of cooperative relaxation properties in glass-forming liquids. *J. Chem. Phys.*, **43**, 139.

Aharoni, S.M. (1998) Increased glass transition temperature in motionally constrained semicrystalline polymers. *Polym. Adv. Technol.*, **9**, 169.

Arrhenius, Z. (1889) Über die Reaktionsgeschwindigkeit bei der Inversion von Rohrzucker durch Säuren. *J. Physik. Chem.*, **4**, 226.

Bueche, F. (1953) Segmental mobility of polymers near their glass temperature. *J. Chem. Phys.*, **21**, 1850.

Bueche, F. (1954) The viscoelastic properties of plastics. *J. Chem. Phys.*, **22**, 603.

Cohen, M.H. and Turnbull, D. (1959) Molecular transport in liquids and glasses. *J. Chem. Phys.*, **31**, 1164.

Debye, P. (1945) *Polar Molecules*, Dover, New York.

Doi, M. (1983) Explanation for the 3.4-power law for viscosity of polymeric liquids on the basis of the tube model. *J. Polym. Sci. Polym. Phys. Ed.*, **21**, 667.

Doi, M. and Edwards, S.F. (1986) *The Theory of Polymer Dynamics*, Oxford University Press, New York.

Doolittle, A.K. (1951) Studies in Newtonian flow. 2. The dependence of the viscosity of liquids on free-space. *J. Appl. Phys.*, **22**, 1471.

Ferry, J.D. (1980) *Viscoelastic Properties of Polymers*, 3rd edn, John Wiley & Sons, New York. Chapter 11

Ferry, J.D. (1991) Some reflections on the early development of polymer dynamics: viscoelasticity, dielectric dispersion, and self-diffusion. *Macromolecules*, **24**, 5237–5245.

Fröhlich, H. (1968) *Theory of Dielectrics*, 2nd edn, Oxford University Press, Oxford.

de Gennes, P.G. (1971) Reptation of a polymer chain in presence of fixed obstacles. *J. Chem. Phys.*, **55**, 572.

Gibbs, J.H. and di Marzio, E.A. (1958a) Chain stiffness and the lattice theory of polymer phases. *J. Chem. Phys.*, **28**, 807.

Gibbs, J.H. and di Marzio, E.A. (1958b) Nature of the glass transition and the glassy state. *J. Chem. Phys.*, **28**, 373.

Glasstone, S., Laidler, K.J. and Eyring, H. (1941) *The Theory of Rate Processes*, McGraw-Hill, New York.

Hansel, D.R., Williams, M.C. and Shen, M. (1976) Viscoelastic properties of entangled polymers. I. The elastically coupled entanglement model. *Macromolecules*, **9**, 345–354.

Hoffman, J.D., Williams, G. and Passaglia, E. (1966) Analysis of alpha beta and gamma relaxations in polychlorotrifluoroethylene and polyethylene – dielectric and mechanical properties. *J. Polym. Sci. C*, **14**, 173.

Ianniruberto, G. and Marrucci, G. (1996) On compatibility of the Cox-Merz rule with the model of Doi and Edwards. *J. Non-Newtonian Fluid Mech.*, **65**, 241.

Janeschitz-Kriegl, H. (1982) Some pending problems in polymer melt rheology, as seen from the point of view of Doi slip-link model. *Rheol. Acta.*, **21**, 388.

Kovacs, A. (1958) La Contraction Isotherme Du Volume Des Polymeres Amorphes. *J. Poym. Sci.*, **30**, 131.

Lamb, J. and Matheson, A.J. (1964) Viscoelastic properties of polystyrene solutions. *Proc. Roy. Soc. A*, **281**, 207.

Lee, E.H. (1960) Viscoelastic stress analysis, in *Proceedings of the First Symposium on Naval Structural Mechanics (London)*, (eds J.N. Goodier and J.F. Hoff), Pergamon Press, Oxford, p. 456.

Likhtman, A.E. and McLeish, T.C.B. (2002) Quantitative theory for linear dynamics of linear entangled polymers. *Macromolecules*, **35**, 6332.

Manero, O., Soltero, J.F.A., Puig, J.E. and Gonzalez-Romero, V.M. (1997) On the application of the modeling of the relaxation spectrum to the prediction of linear viscoelastic properties of surfactant systems. *Colloid Polym. Sci.*, **275**, 979–985.

Marrucci, G. (1996) Dynamics of entanglements: a nonlinear model consistent with the Cox-Merz rule. *J. Non-Newtonian Fluid Mech.*, **62**, 279.

Marvin, R.S. (1954) The dynamic mechanical properties of polyisobutylene, in *Proceedings of the Second International Congress of Rheology*, (ed. V.G.W. Harrison), Butterworths, London, pp. 156–164.

Marvin, R.S. and Oser, H. (1962) Model for the viscoelastic behavior of rubberlike polymers including entanglement effects. *J. Res. Natl Bur. Stand. B*, **66**, 171.

Matsuoka, S., Bair, H.E., Bearder, S.S., Kern, H.E. and Ryan, J.T. (1978) Analysis of non-linear stress relaxation in polymeric glasses. *Polm. Eng. Sci.*, **18**, 1073–1080.

McCrum, N.G. and Morris, E.L. (1964) On measurement of activation energies for creep and stress relaxation. *Proc. R. Soc.*, **A281**, 258.

McLeish, T.C.B. (2002) Tube theory of entangled polymer dynamics. *Adv. Phys.*, **51**(6), 1379.

Mead, D.W., Larson, R.G. and Doi, M. (1998) A molecular theory for fast flows of entangled polymers. *Macromolecules*, **31**, 7895.

Milner, S.T. and McLeish, T.C.B. (1998) Reptation and contour-length fluctuations in melts of linear polymers. *Phys. Rev. Lett.*, **81**, 725.

Milner, S.T., McLeish, T.C.B. and Likhtman, A.E. (2001) Microscopic theory of convective constraint release. *J. Rheol*, **45**, 539.

Rouse, P.E. (1953) A theory of the linear viscoelastic properties of dilute solutions of coiling polymers. *J. Chem. Phys.*, **21**, 1272.

Schausberger, A., Schinlauer, G. and Janeschitz-Kriegl, H. (1985) Linear elasticoviscous properties of molten standard polystyrenes. 1. Presentation of complex modulirole of short-range structural parameters. *Rheol. Acta.*, **24**, 220.

Schindlauer, G., Schausberger, A. and Janeschitz-Kriegl, H. (1985) Linear elasticoviscous properties of molten standard polystyrenes. 2. Improvement of conventional characterization on molar mass-distribution. *Rheol. Acta.*, **24**, 228.

Schmieder, K. and Wolf, K. (1953) Mechanische Relaxationserscheinungen an Hochpolymeren - Beziehungen Zur Struktur. *Kolloidzeitschrift*, **134**, 149.

Staverman, A.J. and Schwarzl, F. (1956) *Die Physik der Hochpolymeren*, Springer-Verlag, Berlin. Chapter 1

Thompson, A.B. and Woods, D.W. (1956) The transitions of polyethylene terephthalate. *Trans. Faraday Soc.*, **52**, 1383.

Tschoegl, N.W. and Ferry, J.D. (1963) Viskoelastische Eigenschaften von Polyisobutylenen in Verdunnten Losungen. *Kolloidzeitschrift*, **189**, 37.

Viovy, J.L., Rubinstein, M. and Colby, R.H. (1991) Constraint release in polymer melts - tube reorganization versus tube dilation. *Macromolecules*, **24**, 3587.

Wachtman, J.B. (1963) Mechanical and electrical relaxation in ThO_2 containing CaO. *Phys. Rev.*, **131**, 517.

Williams, G. (1962) Upper limit of the shear modulus calculated from the Rouse theory. *J. Polym. Sci.*, **62**, 87.

Williams, G. (1964) Complex dielectric constant of dipolar compounds as function of temperature, pressure and frequency. 1. Alpha-relaxation of polymethylacrylate. *Trans. Faraday Soc.*, **60**, 1548.

Williams, M.L., Landel, R.F. and Ferry, J.D. (1955) Mechanical properties of substances of high molecular weight. 19. The temperature dependence of relaxation mechanisms in amorphous polymers and other glass-forming liquids. *J. Am. Chem. Soc.*, **77**, 3701.

Zimm, B.H. (1956) Dynamics of polymer molecules in dilute solution - viscoelasticity, flow birefringence and dielectric loss. *J. Chem. Phys.*, **24**, 269.

8

Anisotropic Mechanical Behaviour

8.1 The Description of Anisotropic Mechanical Behaviour

An oriented polymer is in the strictest terms an anisotropic non-linearly viscoelastic material. A comprehensive understanding of anisotropic mechanical behaviour is therefore a very considerable task. In this chapter, we will restrict the discussion to cases where the strains are small.

The mechanical properties of an anisotropic elastic solid for small strains are defined by the generalized Hooke's law:

$$e_{ij} = s_{ijk\ell}\sigma_{k\ell} \quad \sigma_{ij} = c_{ijk\ell}e_{k\ell},$$

where the $s_{ijk\ell}$ are the compliance constants and the $c_{ijk\ell}$ are the stiffness constants. This has been discussed in Chapter 2. The use of this representation does not necessarily restrict the discussion to time-independent behaviour. The compliance and stiffness constants could be time dependent, defining creep compliances and relaxation stiffnesses in step function loading experiments, or complex compliances and complex stiffnesses in dynamic mechanical measurements. For simplicity, the methods of measurement are usually carefully standardized, for example by measuring each creep compliance after the same loading programme and the same time interval. It will be assumed that for such measurements, there is an exact equivalence between elastic and linear viscoelastic behaviour, as proposed by Biot (1955).

In an elastic material, the presence of symmetry elements leads to a reduction in the number of independent elastic constants and corresponding reductions will be assumed for anisotropic linear viscoelastic behaviour, although there is not enough experimental evidence to confirm that exactly the same rules hold in every case (Rogers and Pipkin 1963).

There are two important points to emphasize regarding the description of data:

1) In practice an abbreviated notation is often used in which

$$e_p = s_{pq}\sigma_q \quad \sigma_p = c_{pq}e_q.$$

As explained in Section 2.5, σ_p represents σ_{11}, σ_{22}, ... , σ_{12}, etc., and e_p represents e_{11}, e_{22}, ..., e_{12}, etc., and in the compliance and stiffness matrices s_{pq} and c_{pq}, p, q take the values 1, 2, ..., 6.

Mechanical Properties of Solid Polymers, Fourth Edition. John Sweeney and Peter Hine.

The conversion rules from the $s_{ijk\ell}$ and $c_{ijk\ell}$ notation to the abbreviated notation are given in Section 2.5. It is important to remember that the engineering strains e_p are not the components of a tensor. Similarly, the 6×6 compliance matrix s_{pq} does not represent a tensor, and therefore tensor manipulation rules do not apply. As will be demonstrated, for working out problems involving transformation of coordinates from one system of axes to another, it is always desirable to use the original tensor notation in terms of e_{ij}, $\sigma_{k\ell}$ and $s_{ijk\ell}$ or $c_{ijk\ell}$.

2) It is usually more convenient to work in terms of compliance constants than stiffness constants. This is because in the experimental procedures it is easier to apply a simple stress of a given type, e.g., a tensile stress or a shear stress, and measure the corresponding strains, e.g.,

$$e_1 = s_{11}\sigma_1 + s_{12}\sigma_2 + s_{13}\sigma_3 + s_{14}\sigma_4 + s_{15}\sigma_5 + s_{16}\sigma_6.$$

The compliance constants s_{11}, s_{12}, s_{13}, etc., can be found by applying stresses $\sigma_1, \sigma_2, \sigma_3$, etc., and measuring e_1 in each case. The procedures will become clearer as the various experimental methods are discussed.

8.2 Mechanical Anisotropy in Polymers

8.2.1 The Elastic Constants for Specimens Possessing Fibre Symmetry

Studies of mechanical anisotropy in polymers have for the most part been restricted to drawn fibres and uniaxially drawn films, both of which show isotropy in a plane perpendicular to the direction of drawing. The number of independent elastic constants is reduced to five (Lekhnitzkii 1981, pp. 37–38). Choosing the 3 direction as the axis of symmetry, the compliance matrix s_{ij} reduces to

$$\begin{pmatrix} s_{11} & s_{12} & s_{13} & 0 & 0 & 0 \\ s_{12} & s_{11} & s_{13} & 0 & 0 & 0 \\ s_{13} & s_{13} & s_{33} & 0 & 0 & 0 \\ 0 & 0 & 0 & s_{44} & 0 & 0 \\ 0 & 0 & 0 & 0 & s_{44} & 0 \\ 0 & 0 & 0 & 0 & 0 & 2(s_{11}-s_{12}) \end{pmatrix}.$$

The various compliance constants are illustrated diagrammatically in Figure 8.1. The situation is most easily appreciated for a fibre specimen, but a uniaxially oriented sheet possesses identical symmetry.

The relationships of these compliance constants to the better-known Young's moduli and Poisson's ratios are as follows:

1) Consider application of a stress along the three direction, that is along the fibre axis, or the draw direction for the polymer film. Then $e_{33} = s_{33}\sigma_{33}$ and

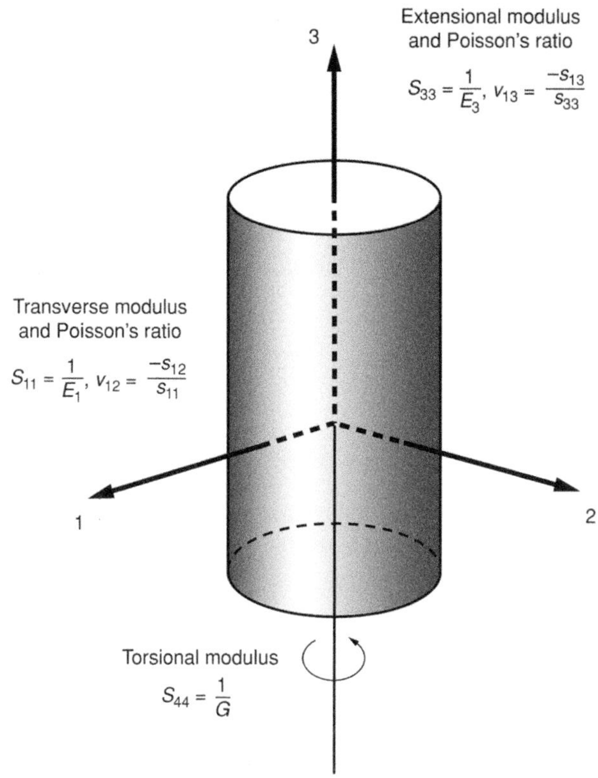

Figure 8.1 The fibre compliance constants.

$$\text{Young's modulus:} \quad E_3 = \frac{\sigma_{33}}{e_{33}} = \frac{1}{s_{33}}, \quad \text{giving } s_{33} = \frac{1}{E_3}.$$

Similarly the strain in the plane transverse to the fibre axis for a stress σ_{33} along the fibre axis is given by $e_{11} = e_{22} = s_{13}\sigma_{33}$ and

$$\text{Poisson's ratio:} \quad \nu_{13} = -\frac{e_{11}}{e_{33}} = -\frac{s_{13}}{s_{33}}.$$

(The negative sign ensures that Poisson's ratio is the conventionally positive quantity since e_{11} is negative, i.e., a contraction.)

3) In a similar manner, s_{11}, s_{12} and s_{13} are related to the modulus E_1 (the transverse modulus) and the corresponding Poisson's ratios $\nu_{21} = \nu_{12}$, ν_{31} and ν_{13} for application of a stress in a plane perpendicular to the fibre axis, i.e.,

$$s_{11} = \frac{1}{E_1} \quad \text{and} \quad \nu_{21} = -\frac{s_{12}}{s_{11}} = -\frac{s_{12}}{s_{22}}, \quad \nu_{31} = -\frac{s_{13}}{s_{11}}, \quad \nu_{13} = -\frac{s_{13}}{s_{33}}.$$

The remaining Poisson's ratios ν_{23} and ν_{32} are not independent quantities, since

$$\nu_{23} = -\frac{s_{23}}{s_{33}} = -\frac{s_{13}}{s_{33}} = \nu_{13} \quad \text{and} \quad \nu_{32} = -\frac{s_{23}}{s_{22}} = -\frac{s_{13}}{s_{11}} = \nu_{31}.$$

4) The shear compliance is the reciprocal of the shear or torsional modulus G. There are two equivalent shear compliances $s_{44} = s_{55} = 1/G$. These relate to torsion about the symmetry axis 3, i.e., shear in the 23 or 13 planes.

   The shear compliance s_{66} relates to shear in the 12 plane and is related to the compliance constants s_{11} and s_{12}, such that $s_{66} = 2(s_{11} - s_{12})$. This relationship expresses the fact that these specimens are isotropic in a plane perpendicular to the symmetry axis, i.e., that the elastic behaviour in this plane is specified by only two elastic constants as for an isotropic material. It will be seen that this property is very important in determining the elastic constants for fibres.

8.2.2 The Elastic Constants for Specimens Possessing Orthorhombic Symmetry

Oriented polymer films that are prepared by either rolling, rolling and annealing, or some commercial one-way draw processes, may possess orthorhombic rather than transversely isotropic symmetry. For such films, the elastic behaviour is specified by nine independent elastic constants. Choose the initial drawing or rolling direction as the 3 axis for a system of rectangular Cartesian coordinates; the 1 axis to lie in the plane of the film and the 2 axis normal to the plane of the film (Figure 8.2). The compliance matrix is

$$\begin{pmatrix} s_{11} & s_{12} & s_{13} & 0 & 0 & 0 \\ s_{12} & s_{22} & s_{23} & 0 & 0 & 0 \\ s_{13} & s_{23} & s_{33} & 0 & 0 & 0 \\ 0 & 0 & 0 & s_{44} & 0 & 0 \\ 0 & 0 & 0 & 0 & s_{55} & 0 \\ 0 & 0 & 0 & 0 & 0 & s_{66} \end{pmatrix}.$$

There are three Young's moduli,

$$E_1 = \frac{1}{s_{11}}, \quad E_2 = \frac{1}{s_{22}} \quad \text{and} \quad E_3 = \frac{1}{s_{33}},$$

and six Poisson's ratios,

$$\nu_{21} = -\frac{s_{21}}{s_{11}}, \quad \nu_{31} = -\frac{s_{31}}{s_{11}}, \quad \nu_{32} = -\frac{s_{32}}{s_{22}}$$
$$\nu_{12} = -\frac{s_{21}}{s_{22}}, \quad \nu_{13} = -\frac{s_{31}}{s_{33}}, \quad \nu_{23} = -\frac{s_{32}}{s_{33}}.$$

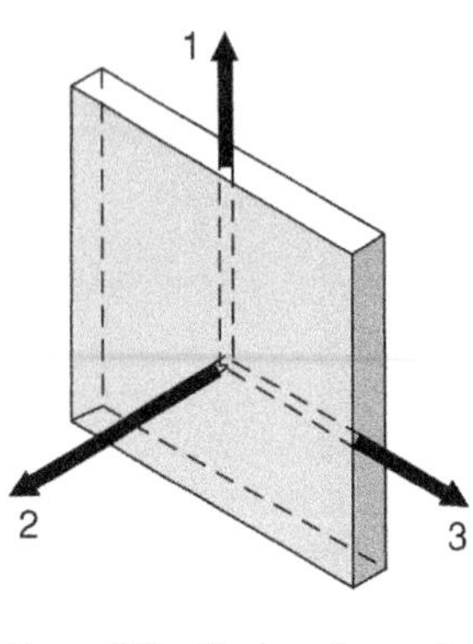

Figure 8.2 Choice of axes for a polymer sheet possessing orthorhombic symmetry.

There are three independent shear moduli $G_1 = 1/s_{44}$, $G_2 = 1/s_{55}$ and $G_3 = 1/s_{66}$ corresponding to shear in the 23, 13 and 12 planes, respectively. For a sheet of general dimensions, torsion experiments where the sheet is twisted about the 1, 2 or 3 axis will involve a combination of shear compliances. This will be discussed in greater detail later, when methods of obtaining the elastic constants are described.

8.3 Measurement of Elastic Constants

The measurement of elastic constants is a very different undertaking for the two situations of a sheet and a fibre. The experimental methods employed for these two cases will therefore be discussed separately.

8.3.1 Measurements on Films or Sheets

8.3.1.1 Extensional Moduli

The simplest measurement on a polymer film is to determine Young's modulus in various directions in the film by cutting long thin strips in the selected directions.

For anisotropic materials, it is important to recognize that it is generally necessary to measure samples of very high aspect ratio[1] to minimize 'end effects'. These end effects arise from non-uniform stress conditions near the clamps that are much more severe than would be anticipated on the basis of St Venant's principle. The situation has been discussed in detail by Horgan (1972, 1974) and by Folkes and Arridge (1975).

We will consider a film of orthorhombic symmetry, the 1 and 3 axes lying in the plane of the film and the 2 axis normal to the film as in Section 8.2.2.

Consider a long strip cut in a direction making an angle θ with the 3 direction (Figure 8.3(a)).

Young's modulus for this strip $E_\theta = 1/s_\theta$, where s_θ is the compliance in a direction making an angle θ with the 3 direction.

To calculate s_θ in terms of the compliance constants, we will use the full tensor notation.

The compliance constants s_{ijkl} referred to one system of Cartesian axes are related to those s'_{pqmn} referred to a second system of Cartesian axes by the tensor transformation rule:

$$s'_{pqmn} = a_{pi}\, a_{qj}\, a_{mk}\, a_{n\ell}\, s_{ijk\ell},$$

where a_{pi}, a_{qj}, ... define the cosines of the angles between the p axis in the second system and the i axis in the first, the q axis in the second system and the j axis in the first, ... and p, q, m, n

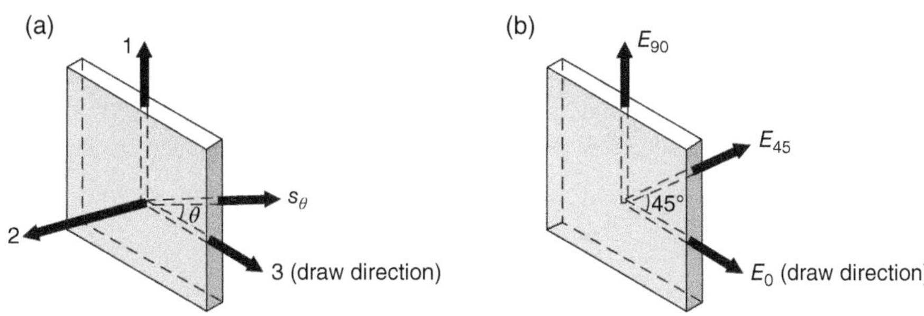

Figure 8.3 (a) The compliance s_θ is at an angle θ to the initial draw direction in the plane of the sheet. (b) The E_0, E_{45} and E_{90} moduli.

1 Aspect ratio is the ratio of length to width or thickness.

take the values 1, 2, 3 in the second system of axes and *i, j, k, ℓ,* take the values 1′, 2′, 3′ in the first system of axes.

We will take the direction of the strip to be the 3′ direction in a second system of Cartesian axes. Then $s_\theta = s_{3'3'3'3'}$ is given by

$$\begin{aligned} s_{3'3'3'3'} &= a_{3'1}a_{3'1}a_{3'1}a_{3'1}s_{1111} + a_{3'3}a_{3'3}a_{3'3}a_{3'3}s_{3333} \\ &+ a_{3'1}a_{3'1}a_{3'3}a_{3'3}s_{1133} + a_{3'3}a_{3'3}a_{3'1}a_{3'1}s_{3311} \\ &+ a_{3'1}a_{3'3}a_{3'3}a_{3'1}s_{1331} + a_{3'3}a_{3'1}a_{3'1}a_{3'3}s_{3113} \\ &+ a_{3'3}a_{3'1}a_{3'3}a_{3'1}s_{3131} + a_{3'1}a_{3'3}a_{3'1}a_{3'3}s_{1313}. \end{aligned}$$

Note that all compliance terms containing the subscript 2 will vanish, because $a_{3'2} = 0$.

The change in coordinate systems corresponds to a rotation of the coordinate axes through an angle θ about the 2 direction as axis.

We therefore put $a_{3'1} = \sin\theta$ and $a_{3'3} = \cos\theta$ and

$$s_{3'3'3'3'} = \sin^4\theta s_{1111} + \cos^4\theta s_{3333} + 2\sin^2\theta \;\; \cos^2\theta s_{1133} + 4\sin^2\theta \cos^2\theta s_{1313}.$$

In the abbreviated notation,

$$s_\theta = s_{3'3'} = \sin^4\theta s_{11} + \cos^4\theta s_{33} + \sin^2\theta \cos^2\theta \,(2s_{13} + s_{55}). \tag{8.1}$$

(Note factor 4 in converting from $s_{ijk\ell}$ to s_{pq}, when p and $q = 4, 5, 6$, i.e., 23, 13, 12.)

It is thus possible to undertake three independent measurements on these sheets. For convenience, choose these to be Young's modulus on strips at 0°, 45° and 90° to the initial draw direction and denote these by E_0, E_{45} and E_{90}, respectively (see Figure 8.3(b)). From Equation (8.1)

$$E_0 = \frac{1}{s_{33}}, \quad E_{90} = \frac{1}{s_{11}} \quad \text{and} \quad \frac{1}{E_{45}} = \frac{1}{4}\left[s_{11} + s_{33} + (2s_{13} + s_{55})\right]. \tag{8.2}$$

Such measurements yield immediately two of the nine independent elastic constants, s_{11} and s_{33}, and give the combination $(2s_{13} + s_{55})$ but do not involve s_{12}.

For a transversely isotropic sheet where 3 is the symmetry axis, there are only five independent elastic constants, and $s_{55} = s_{44}$.

8.3.1.2 Use of Imaging for Full-field Strain Fields

In recent years the ease with which visual images can be captured and analysed has revolutionized the measurement of strain. Dot or speckle patterns can be applied to specimen surfaces and tracked by video capture during testing; images are then processed and strain fields derived as a function of time. This method is known as Digital Image Correlation (DIC); Jerabek *et al.* (2010) have given a description of its application to polymers. A review of the current state of the art in DIC measurement was given by Zhao *et al.* (2019). An example of a DIC setup is shown in Figure 8.4.

Full-field strain measurements are advantageous for efficient determination of elastic constants as they enable measurement of both longitudinal and transverse strains in a uniaxial test. For testing of the solid of Figure 8.2 by stretching or compressing along the 1 direction and observing the strains on the surface normal to the 3 axis, both the compliance s_{11} and the Poisson's ratio ν_{12}, and equivalently the compliance s_{12}, can be found. In the same

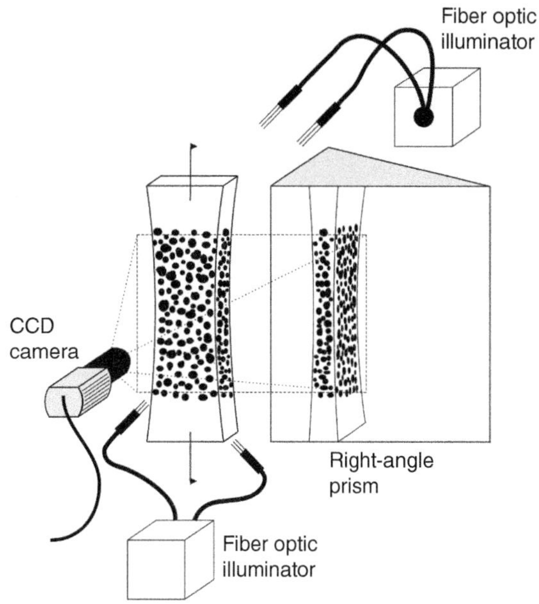

Figure 8.4 DIC setup showing measurement of two surfaces simultaneously. (Parsons E.M., Boyce, M.C., Parks, D.M., Weinberg, M. (2005) Three-dimensional large-strain tensile deformation of neat and calcium carbonate-filled high-density polyethylene. Polymer, **46**, 2257–2265.)

experiment, observation of strains in the plane normal to the 2 axis gives values of both s_{11} and s_{13}. Thus, testing along the 1 axis can provide s_{11}, s_{12} and s_{13}; this could be done with a single experiment, as in Figure 8.4. Similarly, testing along the 2 and 3 axes gives the constants s_{12}, s_{22}, s_{23} and s_{13}, s_{23}, s_{33}, respectively, to yield six independent compliances. The method has been used in compression to obtain these data for composite material (Orell *et al.* 2018). It provides no information on the shear compliances s_{44}, s_{55} and s_{66}.

Attar *et al.* (2022) have experimented upon sheets of orthotropic PET cut from stretch blow-moulded bottles. Operating in the 1–2 plane, they required to determine the values of the compliance matrix entries

$$\begin{pmatrix} s_{11} & s_{12} & 0 \\ s_{12} & s_{22} & 0 \\ 0 & 0 & s_{66} \end{pmatrix}$$

While s_{11}, s_{12} and s_{22} can be readily determined by methods such as described in the preceding paragraph, some shear strain must be applied to determine s_{66}. Attar *et al.* used a biaxial testing machine to apply equibiaxial extensions to cruciform specimens and created shear strains in the specimen by introducing a hole in it. The strain field that results is non-uniform, and the experiments are analysed by comparing DIC measured strains with those calculated using finite element models of the specimens. They used an established method for extracting mechanical parameters from full-field strains, known as the Virtual Fields Method (Pierron and Grédiac 2012). Figure 8.5 shows an example of measured strain fields. A single experiment provided values for the four compliances which were found to be consistent with values provided using alternative methods.

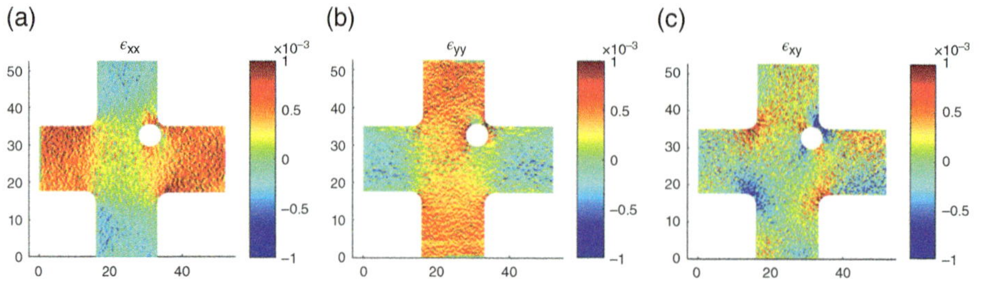

Figure 8.5 Strain values of equibiaxially stretched orthotropic PET cruciform specimen, showing strains $\varepsilon_{11}, \varepsilon_{22}$ and ε_{12} in (a), (b) and (c) respectively. (Attar H., Luo Y-M, Chevalier L., Nguyen T.T., Detrez F. (2022). Identification of anisotropic properties of polymer sheets from heterogeneous biaxial tests. Polym. Test., **115** 107721.)

8.3.1.3 Measurement of Shear Compliance

The need for knowledge of the elastic constants of oriented polymer precedes the availability of DIC by many decades. Historically, mechanical measurements have been devised to measure the shear compliances. Currently, torsion measurements can be most conveniently made using DMTA apparatus with torsional capability (see Chapter 6). However, the analyses outlined later would be needed for interpretation of results with anisotropic material. Raumann (1962) measured torsion of oriented polymer sheets to determine the shear compliances s_{44} and s_{66} for low-density polyethylene oriented uniaxially (transversely isotropic) along the 3 axis. Torsion of oriented sheets can also be used to determine the shear compliances s_{44}, s_{55} and s_{66} for sheets possessing orthorhombic symmetry. As this situation is more general than that of transverse isotropy, it will be considered first.

This problem is known as the St Venant torsion problem. Ecsedi and Baksa (2023) have recently provided closed-form solutions for cylindrical specimens. For the orthorhombic sheets, a solution can only be found when the sheets are cut as rectangular prisms with their surfaces normal to the three axes of orthorhombic symmetry, and where the torsion axis coincides with one of these three axes. This is illustrated in Figure 8.6. Torsion about the 3 axis involves the shear compliances in the 23 and 13 planes, which are s_{44} and s_{55}, respectively.

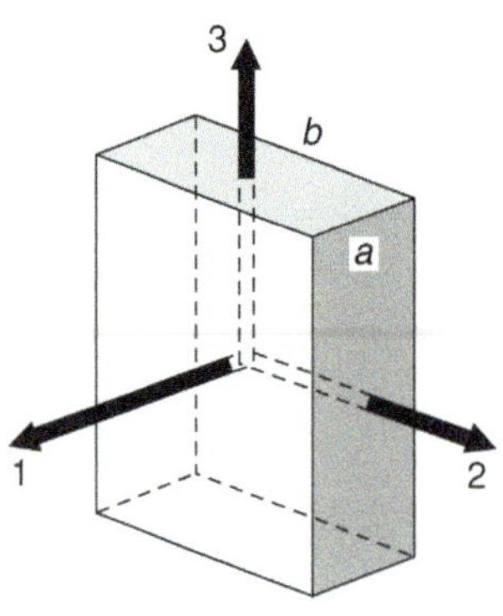

Figure 8.6 The orthorhombic sheet.

The problem has been considered by Lekhnitskii (1981, pp. 283–287). The torque Q_z required to produce the twist T in a specimen of length ℓ, thickness a and width b is

$$Q_z = \frac{ab^3 T}{s_{55}\ell}\beta(c_z) = \frac{ba^3 T}{s_{44}\ell}\beta(c_z^+)$$

to give Q_z/T, the torsional rigidity of the specimen, where

$$c_z = \frac{1}{c_z^+} = \frac{a}{b}\left(\frac{s_{55}}{s_{44}}\right)^{1/2}.$$

The dependency of the function β means that these measurements do not give separate values for s_{44} and s_{55}, so that measurements about different axes are required to fully define them.

For a transversely isotropic sheet (with 3 direction as axis of symmetry), a similar expression describes torsion about an axis perpendicular to the symmetry axis. In this case, $s_{44} = s_{55}$ and the torque Q_z is given by

$$Q_z = \frac{bc^3T}{s_{66}\ell}\beta(c^*) = \frac{cb^3T}{s_{44}\ell}\beta(c^+),$$

where

$$c^+ = \frac{1}{c^*} = \frac{c}{b}\left(\frac{s_{44}}{s_{66}}\right)^{1/2}$$

where b is the thickness, c is the width and ℓ is the length of the specimen.

These formulae show that the relative contribution of the various shear compliances to the torque depends on their relative magnitude and the aspect ratio a/b or b/c. In principle, therefore, both compliances can be obtained from measurements on sheets of different aspect ratios, and this has been done for several polymers (Ladizesky and Ward 1971, 1974; Lewis and Ward 1980). A torsional pendulum method was used by Feng and Farris (2002) to measure shear moduli in a full characterization of oriented PET sheet, where the other orthotropic elastic constants were evaluated using optical strain measurements.

For transversely isotropic sheets, a much simpler formula applies for torsion around the symmetry axis 3, where the torque Q_z is given by

$$Q_z = \left(ab^3/s_{44}\ell\right)\beta(c),$$

where $c = a/b$. $\beta(c)$ is now the same function of $c = a/b$ only.

In practice, the torsion of oriented polymer sheets is complicated by several effects that cause an apparent increase in the stiffness of the sample (Lewis and Ward 1980).

First, there is the extension of lines parallel to the twist axis at the sample edge with respect to similar lines near the centre. This is analogous to the bifilar or multifilar effect in suspensions. When the twist axis is parallel to the symmetry axis in transversely isotropic sheets, the effect of such small axial stresses can be dealt with by the theoretical treatment of Biot (1939). In general, however, it is necessary to carry out experiments over a range of axial stresses, and extrapolate to zero axial stress.

Secondly, planes normal to the twist axes warp into characteristic patterns, but the grips at each end prevent such warping locally. Although Timoshenko and Goodier (2008) have given a theoretical treatment for the effective increase in stiffness, it is again satisfactory to adopt a more empirical procedure. Following Folkes and Arridge (1975), it is considered that such end effects are confined to a block with sample compliance s' and length p at each end of the sample of total length ℓ, the central region of homogeneous stress having the true sample compliance s^0. This gives a linear variation in measured overall sample compliance s with reciprocal length

$$s = s^0 + (2p/\ell)\left(s' - s^0\right).$$

s^0 can be found by taking measurements on samples of different length and extrapolating to zero reciprocal length (Lewis and Ward 1980).

Both these effects could be evaluated using numerical techniques such as the finite element method. Chen *et al.* (2020) have recently published a treatment of the St Venant torsion problem using the finite volume method.

In view of the complications arising in the torsion of oriented polymer sheets due to the effects of axial stress and end effects associated with the grips, there is considerable incentive to determine the shear compliances s_{44} and s_{66} by simple shear. Lewis *et al.* (1979) created an experimental device in which the strains induced by weights were measured by a device using the Hall effect. Zihlif *et al.* (1982) used this method to obtain a comprehensive set of elastic constants for highly oriented polyethylene sheet.

More recently DIC has been used to measure the shear strains in lap-joint tests on elastomers (Nunes 2010; Pal and Bhattacharyya 2025); this could provide an alternative method for oriented polymer.

8.3.2 Measurements on Fibres and Monofilaments

8.3.2.1 Extensional and Torsional Moduli

The extensional modulus $E_3 = 1/s_{33}$ can be conveniently measured in tension using tensile testing machines or, if dynamic data is required, DMTA (see Chapter 6). As with orthotropic sheet, the torsional modulus $G = 1/s_{44}$ can most readily be measured using DMTA with a torsional testing head. This elastic constant is also one of those available via the ultrasonic immersion technique (Darras *et al.* 1995), as described in the next section.

8.3.2.2 The Extensional Poisson's Ratio $\nu_{13} = -s_{13}/s_{33}$

Measurements are made of the diametrical contraction of the fibre under tension. Early work attempted this using optical diffraction (Davis 1959) and mercury-displacement techniques (Frank and Ruoff 1958). Satisfactory data were only obtained for nylon because this fibre is a particularly favourable case, showing no permanent deformation up to 5% extension.

Later measurements have been made by observing in a microscope the radial contraction, together with the corresponding lateral extension of a fibre monofilament (Hadley *et al.* 1965). The monofilament was extended between two moveable grips, which were part of a specially constructed microscope stage. Two ink marks were placed on the monofilament to act as reference points for the measurement of length and changes in length. An immersion liquid was used to reduce diffraction effects at the edges of the monofilament. The method was of limited accuracy, and errors of at least 10% were reported for 95% confidence limits on the mean value.

More recently, a multiple-beam Fizeau fringe technique has been used to measure Poisson's ratio during creep (Hamza *et al.* 2010). This was a study of viscoelasticity of initially isotropic fibre that became significantly oriented during testing and is clearly applicable to preoriented material.

Darras *et al.* (1995) used the ultrasonic immersion method (see Chapter 6) to measure the elastic constants of a range of orthotropic polyethylene-based materials. Some of the specimens were made of compacted oriented polymer fibres, essentially as unidirectional fibre composites with 100% reinforcement. These are transversely isotropic materials with symmetry direction along the fibre (3) axis, to give five independent compliances s_{11}, s_{12}, s_{13}, s_{33}

and s_{44} with $s_{66} = 2(s_{11} - s_{12})$. The compliance values for these sheet specimens gave good agreement with those of the original polymer fibres.

8.3.2.3 The Transverse Modulus $E_1 = 1/s_{11}$

The two remaining elastic constants for fibres, the compliances s_{11} and s_{12}, have to be determined by more sophisticated methods. Both can be obtained from the compression of fibre monofilaments between parallel plates under conditions of plane strain. The transverse modulus is involved in the contact width $2b$ shown in Figure 8.7(a) (Hadley *et al.* 1965; Pinnock *et al.* 1966).

The monofilament is a transversely isotropic solid; thus, it is isotropic in a plane perpendicular to the fibre axis. This implies that under compressive loading normal to the fibre axis the stresses in the transverse plane will be identical in form to those for the compression of an isotropic cylinder. As the length of monofilament under compression is comparatively long, friction ensures that the compression occurs under plane strain conditions. There is, therefore, no change in dimension along the fibre axis ($e_{33} = 0$) and only a normal stress acts along the fibre axis σ_{33} that can be found in terms of the normal stresses σ_{11} and σ_{22} in the plane perpendicular to the fibre axis. We have

$$\sigma_{33} = -\frac{s_{13}}{s_{33}}(\sigma_{11} + \sigma_{22}).$$

All the stresses can therefore be obtained from the solution to the problem of compression of an isotropic cylinder. The corresponding strains can then be obtained using the constitutive equations $e_p = s_{pq}\sigma_q$.

The contact zone is arranged to be small compared with the radius of the monofilament. It is therefore adequate to assume that we are dealing with the contact between two semi-infinite solids and follow Hertz's classic solution for the compression of an isotropic cylinder (Hertz 1896). In this solution, the displacement of the cylinder within the contact zone is assumed to be parabolic, and the boundary conditions are satisfied along the boundary plane only. For purely algebraic reasons, it is most convenient to use the complex variable

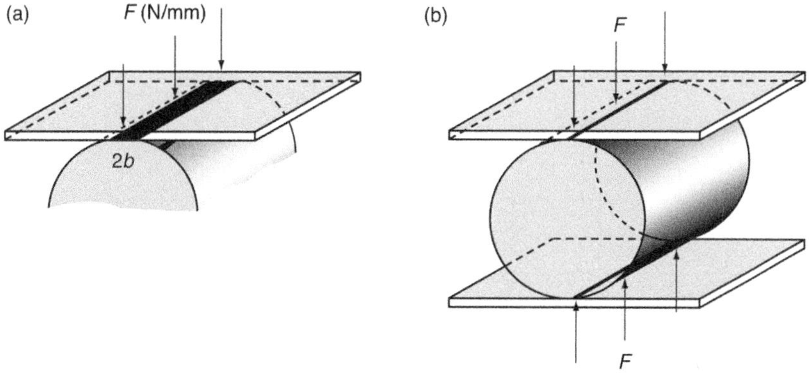

Figure 8.7 The contact zone in the compression of a fibre monofilament (a) and for consideration of deformation in the central zone of the compressed monofilament it is sufficient to assume line contacts (b).

method of McEwen (1949) to obtain an analytical solution for b. It was shown by Hadley *et al.* (1965) that

$$b^2 = \frac{4FR}{\pi}\left(s_{11} - \frac{s_{13}^2}{s_{33}}\right),$$

where F is the load per unit length of monofilament and R is the radius of the monofilament. This expression may be written as

$$b^2 = \frac{4FR}{\pi}\left(s_{11} - \nu_{13}^2 s_{33}\right).$$

Highly oriented polymers are usually much stiffer along their axis than transverse to it. The quantity s_{33} is therefore usually very small compared with s_{11}. Since Poisson's ratio ν_{13} is typically near to 0.5, it follows that the term $\nu_{13}^2 s_{33}$ is only a small correction factor and that the contact width depends primarily on s_{11}. Thus, the contact problem provides a good method in principle for determining s_{11}.

After pioneering work by Hadley *et al.* (1965), further experimental developments by Kotani *et al.* (1994) resulted in the apparatus shown schematically in Figure 8.8. The

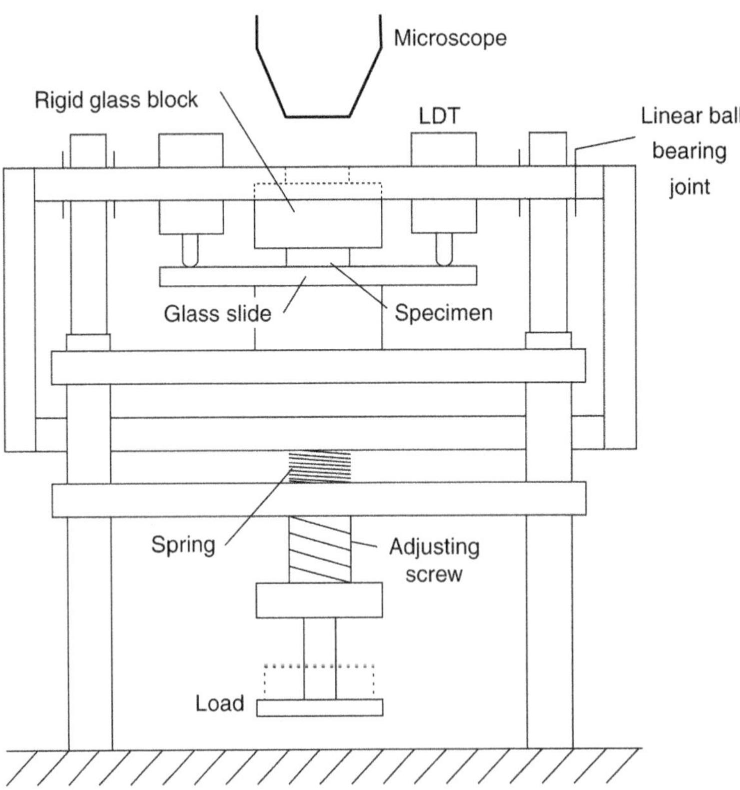

Figure 8.8 Transverse fibre compression apparatus. Kotani et al., 1994 / Springer Nature.

technique has been adopted by Stamoulis *et al.* (2005). The monofilament is compressed between two parallel glass plates on a microscope stage. Illumination is axial through the microscope normal to the visual field, resulting in the contact zone appearing as a dark rectangle. Displacement transducers were added to measure diametral compression.

It can be shown (Abdul Jawad and Ward 1978) that the total diametral compression in the direction u_1 (i.e., parallel to the direction of the applied load) is

$$u_1 = -\frac{4F}{\pi}\left(s_{11} - \frac{s_{13}^2}{s_{33}}\right)\left(0.19 + \sinh^{-1}(R/b)\right). \tag{8.3}$$

Measurements of this type were undertaken simultaneously with the contact zone observations (Kotani *et al.* 1994) and produced consistent values of $\left(s_{11} - s_{13}^2/s_{33}\right)$ for a range of polymer fibres, namely polyethylene terephthalate, polyethylene and a thermotropic liquid crystalline polymer based on hydroxybenzoic and hydroxynaphthoic acid in the ratio 73:27 (Vectra).Values for E_1 were in the range 1.94–2.34 GPa for polyethylene terephthalate, 0.63–1.50 GPa for polyethylene and 0.96–1.01 GP for Vectra. Typical contact zone observations are shown in Figure 8.9.

Similar work has been applied to oriented poly(*para*-phenyleneterephthalamide) (PPTA) fibres (Singletary *et al.* 2000) and nylon fibres (Stamoulis *et al.* 2005). The method has been reviewed and different analyses compared by Hillbrick *et al.* (2019).

8.3.2.4 The Transverse Poisson's Ratio $\nu_{12} = s_{12}/s_{11}$

The transverse Poisson's ratio can be determined by measuring u_2 the change in diameter parallel to the plane of contact in the compression of the monofilament under conditions of plane strain as described in the section on the transverse modulus above.

A simple analysis of this problem follows from the condition that the contact zone can be arranged to be small compared with the radius of the monofilament. To calculate the

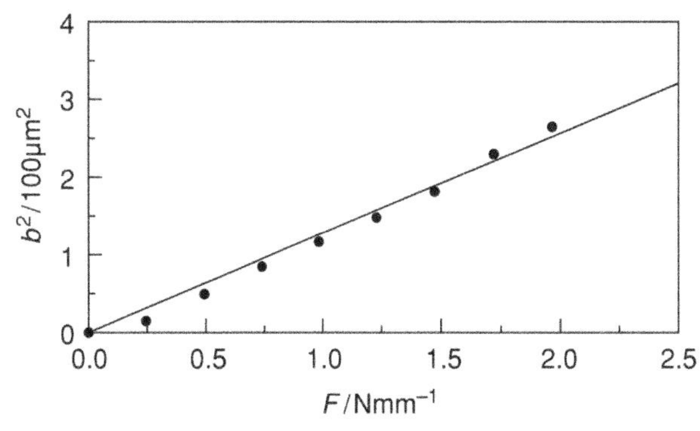

Figure 8.9 Measurement of contact width as a function of the applied force for polyethylene fibres. Kotani et al., 1994 / Springer Nature.

deformations in the diametral plane, it is then adequate to consider the problem as the compression of a cylinder under concentrated loads (Figure 8.7(b)). For an isotropic cylinder, this is a well-known problem to be found in textbooks on elasticity (see Timoshenko and Goodier 2008, p. 122). It is necessary to satisfy the boundary conditions on the surface of the cylinder, and this is done by addition of an isotropic tension in the plane perpendicular to the fibre axis.

The stresses for the transversely isotropic monofilament correspond exactly to those for the isotropic case. It is therefore very straightforward to calculate the strains and hence evaluate the diametral expansion u_2.

It is found that

$$u_2 = F\left\{\left(\frac{4}{\pi} - 1\right)\left(s_{11} - \frac{s_{13}^2}{s_{33}}\right) - \left(s_{12} - \frac{s_{13}^2}{s_{33}}\right)\right\}.$$

For most oriented monofilaments, s_{13}^2/s_{33} is small compared with s_{11}, as discussed previously. Hence, u_2 will depend primarily on s_{12} with a substantial term in s_{11}, which is about $s_{11}/4$. Thus, the measurement of the diametral expansion provides a method for determining s_{12} provided that s_{11} is determined from a measurement of the contact width b, as described in 8.3.2.4.

Measurements of u_2 have been carried out with the monofilament surrounded by an immersion liquid (Hadley *et al.* 1965), and the diameter is measured directly with a calibrated eyepiece. The immersion liquid was chosen to have refractive index approximately equal to that of the monofilament, hence reducing diffraction effects without making the monofilament invisible. Very careful focusing of the microscope was necessary in these experiments. Inaccuracy in focusing can cause errors in the diameter measurements of the order of u_2 itself.

A typical set of results for polyethylene terephthalate is shown in Figure 8.10. It can be seen that the change in diameter is proportional to the applied load, as predicted theoretically.

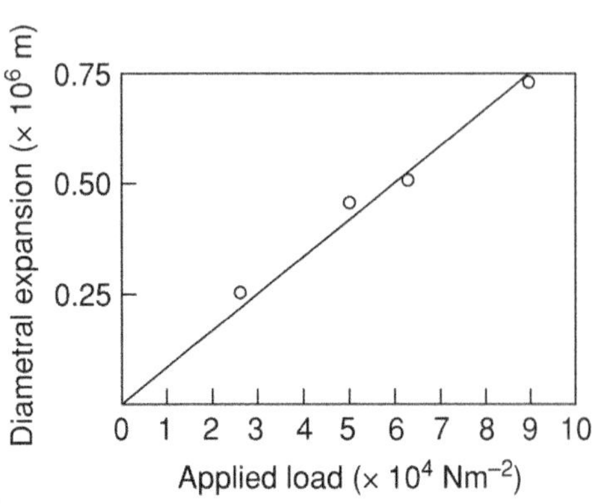

Figure 8.10 Compression of polyethylene terephthalate monofilament (diameter: 0.282 mm): measurement of diametral expansion as a function of applied load.

8.4 Development of Mechanical Anisotropy in Oriented Polymers

The review of experimental studies has two principal aims. First, we will describe early studies on low-density polyethylene and oriented monofilaments because these highlighted several unexpected features of the anisotropy, indicated the differences between individual polymers and provided the basis for the aggregate model against which many subsequent measurements have been tested. Secondly, we will provide very brief summaries of further work in this area and indicate its relevance to our present understanding.

8.4.1 Sheets of Low-Density Polyethylene

Raumann and Saunders (1961) uniaxially stretched isotropic sheets of low-density (i.e., branched) polyethylene to varying final extensions and measured the tensile modulus in directions over a range of angles to the initial draw direction. For a highly oriented sample, the plot of Young's modulus ($1/s_\theta$) against angle with the draw direction (Figure 8.11) shows the lowest stiffness at an angle close to 45° to that direction.

The general compliance equation for transverse isotropy is

$$s_\theta = s_{11}\sin^4\theta + s_{33}\cos^4\theta + (2s_{13} + s_{44})\sin^2\theta\cos^2\theta.$$

The experimental result implies that $(2s_{13} + s_{44})$ is much greater than either s_{11} or s_{33}, because when $\theta = 45°$ the terms will be equally weighted.

Replotting to obtain the modulus at a given angle as a function of draw ratio (Figure 8.12), the results are again somewhat unexpected: E_0 initially falls with increasing draw ratio, so that at low draw ratios $E_{90} > E_0$. Subsequently, Gupta and Ward (1967) showed that this unusual behaviour was specific to room-temperature measurements, and at a sufficiently low temperature, the behaviour resembled that of most other polymers (Figure 8.13).

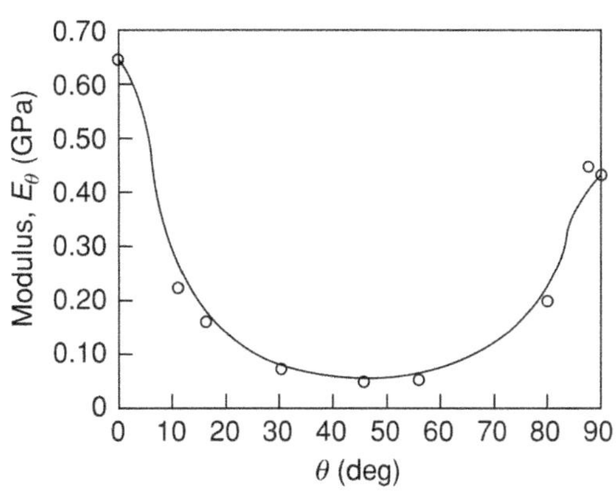

Figure 8.11 Comparison of the observed variation in modulus E_θ with angle θ to draw direction and the theoretical relation (i.e., full curve) calculated from E_0, E_{45} and E_{90} for low-density polyethylene sheet drawn to a draw ratio of 4.65. Raumann and Saunders, 1961 / IOP Publishing.

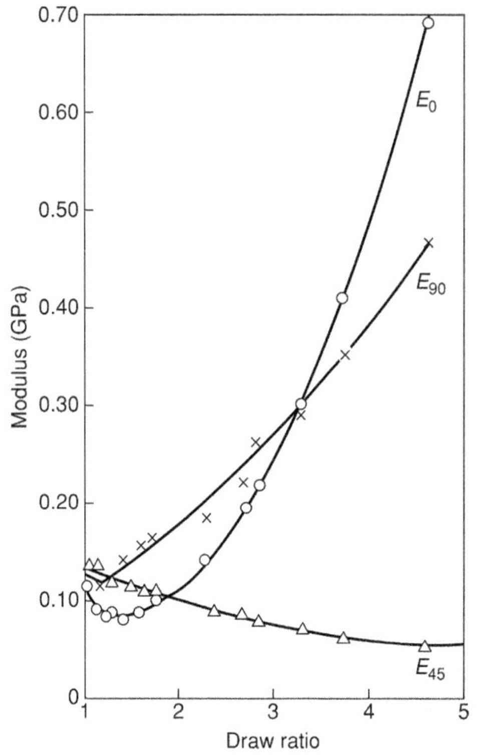

Figure 8.12 The variation in E_0, E_{45} and E_{90} with draw ratio in cold-drawn sheets of low-density polyethylene. Modulus measurements taken at room temperature. Raumann and Saunders, 1961 / IOP Publishing.

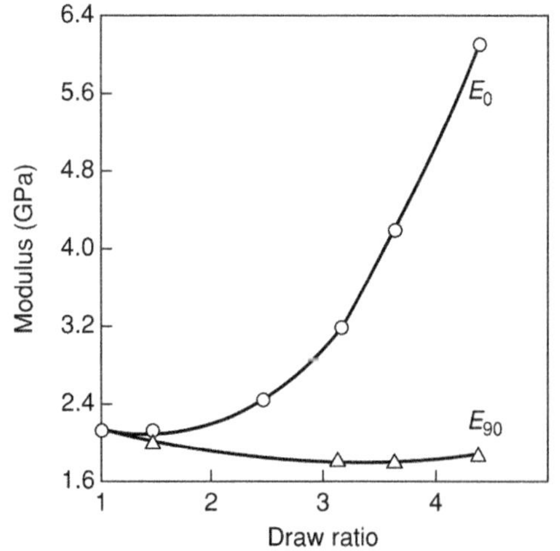

Figure 8.13 The variation in E_0 and E_{90} with draw ratio in cold-drawn sheets of low-density polyethylene. Modulus measurements taken at −125°C.

8.4.2 Filaments Tested at Room Temperature

In a comprehensive study at room temperature, Hadley *et al.* (1969) determined the five independent elastic constants for oriented filaments of polyethylene terephthalate, nylon 6:6, low- and high-density polyethylene and polypropylene. The orientation was determined in terms of draw ratio and optical birefringence. Subsequent studies indicated that it would have been appropriate to record not only the overall orientation, as derived from birefringence, but also the crystal orientation, obtainable from X-ray measurements. The results are summarized in Table 8.1 and Figures 8.14–8.18 (see Section 8.6.2 for discussion of the aggregate theory predictions).

Although the detailed development of mechanical anisotropy in these particular filaments must depend on their exact chemical composition and subsequent processing, several general features can be distinguished. The principal effect of drawing (i.e., increasing molecular orientation) is to increase Young's modulus E_3 measured along the filament axis. In nylon 6:6 and polyethylene terephthalate, there is a corresponding but small decrease in the transverse modulus E_1. For polypropylene and high-density polyethylene E_1 is almost independent of draw ratio and for low-density polyethylene E_1 increases significantly, in agreement with the results of Raumann and Saunders. Note here, too, the anomalous behaviour of this polymer at low draw ratios. Overall, E_3 for highly oriented filaments is greater than E_1, with the anisotropy being greatest for polyethylene terephthalate:

$$\frac{E_3}{E_1} = \frac{s_{11}}{s_{33}} \approx 27.$$

The shear modulus G of low-density polyethylene, and its change with orientation, provides another striking contrast with the other materials examined. A decrease in G by more than a factor of 3 over the range of orientation used compares with only small changes for the other filaments. In polyethylene terephthalate, high-density polyethylene and polypropylene $s_{44} \sim s_{11}$; in nylon $s_{44} \sim 2s_{11}$. By contrast, low-density polyethylene, at least as regards room-temperature behaviour, is exceptional, with the extensional compliance s_{33} having the same order of magnitude as the transverse compliance s_{11}, and with the shear compliance s_{44} being more than an order of magnitude greater than either s_{11} or s_{33}. Such measurements provide the basis for the discussion of relaxation transitions in Chapter 10.

In all cases, the compliance s_{13} is low and appears to decrease fairly rapidly with increasing draw ratio, in a manner comparable with s_{33}. Hence, the extensional Poisson's ratio $\nu_{13} = -s_{13}/s_{33}$ is rather insensitive to draw ratio and, with the exception of high-density polyethylene, does not differ significantly from 0.5. It is thus generally a valid approximation to assume that the filaments are incompressible. Note that for anisotropic bodies ν_{13} is not confined to a maximum value of 0.5, but is limited solely by the inequalities necessary for a positive strain energy (Mentrasti *et al.* 2021)

$$s_{12}^2 < s_{11}^2; \quad s_{13}^2 < \frac{1}{2}s_{33}(s_{11} + s_{12}).$$

Hine and Ward (1996) have used the ultrasonic immersion method (Section 6.3.2) to determine a full set of elastic constants for a range of fibres, by making uniaxially oriented fibre composites. In the first method, the oriented composites were produced by the Leeds

Table 8.1 Elastic compliances of oriented fibres (units of compliance are 10^{-10} Pa^{-1}, errors quoted are 95% confidence limits) Adapted from Hadley et al. (1969).

Material	Birefringence (Δn)	s_{11}	s_{12}	s_{33}	s_{13}	s_{44}	$\nu_{13} = -s_{13}/s_{33}$	$\nu_{12} = -s_{12}/s_{11}$
Low-density polyethylene film Raumann (1962)		22	−15	14	−7	680	0.50	0.68
Low-density polyethylene 1	0.0361	40 ± 4	-25 ± 4	20 ± 2	-11 ± 2	878 ± 56	0.55 ± 0.08	0.61 ± 0.20
Low-density polyethylene 2	0.0438	30 ± 3	-22 ± 3	12 ± 1	-7 ± 1	917 ± 150	0.58 ± 0.08	0.73 ± 0.20
High-density polyethylene 1	0.0464	24 ± 2	-12 ± 1	11 ± 1	-5.1 ± 0.7	34 ± 1	0.46 ± 0.15	0.52 ± 0.08
High-density polyethylene 2	0.0594	15 ± 1	-16 ± 2	2.3 ± 0.3	-0.77 ± 0.3	17 ± 2	0.33 ± 0.12	1.1 ± 0.14
Polypropylene I	0.0220	19 ± 1	-13 ± 2	6.7 ± 0.3	-2.8 ± 1.0	18 ± 1.5	0.42 ± 0.16	0.68 ± 0.18
Polypropylene 2	0.0352	12 ± 2	-17 ± 2	1.6 ± 0.04	-0.73 ± 0.3	10 ± 2	0.47 ± 0.17	1.5 ± 0.3
Polyethylene terephthalate 1	0.153	8.9 ± 0.8	-3.9 ± 0.7	1.1 ± 0.1	-0.47 ± 0.05	14 ± 0.5	0.43 ± 0.07	0.44 ± 0.09
Polyethylene terephthalate 2	0.187	16 ± 2	-5.8 ± 0.7	0.71 ± 0.04	-0.31 ± 0.03	14 ± 0.2	0.44 ± 0.07	0.37 ± 0.06
Nylon 6:6	0.057	7.3 ± 0.7	-1.9 ± 0.4	2.4 ± 0.3	-1.1 ± 0.15	15 ± 1	0.48 ± 0.05	0.26 ± 0.08

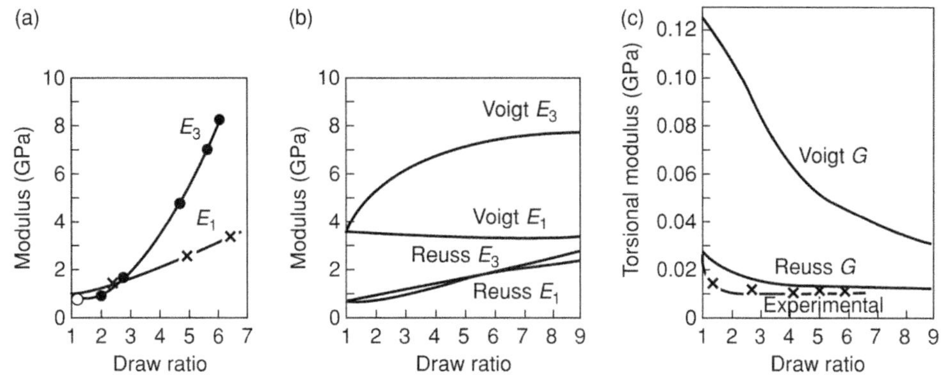

Figure 8.14 Low-density polyethylene filaments: extensional (E_3), transverse (E_1) and torsional moduli (G); comparison between experimental results and simple aggregate theory for E_3 and E_1 ((a) and (b)) and for G (c).

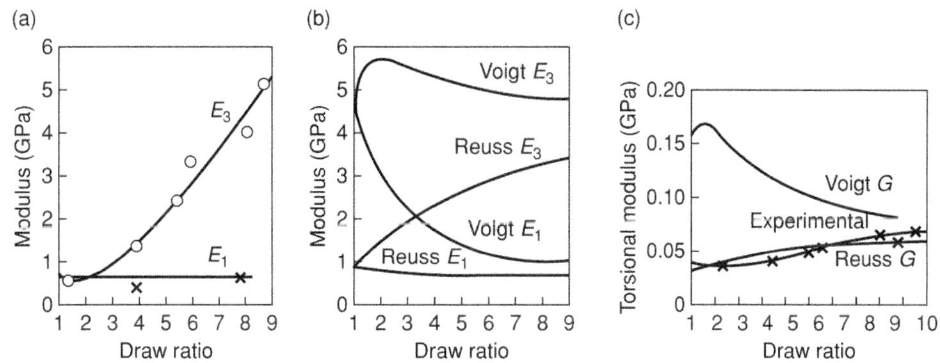

Figure 8.15 High-density polyethylene filaments: extensional (E_3), transverse (E_1) and torsional moduli (G); comparison between experimental results and simple aggregate theory for E_3 and E_1 ((a) and (b)) and for G (c).

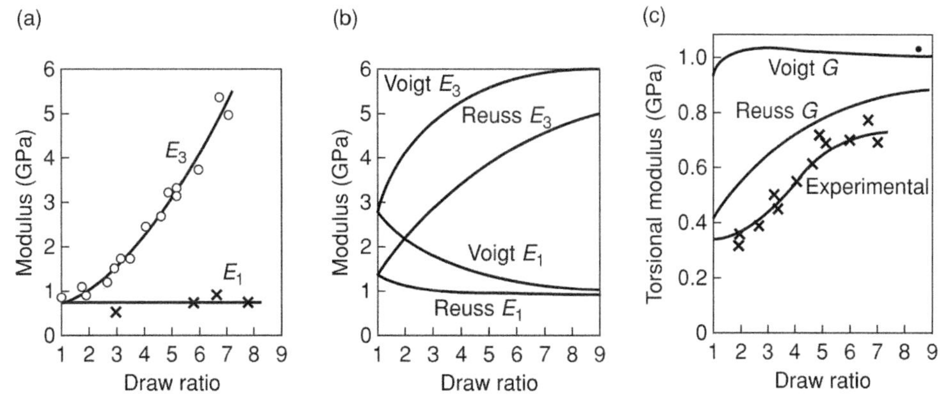

Figure 8.16 Polypropylene filaments; extensional (E_3), transverse (E_1) and torsional moduli (G); comparison between experimental results and simple aggregate theory for E3 and E1 ((a) and (b)) and for G (c).

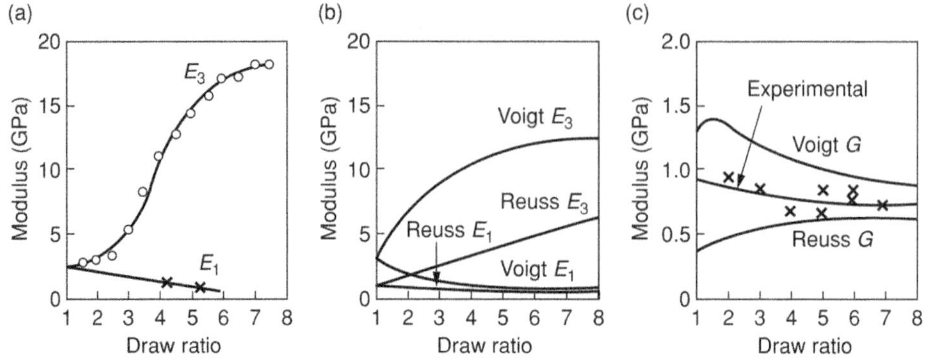

Figure 8.17 Polyethylene terephthalate filaments: extensional (E_3), transverse (E_1) and torsional moduli (G); comparison between experimental results and simple aggregate theory for E_3 and E_1 ((a) and (b)) and for G (c).

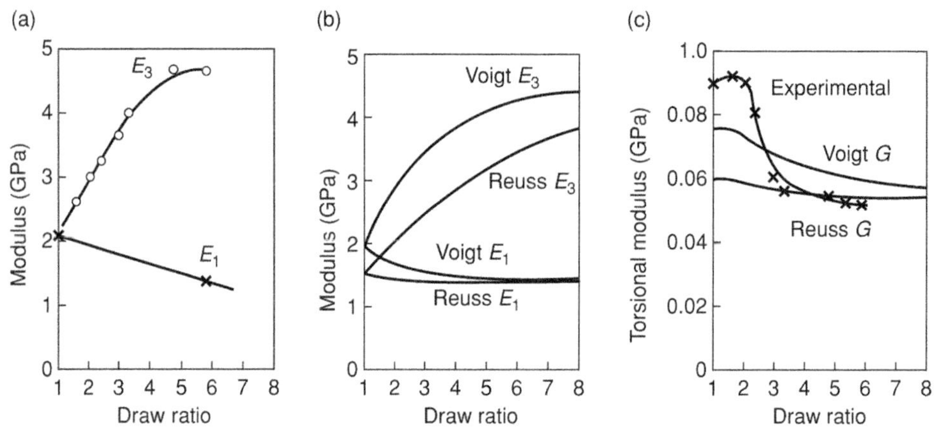

Figure 8.18 Nylon filaments: extensional (E_3), transverse (E_1) and torsional moduli (G); comparison between experimental results and simple aggregate theory for E_3 and E_1 ((a) and (b)) and for G (c).

hot compaction process. Here fibres are compacted under suitable conditions of temperature and pressure to form an homogeneous oriented material in which only a small fraction of the original fibre is melted and re-crystallized to form the matrix of the fibre composite. This matrix fraction can be removed and results extrapolated to 100% fibre fraction. The measured elastic properties for a range of fibres are shown in Table 8.2. The overall patterns of anisotropy determined by these ultrasonic measurements are similar to those found earlier by Hadley *et al.* (1969). It is particularly notable that ν_{12} is greater than 0.5, which is consistent with the view that in highly oriented fibres the high axial stiffness means that transverse compression becomes close to pure shear in the transverse plane.

Table 8.2 The elastic properties of the plates of compacted fibres.

Fibre type	E_{33} (GPa)	E_{11} (GPa)	ν_{13}	ν_{12}	G_{13} (GPa)
Tenfor polyethylene	57.7	4.68	0.45	0.55	1.63
Dyneema polyethylene	74.3	4.31	0.47	0.57	1.36
Polyethylene terephthalate	14.9	3.70	0.39	0.65	1.62
Polypropylene	11.0	2.41	0.39	0.58	1.52
Liquid crystal polymer	97.2	3.24	0.48	0.73	1.3

Standard deviation: $E_{33} \pm 3\%$; all others, $\pm 2\%$.

More recently, Wilczynski *et al.* (1995) have proposed an 'inverse calculation' method where the elastic constants of a fibre can be estimated from fibre resin composite and the elastic constants of the resin. The method was confirmed by measurements on polyethylene/epoxy and carbon fibre/epoxy resin composites. It has been applied (Brew *et al.* 1999) to the determination of the elastic constants of an organic fibre, poly {2,6-dimidazo [4,5-6′. 4′5′-e] pyridinylene – 1,4(2,5 dihydroxy)phenylene} (PIPD). This fibre is a lyotropic liquid crystalline fibre with very high Young's modulus of 285 GPa and a much higher tensile strength (5.21 GPa) and compressive strength (500 MPa) than other polyaramid fibres such as Kevlar.

The technique of Brillouin spectroscopy (Section 6.3.3) has been applied to determine the elastic constants of oriented polymer fibres. Early studies of this nature were undertaken by Krüger *et al.* (1986) and Krüger *et al.* (1991) on oriented polycarbonate films, also determining the third-order constants, which define the elastic non-linear behaviour. Wang *et al.* (1987) and Cavanagh and Wang (1982) have described measurements on oriented polyvinylidene fluoride and polychlorotrifluoroethylene films. In the latter case the results were interpreted using an aggregate model differing in detail from that of Ward discussed in Section 8.6.2.

Reported Brillouin spectroscopy measurements include those of Kumar *et al.* (2000) on uniaxially and biaxially stretched polypropylene films. Using the Ward aggregate model with a modified definition of molecular alignment, enabled elastic anisotropy and refractive index data to be quantitatively related to molecular orientation.

Choy, Leung and his colleagues, initially in collaboration with Ward, have undertaken very extensive measurements of the elastic constants of oriented polymers using the ultrasonic techniques described in Section 6.3.2. Results have been obtained for hydrostatically extruded polypropylene and polyethylene terephthalate (Chan *et al.* 1978), die-drawn polyethylene (Leung *et al.* 1984), hot-rolled nylon 66 (Choy *et al.* 1988) and polyoxymethylene (Choy *et al.* 1983), high modulus polyethylene (Choy and Leung 1985) and polypropylene (Leung and Choy 1983) produced by tensile drawing. The results have been analysed in terms of the aggregate model, which has proved a fair approximation for the low-temperature data, and the Takayanagi model with tie molecules or crystalline bridges that is more satisfactory for high-temperature data.

8.5 Interpretation of Mechanical Anisotropy: General Considerations

The mechanical anisotropy of oriented polymers is determined by the following factors, which will be discussed in turn: (1) the structure of the molecular chain and, where the polymer crystallizes, the crystal structure; (2) the molecular orientation and, in a crystalline polymer, the morphology; (3) thermally activated relaxation processes in both crystalline and non-crystalline regions.

8.5.1 Theoretical Calculation of Elastic Constants

Major developments in fibre technology have led to ultra-high modulus fibres such as poly-paraphenylene terephthalamide (PPTA) (Kevlar and Twaron) and polyethylene (Dyneema, Spectra and Certran). There has therefore been an increased interest in theoretical calculations of the elastic constants, of oriented polymers. This has been given added impetus by the development of more sophisticated computational techniques, some of which are available as commercial software packages (e.g., Accelrys (2025).

Historically, this area developed in three stages:

1) Calculation of the elastic modulus of a single polymer chain. This method has provided useful estimates of Young's modulus of an ideal perfectly oriented polymer. Only the intramolecular interactions within the polymer chain are taken into account, and only the stiffness in the chain direction can be estimated, none of the other elastic constants.

   In the simplest case, consider a planar zigzag chain of identical bonds, of which the classic example is polyethylene. There are two modes of deformation: bond stretching and valence angle opening.

   If these are defined by force constants k_s and k_v that can be estimated from infrared and Raman spectroscopy, it can be shown that the chain modulus $E_3 = \dfrac{1}{s_{33}}$ is proportional to $\left[\dfrac{\cos^2\theta}{k_s} + \dfrac{\ell^2 \sin^2\theta}{4k_v}\right]^{-1}$, where ℓ is the calculated bond length and θ is the inclination of the bond to the chain axis.

   Following the treatment presented by Treloar (1960), consider an idealized polyethylene chain with n bonds of length ℓ, each bond inclined at an angle θ to the chain axis, along which the force F is applied.

   The change in length due to the force F is given by

$$\delta L = n\delta(\ell \cos\ \theta) = n\,(\delta\ell \cos\ \theta - \ell \sin\ \theta\,\delta\theta) \tag{8.4}$$

$$\delta\ell = F \cos\ \theta / k_s. \tag{8.5}$$

   The force F also applies a torque $\frac{1}{2}F\ell \sin\theta$ that causes valence angle deformation

$$\delta\alpha = F\ell\ \sin\theta / 2k_v. \tag{8.6}$$

Since $\theta = 90^0 - \alpha/2$,

$$\delta\theta = -\delta\alpha/2 = -F\ell \ \sin\theta/4k_\nu. \tag{8.7}$$

Substituting Equations (8.5) and (8.7) into (8.4),

$$\text{we have } \delta L = nF\left[\frac{\cos^2\theta}{k_s} + \frac{\ell^2\sin^2\theta}{4k_\nu}\right]. \tag{8.8}$$

The longitudinal modulus E_3 is given by

$$E_3 = (F/A)/(\delta L/L), \tag{8.9}$$

where A is the cross-sectional area of the chain and the initial chain length $L = n\ell\cos\theta$. This gives

$$E_3 = \frac{\ell\cos\ \theta}{A}\left[\frac{\cos^2\theta}{k_s} + \frac{\ell^2\sin^2\theta}{4k_\nu}\right]^{-1}. \tag{8.10}$$

Treloar found that spectroscopic data for k_s and k_ν indicated that bond stretching and bond bending make rather similar contributions to the deformation, bond bending about 60% and bond stretching 40%. The chain modulus was estimated as 182 GPa, i.e., of a similar order to that determined experimentally by X-ray diffraction or neutron diffraction.

2) The lattice dynamical method, which was originally proposed by Born and Huang (1954), to calculate the velocity of sound waves in elastic solids and hence their elastic constants. The equation of motion of an atom is given by

$$m\frac{\partial^2 u}{\partial t^2} + fu = 0, \tag{8.11}$$

where u is the displacement from the equilibrium position and f is the force constant and is given by the second derivative of the interatomic potential energy, i.e.,

$$f = \frac{\partial^2 U}{\partial x^2}.$$

The solution to Equation (8.11) is a sound wave

$$u = u_0\exp[i\,(kx - \omega t)],$$

where $k = 2\pi/\lambda$ is the wave vector and $\omega = 2\pi\nu$ is the angular frequency. Note the similarity of Equation (8.11) to the macroscopic equation for the propagation of a plane elastic wave. In the simplest terms this is

$$\rho\frac{\partial^2 u}{\partial t^2} = \frac{\partial\sigma}{\partial x} = c\frac{\partial^2\varepsilon}{\partial x^2}.$$

The solution to this equation is also given by a wave equation with longitudinal wave of velocity w

$$w = \nu\lambda = \left(\frac{c}{\rho}\right)^{1/2},$$

where c is the appropriate stiffness constant (c_{11}, c_{22}, c_{33}, etc.)

This method is the subject of a classic textbook by Born and Huang (1954) and has been followed extensively by Odajima and Maeda (1967) (for a review see Tashiro 1993) to calculate full sets of elastic constants for polymer crystals.

We will describe the principles of this method by considering the calculation of the chain modulus E_c. For an external strain ε_i applied along the chain axis, the displacement Δx_i of each atom in the chain can be written as

$$\Delta x_i = \rho_i + W\varepsilon_i. \tag{8.12}$$

The total displacement Δx_i is the sum of two terms, one of which (the second term on the right-hand side of Equation (8.12)) is proportional to the external strain ε_i and the term ρ_i that is an internal strain. This additional displacement ρ_i occurs because the atoms in the chain can change their position when the elastic deformation energy V is minimized for a given macroscopic deformation, expressed formally in Equation (8.12) by the strain ε_i.

V is calculated on the basis of interval displacement vectors $\Delta\mathbf{R}$ given formally by

$$\Delta\mathbf{R} = \mathbf{B}_\rho\rho + \mathbf{B}_\varepsilon\varepsilon, \tag{8.13}$$

where $\mathbf{B}_\rho$ and $\mathbf{B}_\varepsilon$ are appropriate geometrical matrices. The elastic deformation energy V is minimized with respect to the internal strain ρ using an appropriate force constant matrix, to give the relationships between $\Delta\mathbf{R}$ and the macroscopic strain ε. Hence, we have V in terms of an equation, which for the one-dimensional calculation of the chain modulus E_c , can also be written as

$$V = \frac{1}{2}vE_c\varepsilon^2,$$

where v is an effective volume.

3) The energy minimization method

Many estimates of the crystal elastic constants for polymers follow the method of calculating the second derivatives of the Helmholtz Free Energy A with respect to deformation of a small volume element v_0 (Rutlede and Suter 1991; Nicholson *et al.* 1994). The stiffness constants $c_{\ell mnk}$ are defined as

$$\frac{\partial^2 A}{\partial\varepsilon_{\ell m}\partial\varepsilon_{nk}} = v_0 c_{\ell mnk}, \tag{8.14}$$

where

$$\varepsilon_{\ell m} = \frac{1}{2}\left\{\frac{\partial u_\ell}{\partial x_m} + \frac{\partial u_m}{\partial x_\ell}\right\} + \text{second-order terms.}$$

The Helmholtz Free Energy includes both intermolecular forces and intramolecular forces and also entropic contributions. The intramolecular contributions are the same as those required for the single-chain calculation, but there is more difficulty in producing force fields that include intermolecular contributions. The intermolecular contributions are typically Lennard-Jones type interactions and to obtain plausible values that are satisfactory for a range of different chemical compositions is often debatable. It is, however, possible to obtain some confirmation of their validity in a particular instance by verifying that the calculations predict the correct crystal structure, and this must be regarded as a the first step to calculating the elastic constants.

Rutledge and Suter (1991) have shown that for Kevlar the intermolecular contributions contribute significantly to the free energy, changing the minimum energy conformation of the chain and leading to significantly higher values for the tensile modulus. They also showed that entropy contributions to the free energy amounted to about 20% of the total free energy. It was concluded that for Kevlar, the tensile modulus E_3, calculated on the basis of internal energy contributions alone was about 20% higher than the true value, and ignoring the intermolecular interactions gave a value too low by a similar amount. This explained why the simple single-chain calculations produced results that were rather close to experimental values obtained by X-ray diffraction, for example.

8.5.1.1 Theoretical Values for Elastic Constants

The lattice dynamical theory and the energy minimization method are very similar because in both cases the elastic constants are determined by imposing appropriate deformations to the unit cell and reminimizing the energy, permitting the components of the structure to deform individually in a non-affine manner. In both cases, the energy involves intramolecular and intermolecular terms. The intramolecular terms involve bond stretching and bond angle bending as described above for the single-chain calculation for the linear zigzag chain of polyethylene. For a more general chain conformation such as a helix, there will also be the possibility of bond torsion. The forms chosen for these terms can differ, but the numerical values for the parameters are always optimized on the basis of infrared and Raman spectra (see, for example, Schachtschneider and Snyder 1963; Snyder and Schachtschneider 1965; Snyder and Zerbi 1967).

For the intermolecular terms, the Lennard-Jones 6–12 potential

$$V(r) = -Dr^{-6} + Er^{-12}$$

is often chosen, but the 6- exponential Buckingham potential

$$V(r) = -Ar^{-6} + B\ \exp(-cr)$$

can also be used.

The optimization of the intermolecular terms is less straightforward because these depend on the details of the crystal structure and especially the intermolecular distances. A customary procedure is therefore to use the computation modelling first to give predictions for the crystal structure based on assumptions of the structure from structure information (X-ray diffraction, infrared and Raman spectroscopy, nuclear magnetic resonance). This is then followed by the estimation of the elastic constants assuming that entropic considerations can be ignored. Essentially these give predictions for 0 K that should be corrected for thermal motion to give predictions for ambient temperature.

It is instructive to compare the results obtained by different researchers, and some of these are shown in Table 8.3 for polyethylene.

It can be seen that a rather wide range of predicted values is obtained that is partly due to choice of different force constants. The results are also sensitive to the details of the assumed

anisotropy arises through the orientation of anisotropic units of structure; and a microscopic model, proposed by Takayanagi in which the crystalline and amorphous regions are considered as two distinct phases. It is appropriate to discuss the mechanical anisotropy of polymers in terms of one or other of these models. In this chapter, those polymers whose behaviour best approximates to the Ward aggregate model will be discussed. In the following chapter, those polymers that are best considered as composite solids will be discussed, following an introductory presentation to the theoretical understanding of the mechanical behaviour of composite materials.

8.6.1 The Aggregate Model and Mechanical Anisotropy

In this model, it is proposed that the polymer can be considered as an aggregate of identical units, which in the unstretched state are oriented randomly. As orientation develops, the units rotate and become completely aligned at the maximum achievable orientation. The elastic properties of the units are those of the mostly highly aligned structure, that could be a fibre with transverse isotropy or a film with orthorhombic symmetry (Ward 1962). In this chapter, the theory will be developed for a fibre, in the first instance for the case where both the fibre and the structural units possess transverse isotropy.

It is instructive to test the appropriateness of the model in two stages:

1) Can the elastic constants of the isotropic polymer be deduced from measurements on the most highly oriented sample?
2) Does the mechanical anisotropy develop with orientation in the predicted manner? We shall consider each stage in turn.

8.6.2 Correlation of the Elastic Constants of an Oriented Polymer with Those of an Isotropic Polymer: The Aggregate Model

It is to be expected that the mechanical properties of polymers will depend on the exact details of the molecular arrangements, i.e., both the crystalline morphology and the molecular orientation, these being intimately related so that any attempt to separate their influence must be an artificial one to a greater or lesser degree. In the case of polyethylene terephthalate, it was found that the degree of molecular orientation (as measured from birefringence, for example) was the primary factor in determining the mechanical anisotropy. Table 8.6 shows some extensional and torsional modulus results for a number of polyethylene terephthalate fibres measured at room temperature. It can be seen that the influence of crystallinity on these moduli is small compared with the effect of molecular orientation on the extensional modulus. It has therefore been proposed that to a first approximation the unoriented fibre or polymer can be regarded as an aggregate of anisotropic elastic units whose elastic properties are those of the highly oriented fibre or polymer (Ward 1962; Kausch 1970). The average elastic constants for the aggregate can be obtained in two ways: either by assuming uniform stress throughout the aggregate (which will imply a summation of compliance constants) or uniform strain (which will imply a summation of stiffness constants). Because in general the principal axes of stress and strain do not coincide for an anisotropic solid, these two approaches both involve an approximation. With the first assumption of uniform stress, the strains throughout the aggregate are not uniform; with

Table 8.6 Physical properties of polyethylene terephthalate fibres at room temperature. Adapted from Pinnock and Ward, 1963.

Birefringence	X-ray crystallinity (%)	Extensional modulus (GPa)	Torsional modulus (GPa)
0	0	2.0	0.77
0	33	2.2	0.89
0.142	31	9.8	0.81
0.159	30	11.4	0.62
0.190	29	15.7	0.79

the alternative assumption of uniform strain, non-uniformity of stress occurs. It was shown by Bishop and Hill (1951) that for a random aggregate the correct value lies between the two extreme values predicted by these alternative schemes.

Consider the case of uniform stress. This can be imagined as a system of N elemental cubes arranged end-to-end forming a 'series' model (Figure 8.19). Assume that each elemental cube is a transversely isotropic elastic solid, the direction of elastic symmetry being defined by the angle θ, which its axis makes with the direction of applied external stress σ. The strain in each cube e_1 is then given by the compliance formula

$$e_1 = \left[s_{11}\sin^4\theta + s_{33}\cos^4\theta + (2s_{13} + s_{44})\sin^2\theta\cos^2\theta\right]\sigma,$$

where s_{11}, s_{33}, etc. are the compliance constants of the cube. We ignore the fact that the cubes in general distort under the applied stress and do not satisfy compatibility of strain throughout the aggregate. Then, the average strain e is

$$e = \frac{\sum e_1}{N} = \left[s_{11}\overline{\sin^4\theta} + s_{33}\overline{\cos^4\theta} + (2s_{13} + s_{44})\overline{\sin^2\theta\cos^2\theta}\right]\sigma,$$

where $\overline{\sin^4\theta}$, etc. now define the average values of $\sin^4\theta$, etc. for the aggregate of units. For a random aggregate, it is found that

$$\begin{aligned} e/\sigma &= \text{average extensional compliance} \\ &= \overline{s'_{33}} = \frac{8}{15}s_{11} + \frac{1}{5}s_{33} + \frac{2}{15}(2s_{13} + s_{44}). \end{aligned} \tag{8.15}$$

In a similar manner, the case of uniform strain can be imagined as a system of N elemental cubes stacked in a 'parallel' model (Figure 8.19(b)). For this case, the stress in each cube σ_1 is given by the stiffness formula:

$$\sigma_1 = \left[c_{11}\sin^4\theta + c_{33}\cos^4\theta + 2(c_{13} + 2c_{44})\sin^2\theta\cos^2\theta\right]e,$$

where c_{11}, c_{33}, etc. are the stiffness constants of the cube. The average stress σ is then

$$\sigma = \frac{\sum \sigma_1}{N} = \left[c_{11}\overline{\sin^4\theta} + c_{33}\overline{\cos^4\theta} + 2(c_{13} + 2c_{44})\overline{\sin^2\theta\cos^2\theta}\right]e,$$

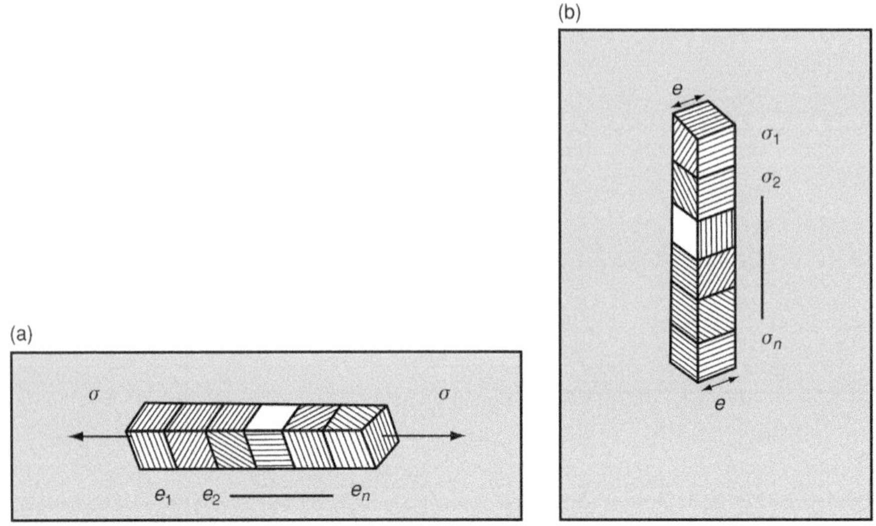

Figure 8.19 The aggregate model (a) for uniform stress and (b) for uniform strain.

where $\overline{\sin^4\theta}$, etc. are the average values of $\sin^4\theta$. For a random aggregate

$$\frac{\sigma}{e} = \overline{c'_{33}} = \frac{8}{15}c_{11} + \frac{1}{5}c_{33} + \frac{4}{15}(c_{13} + 2c_{44}). \tag{8.16}$$

Equations (8.15) and (8.16) define one compliance constant and one stiffness constant for the isotropic polymer. For an isotropic polymer, there are two independent elastic constants, and these two schemes predict a value for the isotropic shear compliance $\overline{s'_{44}}$ and the isotropic shear stiffness $\overline{c'_{44}}$, respectively. These are

$$\overline{s'_{44}} = \frac{14}{15}s_{11} - \frac{2}{3}s_{12} - \frac{8}{15}s_{13} + \frac{4}{15}s_{33} + \frac{2}{5}s_{44} \tag{8.17}$$

$$\overline{c'_{44}} = \frac{7}{30}c_{11} - \frac{1}{6}c_{12} - \frac{2}{15}c_{13} + \frac{1}{15}c_{33} + \frac{2}{5}c_{44}. \tag{8.18}$$

Averaging the compliance constants defines the elastic properties of the isotropic aggregate in terms of $\overline{s'_{33}}$ and $\overline{s'_{44}}$. This is called the 'Reuss average' (Reuss 1929). Averaging the stiffness constants defines the elastic properties of the aggregate in terms of $\overline{c'_{33}}$ and $\overline{c'_{44}}$. This is called the 'Voigt average' (Voigt 1928). In the latter case, it is desirable to invert the matrix and obtain the $\overline{s'_{33}}$ and $\overline{s'_{44}}$ corresponding to these values of $\overline{c'_{33}}$ and $\overline{c'_{44}}$ in order to compare directly the values obtained by the two averaging procedures.

The results of such a comparison are summarized in Table 8.7 for five polymers. For polyethylene terephthalate and low-density polyethylene, the measured isotropic compliances lie between the calculated bounds, suggesting that in these polymers the molecular orientation is indeed the primary factor determining the mechanical anisotropy. In nylon, the measured compliances lie just outside the bounds, suggesting that although molecular

Table 8.7 Comparison of calculated and measured extensional and torsional compliances (in units of 0.1 $(\text{GPa})^{-1}$) for unoriented fibres.

	Extensional compliance			Torsional compliance		
	$\overline{s'_{11}} = \overline{s'_{33}}$			$\overline{s'_{44}}$		
	Calculated			Calculated		
	Reuss average	Voigt average	Measured	Reuss average	Voigt average	Measured
Low-density polyethylene	139	26	81	416	80	238
High-density polyethylene	10	2.1	17	30	6	26
Polypropylene	7.7	3.8	14	23	11	2.7
Polyethylene terephthalate	10.4	3.0	4.4	25	7.6	11
Nylon	6.6	5.2	4.8	17	13	12

orientation is important in determining the mechanical anisotropy, other structural factors play an important part. Finally, in high-density polyethylene and polypropylene, measured values for the isotropic compliances $\overline{s'_{11}} = \overline{s'_{33}}$ lie well outside the calculated bounds, suggesting that factors other than orientation play a major role in the mechanical anisotropy. In polypropylene, Pinnock and Ward (1966) suggested that simultaneous changes occur in morphology and molecular mobility, both of which affect the mechanical properties.

8.6.3 The Development of Mechanical Anisotropy with Molecular Orientation

Fibres and films of intermediate molecular orientation are often produced by a two-stage process in which the first stage consists of making an approximately isotropic specimen, which is then uniaxially stretched or drawn. The aggregate model can be extended to determine the mechanical anisotropy as a function of the draw ratio.

The starting point for such a theory was the observation that in general terms the birefringence–draw ratio curves for several crystalline polymers take a similar form, as noted previously by several workers (Crawford and Kolsky (1951) for low-density polyethylene and Cannon and Chappel (1959) for nylon), with the birefringence increasing rapidly at low draw ratios, but approaching the maximum value asymptotically at draw ratios greater than about five. Results for low-density polyethylene are shown in Figure 8.20(a).

Crawford and Kolsky (1951) concluded that the birefringence was directly related to the permanent strain, and they proposed a model of rod-like units rotating towards the draw direction on drawing. The essential mathematical step in the theory is illustrated in Figure 8.21. Each unit is considered to be transversely isotropic. The orientation of a single unit is therefore defined by the angle θ between its symmetry axis and the draw direction, and the angle φ which is the angle between the projection of the symmetry axis on a plane

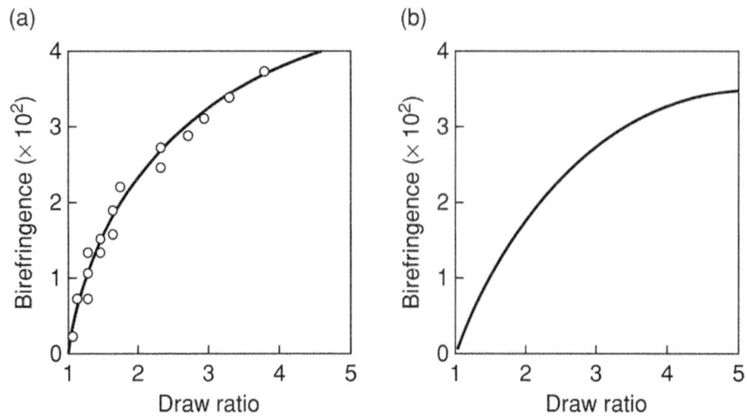

Figure 8.20 Experimental (a) and theoretical (b) curves for the birefringence of low-density polyethylene as a function of draw ratio.

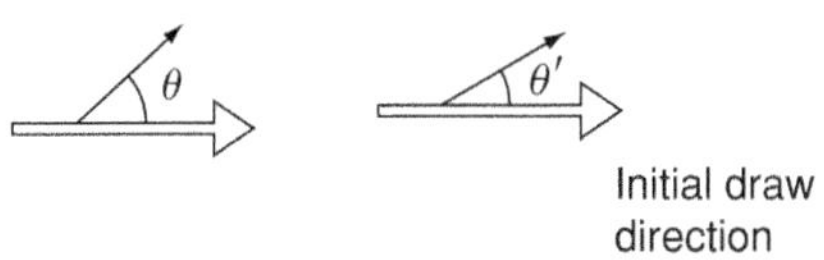

Figure 8.21 The pseudo-affine deformation scheme.

perpendicular to the draw direction and any direction in this plane. It is assumed that the symmetry axes of the anisotropic units rotate in the same manner as lines joining pairs of points in the macroscopic body, which deforms uniaxially at constant volume. This assumption is similar to the 'affine' deformation scheme of Kuhn and Grün (1942) for the optical anisotropy of rubbers (see Section 4.3.4 for a definition of 'affine'), but ignores the required change in length of the units on deformation. We will therefore call it the 'pseudo-affine' deformation scheme. Kuhn and Grün did in fact consider this scheme and rejected it in their discussion of rubber-like behaviour. The angle θ in Figure 8.21 thus changes to θ', $\phi = \phi'$, and it can be shown that

$$\tan\theta' = \frac{\tan\theta}{\lambda^{3/2}},$$

where λ is the draw ratio. This relationship can be used to calculate the orientation distribution function for the units in terms of the draw ratio.

On this model, the birefringence Δn of a uniaxially oriented polymer is given by where $\overline{\sin^2\theta}$ is the average value of $\sin^2\theta$ for the aggregate of units and $\Delta n_{\max}$ is the maximum birefringence observed for full orientation.

This pseudo-affine deformation scheme gives a reasonable first-order fit to the birefringence data for low-density polyethylene (Crawford and Kolsky 1951), nylon (Cannon and Chappel 1959), polyethylene terephthalate (Pinnock and Ward 1964) and polypropylene (Pinnock and Ward 1966). Figure 8.20(b) shows the first case. It is to be noted that this formulation of the birefringence equation ignores the distinction between different structural elements in the polymer (e.g., crystalline regions and disordered regions).With this reservation in mind, the aggregate model is now extended to predict the mechanical anisotropy in the manner outlined in Section 8.6.2. This gives the following equations for the compliance constants s'_{11}, s'_{12}, s'_{13}, s'_{33} and s'_{44} and the stiffness constants c'_{11}, c'_{12}, c'_{13}, c'_{33}, and c'_{44} of the partially oriented polymer:

$$
\begin{aligned}
s'_{11} &= \frac{1}{8}(3I_2 + 2I_5 + 3)s_{11} + \frac{1}{4}(3I_3 + I_4)s_{13} + \frac{3}{8}I_1\, s_{33} + \frac{1}{8}(3I_3 + I_4)s_{44}\\
c'_{11} &= \frac{1}{8}(3I_2 + 2I_5 + 3)c_{11} + \frac{1}{4}(3I_3 + I_4)c_{13} + \frac{3}{8}I_1\, c_{33} + \frac{1}{2}(3I_3 + I_4)c_{44}\\
s'_{12} &= \frac{1}{8}(I_2 + 2I_5 + 1)s_{11} + I_5\, s_{12} + \frac{1}{4}(I_3 + 3I_4)s_{13} + \frac{1}{8}I_1\, s_{33} + \frac{1}{8}(I_3 + I_4)s_{44}\\
c'_{12} &= \frac{1}{8}(I_2 + 2I_5 + 1)c_{11} + I_5\, c_{12} + \frac{1}{4}(I_3 + 3I_4)c_{13} + \frac{1}{8}I_1\, c_{33} + \frac{1}{2}(I_3 + I_4)c_{44}\\
s'_{13} &= \frac{1}{2}I_3\, s_{11} + \frac{1}{2}I_4 s_{12} + \frac{1}{2}(I_1 + I_2 + I_5)\, s_{13} + \frac{1}{2}I_3\, s_{33} - \frac{1}{2}I_3\, s_{44}\\
c'_{13} &= \frac{1}{2}I_3\, c_{11} + \frac{1}{2}I_4 c_{12} + \frac{1}{2}(I_1 + I_2 + I_5)\, c_{13} + \frac{1}{2}I_3\, c_{33} - 2I_3\, c_{44}\\
s'_{33} &= I_1\, s_{11} + I_2 s_{33} + I_3(2\, s_{13} - s_{44})\\
c'_{33} &= I_1\, c_{11} + I_2 c_{33} + 2I_3(c_{13} - 2c_{44})\\
s'_{44} &= (2I_3 + I_4)s_{11} - I_4 s_{12} - 4I_3 s_{13} + 2I_3\, s_{33} + \frac{1}{2}(I_1 + I_2 - 2I_3 + I_5)s_{44}\\
c'_{44} &= \frac{1}{4}(2I_3 + I_4)c_{11} - \frac{1}{4}I_4 c_{12} - I_3 c_{13} + \frac{1}{2}I_3\, c_{33} + \frac{1}{2}(I_1 + I_2 - 2I_3 + I_5)c_{44}.
\end{aligned}
\tag{8.19}
$$

In these equations, s_{11}, s_{12}, etc. are the compliance constants and c_{11}, c_{12}, etc. are the stiffness constants for the anisotropic elastic unit, which in practice means those of the most highly oriented specimen obtained. The terms I_1, I_2, I_3, I_4 and I_5, are the orientation functions, defining the average values of

$$\sin^4\theta\, (I_1),\ \cos^4\theta\, (I_2),\ \cos^2\theta\, \sin^2\theta\, (I_3),\ \sin^2\theta\, (I_4),\ \cos^2\theta\, (I_5)$$

for the aggregate. Note that only two of these orientation functions are independent parameters (e.g., $I_4 = I_1 + I_3$, $I_5 = I_2 + I_3$, $I_4 + I_5 = 1$).

The orientation functions can be calculated on the pseudo-affine deformation scheme, and Figures 8.14–8.18 show that the aggregate model then predicts the general form of the mechanical anisotropy. It is particularly interesting that the predicted Reuss average curves for low-density polyethylene show the correct overall pattern, including the minimum in the extensional modulus. This arises as follows. On the pseudo-affine deformation scheme, $\overline{\sin^4\theta}$ and $\overline{\cos^4\theta}$ decrease and increase monotonically, respectively, with increasing draw ratio, whereas $\overline{\sin^2\theta \cos^2\theta}$ shows a maximum value at a draw ratio of about 1.2. Thus, s'_{33} can pass through a maximum with increasing draw ratio (giving a minimum in Young's modulus E_0) provided that $(2s_{13} + s_{44})$ is sufficiently large compared with s_{11} and s_{33}, which should be approximately equal. The theory assumes elastic constants for the units that are identical with those measured for the highly oriented polymer.

In low-density polyethylene, s_{44} is much larger than s_{11} and s_{33}, which are fairly close in value; hence, these conditions are fulfilled and the anomalous mechanical anisotropy is predicted.

At low temperatures, as discussed earlier (see Figure 8.17), a more conventional pattern of mechanical anisotropy is observed for low-density polyethylene. These results are thus consistent with the aggregate model.

The theoretical curves of Figures 8.14–8.18 differ from those obtained experimentally in two ways. Firstly, there are features of detail (a small minimum in the transverse modulus of low-density polyethylene; a small minimum in the extensional modulus of high-density polyethylene), which are not predicted at all. It has been shown elsewhere (Frank *et al.* 1970) that such effects may be associated with mechanical twinning. Secondly, the predicted development of mechanical anisotropy with increasing draw ratio is much less rapid than is observed in practice. Deficiencies in the pseudo-affine deformation scheme are not unexpected due to the simplifying nature of the assumptions made. The quantities $\overline{\sin^4\theta}$, $\overline{\cos^4\theta}$ and $\overline{\sin^2\theta\cos^2\theta}$ can also be determined experimentally by wide-angle X-ray diffraction and nuclear magnetic resonance (Gupta *et al.* 1968; McBrierty and Ward 1968; Gupta and Ward 1970). In low-density polyethylene, a considerably improved fit was obtained in this manner (Figure 8.22). The conclusion from these results is that the mechanical anisotropy of low-density polyethylene relates to the orientation of the crystalline regions and that it is predicted to a very good degree of approximation by the Reuss averaging scheme.

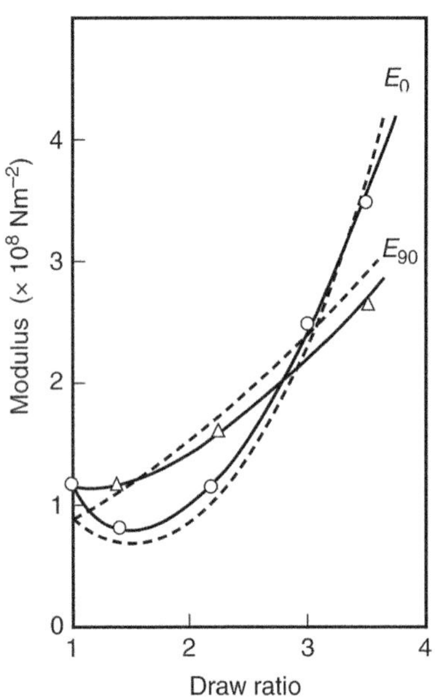

Figure 8.22 Comparison of experimental (–) variation in E_0 and E_{90} for cold-drawn low-density polyethylene with those predicted by the aggregate model using orientation functions from nuclear magnetic resonance (- - -).

The aggregate model predicts only that the elastic constants should lie between the Reuss and Voigt average values. In polyethylene terephthalate, it is clear that the experimental compliances lie approximately midway between the two bounds. For cold-drawn fibres, it has been shown that this median condition applies almost exactly (Allison and Ward 1967).

For low-density polyethylene, the Voigt averaging scheme does not predict the anomalous behaviour. However, the Reuss average does and therefore appears to describe the physical situation more closely. A similar conclusion was reached by Odajima and Maeda (1967) who compared theoretical estimates of the Reuss and Voigt averages of single crystals of polyethylene with experimental values.

In nylon, the Voigt average is closest to the experimentally observed data. It is interesting to note that both averaging schemes predict a maximum in the torsional modulus as a function of draw ratio.

The aggregate model would not appear to be generally applicable to high-density polyethylene and polypropylene. It appears that for polypropylene the aggregate model is applicable only at low draw ratios (Pinnock and ward 1966). As discussed above, there are simultaneous changes in morphology and molecular mobility at higher draw ratios.

It is interesting that in different polymers the Reuss or Voigt averages or a mean of these is closest to the measured values. It is likely that these conclusions will relate to the detailed nature of the stress and strain distributions at a molecular level in the polymers and should in turn be related to the structure.

Kausch, who has applied the aggregate model to a range of both crystalline and amorphous polymers (Kausch 1967, 1969), also noted that for some materials the experimental increase in extensional modulus with stiffness was significantly greater than predicted, and has suggested that the effect may be due to an additional orientation of segments within each unit of the aggregate. Another possibility is that the increase in stiffness of crystalline polymers may be enhanced at the higher draw ratios through the competing process of pulling out more intercrystalline tie molecules. Subsequently Kausch (1971) emphasized the value of reformulating the model in terms of a molecular network rather than orienting rods, which would allow the representation of high strain properties. In summary, it is evident that despite the highly simplistic nature of the assumptions made the aggregate model provides an appropriate model for a number of important polymers. For such materials, details of the crystal structure can play no more than a subsidiary role in the development of mechanical anisotropy, and the deformation is essentially that expected for a single-phase texture or a distorted network.

8.6.4 The Sonic Velocity

It has been suggested by Morgan (1962) and others (Charch and Moseley 1959) that the sonic modulus (i.e., the extensional modulus measured at high frequencies by a wave-propagation technique) can be used to obtain a direct measure of molecular orientation in a manner analogous to the derivation of the so-called optical orientation function $f_0 = \left(1 - \frac{3}{2}\overline{\sin^2\theta}\right)$ from the birefringence.

Consider the equations for the extensional modulus of the aggregate

$$s'_{33} = \overline{\sin^4\theta}\, s_{11} + \overline{\cos^4\theta}\, s_{33} + \overline{\sin^2\theta \cos^2\theta}\,(2s_{13} + s_{44}).$$

Table 8.1 summarizes the measured values of s_{11}, s_{33} and s_{44} for a number of polymers, as obtained from the monofilament data of Hadley *et al.* (1969). It can be seen that in all cases except that of low-density polyethylene, s_{11} and s_{44} are of approximately the same value, and that s_{33} is comparatively small. Remembering that Poisson's ratio is usually close to 0.5, this implies that s_{13} will also be comparatively small.

This suggests that, except for high degrees of orientation, both the terms $\overline{\cos^4\theta}\, s_{33}$ and $\overline{\sin^2\theta \cos^2\theta}\, s_{13}$ will be small and we can approximate to

$$s'_{11} = \overline{\sin^4\theta}\, s_{11} + \overline{\sin^2\theta \cos^2\theta}\, s_{44} \tag{8.20}$$

$$= \overline{\left(\sin^4\theta + \sin^2\theta \cos^2\theta\right)}\, s_{11} \quad = \overline{\sin^2\theta}\, s_{11}. \tag{8.21}$$

Remembering that the birefringence is given by $\Delta n = \Delta n_{\max}\left(1 - \frac{3}{2}\overline{\sin^2\theta}\right)$, it can be seen that the extensional compliance, the reciprocal of the extensional modulus, should be directly related to the birefringence through $\overline{\sin^2\theta}$ independent of the mechanism of molecular orientation (Ward 1964). To this degree of approximation, it then follows that

$$s'_{33} = \frac{2}{3}s_{11}(\Delta n_{\max} - \Delta n). \tag{8.22}$$

We would therefore predict a linear relationship between the extensional compliance s'_{33} and the birefringence Δn, which extrapolates to zero extensional compliance at the maximum birefringence value.

Figure 8.23 shows results for polyethylene terephthalate and polypropylene, which suggest that this is a reasonable approximation. But the values of s_{11} obtained from these plots do not agree with that measured experimentally for the most highly oriented fibre monofilament, suggesting that this approximate treatment is not very soundly based.

Samuels (1974) has carried the sonic velocity analysis one stage further by recognizing the two-phase nature of a crystalline polymer. The natural extension of Equation (8.21) would then be

$$\frac{1}{E} = s'_{33} = \frac{\beta}{E^0_{t,c}}\overline{\sin^2\theta_c} + \frac{1-\beta}{E^0_{t,\mathrm{am}}}\overline{\sin^2\theta_{\mathrm{am}}}, \tag{8.23}$$

where E is the sonic modulus of the sample; $E^0_{t,c}$, $E^0_{t,\mathrm{am}}$ are the lateral moduli of the crystalline and amorphous regions, respectively; $\overline{\sin^2\theta_c}$, $\overline{\sin^2\theta_{\mathrm{am}}}$ are orientation functions for the crystalline and amorphous regions, respectively and β is the fraction of crystalline material.

For an isotropic sample $\overline{\sin^2\theta_c} = \overline{\sin^2\theta_{\mathrm{am}}} = \frac{2}{3}$ and the isotropic sonic modulus E_u is given by

$$\frac{3}{2E_u} = \frac{\beta}{E^0_{t,c}} + \frac{1-\beta}{E^0_{t,\mathrm{am}}}. \tag{8.24}$$

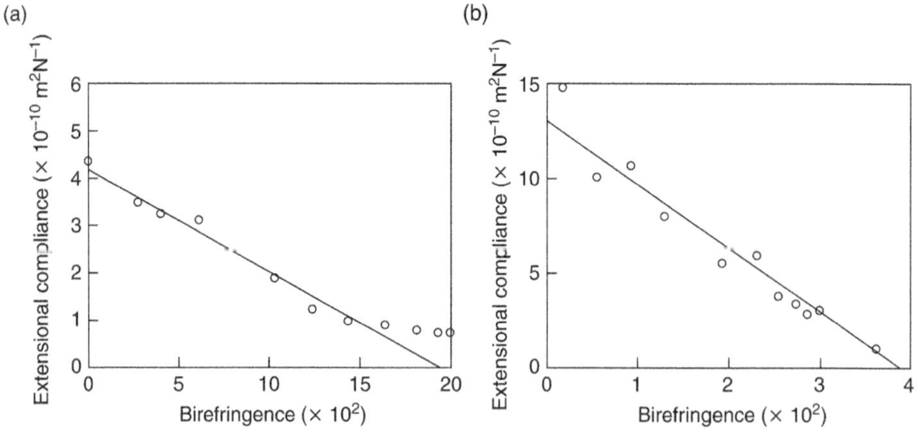

Figure 8.23 Experimental curves showing the relationship between the extensional compliance and birefringence for fibres of (a) polyethylene terephthalate and (b) polypropylene.

If we define orientation averages

$$f_c = \frac{1}{2}\left(3\overline{\cos^2\theta_c} - 1\right), \quad f_{\text{am}} = \frac{1}{2}\left(3\overline{\cos^2\theta_{\text{am}}} - 1\right)$$

for the crystalline and amorphous regions, respectively, we can combine Equations (8.23) and (8.24) to give

$$\frac{3}{2}\left\{\frac{1}{E_u} - \frac{1}{E}\right\} = \frac{\beta f_c}{E^0_{t,c}} + \frac{(1-\beta) f_{\text{am}}}{E^0_{t,\text{am}}}. \tag{8.25}$$

Measurement of the dependence of the sonic modulus on crystallinity in isotropic samples gives through Equation (8.24) a method of determining $E^0_{t,c}$ and $E^0_{t,\text{am}}$. Measurements of the sonic modulus on oriented samples then give through Equation (8.25), a method of determining the orientation function of the amorphous regions f_{am}, providing that f_c can be obtained from another technique, i.e., wide-angle X-ray diffraction.

Samuels (1974) obtained justification for this argument by combining such sonic modulus and X-ray diffraction measurements with the measurement of birefringence. Now the birefringence of a polymer on the two-phase model (ignoring form birefringence) is given by

$$\Delta n = \Delta n^0_c f_c + (1-\beta)\Delta n^0_{\text{am}} \left(\frac{1-\beta}{\beta}\right) \frac{f_{\text{am}}}{f_c}. \tag{8.26}$$

Samuels (1974) showed that plots of $\Delta n/\beta f_c$ as a function of

$$\left(\frac{1-\beta}{\beta}\right) \frac{f_{\text{am}}}{f_c}$$

gave good straight line fits for a range of polypropylene samples. This provided support for his analysis and enabled values to be deduced for Δn^0_c, Δn^0_{am}, the maximum birefringence (i.e., for a completely oriented phase) of the crystalline and amorphous regions, respectively. However, it should be noted that this treatment involves several approximations, as well as the basic assumption of homogeneous stress.

8.6.5 Amorphous Polymers

There are relatively few measurements on amorphous polymers, where the degree of mechanical anisotropy is much less than in crystalline polymers. Early studies include those of Hennig (1964) on polyvinyl chloride, polymethylmethacrylate and polystyrene and Robertson and Buenker (1964) on bisphenol A polycarbonate. The results are summarized in Table 8.8. Hennig's measurements on s_{33} and s_{11} were obtained from dynamic testing at 320 Hz, and the s_{44} measurements at 1 Hz. Robertson and Buenker used the vibrating reed technique to obtain values in the range 100–400 Hz.

A more comprehensive investigation on uniaxially oriented sheets of polymethylmethacrylate and polystyrene was undertaken by Wright *et al.* (1971) using ultrasonic measurements. The results are summarized in Figure 8.24(a) and (b), where the stiffness constants are shown as a function of the birefringence. Rawson and Rider (1974) have also reported ultrasonic data for oriented polyvinyl chloride and observed a similar degree of anisotropy to that seen in Table 8.8 from Hennig's work.

Table 8.8 Elastic compliances of oriented amorphous polymers (units of compliance are GPa^{-1}).

Material	Draw ratio	s_{33}	s_{11}	s_{44}
Polyvinyl chloride	1	0.313	0.313	0.820
	1.5	0.276	0.319	0.794
	2.0	0.255	0.328	0.781
	2.5	0.243	0.337	0.769
	2.8	0.238	0.341	0.763
	∞	0.204	0.379	0.730
Polymethyl methacrylate	1	0.214	0.214	0.532
	1.5	0.208	0.215	0.524
	2.0	0.204	0.215	0.518
	2.5	0.200	0.216	0.510
	3.0	0.196	0.217	0.505
Polystyrene	1	0.303	0.303	0.769
	2.0	0.296	0.304	0.769
	3.0	0.289	0.305	0.769
Polycarbonate	1	0.376	0.376	1.05
	1.3	0.314	0.408	0.980
	1.6	0.268	0.431	0.926

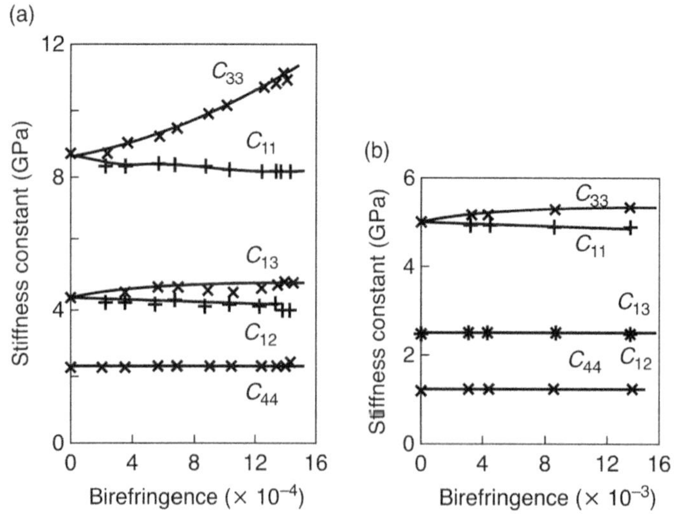

Figure 8.24 Stiffness constants of uniaxially drawn amorphous polymers, measured at room temperatures, as a function of birefringence: (a) polymethylmethacrylate, (b) polystyrene. Wright et al. (1971) / IOP Publishing.

For amorphous polymers, Kashiwigi *et al.* (1971) and Kausch (1967, 1969) and later Rawson and Rider (1974) are in agreement that the mechanical anisotropy can be discussed very satisfactorily by the aggregate model. Moreover, the development of anisotropy with draw ratio can often be described by the pseudo-affine deformation scheme (Wright *et al.* 1971).

8.6.6 Oriented Polyethylene Terephthalate Sheet with Orthorhombic Symmetry

All nine independent elastic constants have been determined for one-way drawn oriented polyethylene terephthalate sheet. The sheet was prepared by drawing isotropic sheet at constant width. It has been shown that there is then both a high degree of chain orientation in the draw direction and that the (100) crystal planes (which mainly reflect preferential orientation of the terephthalate residues in the chain) are preferentially oriented in the plane of the sheet. This type of orientation has been termed uniplanar axial. From the viewpoint of elastic anisotropy, the sheet possesses three orthogonal planes of symmetry and can be described as possessing orthorhombic symmetry.

Collected results for the nine compliance constants are shown in Table 8.9. The 3 axis is the initial draw direction and the 1 axis lies in the plane of the sheet, following the convention indicated in Figure 8.2. s_{11} and s_{33} were obtained from measurements of extensional creep in a dead loading creep machine and refer to the 10s response at 0.1% strain. s_{13} was obtained from the deformation of an electron microscope grid printed on the surface of the sample (Wilson *et al.* 1976b), and s_{12} and s_{23} by the Michelson interferometer method (Wilson *et al.* 1976a). s_{22} was determined by increasing the compressive strain of strips under load in a compressional creep apparatus (Wilson *et al.* 1976c). s_{55} was determined by the torsion of rectangular samples cut with their long axes parallel to 3 and 1, respectively; s_{44} and s_{66} were also determined in this way by making measurements on samples of different aspect ratio (Lewis and Ward 1980). In addition, s_{44} and s_{66} were determined by

Table 8.9 Full set of compliances for the oriented polyethylene terephthalate sheet with orthorhombic symmetry. Adapted from Lewis and Ward, 1980.

Compliance	Value (GPa^{-1})
s_{11}	0.361 ± 0.012
s_{22}	0.9 ± 0.16
s_{33}	0.066 ± 0.001
s_{12}	-0.38 ± 0.04
s_{13}	-0.018 ± 0.001
s_{23}	-0.037 ± 0.005
s_{44}	9.7 ± 0.3
s_{55}	0.564 ± 0.025
s_{66}	14.1 ± 0.8

the simple shear technique (Lewis *et al.*1979; Lewis and Ward 1980), and the values quoted in Table 8.9 are weighted means for the two methods.

It can be seen from Table 8.9 that the sheet shows a very high degree of mechanical anisotropy, and it is interesting to consider how this relates to the two major structural features: the high-chain axis orientation and the preferential orientation of the terephthalate residues. Infrared measurements have shown that the high degree of orientation in the crystalline regions is accompanied by a high proportion of glycol residues in the extended chain *trans* conformation and that these are also highly oriented along the draw direction (Cunningham *et al.* 1974). The low value of the extensional compliance s_{33} can then be explained by supposing that the deformation involves bond stretching and bending in these extended chain molecules. These molecules could be the taut tie molecules proposed by Peterlin (1979). The transverse compliances s_{33} and s_{33} are approximately an order of magnitude greater, which is consistent with these relating primarily to dispersion forces. It also follows from such considerations that if the polymer is stressed in a direction perpendicular to the draw direction, the major contraction is likely to take place in the direction perpendicular to the draw direction (the 2 direction) rather than parallel to it (the 3 direction). Thus, the magnitude of s_{12} would be expected to be much greater than that of s_{23}, as is observed. The value of s_{13} is similar to that of $s_{23,}$ which is consistent with this line of argument, namely the comparative difficulty of deformation in the 3 direction compared with the 1 and 2 directions. Poisson's ratios reflect the same argument and in particular ν_{13} has the very small value of 0.05.

The anisotropy of the shear compliances is also very remarkable. Both s_{44} and s_{66} are large compared with s_{55} and reflect easy shear in the 23 and 12 planes, respectively, presumably where the planar terephthalate chains are sliding over each other constrained only by weak dispersion forces. The compliance s_{55}, which geometrically involves distortion of the plane of the polyester molecule, is of a similar order of magnitude to s_{11} and s_{22}.

It is interesting to apply the aggregate model to these data, calculating bounds for the elastic constants of an 'equivalent fibre' by averaging the sheet constants in the plane normal to the sheet draw direction. This requires an extension of the mathematical treatment of Section 8.6.2 to deal with the case of a transversely isotropic aggregate of orthorhombic units. The basic equations have been given in detail elsewhere (Cunninham 1974), so only the key results will be summarized here. If the orthorhombic unit constants are s_{11}, s_{13}, ..., s_{66} (Section 8.2.2), the Reuss average fibre constants s'_{33}, s'_{13}, ..., s'_{44} obtained by averaging in the 12 plane are given by

$$s'_{33} = s_{33}$$

$$s'_{11} = \frac{3}{8}s_{11} + \frac{1}{4}s_{12} + \frac{3}{8}s_{22} + \frac{1}{8}s_{66}$$

$$s'_{12} = \frac{1}{8}s_{11} + \frac{3}{4}s_{12} + \frac{1}{8}s_{22} - \frac{1}{8}s_{66}$$

$$s'_{13} = \frac{1}{2}(s_{13} + s_{23})$$

$$s'_{44} = \frac{1}{2}(s_{44} + s_{55})$$

and the Voigt average fibre constants in similar terms are c'_{33}, c'_{13}, ..., c'_{44}, where

$$c'_{33} = c_{33}$$

$$c'_{11} = \frac{3}{8}c_{11} + \frac{1}{4}c_{12} + \frac{3}{8}c_{22} + \frac{1}{2}c_{66}$$

$$c'_{12} = \frac{1}{8}c_{11} + \frac{3}{4}c_{12} + \frac{1}{8}c_{22} - \frac{1}{2}c_{66}$$

$$c'_{13} = \frac{1}{2}(c_{13} + c_{23})$$

$$c'_{44} = \frac{1}{2}(c_{44} + c_{55}).$$

The results of this calculation are shown in Table 8.10 together with the experimental value obtained for a highly oriented fibre monofilament. Although the experimental values do not always lie exactly within the predicted bounds, they are always in the correct range. In Table 8.10, a comparison is given between the calculated and measured compliance constants for isotropic polyethylene terephthalate based on the sheet data. Again the measured values lie between the Reuss and Voigt bounds. Taking into account the very large degree of anisotropy and the very simplistic nature of these calculations, it is considered that these results afford good support for the contention that to a first approximation the mechanical anisotropy can be considered in terms of the single-phase aggregate model.

Table 8.10 Comparison of calculated and measured compliance constants (GPa^{-1}) for polyethylene terephthalate fibres based on the sheet compliances.

	Calculated bonds		
Compliance constant	**Reuss**	**Voigt**	**Experimental value**
Highly oriented fibres			
s_{11}	2.1	0.73	1.61
s_{12}	−1.9	−0.55	−0.58
s_{13}	−0.028	−0.025	−0.031
s_{33}	0.066	0.066	0.071
s_{44}	5.1	1.07	1.36
Isotropic fibres			
s_{33}	1.8	0.24	0.44
s_{44}	5.3	0.64	1.1

8.7 The Aggregate Model for Chain-Extended Polyethylene and Liquid Crystalline Polymers

Annealing polyethylene at high temperatures (~230–240°C) and high pressures (~200 MPa) produces a chain-extended structure consisting of small domains where the crystal thicknesses are ~2 μm, i.e., an aggregate of small, highly aligned units. This material can be aligned by hydrostatic extrusion, where a billet is extruded in the solid phase through a conical die to give a product with moderately high orientation and Young's moduli of ~40 GPa in the alignment direction. Not surprisingly, the aggregate model can be applied to the development of orientation, which follows the pseudo-affine deformation scheme, leading to an understanding of the mechanical anisotropy (Maxwell *et al.* 1996).

On the compliance averaging scheme, the tensile modulus of the oriented polymer E_3 is given by

$$\frac{1}{E_3} = s'_{33} = s_{11}\,\overline{\sin^4\theta} + s_{33}\,\overline{\cos^4\theta} + (2s_{13} + s_{44})\,\overline{\sin^2\theta\cos^2\theta}. \tag{8.27}$$

As before, θ represents the angle between the unit of the aggregate and the axis representing total alignment, and $\overline{\sin^4\theta}$ and so on represent average values.

For the very high degree of molecular orientation found in these polymers $\sin^4\theta \ll 1$ and $\overline{\cos^4\theta} \approx \overline{\cos^2\theta} \approx 1$. We can then rewrite Equation (8.27) as

$$\frac{1}{E_3} = s'_{33} = \frac{1}{E_c} + s_{44}\,\overline{\sin^2\theta}, \tag{8.28}$$

where E_c is the tensile modulus of the aggregate unit. To a similar degree of approximation, it can be shown that

$$s'_{44} = s_{44} = \frac{1}{G}, \tag{8.29}$$

where G is the shear modulus of the polymer.

Combining Equations (8.28) and (8.29)

$$\frac{1}{E_3} = \frac{1}{E_c} + \frac{\overline{\sin^2\theta}}{G}. \tag{8.30}$$

Figure 8.25 show results for a plot of $1/E_3$ versus $1/G$ for hydrostatic extrusion of pressure-annealed polyethylene taken to a series of extrusion ratios (equivalent to draw ratios). It can be seen that the gradients of the fitted lines decrease with increasing extrusion ratio as the orientation parameter $\overline{\sin^2\theta}$ decreases, and that the lines converge to a value of $E_c \sim 250$ GPa, in the correct range for the chain modulus of polyethylene.

The aggregate model also has been used with success to describe the mechanical anisotropy of several liquid crystalline polymers. Ward and co-workers (Davies and Ward 1988; Troughton *et al.* 1989; Green *et al.* 1990) examined the dynamic mechanical behaviour of several thermotropic polyesters in tension and shear over a wide temperature range, and used the single-phase aggregate model to relate quantitatively the fall in tensile modulus with temperature to the corresponding fall in shear modulus.

Figure 8.26 shows data for the temperature dependence for the tensile modulus and the shear modulus for a highly oriented thermotropic copolyester, on a plot of $1/E_3$ versus $1/G$.

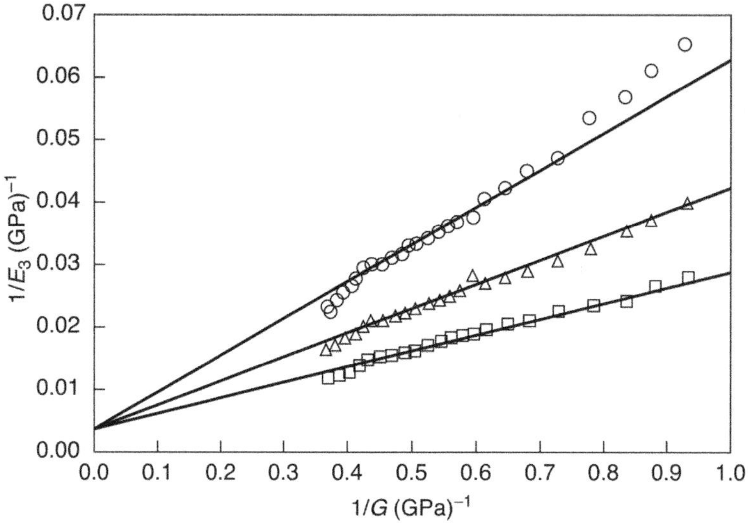

Figure 8.25 A plot of $1/E_3$ against $1/G$ showing the validity of the aggregate model analysis for a range of R006-60 extrudates previously pressure annealed at 238°C for 15 min at 450 MPa pressure: (□)10:1; (Δ) 7:1; (○) 5:1. (Reproduced from Powell et al. (1990) The structure and properties of oriented chain-extended polyethylene J. Mater. Sci., **25**, 3990. Copyright (1990) Springer Science and Business Media.)

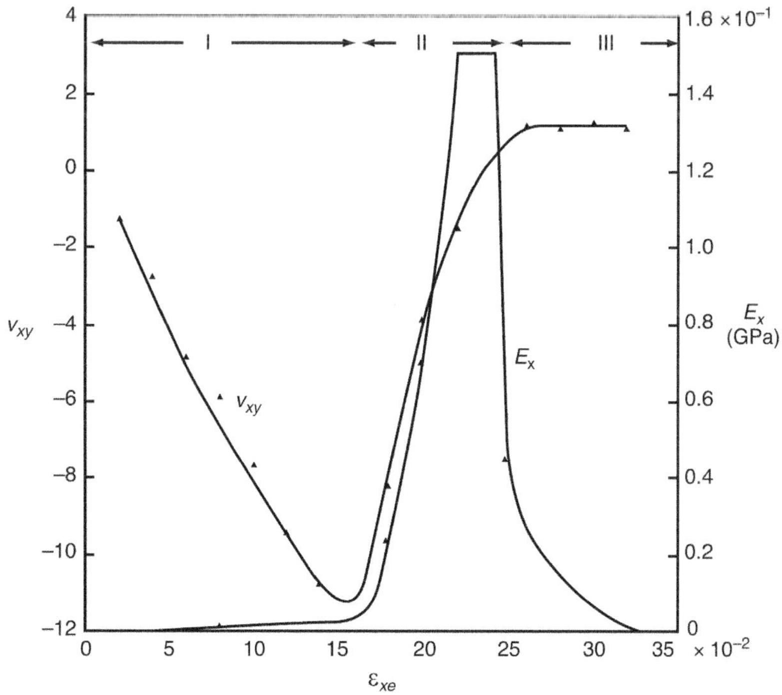

Figure 8.26 Plot of Poisson's ratio ν_{xy} (= ν_{12}) and Young's modulus E_x (= E_1) against engineering strain ε_{xe} showing the three regions of behaviour: I, low modulus; II, increasing modulus; and III plastic. (Reproduced from Caddock, B.D. and Evans, K.E. (1989) Microporous materials with negative Poisson's ratios. I. Microstructure and mechanical properties. J. Phys. D. Appl. Phys., **22**, 1877. Copyright (1989) Institute of Physics.)

The results lie on a reasonable straight line extrapolating to a value of 173 GPa for the tensile modulus of a unit of aggregate. This value compares favourably with theoretical estimates based on bond stretching and bond bending modes of deformation.

In a second version of the aggregate model, Ward and co-workers assumed, on the basis of observation of the X-ray diffraction pattern, that the aggregate unit averages the deformation over a length of 8–10 monomer units. The chain modulus in this case can be determined experimentally by measuring the change in the X-ray diffraction pattern under stress, and a temperature-dependent E_c was observed. By rearrangement of Equation (8.30) a further plot of

$$\left(\frac{1}{E_3} - \frac{1}{E_c}\right) \text{ versus } \frac{1}{G}$$

was obtained (Figure 8.27), and this gave a good straight line through the origin.

The value of $\overline{\sin^2\theta}$ obtained from the slope of this line is less than that found from the fitting procedure of Figure 8.28, which reflects the misalignment along the length of the liquid crystalline polymer chain. Because of the sinuous nature of the polymer chains, the authors have called the effect the 'sinuosity'.

Northolt and Van Aartsen (1977) used the aggregate model to interpret the rapid increase in Young's modulus with increasing crystalline orientation for PPTA fibres. PPTA is a

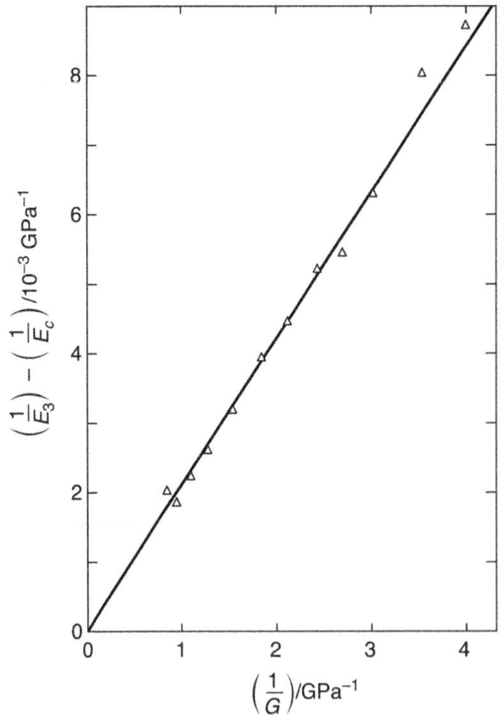

Figure 8.27 Plot of $(1/E_3 - 1/E_c)$ versus $1/G$ for a highly oriented thermotropic copolyester.

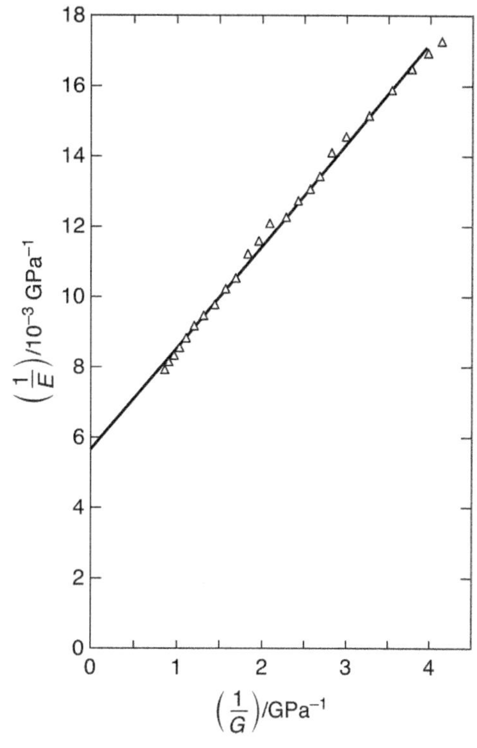

Figure 8.28 Plot of $1/E_3$ versus $1/G$ for a highly oriented thermotropic copolyester.

lytropic liquid crystalline polyester whose X-ray diffraction pattern suggests a crystallite size of about 70 nm in the axial direction and 5 nm transverse.

The sonic modulus E_{sonic} and the X-ray data for $\overline{\sin^2\theta}$ were fitted on the assumption that PPTA fibres can be considered an aggregate of small crystals (similar to chain-extended polyethylene discussed above). We have

$$\frac{1}{E_{sonic}} = \frac{1}{E_c} + A\overline{\sin^2\theta}. \tag{8.31}$$

It was shown that the measured value for E_c and A agreed well with those calculated theoretically. For high orientation $A = 1/G$, so that Equation (8.31) is exactly equivalent to Equation (8.30).

In more recent research, Northolt and van der Hout (1985) have proposed explicit theoretical models for the elastic modulus of a fibre consisting of fibrils containing rigid-rod chains. These models take as their starting point a planar zigzag chain. Deformation can occur by extensional strain along the axes of the rigid-chain segments (e.g., PQ in Figure 8.29) and by shear. The extensional compliance of the fibre s_{33} is given by

$$s_{33} = \frac{1}{E_c} + \frac{\langle \sin^2\theta \rangle}{2G},$$

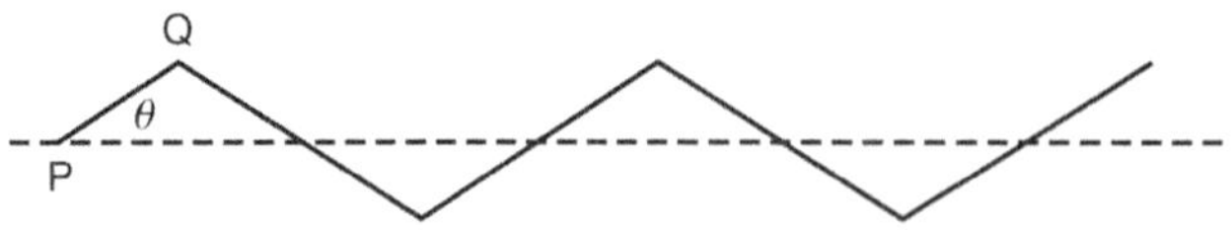

Figure 8.29 Deformation of a rigid chain on the Northolt/Van Aartsen model.

where E_c is the axial Young's modulus of each segment, G is the shear modulus of the fibre and

$$\langle \sin^2\theta \rangle = \frac{\int_0^{\pi/2} R'(\theta)\sin^2\theta \cos\theta \, d\theta}{\int_0^{\pi/2} R(\theta_0)\cos\theta_0 d\theta_0}.$$

Note that this equation differs from the classical aggregate model equation in two respects:

1) The denominator of the second term as the left-hand side is $2G$ and not G.
2) The definition of $<\sin^2\theta>$ differs from that for $\sin^2\theta$ in that the average is carried out in the plane of the chains and is not a three-dimensional average; $R(\theta_0)$ refers to the initial chain distribution and $R'(\theta)$ the final distribution under stress.

8.8 Auxetic Materials: Negative Poisson's Ratio

Although the possibility of materials with a negative Poisson's ratio had been recognized theoretically, it is only comparatively recently, since the mid-1980s, that examples of such materials have become available. These materials expand laterally when stretched and contract laterally when compressed, and are called auxetic from the classical Greek word auxetos meaning increase.

A simple example of a negative Poisson's ratio material is a cellular foam based on a re-entrant honeycomb, shown in Figure 8.30. On stretching, the axial fibrils straighten to produce lateral expansion. Other theoretical auxetic structures have since been studied, including rotating rhombi (Attard and Grima 2008), rotating rigid parallelograms (Attard *et al.* 2009), rigid rectangles of differing sizes (Grima *et al.* 2011), deformable squares (Grima *et al.* 2008) and interlocking hexagons (Ravira *et al.* 2007). Fozdar *et al.* (2011) used a form of three-dimensional printing to produce regular structural geometries, both single- and multilayer, and investigate their auxetic properties experimentally. The subject of geometrical structures and models in relation to auxetic polymers has been reviewed by Liu and Hu (2010).

Since the pioneering work of Lakes (1987), there have been a number of instances of auxetic polymer foams, for instance Binchi *et al.* (2008), Binchi *et al.* (2010) and Xu and Li (2011). Caddock and Evans (1989) have shown how an anisotropic microstructure consisting of nodules and fibrils can be produced in polytetrafluoroethylene (PTFE), and give rise to a very large negative Poisson's ratio. Figure 8.31 is a schematic representation of the deformation of microporous PTFE.

Figure 8.30 Re-entrant honeycomb with negative Poisson's ratio made of bendable ligaments. A similar structure can be made with rigid ligaments if a spring is placed between points of types A and B. Evans and Caddock, 1989 / IOP Publishing.

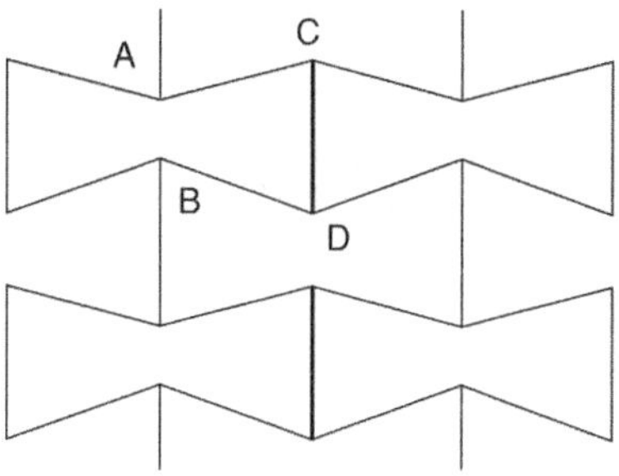

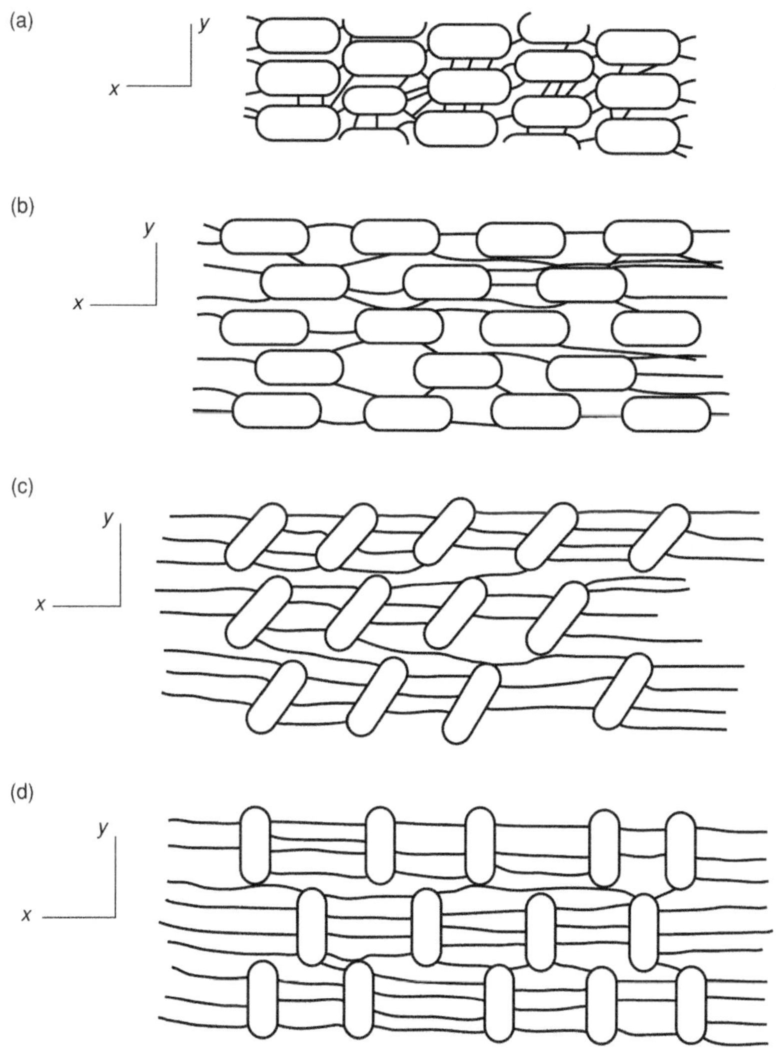

Figure 8.31 A schematic representation of the structural changes observed in microporous polytetrafluoroethylene undergoing tensile loading in the x direction: (a) initial dense microstructure, (b) tension in fibrils causing transverse displacement of anisotropic nodal particles with lateral expansion, (c) rotation of nodes producing further lateral expansion and (d) fully expanded structure prior to further, plastic deformation due to node break-up. Evans and Caddock, 1989 / IOP Publishing.

Initially, there is expansion in both the axial and lateral directions as the fibrils are stretched, similar to the deformation of the re-entrant structure of Figure 8.30. This is followed by rotation of the nodules with a particular handedness as shown in the figure, to give further lateral expansion. Evans and Caddock (1989) showed how two theoretical models, a translational model for stages (a) to (b) of Figure 8.31, and a rotational model for stages (b), (c) and (d), could provide a good confirmation of the observed changes in Poisson's ratio with strain. Their results are illustrated in Figure 8.26, from which it can be seen that experimental values of Poisson's ratio approaching −12 can be observed.

Further studies (Alderson and Evans 1992; Evans and Alderson 1992; Neale et al. 1995) showed that in both PTFE and ultrahigh-molecular-weight polyethylene (UHMPE) an isotropic microstructure of nodules and fibrils shown schematically in Figure 8.32 can give rise to a negative Poisson's ratio. Essentially this arises because when the materials are stretched, the extension of the fibrils causes the nodules to move apart.

The work on auxetic polymers arising from microporous structure has continued. Alderson *et al.* (2002), in an attempt to produce auxetic material in a more easily usable form, have used melt-spinning to produce auxetic polypropylene fibres. This work was developed further, by way of a study of the processing parameters for melt-spinning of auxetic polypropylene, polyester and nylon fibres (Ravirala *et al.* 2005; Alderson *et al.* 2007) have produced auxetic polypropylene film using melt extrusion. Less conventionally, Alderson *et al.* (2005) have produced auxetic polyethylene by a combination of powder compaction and sintering, without an extrusion step.

As a practical means of creating material with negative Poisson's ratio, it has been proposed to use auxetic networks of polymer fibres within a conventional matrix to produce composites that are themselves auxetic. Here the individual fibres are not in general auxetic, but the overall network structure is. Material in sheet form, consisting of fibres compressed together, has been shown by experiment to possess negative out-of-plane Poisson's ratios, an example being paper. Tatlier and Berhan (2009) used finite element modelling of three-dimensional compressed fibre systems and found that negative Poisson's ratios were predicted. The physical basis of the effect is that, during the compressive forming process, initially straight fibres become bent over other non-parallel fibres at contact points; then, on

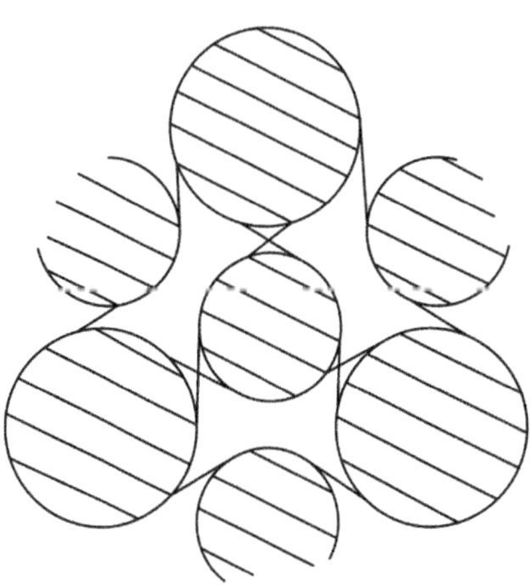

Figure 8.32 Schematic diagram of an isotropic microstructure consisting of nodules and fibrils that causes a negative Poisson's ratio when pulled in any direction. Evans and Alderson, 1992 / Springer Nature.

the application of tension, the fibres are deformed back to their initial straight configurations, producing a lateral expansion. Following the modelling work, Jayanty *et al.* (2011) infused auxetic sintered metal fibre networks with polymer and successfully produced an auxetic composite. Also, they produced polymer composites using carbon nanofibres that also proved to be auxetic, even though the fibres were not originally in a fused network. This was attributed to the fibres forming an effective network as a result of proximity rather than true contact.

References

Abdul Jawad, S. and Ward, I.M. (1978) Transverse compression or oriented nylon and polyethylene extrudates. *J. Mater. Sci.*, **13**, 1381.

Accelrys. (2025) *Materials Studio 7*, Accelrys Inc., San Diego, CA.

Alderson, K.L. and Evans, K.E. (1992) The fabrication of microporous polyethylene having a negative Poisson's ratio. *Polymer*, **33**, 4435.

Alderson, K.L., Alderson, A., Smart, G. *et al.* (2002) Auxetic polypropylene fibres: part 1 – manufacture and characterization. *Plast. Rubber Compos.*, **31**, 344–349.

Alderson, K.L., Webber, R.S., Kettle, A.P. *et al.* (2005) Novel fabrication route for auxetic polyethylene. Part 1. Processing and microstructure. *Polym. Eng. Sci.*, **45**, 568–578.

Alderson, K.L., Alderson, A., Davies, P.J. *et al.* (2007) The effect of processing parameters on the mechanical properties of auxetic polymeric fibers. *J. Mater. Sci.*, **42**, 7991.

Allison, S.W. and Ward, I.M. (1967) The cold drawing of polyethylene terephthalate. *Br. J. Appl. Phys.*, **18**, 1151.

Attard, D. and Grima, J.N. (2008) Auxetic behaviour from rotating rhombi. *Phys. Status Solidi B*, **245**(S1), 2395–2404.

Attard, D., Manicaro, E. and Grima, J.N. (2009) On rotating rigid parallelograms and their potential for exhibiting auxetic behaviour. *Phys. Status Solidi B*, **246**, 2033–2044.

Bianchi, M., Scarpa, F.L. and Smith, C.W. (2008) Stiffness and energy dissipation in polyurethane auxetic foams. *J. Mater. Sci.*, **43**, 5851–5860.

Bianchi, M., Scarpa, F.L. and Smith, C.W. (2010) Shape memory behaviour in auxetic foams: mechanical properties. *Acta Mater.*, **58**, 858–865.

Biot, M.A. (1939) Increase of torsional stiffness of a prismatical bar due to axial tension. *J. Appl. Phys.*, **10**, 860.

Biot, M.A. (1955) *International Union of Theoretical and Applied Mechanics Colloquium (Madrid)*, Springer-Verlag, Berlin, p. 251.

Bishop, J. and Hill, R. (1951) A theory of the plastic distortion of a polycrystalline aggregate under combined stresses. *Philos. Mag.*, **42**, 1248.

Born, M. and Huang, K. (1954) *Dynamical Theory of Crystal Lattices*, Clarendon Press, Oxford.

Brew, B., Hine, P.J. and Ward, I.M. (1999) The properties of pipd-fibre/epoxy composites. *Compos. Sci. Technol.*, **59**, 1109.

Caddock, B.D. and Evans, K.E. (1989) Microporous materials with negative Poisson's ratios. I. Microstructure and mechanical properties. *J. Phys. D. Appl. Phys.*, **22**, 1877.

Cannon, C.G. and Chappel, F.C. (1959) Effect of temperature and moisture content on the orientation produced by the drawing of nylon 66. *Br. J. Appl. Phys.*, **10**, 68.

Cavanagh, D.B. and Wang, C.H. (1982) Studies of the effect of orientation on the elastic constants of stretched and hydrostatically extruded polychlorotrifluoroethylene: a reorientational model in semicrystalline polymers. *J. Polym. Sci. Polym. Phys. Edn.*, **20**, 1647.

Chan, O.K., Chen, F.C., Choy, C.L. *et al.* (1978) Elastic constants of extruded polypropylene and polyethylene terephthalate. *J. Phys. D. Appl. Phys.*, **11**, 617.

Charch, W.H. and Moseley, W.W. (1959) Structure-property relationships in synthetic fibers: part I: structure as revealed by sonic observations. *Text. Res. J.*, **29**, 525.

Chen, H., Gomez, J. and Pindera, M.-J. (2020) Saint Venant's torsion of homogeneous and composite bars by the finite volume method. *Compos. Struct.*, **242**, 112128.

Choy, C.L. and Leung, W.P. (1985) Elastic moduli of ultradrawn polyethylene. *J. Polym. Sci. Polym. Phys. Ed.*, **23**, 1759.

Choy, C.L., Leung, W.P. and Huang, C.E. (1983) Elastic moduli of highly oriented polyoxymethylene. *Polym. Eng. Sci.*, **23**, 910.

Choy, C.L., Leung, W.P., Ong, E.L. *et al.* (1988) Mechanical anisotropy in rolled nylon 66. *Polym. Sci. B, Polym. Phys. Ed.*, **26**, 1569.

Crawford, S.M. and Kolsky, H. (1951) Stress birefringence in polyethylene. *Proc. Phys. Soc. B*, **64**, 119.

Cunningham, A. (1974) The structure and properties of oriented polyethylene terephthalate. Ph. D. thesis. Leeds University.

Cunningham, A., Ward, I.M., Willis, H.A. *et al.* (1974) An infra-red spectroscopic study of molecular orientation and conformational changes in poly(ethylene terephthalate). *Polymer*, **15**, 749.

Darras, O., Ducett, R.A., Hine, P.J. and Ward, I.M. (1995) Anisotropic elasticity of oriented polymer materials. *Compos. Sci. Technol.*, **55**, 131–138.

Davies, G.R. and Ward, I.M. (1988) Structure and properties of oriented thermotropic liquid crystal polymers in the solid state, in *High Modulus Polymers*, (eds A.E. Zachariades and R.S. Porter), Marcel Dekker, New York, pp. Chap. 2, 37–Chap. 70.

Davis, V.V. (1959) A diffraction method for the measurement of diameter changes in strained fibres. *J. Text. Inst.*, **50**, 688.

Ecsedi, I. and Baksa, A. (2023) Saint-Venant torsion of cylindrical orthotropic bar. *Arch. Appl. Mech.*, **93**, 2025–2032.

Evans, K.E. and Alderson, K.L. (1992) The static and dynamic moduli of auxetic microporous polyethylene. *J. Mater. Sci. Lett.*, **11**, 1721.

Evans, K.E. and Caddock, B.D. (1989) Microporous materials with negative Poisson's ratios. II. Mechanisms and interpretation. *J. Phys. D. Appl. Phys.*, **22**, 1883.

Feng, R. and Farris, J. (2002) Linear thermoelastic characterization of anisotropicpoly(ethylene terephthalate) films. *J. Appl. Polym. Sci.*, **86**, 2937–2947.

Folkes, M.J. and Arridge, R.G.C. (1975) The measurement of shear modulus in highly anisotropic materials: the validity of St. Venant's principle. *J. Phys. D*, **8**, 1053.

Fozdar, D.Y., Soman, P., Lee, J.W. *et al.* (2011) Three-dimensional polymer constructs exhibiting a tunable negative Poisson's ratio. *Adv. Funct. Mater.*, **21**, 2712–2720.

Frank, F.I. and Ruoff, A.L. (1958) A method of measuring Poisson's ratio of fibers 1. *Text. Res. J.*, **28**, 213.

Frank, F.C., Gupta, V.B. and Ward, I.M. (1970) Effect of mechanical twinning on tensile modulus of polyethylene. *Philos. Mag.*, **21**, 1127.

Green, D.I., Unwin, A.P., Davies, G.R. and Ward, I.M. (1990) An aggregate model for random liquid crystalline copolyesters. *Polymer*, **31**, 579.

Grima, J.N., Farrugia, P.S., Caruana, C. *et al.* (2008) Auxetic behaviour from stretching connected squares. *J. Mater. Sci.*, **43**, 5962–5971.

Grima, J.N., Manicaro, E. and Attard, D. (2011) Auxetic behaviour from connected different-sized squares and rectangles. *Proc. R. Soc. A*, **467**, 439–458.

Gupta, V.B. and Ward, I.M. (1967) Mechanical and optical anisotropy in low-density polyethylene. *J. Macromol. Sci. B*, **1**, 373.

Gupta, V.B. and Ward, I.M. (1970) Crystallite orientation in low-density polyethylene and its effect on mechanical anisotropy. *J. Macromol. Sci. B*, **4**, 453.

Gupta, V.B., Keller, A. and Ward, I.M. (1968) The effect of crystallite orientation on mechanical anisotropy in low-density polyethylene. *J. Macromol. Sci. B*, **2**, 139.

Hadley, D.W., Ward, I.M. and Ward, J. (1965) Transverse compression of anisotropic fibre monofilaments. *Proc. Roy. Soc. A*, **285**, 275.

Hadley, D.W., Pinnock, P.R. and Ward, I.M. (1969) Anisotropy in oriented fibres from synthetic polymers. *J. Mater. Sci.*, **4**, 152.

Hamza, A.A., Sokkar, T.Z.N. and El-Farahaty, K.A. (2010) On-line opto-viscoelastic analysis of polypropylene fibres using multiple-beam Fizeau fringes in transmission and a modified creep device. *Polym. Int.*, **59**, 1021–1030.

Hennig, J. (1964) Anisotropie des dynamischen Elastizitätsmodulus in einachsig verstreckten amorphen Hochpolymeren. *Kolloidzeitschrift*, **200**, 46.

Hertz, H. (1896) *Miscellaneous Papers*, Macmillan, London, p. 146.

Hillbrick, L.K., Kaiser, J., Huson, M.G., Naylor, G.R.S., Wise, S.W., Miller, A.D. and Lucas, S. (2019) Determination of the transverse modulus of cylindrical samples by compression between two parallel flat plates. *SN Appl. Sci.*, **1**, 724.

Hine, P.J. and Ward, I.M. (1996) Measuring the elastic properties of high-modulus fibres. *J. Mater. Sci.*, **31**, 371.

Horgan, C.O. (1972) On Saint-Venant's principle in plane anisotropic elasticity. *J. Elast.*, **2**, 335.

Horgan, C.O. (1974) The axisymmetric end problem for transversely isotropic circular cylinders. *Int. J. Solids Struct.*, **10**, 837.

Jayanty, S., Crowe, J. and Berhan, L. (2011) Auxetic fibre networks and their composites. *Phys. Status Solidi B*, **248**, 73–81.

Jerabek, M., Major, Z. and Lang, R.W. (2010) Strain determination of polymeric materials using digital image correlation. *Polym. Test.*, **29**, 407–416.

Karasawa, N., Dasgupta, S. and Goddard, W.A. (1991) Mechanical properties and force field parameters for polyethylene crystal. *J. Phys. Chem.*, **95**, 2260.

Kashiwagi, M., Folkes, M.J. and Ward, I.M. (1971) The measurement of molecular orientation in drawn poly(methyl methacrylate) by broad line nuclear magnetic resonance. *Polymer*, **12**, 691–697.

Kausch, H.H. (1967) Elastic properties of anisotropic heterogeneous materials. *J. Appl. Phys.*, **38**, 4213.

Kausch, H.H. (1969). *Kolloidzeitschrift*, **234**, 1148; (1970) 237, 251; (1978) *Polymer Fracture*, Springer, Berlin, p. 33

Kausch, H.H. (1970) Über die Zusammenhänge von makroskopischer und molekularer Anisotropie in Hochpolymeren. *Kolloidzeitschrift*, **237**, 251.

Kausch, H.H. (1971) Elastic and ultimate behavior of amorphous polymer networks. *J. Macromol. Sci. B.*, **5**, 269.

Kotani, T., Sweeney, J. and Ward, I.M. (1994) The measurement of transverse mechanical properties of polymer fibres. *J. Mater. Sci.*, **29**, 5551.

Krüger, J.K., Marx, A., Peetz, L. *et al.* (1986) Simultaneous determination of elastic and optical properties of polymers by high performance Brillouin spectroscopy using different scattering geometries. *Colloid Polym. Sci.*, **264**, 403.

Krüger, J.K., Grammes, C., Stockem, K. *et al.* (1991) Nonlinear elastic properties of solid polymers as revealed by Brillouin spectroscopy. *Colloid Polym. Sci.*, **269**, 764.

Kuhn, W. and Grün, F. (1942) Relations between elastic constants and the strain birefringence of high-elastic substances. *Kolloid-zeitschrift*, **101**, 248.

Kumar, S.R., Rennesch, D.P. and Grimsditch, M. (2000) Effect of molecular orientation on the elastic constants of polypropylene. *Macromolecules*, **33**, 1819.

Ladizesky, N.H. and Ward, I.M. (1971) Measurement of shear moduli of oriented low-density polyethylene. *J. Macromol. Sci. B*, **5**, 745.

Ladizesky, N.H. and Ward, I.M. (1974) Deformation mechanisms of low density polyethylene with parallel lamellar morphology. *J. Macromol. Sci. B*, **9**, 565.

Lakes, R.S. (1987) Foam structures with a negative Poisson's ratio. *Science*, **235**, 1038.

Lekhnitskii, S.G. (1981) *Theory of Elasticity of an Anisotropic Elastic Body*, Mir, Moscow.

Leung, W.P. and Choy, C.L. (1983) The elastic constants of ultradrawn polypropylene. *J. Polym. Sci. Polym. Phys. Ed.*, **21**, 725.

Leung, W.P., Chen, F.C., Choy, C.L. *et al.* (1984) Ultrasonic measurements of the mechanical relaxations and complex stiffnesses in oriented linear polyethylene. *Polymer*, **25**, 447.

Lewis, E.L.V. and Ward, I.M. (1980) The measurement of shear compliances for oriented polyethylene terephthalate sheet. *J. Mater. Sci.*, **15**, 2354.

Lewis, E.L.V., Richardson, I.D. and Ward, I.M. (1979) Hall-effect apparatus for the measurement of simple shear in polymers. *J. Phys. E*, **12**, 189.

Liu, Y. and Hu, H. (2010) A review on auxetic structures and polymeric materials. *Sci. Res. Essays*, **5**, 1052–1063.

Maxwell, A.S., Unwin, A.P. and Ward, I.M. (1996) The mechanical behaviour of oriented high-pressure annealed polyethylene. *Polymer*, **37**, 3283.

McBrierty, V.J. and Ward, I.M. (1968) Investigation of the orientation distribution functions in drawn polyethylene by broad line nuclear magnetic resonance. *J. Phys. D. Appl. Phys.*, **1**, 1529.

McEwen, E. (1949) Stresses in elastic cylinders in contact along a generatrix (including the effect of tangential friction). *Philos. Mag.*, **40**, 454.

Mentrasti, L., Malari, L. and Fabiani, M. (2021) Poisson's ratio bounds in orthotropic materials. Application to natural composites: wood, bamboo and *Arundo donax*. *Compos. Part B Eng.*, **209**, 108612.

Morgan, H.M. (1962) Correlation of molecular orientation measurements in fibers by optical birefringence and pulse velocity methods. *Text. Res. J.*, **32**, 866.

Neale, P.J., Pickles, A.P., Alderson, K.L. *et al.* (1995) The effect of the processing parameters on the fabrication of auxetic polyethylene. *J. Mater. Sci.*, **30**, 4087.

Nicholson, T.M., Davies, G.R. and Ward, I.M. (1994) Conformations in poly(ethylene terephthalate): a molecular modelling study. *Polymer*, **35**, 4259.

Northolt, M.G. and Van Aartsen, J.J. (1977) Chain orientation distribution and elastic properties of poly (p-phenylene terephthalamide), a “rigid rod” polymer. *J. Polym. Sci. Polym. Symp.*, **58**, 283.

Northolt, M.G. and van der Hout, R. (1985) Elastic extension of an oriented crystalline fibre. *Polymer*, **26**, 310.

Nunes, L.C.S. (2010) Shear modulus estimation of the polymer polydimethylsiloxane (PDMS) using digital image correlation. *Mater. Des.*, **31**, 583–588.

Odajima, A. and Maeda, T. (1967) Calculation of the elastic constants and the lattice energy of the polyethylene crystal. *J. Polym. Sci. Part C*, **15**, 55.

Orell, O., Vuorinen, J., Jokinen, J., Kettunun, H., Hytönen, P., Turunen, J. and Kanerva, M. (2018) Characterization of elastic constants of anisotropic composites in compression using digital image correlation. *Compos. Struct.*, **185**, 176–185.

Pal, S. and Bhattacharyya, A. (2025) Measurement of axial and shear mechanical response of PDMS elastomers and determination of Poisson’s ratio using digital image correlation. *Polym. Test.*, **143**, 108687.

Peterlin, A. (1979) Elastic modulus and strength of fibrous structure. *Polymer Engineering and Science*, **19**, 118–124.

Pierron, F. and Grédiac, M. (2012) *The Virtual Fields Method*, Springer, New York.

Pinnock, P.R. and Ward, I.M. (1963) Dynamic mechanical measurements on polyethylene terephthalate. *Proc. Phys. Soc.*, **81**, 260.

Pinnock, P.R. and Ward, I.M. (1964) Mechanical and optical anisotropy in polyethylene terephthalate fibres. *Br. J. Appl. Phys.*, **15**, 1559.

Pinnock, P.R. and Ward, I.M. (1966) Mechanical and optical anisotropy in polypropylene fibres. *Br. J. Appl. Phys.*, **17**, 575.

Pinnock, P.R., Ward, I.M. and Wolfe, J.M. (1966) Compression of anisotropic fibre monofilaments. 2. *Proc. Roy. Soc. A*, **291**, 267.

Raumann, G. (1962) The anisotropy of the shear moduli of drawn polyethylene. *Proc. Phys. Soc.*, **79**, 1221.

Raumann, G. and Saunders, D.W. (1961) The anisotropy of Young’s modulus in drawn polyethylene. *Proc. Phys. Soc.*, **77**, 1028.

Ravirala, N., Alderson, A., Alderson, K.L. *et al.* (2005) Auxetic polypropylene films. *Polym. Eng. Sci.*, **45**, 517–528.

Ravirala, N., Alderson, A. and Alderson, K.L. (2007) Interlocking hexagons model for auxetic behaviour. *J. Mater. Sci.*, **42**, 7433–7445.

Rawson, F.F. and Rider, J.G. (1974) The elastic constants of oriented polyvinyl chloride. *J. Phys. D*, **7**, 41.

Reuss, A. (1929) Berechnung der Fließgrenze von Mischkristallen auf Grund der Plastizitätsbedingung für Einkristalle. *Z. Angew. Math. Mech.*, **9**, 49.

Robertson, R.E. and Buenker, R.J. (1964) Some elastic moduli of bisphenol A polycarbonate. *J. Polym. Sci.*, **A2**, 4889.

Rogers, T.G. and Pipkin, A.C. (1963) Asymmetric relaxation and compliance matrices in linear viscoelasticity. *J. Appl. Maths Phys.*, **14**, 334.

Rutledge, G.C. and Suter, U.W. (1991) Calculation of mechanical properties of poly(pphenylene terephthalamide) by atomistic modeling. *Polymer*, **32**, 2179.

Samuels, R.J. (1974) *Structured Polymer Composites*, John Wiley & Sons, New York.

9

Morphology and Structural Effects

9.1 Evolution of Structures Under Tension: Cavitation

Semi-crystalline polymers have richly varied and complex morphologies. In Chapter 1 we described chain folding to produce lamellae that form crystallites, with potential to form spherulitic morphology (Bassett 2003; Crist and Schultz 2016; see Figure 9.1). Dense hard crystal regions are surrounded by amorphous material that may be much softer, resulting in stress fields that are locally complex. This has important effects, especially when solid polymers are subject to large strains; inhomogeneous stress fields will be the agent for changes in morphology, such as are necessary when producing materials with high levels of molecular chain orientation.

Peterlin (1971) has described the sequence of events that lead from a spherulitic structure to the fibrillar structure observed when high uniaxial tensile strains are imposed on polyethylene and polypropylene. In the initial stages, plastic deformation is enabled by rotation of crystallites and sliding of lamellae. This is followed by more extreme microstructural changes, which is accompanied by a phenomenon termed 'micronecking', where voids are created and lamellae are stretched to become microfibrils. The morphology along the direction of stretch is thus changed out of all recognition, though as shown by Amornsakchai et al. (2000), structure in the transverse plane retains some memory of the unstretched state. Tie-molecules form links between crystallites, become taut on stretching, and are a key factor in the structure's stiffness. Further stretching eventually leads to large-scale fracture. Peterlin presented evidence of the microfibrils from images created by electron microscopy. Galeski (2003) has observed that cavitation is an inherent feature of micronecking, as illustrated in Figure 9.2 from Peterlin (1972).

To understand the evolution of structure, it is necessary to locate the origin of the beginning of cavitation. Pawlak and Galeski (2005) have worked on a range of semicrystalline polymers and proposed that cavitation originated in the amorphous phase, acting in competition with plasticity in the crystal phase. However, Men et al. (2004) located the origin of cavitation of poly(1-butane) in the crystalline phase. The work of Lyu et al. (2023) on polypropylene using X-ray techniques and step-cycling tensile experiments also located cavitation in the crystalline phase. For the particular case of large deformation in polypropylene, cavitation in the amorphous phase via the disentanglement of interfibrillar tie-molecules has been observed by Lu et al. (2015).

Mechanical Properties of Solid Polymers, Fourth Edition. John Sweeney and Peter Hine.

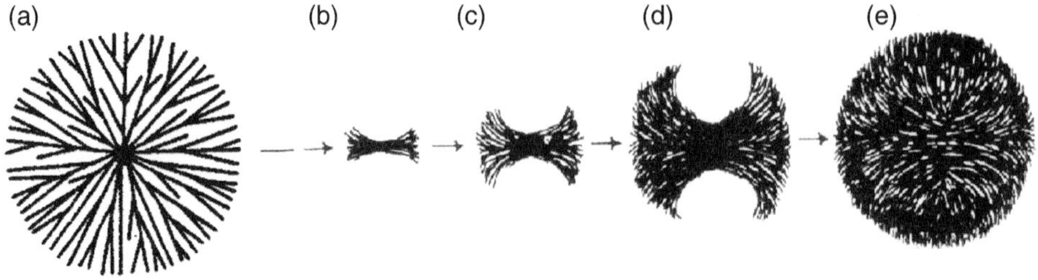

Figure 9.1 Development of spherulite (left) through stages (a–e) as crystals nucleate from a single central site. Gránánsy et al. (2006) / Taylor & Francis Group.

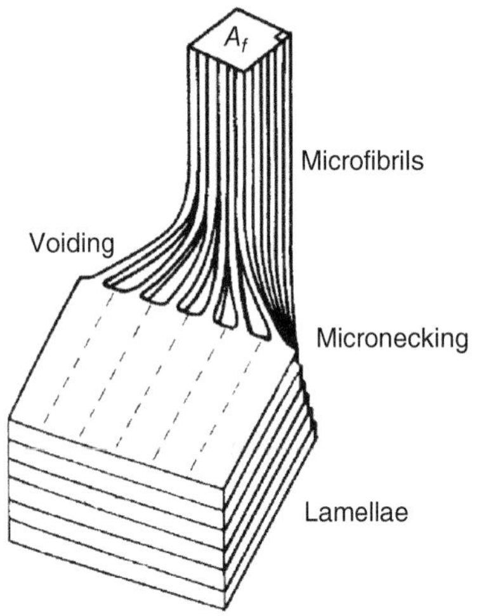

Figure 9.2 Micronecking and the development of voids or cavitation. Peterlin (1972) / Springer Nature.

Another important issue is that tensile deformation of a semicrystalline polymer does not always cause cavitation, and the resulting need to determine the conditions required for it. Humbert et al. (2010) have studied cavitation of a range of polyethylenes using in situ SAXS. They concluded that there was competition between crystallite shearing, characterized by a stress σ_{sh} and cavitation occurring at a stress σ_{cav}. The phenomenon corresponding to the smaller stress then prevails. However, the value σ_{sh} is an increasing function of the crystallite thickness L_c. It follows that there is a critical crystallite thickness L_{cc}, that may depend on temperature and strain rate, above which $\sigma_{cav} < \sigma_{sh}$ and cavitation occurs. Under this regime, Humbert et al. (2010) envisaged three possible scenarios for the stretching of a tensile specimen:

Homogeneous cavitation. $L_c \gg L_{cc}$, $\sigma_{cav} < \sigma_{sh}$. Applied stress increases until σ_{cav} is attained and cavitation occurs homogeneously throughout the specimen. The cavities act as stress concentrators and raise the stress locally to σ_{sh} in many locations adjacent to the crystallites, resulting in widespread necking.

Localized cavitation. $L_c < L_{cc}$, $\sigma_{sh} < \sigma_{cav}$. Applied stress increases until σ_{sh} is attained to initiate crystalline shearing and neck formation. Damage to the crystallites introduces stress concentrations to raise the stress to σ_{cav} producing rapid voiding.

No cavitation. $L_c \ll L_{cc}$, $\sigma_{sh} \ll \sigma_{cav}$. Applied stress increases until σ_{sh} is attained to initiate crystalline shearing and neck formation. However, because of the large difference between σ_{cav} and σ_{sh}, stress concentrations do not raise the stress enough to cause cavitation.

All three scenarios were observed experimentally. It can be seen that cavitation and crystal shear appear to interact in a complex manner to produce qualitatively different results.

9.2 Effects of Stress Field

In the original work by Peterlin and in many subsequent contributions, the emphasis has been on uniaxial tensile stress conditions. Given the importance of cavitation and the expectation that its potential will be heightened in the presence of tension, it seems likely that the evolution of structure and the tendency to tensile failure will depend crucially on the specific nature of the stress field. The suppression of tension was the motivation for the development of the hydrostatic extrusion process for polymers (Gibson et al. 1974; Capaccio et al. 1979). The concept is also behind the effectiveness of processes such as roller drawing (Kaito et al. 1985; Mohanraj et al. 2008) and die drawing (Coates and Ward 1981) shown in Figure 9.3.

In a study of the mechanical effects of cavitation, Pawlak and Galeski (2005) compared the stress–strain behaviour in uniaxial tension with that in channel die compression (see Figure 9.4) of a number of highly crystalline polymers. In channel die compression, the deformation field is the same as that in planar (constant width) tension, but the mean stress is negative rather than positive. They attribute the difference in stress–strain behaviour, shown in Figure 9.4, to the presence of cavitation in the tensile case and its absence in channel die compression. Typical stress–strain curves are shown in Figure 9.5, together with the proposed mechanisms that apply at the various stages of deformation. These resemble observations on polyoxymethylene (POM), nylon (PA6), polypropylene and high density polyethylene. Voids were detected by small-angle X-ray scattering (SAXS), density changes and stress whitening. Similar experimental comparisons on polymers low in crystallinity and yield stress showed little difference in the tensile and channel die compression curves and little evidence of voiding, conforming to the expectation that yielding in the smaller crystallites precluded cavitation.

These observations on the effect of cavitation concern the process of molecular orientation. It is of interest to discover whether the effects of cavitation persist in the mechanical behaviour of the oriented polymer products that result. Mohanraj et al. (2006) produced POM products by both hydrostatic extrusion and die-drawing, with greatly higher pressures applied in the former process. Densities were significantly higher for hydrostatically extruded material at all extension ratios, and SAXS studies showed lower voiding. The strength was found to be higher for hydrostatically extruded material at the higher draw ratios (~7). This effect was attributed to higher levels of voiding in the die-drawn polymer.

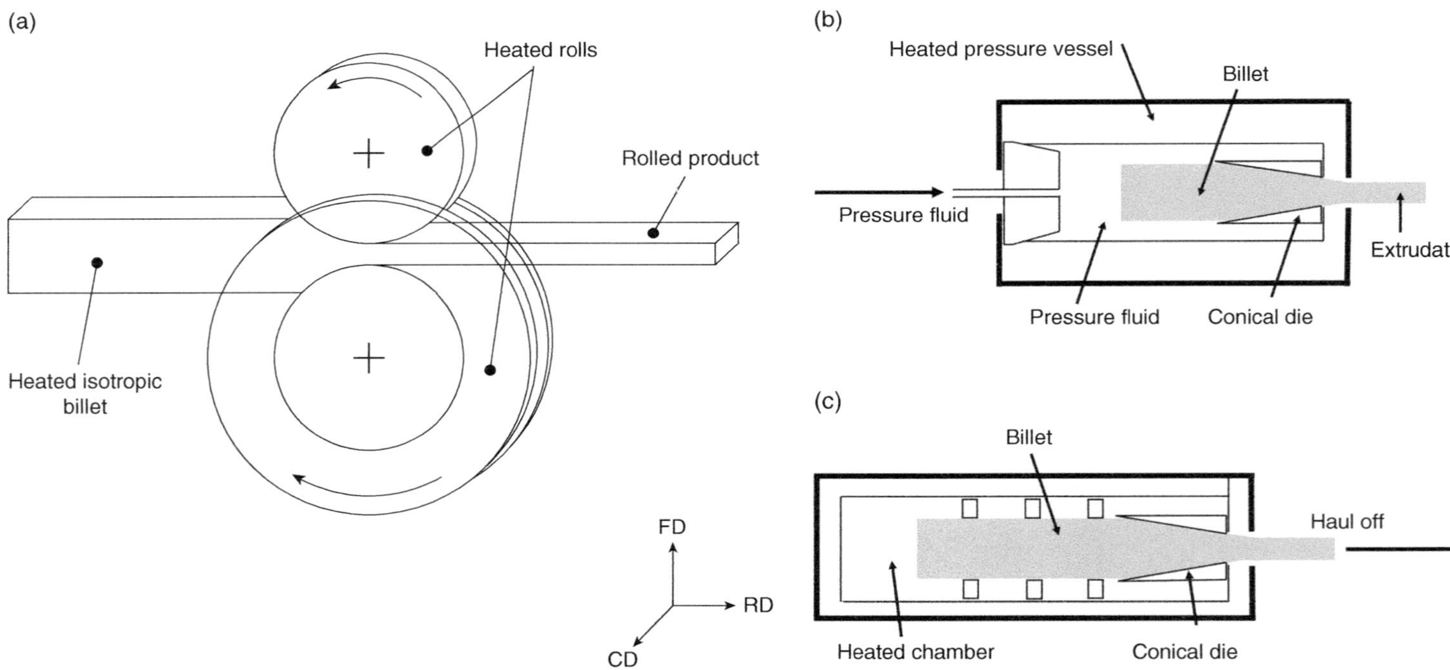

Figure 9.3 Solid-phase deformation processes. (a) Roller drawing with side constraints. (Mohanraj, J., Morawiec, J., Pawlak, A., Barton, D.C., Galeski, A., Ward. I.M. (2008) Orientation of polyoxymethylene by rolling with side constraints. Polymer, **49**, 303–316.) (b) Hydrostatic extrusion. (c) Die drawing.

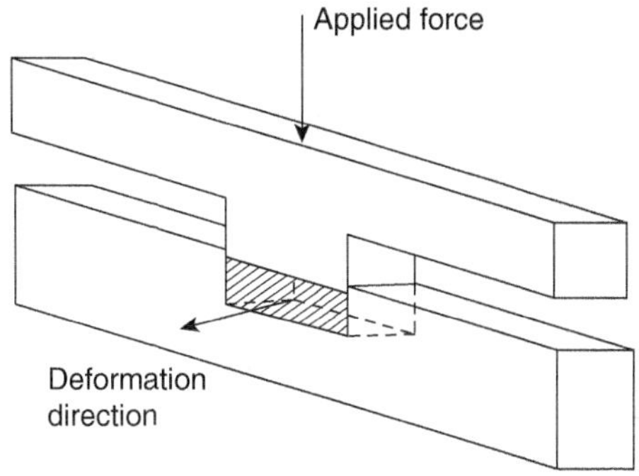

Figure 9.4 Experimental arrangement for channel die compression. Pawlak and Galeski (2005) / American Chemical Society.

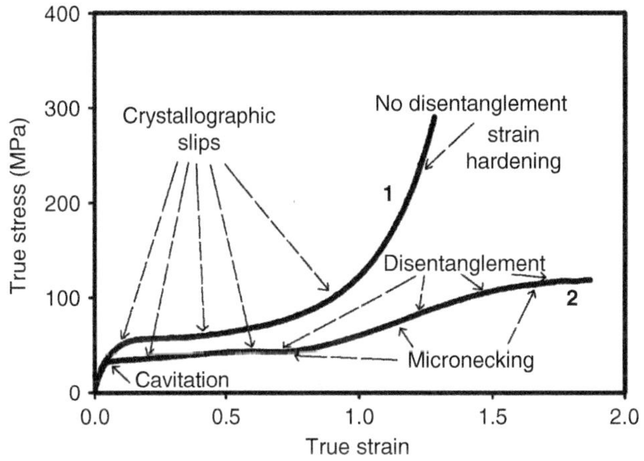

Figure 9.5 Diagram of stress–strain curves for highly crystalline polymers in channel die compression (curve 1) and uniaxial tension (curve 2). Pawlak and Galeski (2005) / American Chemical Society.

Lyu et al. (2020) have made a detailed study of cavitation in polypropylene. Figure 9.6 compares isotactic polypropylene stretched at 130°C both uniaxially and by die drawing, with clear evidence of cavitation in the form of stress whitening in the uniaxial case, and very little whitening in the die-drawn specimen. The lower level of cavitation in the die-drawn product demonstrates the effect of the compressive stress component. However, further experimentation showed that this was not simply the mechanical suppression of cavitation. The die-drawn material was subject to additional stretching by free-drawing, and still no significant whitening was observed. However, cavities could be observed using Ultra-Small-Angle X-ray Scattering (USAXS) observations. These cavities were of roughly elliptical cross-sections, characterized by long and short axes which define an orientation detectable by USAXS. For uniaxially stretched specimens, the long axes of the cavities are initially perpendicular to the axis of stretching, and align progressively along the stretching axis as the test proceeds. On the other hand, in the die-drawn specimens, the long axes of the cavities are aligned along the drawing direction as soon as they appear. Cavities with this orientation to the tensile field are less vulnerable to being propagated than those with long

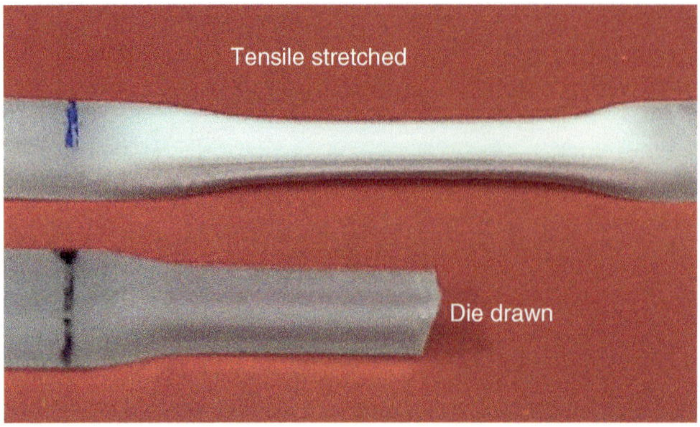

Figure 9.6 Stress whitening in polypropylene stretched in tension at 130°C, compared with the same material stretched by die drawing. Lyu et al. (2020) / American Chemical Society.

axes normal to the stress (see Chapter 13). Lyu et al. concluded that the die-drawing process produced a structure with a reduced number of extendable flaws, to give effects beyond the purely mechanical effect of the compressive stress component. The exploitation of this effect has been explored further by Lyu et al. (2021).

The precise mechanisms behind cavitation in semi-crystalline polymers are still being established. Lu and Men (2018) have reviewed the current state of knowledge and note that there are still different views on issues such as whether cavitation initiates in the amorphous or the crystalline regions. It could be that both are possible, depending on the position of the initiation point within the original spherulite. Figure 9.7 shows cavities originating in the

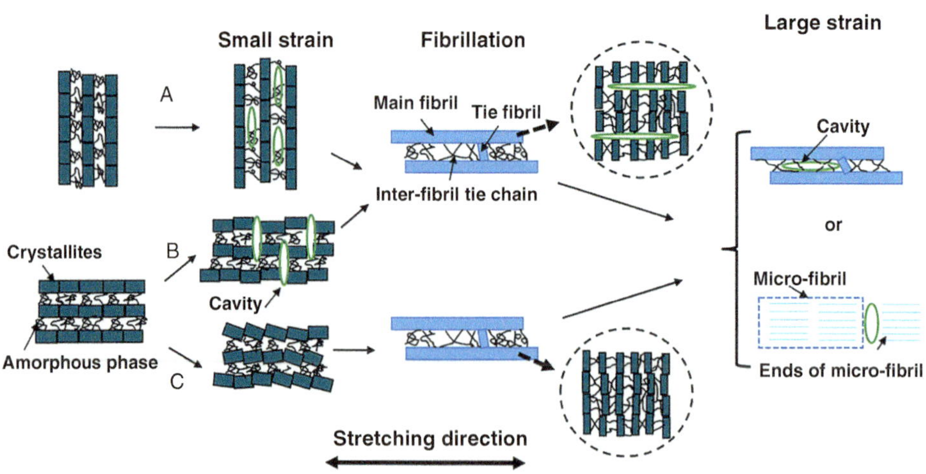

Figure 9.7 Different proposed cavitation mechanisms at different strain levels. Path A shows cavitation (green ellipses) originating in the amorphous phase. Path B shows cavitation occurring at breaks in crystal blocks. In Path C, cavitation is suppressed as crystals yield. After fibrillation, as strains become large the cavities orient along the stress direction. Lu and Men (2018) / John Wiley & Sons.

amorphous region (path A, as suggested by Pawlak and Galeski 2005) and in the breaking crystalline blocks (path B, Men and Rieger 2004). Orientation of cavities in the draw direction at larger strain is indicated.

9.3 One-Dimensional Modelling

9.3.1 Mechanical Anisotropy of Lamellar Structures

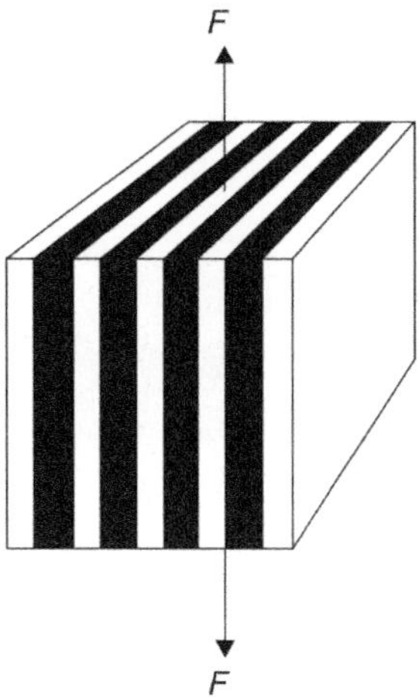

Figure 9.8 Isostrain conditions for layered composite.

It is instructive to start the discussion of the modelling of semi-crystalline polymer using a model of an idealized lamellar composite that consists of a high modulus layer, that can be associated with crystalline material, and a more compliant layer that can be identified with amorphous material. Provided that the bonding between the layers remains intact, the volume fraction of each component and not the thickness of the individual layers is the important factor. As with the aggregate model discussed in Section 8.6, different values of overall stiffness are obtained depending on whether the components are in parallel or in series, yielding the Voigt or Reuss average modulus, respectively.

The maximum stiffness is obtained when a uniaxial stress is applied parallel with the layers, as indicated in Figure 9.8. It is assumed that the strain is the same in all the layers, a form of loading known as the isostrain (or homogeneous strain) condition.

The force acting on the composite (F) is equal to the sum of the forces acting on the crystal (C) and amorphous (A) layers

$$F = F_C + F_A \tag{9.1}$$

Force is equal to stress multiplied by area. Hence for the total stress σ

$$\sigma A = \sigma_C A_C + \sigma_A A_A$$

where A_C and A_A represent the areas of the end faces occupied by each component. As both components are of length l, areas can be represented by volumes, or rather by volume fractions V_C and V_A. The volume fraction of the composite V is given by $V = V_C + V_A = 1$. Hence

$$\sigma V = \sigma_C V_C + \sigma_A V_A \Rightarrow \sigma = \sigma_C V_C + \sigma_A V_A \tag{9.2}$$

Under isostrain conditions, this expression can be rewritten in terms of Young's modulus (E) as

$$E = E_C V_C + E_A V_A \tag{9.3}$$

which is a Voigt average modulus (see Section 8.6.2).

The modulus is, however, much lower in the direction transverse to the layered structure (Figure 9.9). In this case, each layer is subjected to the same force, and hence to the same

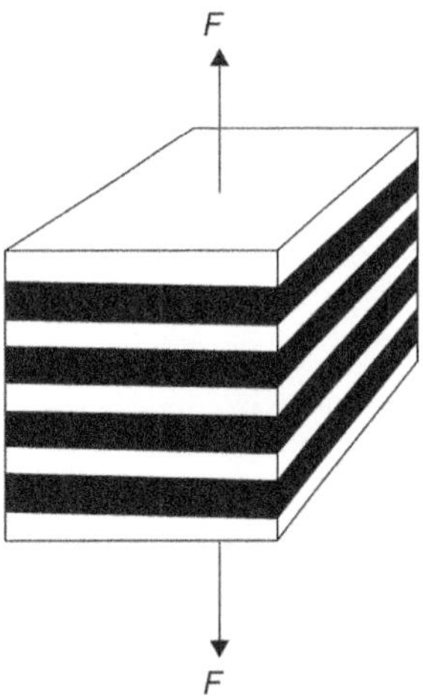

Figure 9.9 Isostress conditions for layered composite.

stress, because the area remains constant through the stack. Loading of this form is known as the isostress (or homogeneous stress) condition.

The total deformation $\delta\ell$ is equal to the sum of the deformations in each component:

$$\delta\ell = \delta\ell_C + \delta\ell_A$$

Length changes can be converted to strains using $\varepsilon = \delta\ell/\ell$:

$$\varepsilon\ell = \varepsilon_C\ell_C + \varepsilon_A\ell_A \tag{9.4}$$

Substituting modulus E as the ratio of (uniform) stress to strain, we obtain

$$\frac{\sigma\ell}{E} = \frac{\sigma\ell_C}{E_C} + \frac{\sigma\ell_A}{E_A}.$$

As the cross section of the composite is assumed to be uniform, the length of a component is proportional to its volume fraction. Again, take V as unity to give

$$\frac{1}{E} = \frac{V_C}{E_C} + \frac{V_A}{E_A} \tag{9.5}$$

This expression can be rewritten as

$$E = \frac{E_C E_A}{E_A V_C + E_C V_A} \tag{9.6}$$

and is a Reuss average modulus (see Section 8.6.2).

The model of Figure 9.8 can be considered to be a parallel model, as it can be represented by a pair of springs with moduli E_C and E_A. Similarly, Figure 9.9 is a series model with the two springs in series. It is clear that either model will show different moduli when strained horizontally rather than vertically; for instance, Figure 9.8 structure will have modulus given by Equation (9.5) when stretched vertically and that given by Equation (9.6) when stretched horizontally. Elastically anisotropic behaviour is thus a feature of these models.

9.3.2 Takayanagi Models

Takayanagi (1963) and Takayanagi et al. (1966) applied the concept of a model with alternating crystalline and amorphous layers to oriented highly crystalline polymers. High-density polyethylene that has been drawn uniaxially (i.e., with fibre symmetry) and then annealed has a distinct lamellar texture, and we will show that the orientation of the lamellae, as distinct from the molecular orientation, plays the dominant role in determining the mechanical anisotropy. The temperature variation in the in-phase (E_1) and out-of-phase (E_2) components of the dynamic modulus, both parallel with ($\|$) and perpendicular to ($\perp$) the draw direction, is shown in Figure 9.10, with the parallel modulus ($E_0 = \|$) crossing the perpendicular modulus ($E_{90} = \perp$) at high temperatures; at low temperatures $E_0 > E_{90}$,

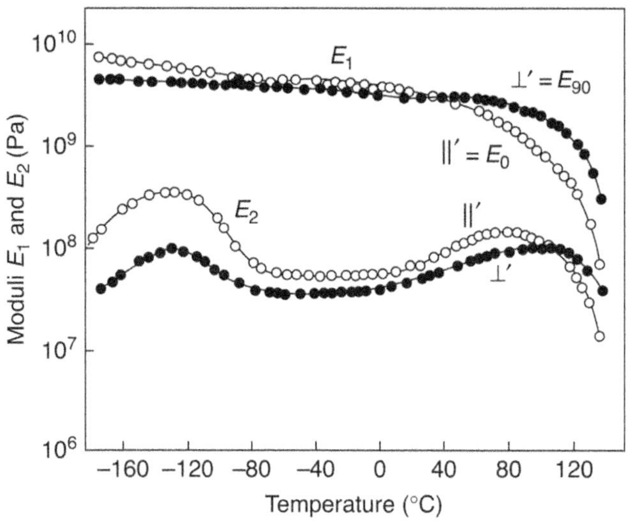

Figure 9.10 Temperature dependence of E_1 and E_2, the components of the dynamic modulus, in directions parallel (||) and perpendicular ($\perp'$) to the initial draw direction for annealed samples of high-density polyethylene. Takayanagi et al., 1966 / John Wiley & Sons.

but at high temperatures $E_{90} > E_0$. The principal features can be explained in terms of a simple model of amorphous (A) and crystalline (C) components, which are in series in the draw direction but in parallel in the transverse direction (Figure 9.11). In the orientation direction, each component is subject to the same stress, so that compliances are added as in the Reuss averaging scheme. The stiffness above the relaxation transition is primarily determined by the compliant amorphous regions, so giving a large fall in modulus as the temperature is increased. In the perpendicular direction, the parallel components are each subject to the same strain, and stiffnesses are added as in the Voigt scheme. The crystalline regions support the applied stress at temperatures above the relaxation transition, and so a comparatively high stiffness is maintained. Takayanagi and colleagues considered that appropriate values of the modulus might be $E^c(\|) = 100\,\text{GPa}$, $E^c(\perp) = 1\,\text{GPa}$, E_A (low T) $= 1\,\text{GPa}$, E_A (high T) $= 0.01\,\text{GPa}$.

9.3.3 The Crystal Bridge Model for Ultra-High-Modulus Polyethylene

In a pair of related papers, Gibson et al. (1978) and Clements et al. (1978) reported a structural model of oriented linear polyethylene (LPE). The model used X-ray-derived measurements on crystal thickness to provide data for the model's predictions of dynamic small strain tensile behaviour (see Chapter 6). X-ray measurements of crystal strain provided data on crystal modulus.

The LPE specimens were uniaxially oriented with draw ratios ranging from 5 to 30; X-ray measurements showed that at a draw ratio of 10 and above, the crystalline phase was

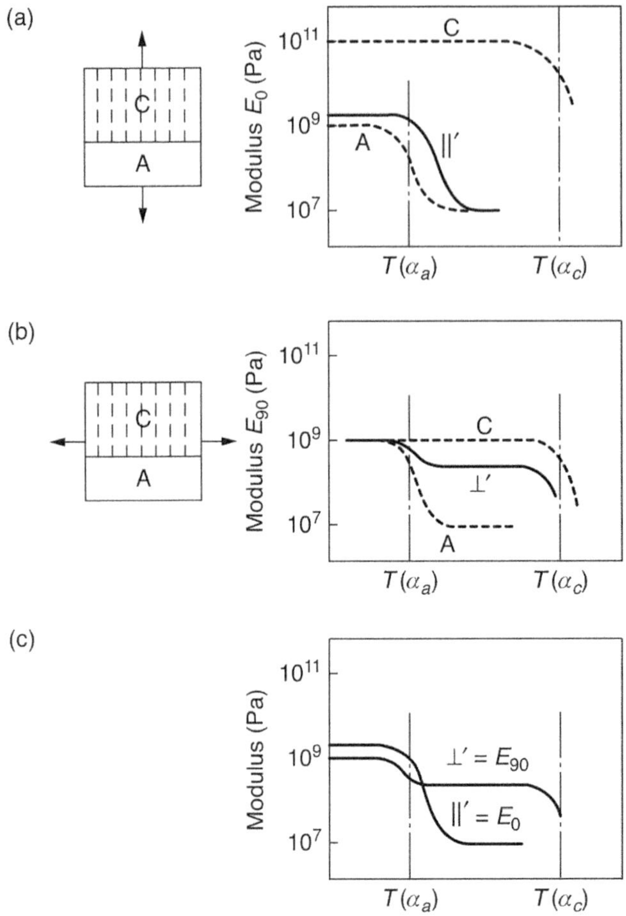

Figure 9.11 Schematic representations of change in modulus E with temperature on the Takayanagi model for (a) the ||′ and (b) the ⊥′ situations corresponding to E_0 and E_{90}, respectively. Calculations assume amorphous relaxation at temperature $T(\alpha_a)$ and crystalline relaxation at temperature $T(\alpha_c)$ and (c) shows combined results. C, crystalline phase; A, amorphous phase. Takayanagi et al., 1966 / John Wiley & Sons.

'almost perfectly' oriented, with c-axes along the stretching direction (Gibson et al. 1978). Density measurements indicated that the crystallinity of the oriented specimens was around 77%. A crucial finding of wide-angle X-ray measurements was that the mean crystal lengths in the c (drawing) direction increased with draw ratio, whereas the transverse crystal dimensions remained constant. This phenomenon lends credibility to the concept that the increase with draw ratio of modulus in the draw direction reflects an increasing continuity of the crystal phase. A diagram of the proposed structural arrangement is given in Figure 9.12(a), where only oriented crystalline polymer is shown and the lines represent backbone chains. The clear areas represent disordered regions which have a periodicity L as measured by small-angle X-ray measurements. The morphology may be interpreted as fibrillar stacks of crystallites joined by crystalline bridges, or as a continuous crystal with

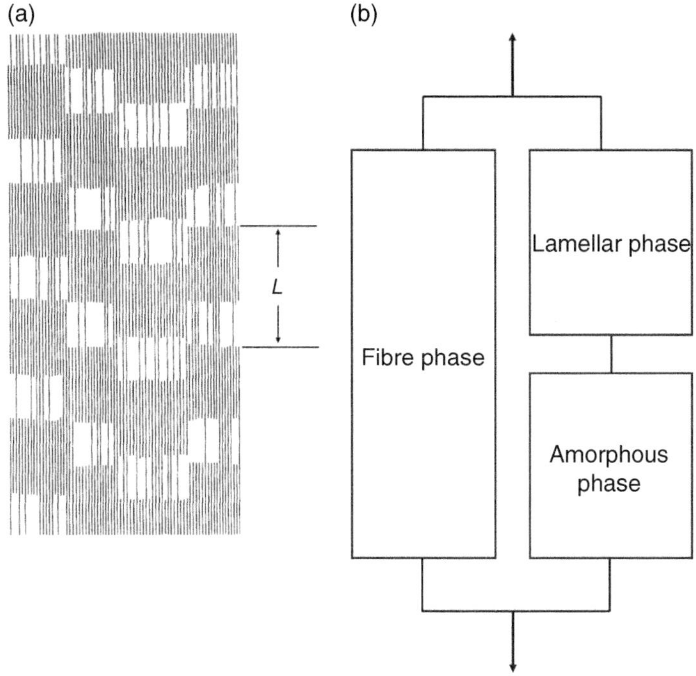

Figure 9.12 Crystal bridge model. (a) Morphology: lines represent backbones of aligned crystals and clear regions are disordered with long period L. Gibson et al. (1978) / with permission of Elsevier. Dynamic mechanical behaviour and longitudinal crystal thickness measurements on ultra-high modulus linear polyethylene: a quantitative model for the elastic modulus. Polymer, **19**, 683–693. (b) Model: fibre phase represents effect of continuous crystallinity.

disordered inclusions. In general, the mean lengths of the crystal c axes exceeded the long period L (Clements et al. 1978).

The model is shown in Figure 9.12(b). The *Fibre phase* arm represents the effect of the continuous crystal network. This was modelled as a fibre composite. Only those crystalline chains that join (bridge) two or more fibrillar stacks will contribute to the overall stiffness of the body, and these are represented as fibres; the matrix then consists of a combination of chain-folded and amorphous material. The composite model used is that of Cox (1952), where the modulus E is given by a variation of Equation (9.3):

$$E = E_f V_f \psi + E_m V_m \tag{9.7}$$

Here ψ is a factor to account for the finite length of the fibres and the subscripts f and m refer to the fibre and matrix, respectively. It was concluded that the whole specimen behaviour was dominated by the fibre component of this equation, and modulus predictions were made on this basis. The model was implemented for conditions at −50°C to eliminate viscoelastic effects. The model produced valid predictions of modulus as draw ratio was varied.

In this approach the increase in modulus was seen solely as a consequence of the increase in crystalline continuity. The hypothesis of increasing crystalline continuity has been supported by observations on thermal conductivity (Gibson et al. 1977); for polyethylene, increasing the draw ratio increased the thermal conductivity along the draw direction, to give an anisotropy as high as 18.5. Gibson et al. (1978), whilst acknowledging the existence of tie molecules, did not consider them to be significant to their work. This is in contrast to Peterlin's description of structural evolution. A review by Breese and Beaucage (2004) gives a comparative account of models of oriented semi-crystalline polymer, encompassing the crystal bridge model, composite models, the Peterlin approach and a model due to Barham and Arridge (1977).

9.3.4 Modelling Structural Development

Deplancke et al. (2019) have produced a model of ultra-high-molecular-weight polyethylene (UHMWPE) that tracks deformation from the undeformed state to a fibrillar structure. The approach still uses two phases, but with more sophisticated mechanical behaviour capable of large strains. The amorphous phase is represented by an eight-chain Arruda–Boyce model of rubber elasticity (see Chapter 4). The crystalline phase is assumed to possess rate-dependent plasticity in the form of an Eyring process (see Chapter 12). At higher strains, when some fibrillar morphology has developed, a second Eyring process begins to act, that represents a combination of possible mechanisms such as fibril slip, fracture of tie molecules and disentanglement. Initially, the morphology is assumed to behave as a combination of parallel and series models such as those in Figures 9.8 and 9.9. As deformation progresses, the structure breaks down as spherulites are disrupted and a morphology develops that becomes closer to the series model of Figure 9.9. This evolution can be represented parametrically. Starting with a parallel model, Equation (9.2) gives

$$\sigma = \sigma_C V_C + \sigma_A V_A.$$

Recalling that $V_C + V_A = 1$ we may rewrite this with $V_C = \phi$ as

$$\sigma = \phi\sigma_C + (1-\phi)\sigma_A \tag{9.8}$$

We may also write

$$\sigma = (1-\phi + k\phi)\sigma_A \tag{9.9}$$

which returns Equation (9.8) when $k = \sigma_C/\sigma_A$. For the series model of Figure 9.9,

$$\sigma = \sigma_C = \sigma_A.$$

This is equivalent to Equation (9.9) with $k = 1$. Thus, as k varies in the range $1 \leq k \leq \sigma_C/\sigma_A$, Equation (9.9) gives a stress that varies smoothly from that of the parallel model to that of the series model. Deplancke et al. set an initial value of k for the undeformed structure based on experimental data, and allowed k to evolve as a function of crystal plastic strain, approaching unity as the fibrillar state was attained. In this way they used a one-dimensional model to simulate the reorganization of a three-dimensional structure. The predictions for stress–strain behaviour under uniaxial compression and tension showed good predictions of strain rate dependence and useful though imperfect representation of macroscopic yielding and strain-hardening.

9.4 Three-Dimensional Models

Two-phase composite elements such as those illustrated in Figure 9.13 have formed the basis of a range of three-dimensional models of semicrystalline polymers. They all include both crystalline and amorphous components. In general models are created by first generating an assembly of composite elements with prescribed orientations and secondly calculating the overall macroscopic stress–strain law. The second stage is achieved using a homogenization procedure that relates the macroscopic stress of the assembly to that of the composite elements. There are several established options for this. The majority of models use a *mean field theory*; this and other options are outlined in the review by Mirkhalaf and Vadizadeh (2024).

In the early stages of this work, Lee et al. (1993a, 1993b) produced models of high-density polyethylene (HDPE) using inclusion elements of the form shown in Figure 9.13(b). The amorphous phase was assumed to be in its rubbery state, as appropriate for polyethylene at room temperature. Its mechanical behaviour was defined by an Argon rate-dependent plasticity (Argon 1973; see Chapter 12); there was no series elastic mechanism and therefore no yield stress associated with the model. However, there was a parallel hyperelastic component in the form of an Arruda–Boyce eight-chain model to represent strain hardening (Arruda and Boyce 1993; see Chapter 4). The crystalline phase was defined by crystal plasticity, with four independent slip systems (see for example Peirce et al. 1983). The parameters defining the crystal plasticity were obtained from the literature. Other significant works in this area are van Dommelen et al. (2004) and Mirkhalaf et al. (2019).

Modelling of this nature automatically creates predictions of crystal orientation distributions (or *texture*). In the model of Lee et al. (1993a, 1993b) no attempt was made to introduce an initial spherulitic morphology. Rather, it comprised 244 randomly oriented composite elements, and in its initial undeformed state the crystal orientation distribution was shown

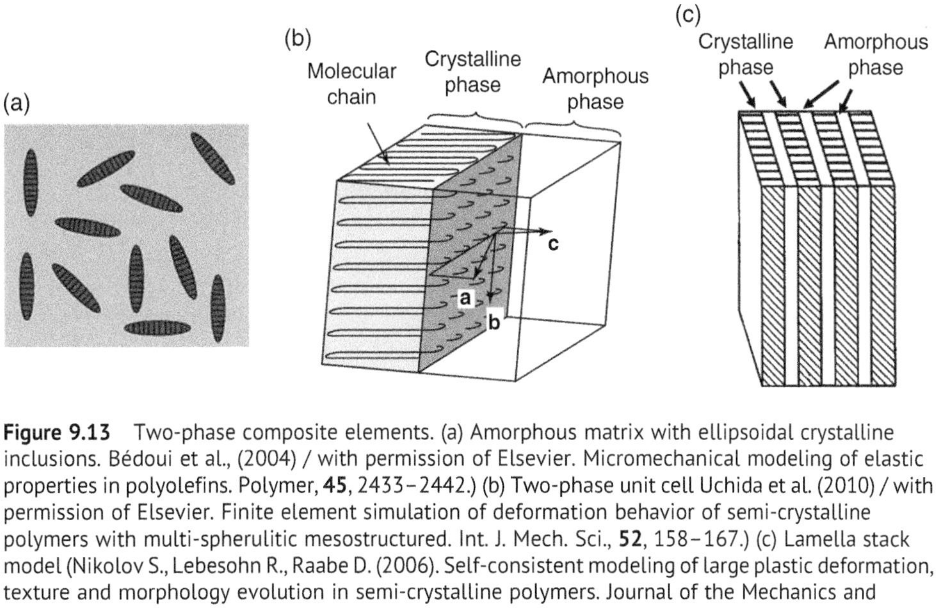

Figure 9.13 Two-phase composite elements. (a) Amorphous matrix with ellipsoidal crystalline inclusions. Bédoui et al., (2004) / with permission of Elsevier. Micromechanical modeling of elastic properties in polyolefins. Polymer, **45**, 2433–2442.) (b) Two-phase unit cell Uchida et al. (2010) / with permission of Elsevier. Finite element simulation of deformation behavior of semi-crystalline polymers with multi-spherulitic mesostructured. Int. J. Mech. Sci., **52**, 158–167.) (c) Lamella stack model (Nikolov S., Lebesohn R., Raabe D. (2006). Self-consistent modeling of large plastic deformation, texture and morphology evolution in semi-crystalline polymers. Journal of the Mechanics and Physics of Solids **54**, 1350–1375.)

to correspond to macroscopic isotropy. The model essentially makes predictions post-yield, in terms of both stress–strain behaviour and texture. The authors compared the model results with those of both uniaxial compression and plane strain compression experiments, showing good predictions of both stress–strain curves and X-ray crystallographic textures.

Bédoui et al. (2004) have compared the performance of models using composite elements shown in Figure 9.13(b) adopted by Lee et al. with those using the inclusion approach typified by Figure 9.13(a). The inclusions are crystalline ellipsoids within an amorphous matrix. They used these approaches to calculate elastic moduli and Poisson's ratios of polypropylene and HDPE as a function of crystallinity. They found the inclusion approach to be substantially inferior to the composite element approach. At crystallinities up to 50%, elastic moduli were greatly over-estimated and Poisson's ratios under-estimated by the inclusion method, with the composite element method generally relatively effective.

Van Dommelen et al. (2003a) used the composite element shown in Figure 9.13(b) to produce models of HDPE of a similar nature to those of Lee et al. (1993a, 1993b) but with the crucial addition of elastic components to both the crystalline and amorphous phases. Anisotropic elasticity and strain-rate-dependent crystal plasticity are combined to represent the crystal phase. For the amorphous phase, rate-dependent plasticity is combined with the parallel Arruda–Boyce eight chain model to provide strain hardening as in the model of Lee et al., with the addition of a series neo-Hookean elastic process (see Chapter 3). Thus, both phases have the potential for yielding behaviour. Realistic strain-rate dependent yield points were achieved. This analysis is also capable of modelling unloading as well as loading behaviour. Both these features are advances on the modelling of Lee et al.

9.4.1 Spherulites

To investigate the detailed effects of a spherulitic undeformed state, van Dommelen et al. (2003b) used multiscale modelling and the finite element method. The lowest level of structure was still the element of Figure 9.13(b), retaining the constitutive behaviour adopted in van Dommelen (2003a). These were incorporated into model spherulites as shown in Figure 9.14. The complex structure of the spherulite was simplified to a model for which lamellae grow radially from the centre along radii, with a specified degree of twisting. Representative Volume Elements (RVEs) were constructed (Figure 9.14(d)) from spherulite models with centres irregularly positioned. The RVEs were stretched in uniaxial and plane strain tension. Figure 9.14(e) shows strain levels for stretching in plane strain. Strain (and stress) concentrations form initially at the spherulite centres, and strains go on to reach values of double the macroscopic level. Uchida et al. (2010) used an RVE-based finite element analysis of spherulitic structure based on Voroni polygons that also showed stress and strain concentration at spherulite centres. Oktay and Gürses (2015) have modelled individual spherulites as two-phase structures using the finite element method, exploring the response of different crystallinities and loading modes.

9.4.2 Shape Evolution of Crystallites

In Section 9.1 we described processes involving gross rearrangements of structure, including breakup of crystalline stacks and destruction of spherulites. So far, the modelling described here falls short of this, with composite elements being rotated but retaining their

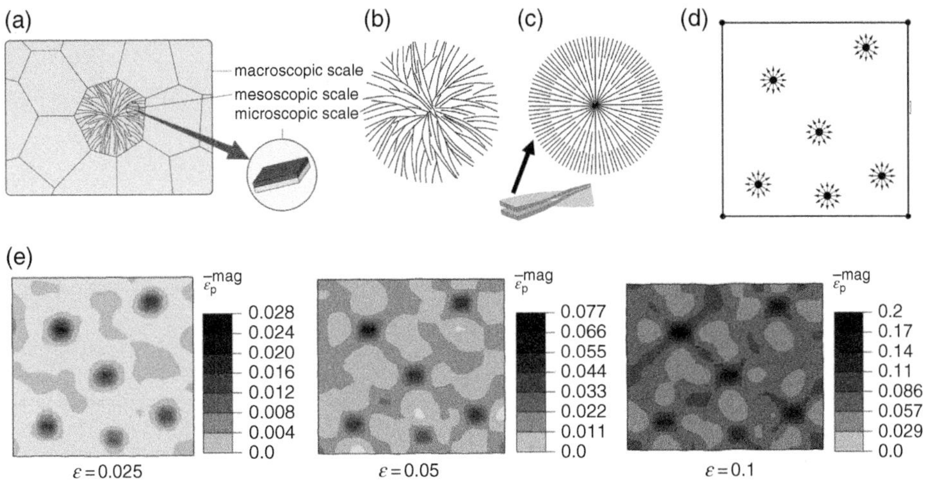

Figure 9.14 Multiscale model. (a) Composite element at microscopic scale. (b) Spherulite, with twisted lamellae. (c) Idealized spherulite, with radial plane lamellae. (d) RVE showing positions of spherulite centres. (e) Contour plots of plastic strain magnitude in plane strain at three levels of macroscopic strain ε, showing strain concentrations at spherulite centres. Brekelmans and Baaijens (2003b) / with permission of Elsevier.

integrity, with overall deformations corresponding to a relatively moderate range of strain. Nikolov et al. (2006) went some way to remedy this situation by introducing model crystalline stacks with the capability of significant change of shape. They used the multilayer composite element type of Figure 9.13(c) for this purpose. One of their motivations for this choice was that it reflects reality; experimental observations show that during large deformations, stacks several laminae thick rotate as a rigid body. For simplicity, they reverted to constitutive relations resembling those of Lee et al. (1993a, 1993b) in that there is no mechanism to represent initial elastic response. The major difference between the Lee and Nikolov constitutive equations is in the origin of strain hardening. In the Nikolov treatment the amorphous phase is assumed to be viscoplastic and the strain hardening originates from an increase in resistance to interlaminar shear, rather than from the Arruda–Boyce model in the Lee treatment.

In the implementation of their model, Nokolov et al. used initially spherical multilayer elements that evolved into prolate ellipsoids in tension and oblate ellipsoids in compression. They modelled HDPE with 70% crystallinity. The evolution with equivalent strain of the aspect ratios of the elements is shown in Figure 9.15. Nikolov et al. noted that, if the spherical elements were not allowed to change shape, results were very similar to those of Lee et al. (1993a, 1993b). The advantage of the Nikolov model over that of Lee can be seen in Figure 9.16, where results are compared with experiments (G'Sell and Jonas 1979; Hiss et al. 1999) in uniaxial tension. The prediction of strain hardening is much improved.

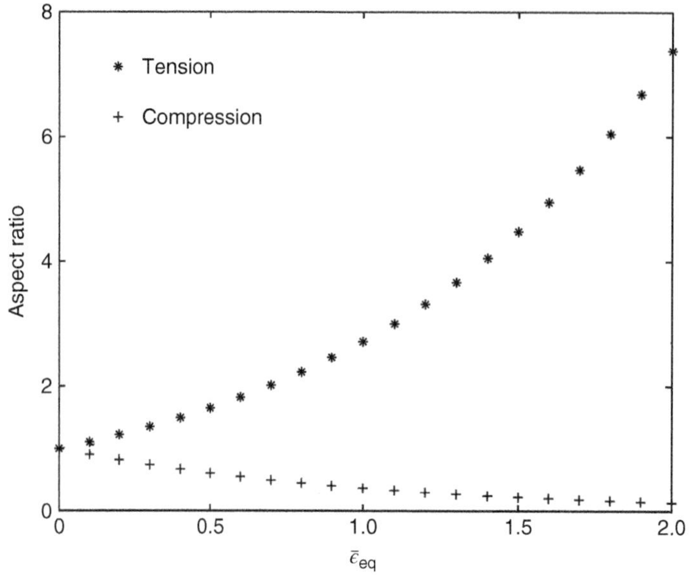

Figure 9.15 Aspect ratios (symmetry axis length/transverse diameter) of ellipsoidal, initially spherical inclusions as strain is increased. (Adapted from Nikolov, S., Lebesohn, R., and Raabe, D. (2006). Self-consistent modeling of large plastic deformation, texture and morphology evolution in semi-crystalline polymers. J. Mech. Phys. Solids, **54**, 1350–1375.)

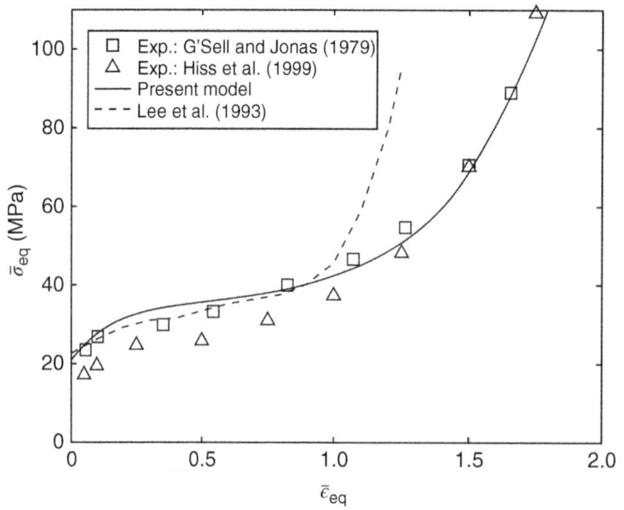

Figure 9.16 Predictions of model of Nikolov et al. (2006) (Present model) and that of Lee et al. (1993a, 1993b) compared with experimental results. (Nikolov, S., Lebesohn, R., and Raabe, D. (2006). Self-consistent modeling of large plastic deformation, texture and morphology evolution in semi-crystalline polymers. J. Mech. Phys. Solids, **54**, 1350–1375.)

9.4.3 Highly Oriented Structures

Each of the three-dimensional simulations reported above in this section started from an undeformed state. There are processes that produce polymer product with much higher draw ratios than have been attained in these models. Major structural breakdown and rearrangement that accompanies very high stretch ratios present a step increase in the challenge to simulation and impose limits to it. It is however of interest to understand the physical origins of the mechanical properties of highly oriented structures, even if we lack a detailed model of how they arise. Taking the oriented structure as given, modelling can provide a means to predict its mechanical properties, as with the crystal bridge model above.

UHMWPE fibres can be produced with fibrillar microstructures by gel-spinning and drawing. Staniszewski et al. (2020) have modelled their mechanical properties. A fibre is treated as an assembly of shorter macrofibrils as shown in Figure 9.17. The fibre is assumed to consist of 83% purely crystalline macrofibrils by volume, with the remaining material amorphous. Axial coupling between macrofibrils is weak as it is transmitted by soft amorphous material, The major mechanical coupling between fibres is by shear, the source of which is interfibrillar tie-chains and microfibrils, acting as shown in Figure 9.18. Model results are compared with experimental strain measurements on stretched fibrils made using an optical method. High-strain regions are detectable, supporting the model approach; predictions and experiments are compared in Figure 9.19.

Highly oriented UHMWPE fibres have attracted the attention of a number of other workers. Strawhecker et al. (2016) used AFM measurements to detect shish-kebab morphologies, and noted that various levels of interaction are possible between neighbouring structures (e.g., standalone or interlocking). Dong et al. (2018) have made detailed models of these phenomena, using a range of RVEs to simulate different modes of interaction.

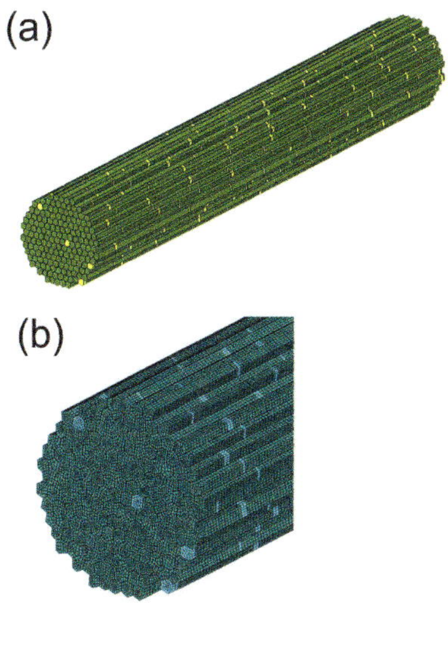

Figure 9.17 Model of fibre consisting of an assembly of macrofibrils. (a) General view (b) one end showing finite element mesh. Lighter regions show higher strain in softer material between fibrils. Staniszewski et al. (2020) / with permission of Elsevier.

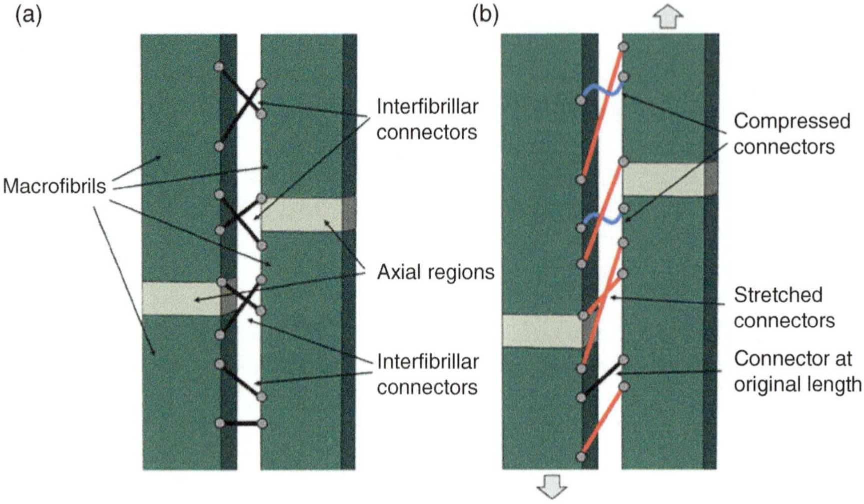

Figure 9.18 Modelling of shear coupling between macrofibrils. In (a), connectors are illustrated; in (b) stretched versus compressed connectors are distinguished. Staniszewski et al. (2020) / with permission of Elsevier.

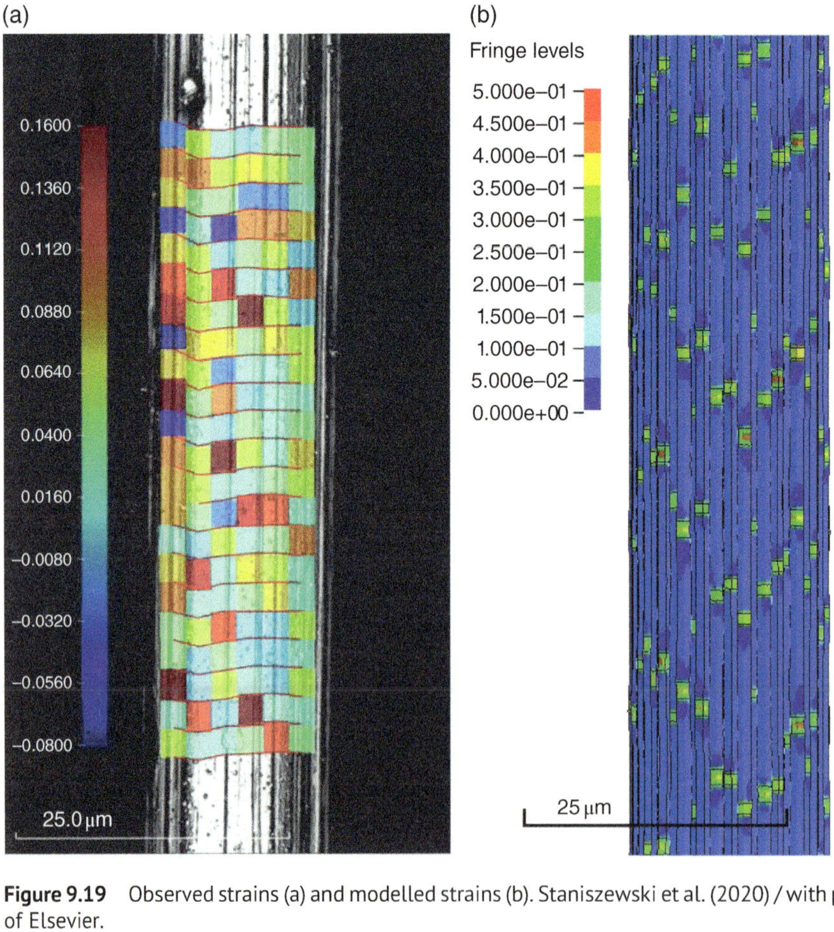

Figure 9.19 Observed strains (a) and modelled strains (b). Staniszewski et al. (2020) / with permission of Elsevier.

References

Amornsakchai, T., Olley, R.H., Bassett, D.C., Al-Hussain, M.O.M., Unwin, A.P. and Ward, I.M. (2000) On the influence of initial morphology on the internal structure of highly drawn polyethylene. *Polymer*, **41**, 8291–8298.

Argon, A.S. (1973) A theory for the low-temperature plastic deformation of glassy polymers. *Philos. Mag.*, **28**, 839.

Arruda, E.M. and Boyce, M.C. (1993) A 3-dimensional constitutive model for the large stretch behavior of rubber elastic-materials. *J. Mech. Phys. Solids*, **41**, 389.

Barham, P.J. and Arridge, R.G.C. (1977) A fibre composite model of drawn crystalline polymers. *J. Polym. Sci. Part B: Polym. Phys.*, **15**, 1177–1188.

Bassett, D.C. (2003) Polymer spherulites: a modern assessment. *J. Macromol. Sci., B: Phys.*, **42**, 227–256.

Bédoui, F., Diani, J. and Régnier, G. (2004) Micromechanical modeling of elastic properties in polyolefins. *Polymer*, **45**, 2433–2442.

Breese, D.R. and Beaucage, G. (2004) A review of modeling approaches for oriented semi-crystalline polymers. *Curr. Opin. Solid State Mater. Sci.*, **8**, 439–448.

Capaccio, G., Gibson, A.G. and Ward, I.M. (1979) Drawing and hydrostatic extrusion of ultra-high modulus polymers, in *Ultra-High Modulus Polymers*, (eds A. Ciferri and I.M. Ward), Applied Science Publishers, London, UK, pp. 1–76.

Clements, J., Jakeways, R. and Ward, I.M. (1978) Lattice modulus and crystallite thickness measurements in ultra-high modulus linear polyethylene. *Polymer*, **19**, 639–643.

Coates, P.D. and Ward, I.M. (1981) Die drawing: solid phase drawing of polymers through a converging die. *Polym. Eng. Sci.*, **21**, 612–618.

Cox, H.L. (1952) The elasticity and strength of paper and other fibrous materials. *Br. J. Appl. Phys.*, **3**, 72.

Crist, B. and Schultz, J.M. (2016) Polymer spherulites: a critical review. *Prog. Polym. Sci.*, **56** (1–63), 103129.

Deplancke, T., Fivel, M. and Lame, O. (2019) 1D strain rate-dependent constitutive model of UHMWPE: from crystalline network to fibrillar structure behavior. *Mech. Mater.*, **137**, 103129.

van Dommelen, J.A.W., Parks, D.M., Boyce, M.C., Brekelmans, W.A.M. and Baaijens, F.P.T. (2003a) Micromechanical modeling of the elasto-viscoplastic behavior of semi-crystalline polymers. *J. Mech. Phys. Solids*, **51**, 519–541.

van Dommelen, J.A.W., Parks, D.M., Boyce, M.C., Brekelmans, W.A.M. and Baaijens, F.P.T. (2003b) Micromechanical modeling of intraspherulitic deformation of semicrystalline polymers. *Polymer*, **44**, 6089–6101.

van Dommelen, J.A.W., Schrauwen, B.A.G., van Breemen, L.C.A. and Govaert, L.E. (2004) Micromechanical modeling of the tensile behavior of oriented polyethylene. *J. Polym. Sci. B Polym. Phys.*, **42**, 2983–2994.

Dong, H., Wang, Z., O'Connor, T.C., Azoug, A., Robbins, M.O. and Nguyen, T. (2018) Micromechanical models for the stiffness and strength of UHMWPE macrofibrils. *J. Mech. Phys. Solids*, **116**, 70–98.

Galeski, A. (2003) Strength and toughness of crystalline polymer systems. *Prog. Polym. Sci.*, **28**, 1643–1699.

Gibson, A.G., Ward, I.M., Cole, B.N. and Parsons, B. (1974) Hydrostatic extrusion of linear polyethylene. *J. Mater. Sci.*, **9**, 1193–1196.

Gibson, A.G., Greig, D., Sahota, M., Ward, I.M. and Choy, C.L. (1977) Thermal conductivity of ultrahigh-modulus polyethylene. *J. Polym. Sci.: Polym. Lett. Ed.*, **15**, 183–192.

Gibson, A.G., Davies, G.R. and Ward, I.M. (1978) Dynamic mechanical behaviour and longitudinal crystal thickness measurements on ultra-high modulus linear polyethylene: a quantitative model for the elastic modulus. *Polymer*, **19**, 683–693.

G'Sell, C. and Jonas, J.J. (1979) Determination of the plastic behavior of solid polymers at constant true strain rate. *J. Mater. Sci.*, **14**, 583–591.

Hiss, R., Hobeika, S., Lynn, C. and Strobl, G. (1999) Network stretching, slip processes, and fragmentation of crystallites during uniaxial drawing of polyethylene and related copolymers. A comparative study. *Macromolecules*, **32**, 4390–4403.

Humbert, S., Lame, O., Chenal, J.M., Rochasw, C. and Vigier, G. (2010) New insight on initiation of cavitation in semicrystalline polymers: in-situ SAXS measurements. *Macromolecules*, **43**, 7212–7221.

Kaito, A., Nakayama, K. and Kanetsuna, H. (1985) Roller drawing of ultrahigh molecular weight polyethylene. *J. Appl. Polym. Sci.*, **30**, 4591–4608.

Lee, B.J., Argon, A.S., Parks, D.M., Ahzi, S. and Bartczak, Z. (1993a) Simulation of large strain plastic deformation and texture evolution in high density polyethylene. *Polymer*, **34**, 3555–3575.

Lee, B.J., Parks, D.M. and Ahzi, S. (1993b) Micromechanical modeling of large plastic deformation and texture evolution in semi-crystalline polymers. *J. Mech. Phys. Solids*, **41**, 1651–1687.

Lu, Y. and Men, Y. (2018) Cavitation-induced stress whitening in semi-crystalline polymers. *Macromol. Mater. Eng.*, **303**, 1800203.

Lu, Y., Wang, Y., Chen, R., Zhao, J., Jiang, Z. and Men, Y. (2015) Cavitation in isotactic polypropylene at large strains during tensile deformation at elevated temperatures. *Macromolecules*, **48**, 5799–5806.

Lyu, D., Sun, Y., Lu, Y. *et al.* (2020) Suppressed cavitation in die-drawn isotactic polypropylene. *Macromolecules*, **53**, 4863–4873.

Lyu, D., Sun, Y., Lu, Y. *et al.* (2021) A simple way to control small-strain cavitation in die-drawn isotactic polypropylenene. *Ind. Eng. Chem. Res.*, **60**, 5151–5160.

Lyu, D., Han, C., Lu, Y. and Men, Y. (2023) The breakage of original crystallites determined small strain cavitation in syndiotactic polypropylene during step-cycle tensile deformation. *Polymer*, **283**, 126294.

Men, Y., Rieger, J. and Homeyer, J. (2004) Synchrotron ultrasmall-angle X-ray scattering studies on tensile deformation of poly(1-butene). *Macromolecules*, **37**, 9481–9488.

Mirkhalaf, M. and Vadizadeh, R. (2024) Micro-mechanical modeling of semi-crystalline polymers: a review. *Int. J. Solids Struct.*, **290**, 112691.

Mirkhalaf, M., van Dommelen, J.A.W., Govaert, L.E., Furmanski, J. and Geers, M.G.D. (2019) Micromechanical modeling of anisotropic behavior of oriented semicrystalline polymers. *J. Polym. Sci. Part B: Polym. Phys.*, **57**, 378–391.

Mohanraj, J., Bonner, M.J., Barton, D.C. and Ward, I.M. (2006) Physical and mechanical characterization of oriented polyoxymethylene produced by die-drawing and hydrostatic extrusion. *Polymer*, **47**, 5897–5908.

Mohanraj, J., Morawiec, J., Pawlak, A., Barton, D.C., Galeski, A. and Ward, I.M. (2008) Orientation of polyoxymethylene by rolling with side constraints. *Polymer*, **49**, 303–316.

Nikolov, S., Lebesohn, R. and Raabe, D. (2006) Self-consistent modeling of large plastic deformation, texture and morphology evolution in semi-crystalline polymers. *J. Mech. Phys. Solids*, **54**, 1350–1375.

Oktay, E.E. and Gürses, E. (2015) Modeling of spherulite microstructures in semicrystalline polymers. *Mech. Mater.*, **90**, 83–101.

Pawlak, A. and Galeski, A. (2005) Plastic deformation of crystalline polymers: the role of cavitation and crystal plasticity. *Macromolecules*, **38**, 9688–9697.

Peirce, D., Asaro, R.J. and Needleman, A. (1983) Material rate dependence and localized deformation in crystalline solids. *Acta Metall.*, **31**, 1951–1976.

Peterlin, A. (1971) Molecular model of drawing of polyethylene and polypropylene. *J. Mater. Sci.*, **6**, 490–508.

Peterlin, A. (1972) Plastic deformation of unoriented polymer under tensile load, in *Advances in Polymer Science and Engineering.*, Plenum Press, New York, pp. 1–19.

Staniszewski, J.M., Bogetti, T.A., Wu, V. and Moy, P. (2020) Interfibrillar behavior in ultra-high molecular weight polyethylene (UHMWPE) single fibers subjected to tension. *Int. J. Solids Struct.*, **206**, 354–369.

Strawhecker, K.E., Sandoz-Rosado, E.J., Stockdale, T.A. and Laird, W.D. (2016) Interior morphology of high-performance polyethylene fibers revealed by modulus mapping. *Polymer*, **103**, 224–232.

Takayanagi, M. (1963) Viscoelastic properties of crystalline polymers. *Mem. Fac. Eng. Kyushu Univ.*, **23**, 41.

Takayanagi, M., Imada, I. and Kajiyama, T. (1966) Mechanical properties and fine structure of drawn polymers. *J. Polym. Sci. Polym. Symp.*, **15**, 263–281.

Uchida, M., Tokuda, T. and Tada, N. (2010) Finite element simulation of deformation behavior of semi-crystalline polymers with multi-spherulitic mesostructure. *Int. J. Mech. Sci.*, **52**, 158–167.

10

Relaxation Transitions: Experimental Behaviour and Molecular Interpretation

We shall discuss the assignment of viscoelastic relaxations in a molecular sense to different chemical groups in the molecule, and in a physical sense to features such as the motion of molecules in crystalline or amorphous regions. Because amorphous polymers exhibit fewer structure-dependent features than those that are semi-crystalline, we shall use these simpler materials to illustrate some general characteristics of relaxation behaviour.

10.1 Amorphous Polymers: An Introduction

It is customary to label relaxation transitions in polymers as α, β, γ, δ and so on in alphabetical order with decreasing temperature. Three of the four transitions in polymethyl methacrylate (PMMA)

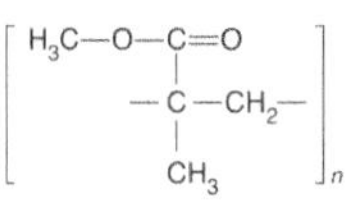

are shown in Figure 10.1, which summarizes data obtained using a low-frequency torsion pendulum. The highest temperature relaxation, the α relaxation, is the glass transition and is associated with a large change in modulus. Comparative studies on similar polymers, together with nuclear magnetic resonance (NMR) and dielectric measurements (Deutsch et al. 1954; Sinnott 1960; Powles et al. 1964; Heijboer 1965, p. 231; McCrum et al. 1967; Roberts and White 1973), have shown the β relaxation to be associated with side-chain motions of the ester group. The γ and δ relaxations involve motion of the methyl groups attached to the main chain and the side chain, respectively.

More extensive recent researches on the β relaxation have been undertaken by Monnerie et al. (2005). It was concluded that for polymers with side chains such as PMMA the β relaxation involved π flips of the ester group around the bond with the main chain. These were isolated in the low-temperature part of the relaxation, but intramolecular cooperativity was involved in the high-temperature part. These ring flips are accompanied by a change in the rotation angles of the main chain. The β transition was also studied in maleimide and

Mechanical Properties of Solid Polymers, Fourth Edition. John Sweeney and Peter Hine.

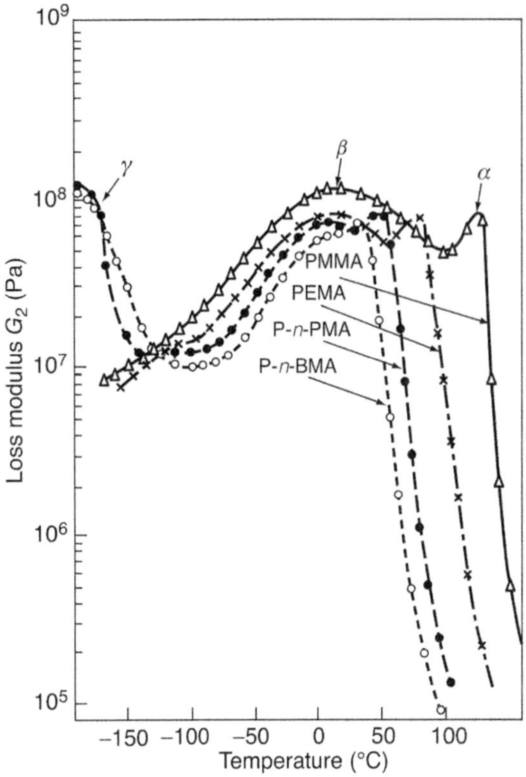

Figure 10.1 Temperature dependence of loss modulus G_2 for polymethyl methacrylate (PMMA), polyethyl methacrylate (PEMA), poly-n-propyl methacrylate (P-*n*-PMA) and poly-*n*-butyl methacrylate (P-*n*-BMA). Adapted from Heijboer, 1965.

glutarimide PMMA copolymers. In all cases, the cooperativity of the molecular motions is intramolecular and in the maleimide copolymers their rigidity hinders the propagation of main-chain adjustment.

The case of amorphous poly(ethylene terephthalate) (PET), where there are no side groups, is considered in Section 10.3.2 in conjunction with semi-crystalline forms of this polymer. Worthy of mention is the use of antiplasticizers to show that the β relaxation is composite (Maxwell et al. 1998a, 1998b). In studies of bisphenol A (and tetramethyl bisphenol A), polycarbonate intramolecular cooperativity associated with the carbonate residue was shown by the introduction of the methyl groups on the phenyl rings (Aoki and Brittain 1977), and intermolecular cooperativity was shown by the dielectric behaviour of blends (Katana et al. 1993). Further evidence for intermolecular contributions to the ring motions was obtained from the observed pressure dependence of the proton NMR transverse relaxation time (Walton et al. 1990). There is a key conclusion for polymers without side chains containing phenyl rings such as PET and bisphenol A polycarbonate. For the linear polymer PET, phenyl ring π flips occurring in the high-temperature parts of the β transition are associated with intramolecular cooperativity only, whereas for polycarbonate there is intermolecular cooperativity.

10.2 Factors Affecting the Glass Transition in Amorphous and Low Crystallinity Polymers

Two distinct models have been used for interpreting the influence of features such as chemical structure, molecular mass, cross-linking and plasticizers on the glass transition in amorphous polymers. The first approach considers changes in molecular flexibility, which modify the ease with which conformational changes can take place. The alternative approach relates all these effects to the amount of free volume, which is assumed to attain a critical value at the glass transition.

10.2.1 Effect of Chemical Structure

Although these factors have been intensively studied because of their importance in selecting polymers for commercial exploitation, much of our knowledge is empirical in nature, due primarily to the difficulty in distinguishing between intra- and intermolecular effects. Some general features are, however, evident.

10.2.1.1 Main-Chain Structure

Flexible groups such as an ether link will enhance main-chain flexibility and reduce the glass transition temperature, with the opposite effect being shown by the introduction of an inflexible group, such as a terephthalate residue.

10.2.1.2 Side Groups (Vincent 1965)

Bulky, inflexible side groups increase the temperature of the glass transition, as is illustrated in Table 10.1 for a series of substituted poly-α-olefins:

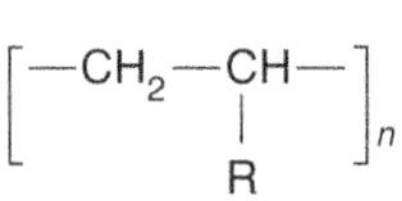

A difference between the effect of rigid and flexible side groups is shown in Table 10.2 for a series of polyvinyl butyl ethers:

Table 10.1 Glass transition of some vinyl polymers.

Polymer	R	Transition temperature in °C at ~1 Hz
Polypropylene	CH_3	0
Polystyrene	C_6H_5	116
Poly-*N*-vinylcarbazole	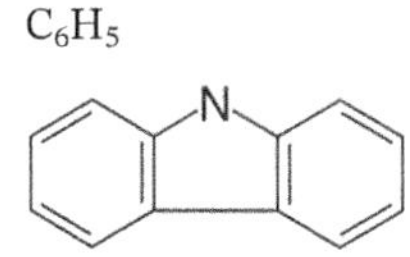	211

Source: Adapted from The Physics of Plastics (ed. P.D. Ritchie).

Table 10.2 Glass transition of some isomeric polyvinyl butyl ethers.

Polymer	R_1	Transition temperature in °C at ~1 Hz
Polyvinyl *n*-butyl ether	$CH_2CH_2CH_2CH_3$	−32
Polyvinyl isobutyl ether	$CH_2CH(CH_3)_2$	−1
Polyvinyl *t*-butyl ether	$C(CH_3)_3$	+83

Source: Adapted from The Physics of Plastics (ed. P.D. Ritchie).

Table 10.3 Glass transition of some polyvinyl n-alkyl ethers.

Polymer	R_2	Transition temperature in °C at ~1 Hz
Polyvinyl methyl ether	CH_3	−10
Polyvinyl ethyl ether	CH_2CH_3	−17
Polyvinyl *n*-propyl ether	$CH_2CH_2CH_3$	−27
Polyvinyl *n*-butyl ether	$CH_2CH_2CH_2CH_3$	−32

Source: Adapted from The Physics of Plastics (ed. P.D. Ritchie).

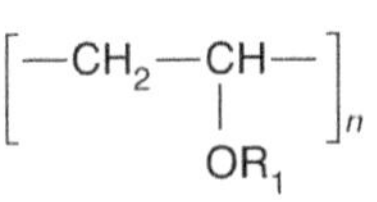

All these polymers contain the same atoms in the side group OR_1 (where R_1 represents the butyl isomeric form), but more compact arrangements reduce the flexibility of the molecule and give a marked increase in the transition temperature.

Increasing the length of flexible side groups reduces the temperature of the glass transition, as is evident from Table 10.3 for a series of polyvinyl ethers:

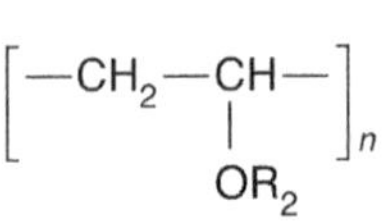

where R_2 represents the *n*-alkyl group. Here, the increase in length is associated with an increase in free volume at a given temperature.

10.2.1.3 Main-Chain Polarity

In a series of polymers of similar main-chain composition, the temperature of the glass transition may be significantly depressed as the number of successive —CH_2 or —CH_3 groups in the side groups is increased (Figure 10.2). It is evident that the temperature of the glass transition increases with main-chain polarity, and it is assumed that the associated reduction in main-chain mobility is due to the increase in intermolecular forces. In particular, it is suggested that the higher values for the polychloracrylic esters arise from the increased valence forces associated with the chlorine molecules.

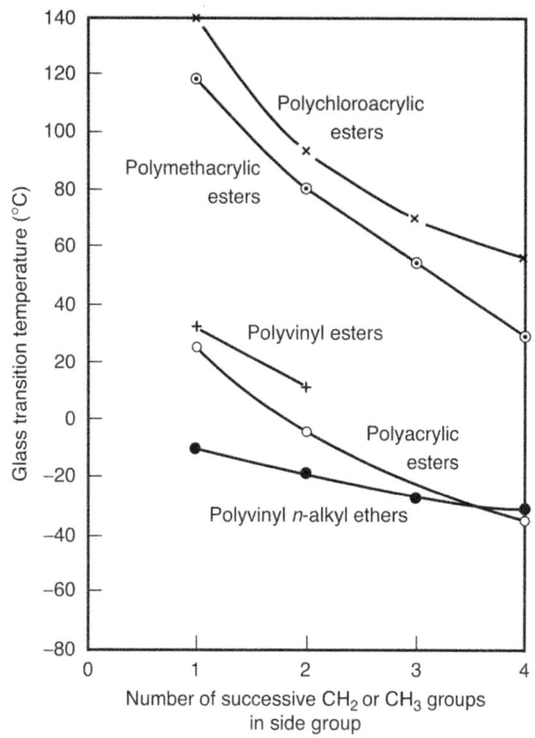

Figure 10.2 The effect of polarity on the position of the glass transition temperature for five polymer series. Adapted from Vincent, 1965.

10.2.2 Effect of Molecular Mass and Cross-Linking

The length of the main chain does not affect the dynamic mechanical properties of polymers in the glassy state, where molecular motions are of restricted extent, but the glass transition temperature is depressed at very low relative molecular masses as a consequence of the additional free volume introduced by the increased proportion of chain ends (Fox and Flory 1950, 1954).

As already discussed (see Section 7.1.1), molecular mass has a large effect in the glass transition range, where viscous flow transforms to a plateau range of rubber-like behaviour due to entanglements between the longer molecular chains.

Chemical cross-links reduce the free volume by bringing adjacent chains close together and so raise the temperature of the glass transition, as is shown in Figure 10.3 for phenol–formaldehyde resin cross-linked with hexamethylene tetramine to different extents. The transition region is greatly broadened (Nielsen 1962), so that in very highly cross-linked materials, where motions of extensive segments of the main chain are not possible, there is no glass transition.

10.2.3 Blends, Grafts and Copolymers

The mechanical properties of blends and graft polymers are determined primarily by the mutual solubility of the two homopolymers. For complete solubility, the properties of the mixture are close to those of a random copolymer of the same composition, as is shown in Figure 10.4, which compares a 50:50 mixture of polyvinyl acetate and polymethyl acrylate

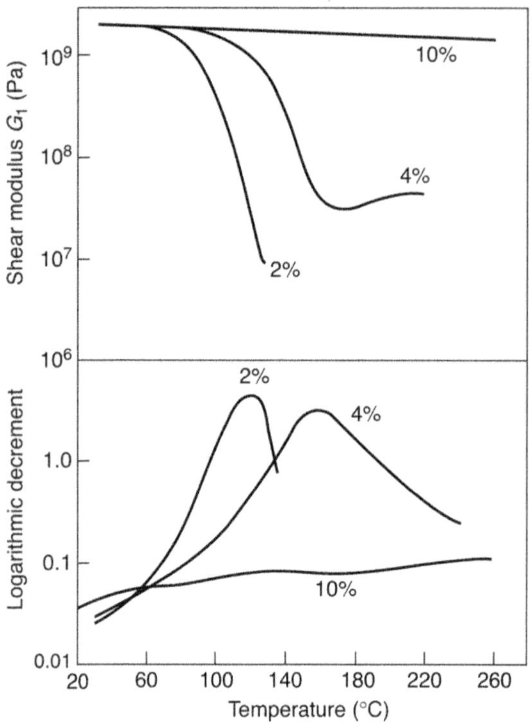

Figure 10.3 Shear modulus G_1 and logarithmic decrement of a phenol–formaldehyde resin cross-linked with hexamethylene tetramine at stated concentrations. Adapted from Drumm et al., 1956.

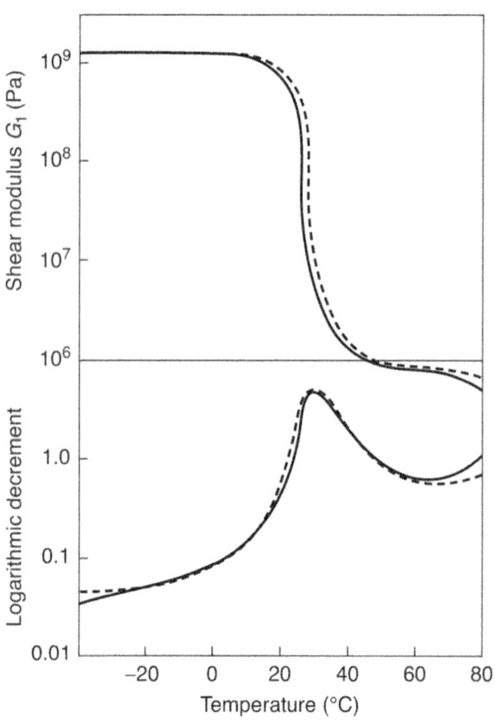

Figure 10.4 Shear modulus G_1 and logarithmic decrement for a miscible blend of polyvinyl acetate and polymethyl acrylate (- - -) and a copolymer of vinyl acetate and methyl acrylate (– – – –). Adapted from Nielsen, 1962.

with the equivalent copolymer (Nielsen 1962). Note that the damping peak occurs at 30°C, compared with 15°C for the polymethyl acrylate and 45°C for polyvinyl acetate.

A theoretical interpretation of the glass transition temperature of a copolymer is based on the assumption that the transition occurs at a constant fraction of free volume. Gordon and Taylor (1952) assume that in an ideal copolymer the partial specific volumes of the two components are constant and equal to the specific volumes of the two homopolymers. The specific volume–temperature coefficients for the two components in the rubbery and glassy states are assumed to remain the same in the copolymer as in the homopolymers, and to be independent of temperature. The glass transition temperature T_g for the copolymer is then given by (Fox and Loshaek 1955; Mandelkern et al. 1957).

$$\frac{1}{T_g} = \frac{1}{(w_1 + Bw_2)} \left[\frac{w_1}{T_{g1}} + \frac{Bw_2}{T_{g2}}\right],$$

where w_1 and w_2 are the mass fractions of the two monomers whose homopolymers have transition temperatures T_{g1} and T_{g2}, respectively, and B is a constant close to unity.

Where the two polymers in a mixture are insoluble they exist as separate phases, so that two glass transitions are observed as shown in Figure 10.5 for a polyblend of polystyrene and styrene–butadiene rubber (Nielsen 1962). The two loss peaks are very close to those for pure polystyrene and pure styrene–butadiene rubber.

10.2.4 Effects of Plasticizers

Plasticizers, which are relatively low-molecular mass organic materials added to soften rigid polymers, must be soluble in the polymer and usually they dissolve it completely at high temperature. The classic example of their application is polyvinyl chloride (PVC) which can be plasticized with di(ethylhexyl)phthalate (Wolf 1951). In this polymer, plasticization

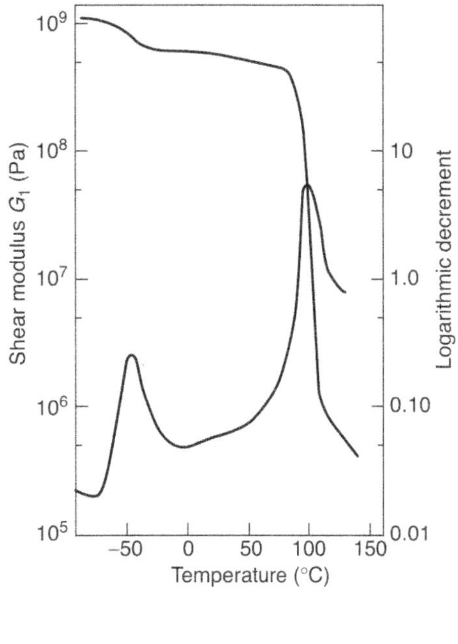

Figure 10.5 Shear modulus G_1 and logarithmic decrement for an immiscible polyblend of polystyrene and a styrene–butadiene copolymer. Adapted from Nielsen, 1962.

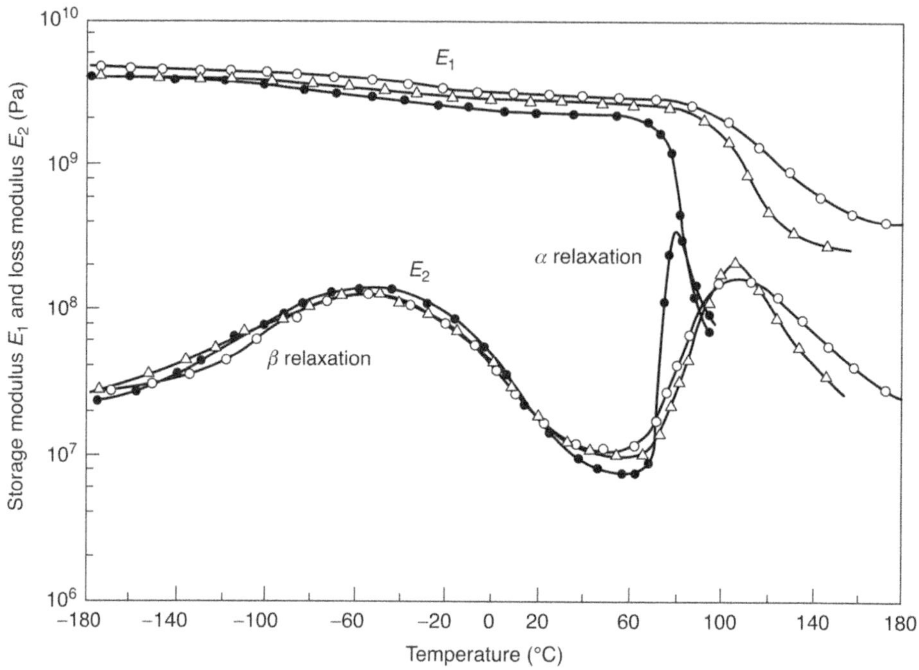

Figure 10.7 Storage modulus E_1 and loss modulus E_2 as a function of temperature at 138 Hz for PET samples of differing degrees of crystallinity (•, 5%; Δ, 34%; ○, 50%). Adapted from Takayanagi, 1964.

the same conclusion. The process is therefore consistent with localized molecular motions, in contrast with the restrained long-range segmental motions involved in the glass–rubber α relaxation.

Boyd (1985b) has analysed the dynamic mechanical behaviour of PET in terms of a composite model of crystalline and amorphous phases. The relaxation strengths for the α and β processes, determined from the shear modulus results of Illers and Breuer, were shown to be related to the crystallinity, indicating that both relaxation processes related to the amorphous regions. For the β processes, the shear modulus for the amorphous phase lies between the upper (Voigt) and lower (Reuss) bounds in both the relaxed and unrelaxed states. The α process, however, appeared to fit best near the lower bound behaviour.

Although the phenomenological approach of Boyd gives some insight into the dynamic mechanical behaviour, it does not provide any significant understanding of the molecular origins of the relaxation processes. Maxwell et al. (1998a, 1998b) have shown how it is necessary to bring together dynamic mechanical, NMR and dielectric measurements to achieve such an understanding, and NMR spectroscopy is probably the most powerful tool in this respect. The main advantage of NMR is that it allows the mobility in different parts of the molecule to be examined. There are several investigations of the molecular relaxations in PET using NMR. Using selectively deuterated samples to distinguish between molecular motions in the aliphatic and aromatic parts of the PET molecule, English (1984) identified significant motion of the ethylene glycol units at the glass transition temperature, which have been assumed to be due to a unique *trans-gauche* movement of these units (see Figure 1.7). It is important to note that the ethylene glycol units do not contribute to the relaxation processes present in the β peak. This result is consistent with much earlier NMR studies on selectively deuterated PET samples by Ward (1960), which showed that the α relaxation in PET involved significant motions of both the aliphatic and aromatic moieties.

Alongside mechanical spectroscopy and NMR, dielectric spectroscopy (Kremer 2003; Popov et al. 2022) is also a very powerful technique for examining relaxation processes within polymers. Dielectric spectroscopy measures the interaction between an oscillating external electric field and internal parts of the polymer that are dielectrically active by having a permanent or induced dipole moment. The movement of these active groups in response to the changing external electric field therefore provides information into a range of relaxation processes, which depends on where these dipole moments are located. For some polymers, for example PMMA, the side group can have a dipole moment and so local relaxation phenomena (such as β and γ relaxations) can be probed, although the direct link between the measured secondary relaxations and chemical groups is not always so clear. For other polymers, for example PPG or PI, there is a dipole moment component along the main polymer backbone, allowing relaxation information to be gathered about whole chain motions (e.g. Rouse dynamics and reptation). Additionally, specific polymers can be synthesized (Kotaka et al. 1992), often based on PI, to contain dielectrically active groups in the position of the chain whose relaxation phenomena are to be studied.

In dielectric spectroscopy, the material to be studied is subject to an alternating electric field by placing it between electrodes. An oscillating voltage is applied to the sample through these electrodes and then the current; hence, the complex impedance is recorded. Analogous to mechanical spectroscopy where a peak in the loss Modulus, E'', or tan δ is used to characterize the temperature or frequency of a particular relaxation process, in dielectric spectroscopy this is characterized by a peak in the dielectric loss, ε''. The advantage of dielectric spectroscopy over mechanical spectroscopy is that it can probe over a very wide range of frequencies (~10^{-4} to ~10^{10} Hz). This is in comparison to mechanical measurements which can probe only four orders of magnitude around 1 Hz due to the necessity to use electric motors to produce the oscillating stress or strain required to probe mechanical relaxations. Additionally, this large frequency range available with dielectric spectroscopy often does not require time–temperature superposition to create a master curve (as with mechanical spectroscopy), although measurements at different temperatures can allow the activation energy of the various probed relaxation phenomena to be determined.

By combining mechanical and dielectric spectroscopy, it can be possible to identify the direct molecular origin of particular mechanical relaxations. The group of Watanabe et al. (2003) have been very active in developing the technique of rheo-dielectric by combining dielectric spectroscopy with rheology. In studies where mechanical and dielectric techniques are combined, it can be advantageous if only one unit of the polymer has a dipole moment, then a comparison of the mechanical and dielectric measurements could pinpoint the contribution of this unit to the overall mechanical relaxation response. Combined dielectric and mechanical dynamic measurements by Maxwell et al. (1998b) on PET used such a combined approach to show that the β peak consists of two different relaxation processes. High-resolution carbon-13 chemical shift and deuterium NMR experiments showed that both the phenyl rings and the carbonyl groups undergo small-angle oscillations at temperatures below the glass transition, and that the phenyl groups also undergo rapid 180° flips. It was confirmed that the ethylene glycol group does not contribute to the β relaxation, and it was concluded that the high-temperature side of the relaxation is due to the 180° flips and the low-temperature side to motion of the carbonyl groups, which has a significantly lower activation energy and activation enthalpy (see Section 7.3).

10.3.3 Relaxation Processes in Polyethylene

Polyethylene is the obvious choice for investigating relaxations in the more highly crystalline polymers. Its structure has been studied in great detail, and the material is readily obtainable in two forms. LDPE typically contains about three short side branches per 100 carbon atoms, together with about one longer branch per molecule. High-density polyethylene (HDPE) is much closer to the pure $(CH_2)_n$ polymer, and the proportion of side branches is often less than 5 per 1000 carbon atoms. The main features of the temperature dependence of tan δ for each material are indicated schematically in Figure 10.8. The LDPE shows clearly resolved α, β and γ loss peaks. In HDPE, the low-temperature γ peak is very similar to that in LDPE, but the β relaxation is hardly resolved and the α relaxation is often considerably modified appearing to consist of at least two processes (α and α') with different activation energies. The high-temperature behaviour is also dependent on whether loss angle or loss modulus is the quantity being measured. At one time, some investigators questioned the existence of the β relaxation in HDPE, but further work involving a range of polymers intermediate between the extremes indicated in Figure 10.8 has established its presence.

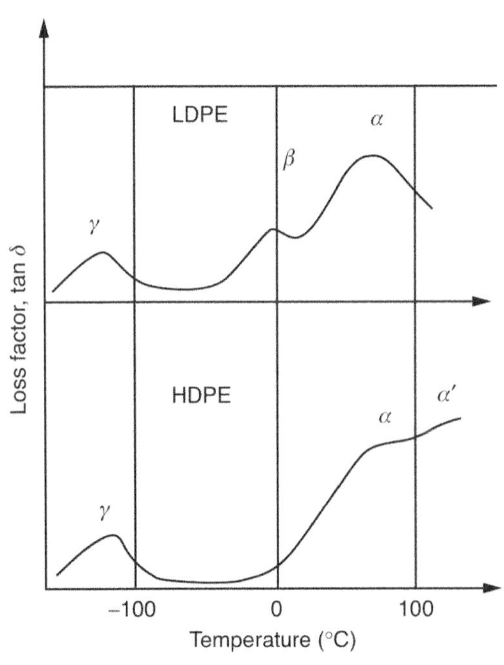

Figure 10.8 Schematic diagram showing α, α', β and γ relaxation processes in LDPE and HDPE.

We must conclude that α, β and γ relaxations occur in all forms of polyethylene. The first stage in the process of analysis is to determine whether a given relaxation is related to the crystalline or the amorphous component, or to an interaction involving both phases. Next, one deduces a process that is able to account for each relaxation. Finally, an attempt must be made to model a molecular mechanism that can cause the process to occur.

In two review articles, Boyd (1985a, 1985b) presents evidence for two major conclusions regarding the origins of the α, β and γ relaxations in polyethylene. First, the mechanical strengths of all three relaxations relate to the amorphous fraction. Secondly, both mechanical and dielectric measurements show that the location of the α relaxation depends on the crystal lamellar thickness. These conclusions might appear to be in conflict regarding the α relaxation, but we will see that recent research by Ward and co-workers (Matthews et al. 1999a, 1999b) resolves this issue and also the relationship between the relaxations and LDPE and HDPE, as to whether they are of similar molecular origin. Two extra ingredients were required to answer this question: dynamic mechanical measurements on specially produced oriented samples and measurements over a wide range of frequencies to determine the activation energies for the processes, as discussed in Section 7.3.

Key results for the mechanical anisotropy of LDPE sheets have already been discussed in Section 9.5.3. Oriented and annealed sheets can be considered as composite solids, where the β relaxation is an interlamellar shear process, consistent with its assignment by Boyd to an amorphous process. Figure 10.9 shows results for cold-drawn and annealed HDPE, where no β relaxation is observed (Stachurski and Ward 1969). In these sheets, the crystal lamellae make an acute angle of about 40° with the initial draw direction (Hay and Keller 1967). Applying the stress along the initial draw dircction then gives the maximum resolved shear stress parallel to lamellar planes. We see from Figure 10.9 that the maximum loss is

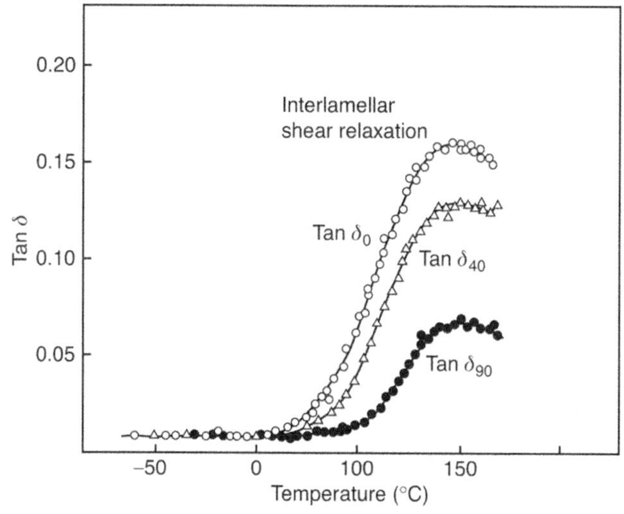

Figure 10.9 Temperature dependence of tan δ in a cold-drawn and annealed HDPE sheet in different directions at 50 Hz. Adapted from Stachurski and Ward, 1969.

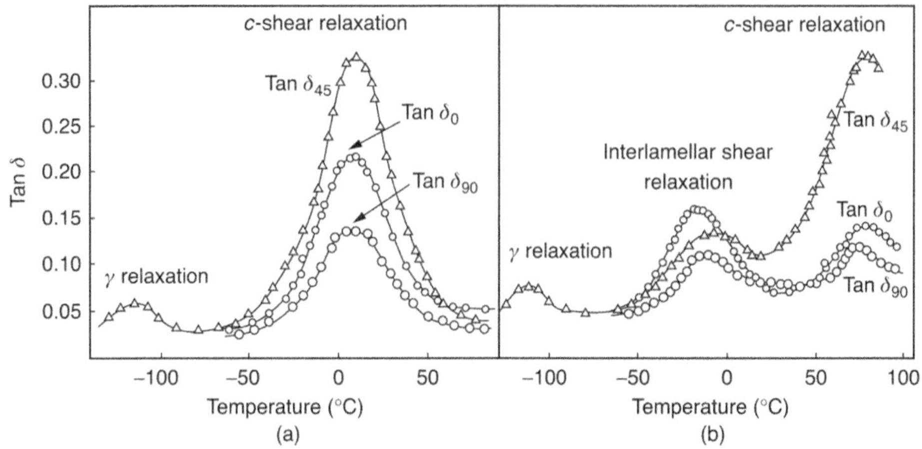

Figure 10.10 Temperature dependence of tan δ in three directions in cold-drawn (a) and cold-drawn and annealed (b) LDPE sheets at approximately 500 Hz. Adapted from Stachurski and Ward, 1969.

tan δ_0, confirming that the α relaxation in HDPE is primarily an interlamellar shear process from a macroscopic mechanical viewpoint.

The situation is apparently made more confusing by comparison of the results of Stachurski and Ward (1968) for cold-drawn (a) and annealed (b) sheets of LDPE illustrated in Figure 10.10. For the cold-drawn sheet, the maximum loss in the 0°C region occurs at 45° to the draw direction. Only one loss process is observed and this shows the anisotropy appropriate for a relaxation that involves shear parallel to the draw direction in a plane containing the draw direction. This material does not show a clear lamellar texture, so it is reasonable to associate this process with the α relaxation. In the cold-drawn and annealed sheet, the process has moved to about 70°C. These results, taken together with the measurements on specially oriented sheets, suggest that the draw or c-axis orientation of the crystalline regions is the governing factor. Because the relaxation involves shear in the c-axis direction in planes containing the c axis it has been termed the 'c-shear relaxation'. The annealed LDPE sheet also shows a β relaxation below 0°C with an anisotropy consistent with interlamellar shear, as discussed in Section 8.4.3.

The identification of the α process as a c-shear relaxation and the β process as interlamellar shear in a drawn and annealed LDPE sheet was nicely confirmed by measurements of the anisotropy of dielectric relaxation (Davies and Ward 1969). Pure polyethylene shows no dielectric response, so experiments were made on specimens that had been lightly decorated with dipoles by means of oxidation, to such a small extent that the overall relaxation behaviour was not significantly affected. The dielectric relaxation data showed marked anisotropy for the relaxation, consistent with its assignment to the c-shear relaxation, but the β relaxation showed no anisotropy, confirming that the mechanical anisotropy observed related to the lamellar texture and not to anisotropy at a molecular level.

We are still left with an apparent paradox that the α relaxation in HDPE relates to interlamellar shear, whereas in LDPE, it is the c-shear relaxation. Recent measurements of the

Table 10.4 Activation energies for α and β relaxations in HDPE and LDPE.

Sample	β Relaxation activation energy (kJ/mol)	α Relaxation activation energy (kJ/mol)
Isotropic LDPE	430	120
Oriented LDPE	500	110
Isotropic HDPE	Not present	120
Oriented HDPE	Not present	80–90

activation energies for these materials by Matthews et al. (1999b) (Table 10.4) show, however, that the α relaxation in both polymers has a comparatively low activation energy consistent with the relaxation being the *c*-shear mechanism at a molecular level, that is the α relaxation in both HDPE and LDPE is associated with identical molecular mechanisms. It appears that interlamellar shear in the mechanical α process in HDPE requires coupled motions of the chains that run through the lamellae (*c*-shear) together with chains on the lamellar surface. In contrast, *c*-shear and interlamellar shear in LDPE are two distinct mechanical relaxations, and interlamellar shear has a much higher activation energy, akin to a glass transition.

Mansfield and Boyd (1978) have proposed that the dielectric α process can be represented by the torsional movement of a segment of chain about 12CH_2 units in length. This motion, which will cause the short-twisted mismatch region to move through the crystal one carbon atom at a time, is consistent with the dependence of activation energy on crystal thickness (Figure 10.11).

In the mechanical situation, the translational component of the crystal process can lead to reorganization of the crystal surface and hence modify the connections of amorphous chains to the crystal surface. An example is shown in Figure 10.12, where translational motion of a decorated dipole (indicated by a horizontal arrow) permits lengthening of the tight tie chain in (a), and so enables further deformation of the amorphous fraction.

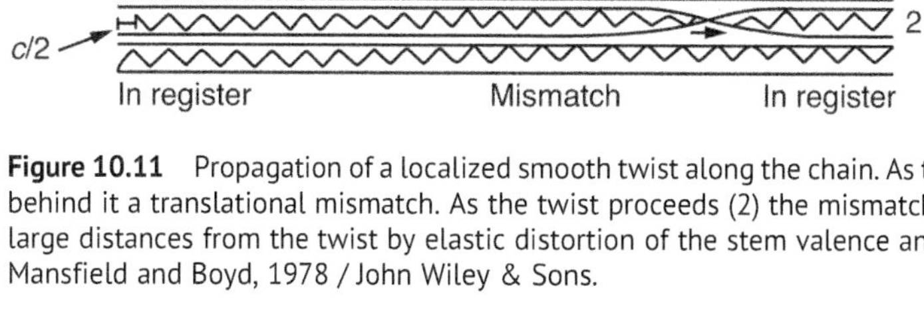

Figure 10.11 Propagation of a localized smooth twist along the chain. As the twist starts (1) it leaves behind it a translational mismatch. As the twist proceeds (2) the mismatch becomes attenuated at large distances from the twist by elastic distortion of the stem valence angles and bond length. Mansfield and Boyd, 1978 / John Wiley & Sons.

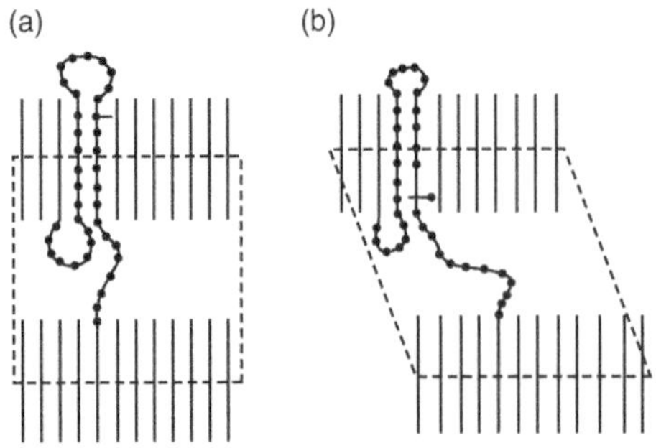

Figure 10.12 Further relaxation of the amorphous fraction resulting from translational mobility in the crystal (the latter acquired in the α process). Illustrated in this case is reorganization of the interface in (a) through shortening of two loops that in (b) permits lengthening of a tight tie chain, which in turn permits more deformation of the amorphous fraction. Also shown is a decorating dipole in the crystal that moves through a number of translational, rotational steps. One such step suffices for dielectric activity. Boyd (1985) / with permission of Elsevier.

The β relaxation is very broad compared with that in completely amorphous polymers due to the immobilizing effect of the crystals on the amorphous fraction. Boyd speculates that the shortest relaxation times may be associated with motions of very loose folds and relatively non-extended tie chains; conversely tight folds are unable to relax. The relative prominence of the β relaxation in LDPE compared with HDPE is enhanced by the lower value of the relaxed β process modulus in LDPE, which will increase the relative intensity of the β and decrease that of the α. On a molecular basis, the branching of LDPE gives a more loosely organized amorphous component, capable of relaxing to a lower limiting rubbery modulus.

As it occurs below the glass transition temperature the γ relaxation will involve simple conformational motions that are relatively short range in character. Such motions must leave the molecular stems adjacent to the bonds undergoing transition relatively undisturbed; they must require only a modest activation energy, and the swept-out volume during the relaxation should be small. Following Willbourn's (1958) suggestion that the γ relaxation in many amorphous and semi-crystalline polymers can be attributed to a restricted motion of the main chain that involves at least four successive —CH_2 groups, both Schatzki (1962) and Boyer (1963) have proposed that subglass relaxations can be modelled in terms of a so-called crankshaft mechanism (Figure 10.13). Shatzki's five-bond mechanism involves the simultaneous rotation about bonds one and seven such that the intervening carbon bonds move as a crankshaft. Boyer's proposal involves only three intermediate

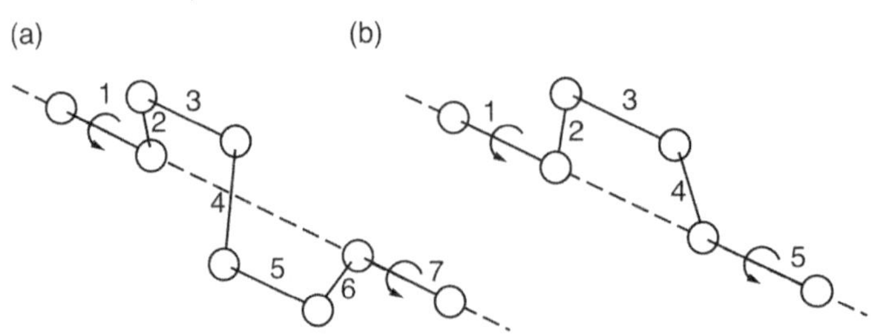

Figure 10.13 The crankshaft mechanisms of Shatzki (a) and Boyer (b). Adapted from McCrum et al., 1967.

carbon bonds. The mechanisms are, as Boyd (1985a, 1985b) has pointed out, the simplest allowed moves of a tetrahedrally bonded chain on a diamond lattice that leave the adjacent stem bonds in place. The internal energetics of the five-bond model are modest, but the swept-out volume is large in the context of a motion in a glassy matrix. For the three-bond transition, a double energy barrier system with an intermediate energy minimum is implied, and the motion associated with one of these barriers requires a significant free volume, and so is inhibited by the matrix. Despite these drawbacks, a crankshaft mechanism has been proposed as being relevant to the γ relaxation in polyethylene.

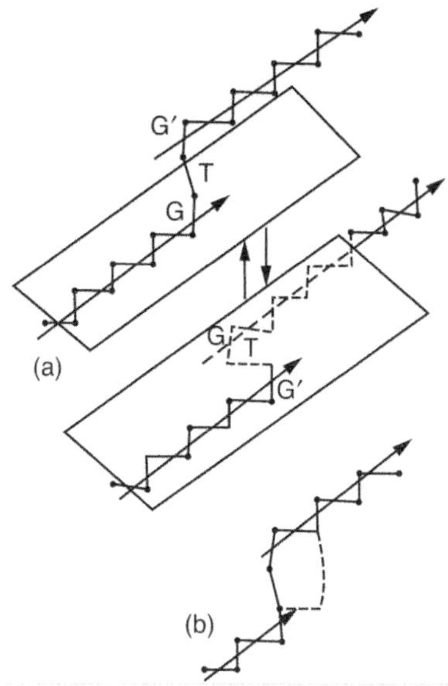

Figure 10.14 Kinks and kink inversion. (a) The conformational sequence ... TTTGTG′TTT ... has parallel offset planar zigzag stems (indicated by arrows) on either side of the GTG′ portion. The transition TGTG′ → TG′TGT (called here kink inversion) creates a mirror image of the kink about the displaced stems. (b) A three-bond crankshaft move is shown at a kink site (as dashed line). This move advances the kink along the chain by $2CH_2$ units. Boyd (1985) / with permission of Elsevier.

Boyd (1985a, 1985b) discusses a motion related to the three-bond mechanism that can accomplish the appropriate shape change without encountering problems associated with free volume. It involves the conformational sequence GTG′ occurring in an otherwise all-*trans* chain (*G* and *G*′ represent alternative *gauche* transformations). From Figure 10.14 it can be seen that this sequence, known as a kink (Pechhold et al. 1963), has the effect of displacing the separated *trans* components of a planar zigzag yet leaving them parallel to one another. Interchanging the senses of the *gauche* bonds

$$...TTGTG'TT... \rightarrow ...TTG'TGTT...$$

converts the kink into a mirror image of itself. The kink inversion process, shown in Figure 10.15, involves only a small swept-out volume and requires a modest activation energy. The stem displacement causes a localized shape change that can propagate through the specimen as a shear strain. The kink inversion process is therefore a possible candidate on which to base a molecular model of the γ relaxation, but it must be emphasized that there is no direct evidence to demonstrate that it is appropriate for modelling the behaviour of polyethylene.

10.3.4 Relaxation Processes in Liquid Crystalline Polymers

Liquid crystalline polymers form another class of polymers from a structural viewpoint and can be produced either from a liquid crystalline solution (termed lyotropic) or a liquid crystalline melt (thermotropic). In this section, we will consider thermotropic liquid crystalline polymers, of which the simplest chemically are those invented by Calundann (1976), which are random copolymers of hydroxybenzoic acid (HBA) and hydroxynaphthoic acid (HNA) (Figure 10.16). The random arrangement of HBA and HNA units along the chain prevents

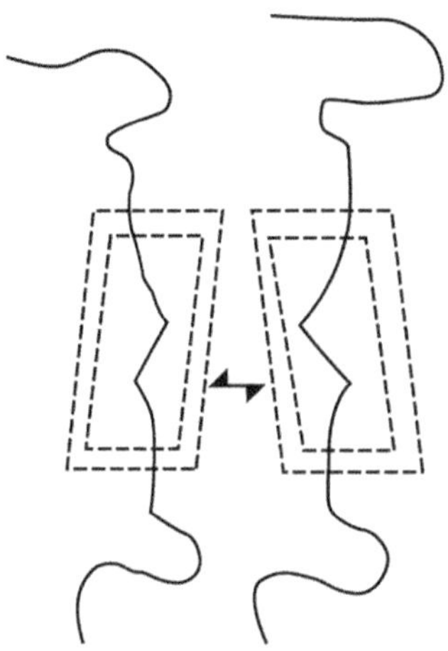

Figure 10.15 Strain fields set up by stem displacement accompanying kink inversion. Boyd (1985) / with permission of Elsevier.

normal crystallization into three-dimensional order (although there is debate about the extent to which there may be some regions of three-dimensional order (Donald and Windle 1992, p. 162)). Molecular alignment of the chains can be achieved readily by melt spinning. This produces an oriented liquid crystalline structure (termed a mesophase) in which there is axial alignment of the chains, which are close packed on hexagonal or orthorhombic lattices without any regularity within the chains along the axial direction.

The dynamic mechanical properties of these oriented HBA/HNA copolymers have been studied by Blundell and Buckingham (1985) and Davies and Ward (1988). It is of particular interest to compare different compositions based on HBA and HNA only, with two other copolymers that incorporate terephthalic acid (TA) dihydroxynaph-thalene (DNA) and biphenol (BP) (Ward 1993). The compositions of the four copolymers to be discussed are shown in Table 10.5. The most instructive comparisons come from consideration of the dynamic mechanical loss factors in shear (Figure 10.17) and the dielectric loss data (Figure 10.18). It can be seen that there are three relaxation processes, labelled α, β and γ. Because these polymers are essentially single phase, it is possible to seek an understanding entirely in terms of molecular relaxation processes. The relatively high intensity of the β relaxation in CO 30/70 identifies this relaxation with the naphthalene residue because this has the highest concentration in all these polymers. It is clear from comparison with CO 2,6

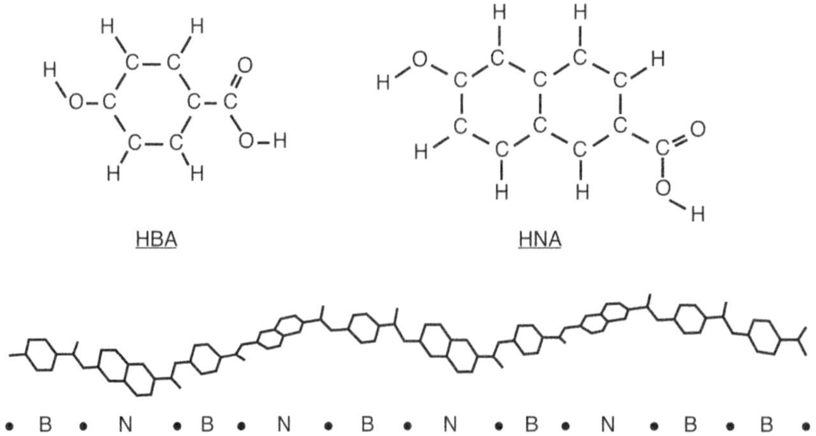

Figure 10.16 Monomers of HBA and HNA and a projection of the random chain. Adapted from Davies and Ward, 1988.

Table 10.5 Chemical compositions of thermotropic liquid crystalline polymers.

	Composition (mole fraction in %)				
Polymer	**HBA**	**HNA**	**TA**	**DHN**	**BP**
CO 73/27	73	27			
CO 30/70	30	70			
CO 2,6	60		20	20	
COTBP	60	5	17.5		17.5

HBA, 4-hydroxybenzoic acid; HNA, 2-hydroxy 6-naphthoic acid; TA, terephthalic acid; DHN, 2,6 dihydroxynaphthalene; BP, 4,4′ biphenyldiol.

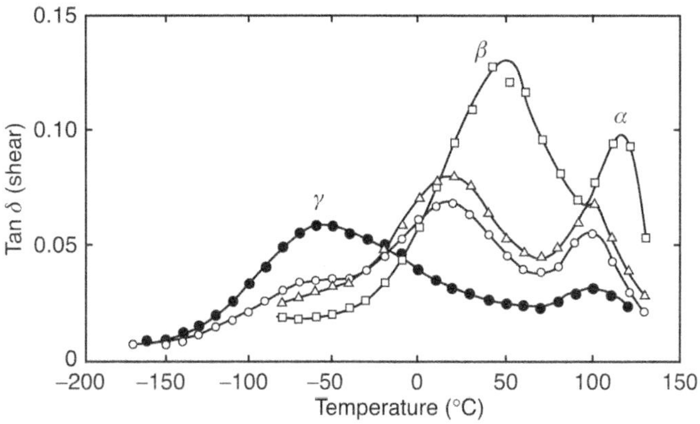

Figure 10.17 Dynamic mechanical loss factor (shear) for CO 30/70 (□), CO 73/27 (Δ), CO 2,6 (○) and COTBP (●). Adapted from Green et al., 1990.

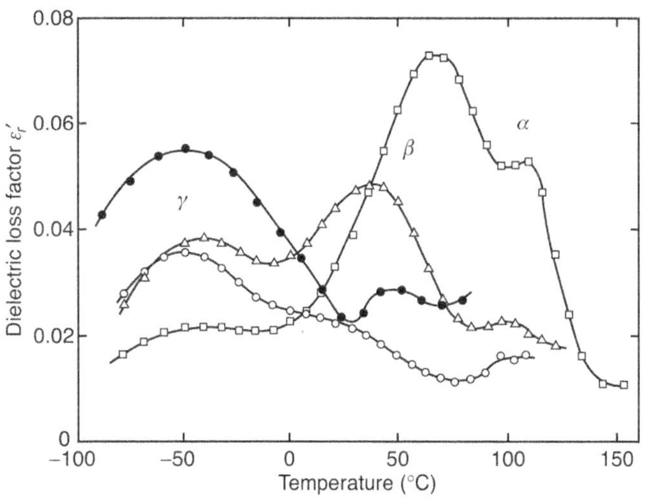

Figure 10.18 Dielectric loss data for CO 73/27(Δ), CO 30/70 (□), CO 2,6 (○) and COTBP (●). Ward, 1993 / John Wiley & Sons.

Boyer, R.F. (1963) The relation of transition temperatures to chemical structure in high polymers. *Rubber Rev.*, **36**, 1303.

Calundann, G.W., 1976 British Patent 1,585,511 (priority 13 May 1976, USA) (US Patent 4067 852).

Coburn, J.C. and Boyd, R.H. (1986) Dielectric relaxation in poly(ethylene terephthalate). *Macromolecules*, **19**, 2338–2245.

Davies, G.R. and Ward, I.M. (1969) The anisotropy of dielectric relaxation in oxidised low density polyethylene. *J. Polym. Sci.*, **B7**, 353.

Davies, G.R. and Ward, I.M. (1988) Structure and properties of oriented thermotropic liquid crystal polymers in the solid state, in *Higin h Modulus Polymers*, (eds A.E. Zachariades and R.S. Porter), Marcel Dekker, New York, pp. 37–69.

Deutsch, K., Hoff, E.A. and Reddish, W. (1954) Relation between the structure of polymers and their dynamic mechanical and electrical properties. 1. Some alpha-substituted acrylic. *J. Polym. Sci.*, **13**, 565.

Donald, A.M. and Windle, A.H. (1992) *Liquid Crystalline Polymers*, Cambridge University Press, Cambridge, p. 162.

English, A.D. (1984) Macromolecular dynamics in solid poly(ethylene-terephthalate)–H-1 and C-13 solid-state NMR. *Macromolecules*, **17**, 2182.

Fox, T.G. and Flory, P.J. (1950) 2nd-order transition temperatures and related properties of polystyrene. 1. Influence of molecular weight. *J. Appl. Phys.*, **21**, 581.

Fox, T.G. and Flory, P.J. (1954) The glass temperature and related properties of polystyrene-influence of molecular weight. *J. Polym. Sci.*, **14**, 315.

Fox, T.G. and Loshaek, S. (1955) Influence of molecular weight and degree of crosslinking on the specific volume and glass temperature of polymers. *J. Polym. Sci.*, **15**, 371.

Gordon, M. and Taylor, J.S. (1952) Ideal copolymers and the 2nd-order transitions of synthetic rubbers. 1. Non-crystalline copolymers. *J. Appl. Chem.*, **2**, 493.

Hay, I.L. and Keller, A. (1967) A study on orientation effects in polyethylene in the light of crystalline texture. *J. Mater. Sci.*, **2**, 538.

Heijboer, J. (1965) *Physics of Non-Crystalline Solids*, North-Holland, Amsterdam.

Illers, K.H. and Breuer, H. (1963) Molecular motions in polyethylene terephthalate. *J. Colloid Sci.*, **18**, 1.

Katana, G., Kremer, F., Fischer, E.W. *et al.* (1993) Broad-band dielectric study on binary blends of bisphenol-a and tetramethylbisphenol-a polycarbonate. *Macromolecules*, **26**, 3075–3080.

Kilian, H.G., Halboth, H. and Jenckel, E. (1960) X-ray photographic examination of the melting and crystallization behavior of poly(ethylene terephthalate) (PET). *Kolloid Z.*, **176**, 166.

Kotaka, T., Yao, M.L., Watanabe, H. and Adachi, K. (1992) A dielectric study on dynamics of tethered chains: styrene-isoprene diblock copolymers in bulk and in selective solvent. *J. Macromol. Sci. A*, **29**, 163–171.

Kremer, F. (2003) *Broadband Dielectric Spectroscopy*, Springer, Berlin, Heidelberg, Germany.

Ljungberg, N. and Wesslén, B. (2001) The effects of plasticizers on the dynamic mechanical and thermal properties of poly(lactic acid). *J. Appl. Polym. Sci.*, **86**, 1227–1234.

Luo, F., Li, J., Ji, F., Weng, Y. and Ren, J. (2022) Preparation of poly(lactic acid)-based shape memory polymers with low response temperature utilizing composite plasticizers. *Polym. Bull.*, **79**, 4761–4781.

Mandelkern, L., Martin, G.M. and Quinn, F.A. (1957) Glassy state transitions of poly-(chlorotrifluoroethylene), poly-(vinylidene fluoride), and their copolymers. *J. Res. Natl. Bur. Stand.*, **58**, 137.

Mansfield, M. and Boyd, R.H. (1978) Molecular motions, alpha-relaxation and chain transport in polyethylene crystals. *J. Polym. Sci. Phys. Ed.*, **16**, 1227.

Matthews, R.G., Ward, I.M. and Capaccio, G. (1999a) Structural heterogeneity and dynamic mechanical relaxations of ethylene α-olefin copolymers. *J. Macromol. Sci. Phys.*, **37**, 51.

Matthews, R.G., Unwin, A.P., Ward, I.M. *et al.* (1999b) A comparison of the mechanical relaxation behavior of linear low and high density polyethylenes. *J. Macromol. Sci. Phys. B*, **38**, 123.

Maxwell, A.S., Ward, I.M., Lauprêtre, F. *et al.* (1998a) Secondary relaxation processes in polyethylene terephthalate-additive blends: 1. N.m.r. investigation. *Polymer*, **39**, 6835–6849.

Maxwell, A.S., Monnerie, L. and Ward, I.M. (1998b) Secondary relaxation processes in polyethylene terephthalate-additive blends: 2. Dynamic mechanical and dielectric investigations. *Polymer*, **39**, 6851.

McCrum, N.G., Read, B.E. and Williams, G. (1967) *Anelastic and Dielectric Effects in Polymeric Solids*, John Wiley & Sons, London.

Mekonnen, R., Mussone, P., Khalil, H. and Bressler, D. (2013) Progress in bio-based plastics and plasticizing modifications. *J. Mater. Chem. A*, **1**, 13379.

Molina, B.G., Ocón, G., Fiorella, M.S., Iribarren, J.I., Armelin, E. and Alemán, C. (2023) Thermally-induced shape memory behavior of polylactic acid/polycaprolactone blends. *Eur. Polym. J.*, **196**, 112230.

Monnerie, L., Lauprêtre, F.O. and Halary, J.L. (2005) Investigation of solid-state transitions in linear and crosslinked amorphous polymers. *Adv. Polym. Sci.*, **187**, 35–213.

Nielsen, L.E. (1962) *Mechanical Properties of Polymers*, Van Nostrand-Reinhold, New York.

Pechhold, W., Blasenbrey, S. and Woerner, S. (1963) Eine Niedermolekulare Modell-substanz fur Lineares Polyathylen – Vorschlag des Kinkenmodells zur Deutung des Gamma- und Alpha-Relaxationsprozesses. *Kolloid-Z, Z. Polym.*, **189**, 14.

Popov, I., Cheng, S., Alexei, P. and Sokolov, A.P. (2022) Broadband dielectric spectroscopy and its application in polymeric materials, structure-properties and characterization, in *Macromolecular Engineering*, (eds N. Hadjichristidis, Y. Gnanou, K. Matyjaszewski and M. Muthukumar).

Powles, J.G., Hunt, B.I. and Sandiford, D.J.H. (1964) Proton spin lattice relaxation and mechanical loss in a series of acrylic polymers. *Polymer*, **5**, 505–515.

Roberts, G.E. and White, E.F.T. (1973) *The Physics of Glassy Polymers*, Applied Science Publishers Ltd, London, p. 212.

Schatzki, T.F. (1962) Statistical computation of distribution functions of dimensions of macromolecules. *J. Polym. Sci.*, **57**, 337.

Senatova, F.S., Zadorozhnyya, M.Y., Niazaa, K.V. *et al.* (2017) Shape memory effect in 3D-printed scaffolds for self-fitting implants. *Eur. Polym. J.*, **93**, 222–231.

Sinnott, K.M. (1960) Nuclear magnetic resonance and molecular motion in polymethyl acrylate, polymethyl methacrylate and polyethyl methacrylate. *J. Polym. Sci.*, **42**, 3.

Song, J.J., Chang, H.H. and Naguib, H.E. (2015) Biocompatible shape memory polymer actuators with high force capabilities. *Eur. Polym. J.*, **67**, 186–198.

Stachurski, Z.H. and Ward, I.M. (1968) β relaxations in polyethylenes and their anisotropy. *J. Polym. Sci. A2*, **6**, 1817.

Stachurski, Z.H. and Ward, I.M. (1969) Mechanical relaxations in polyethylene. *J. Macromol. Sci. Phys. B*, **3**, 445.

Takayanagi, M. (1963) Viscoelastic properties of crystalline polymers. *Mem. Fac. Eng, Kyushu Univ.*, **23**, 1.

Vincent, P.I. (1965), in *The Physics of Plastics*, (ed. P.D. Ritchie), Iliffe, London.

Walton, J.H., Lizak, M.J., Conradi, M.S. *et al.* (1990) Hydrostatic pressure dependence of molecular motions in polycarbonates. *Macromolecules*, **23**, 416–422.

Ward, I.M. (1960) Nuclear magnetic resonance studies of polyethylene terephthalate and related polyesters. *Trans. Faraday Soc.*, **56**, 648.

Ward, I.M. (1993) Relaxation processes in oriented liquid crystalline polymers. *Makromol. Chem. Macromol. Symp.*, **69**, 75.

Ward, I.M. and Coates, P.D. (2001), Introduction in *Solid Phase Processing of Polymers*, (eds I.M. Ward, P.D. Coates and M.M. Dumoulin), Hanser, Munich, pp. 1–10.

Watanabe, H. *et al.* (2003) Rheo-dielectrics in oligomeric and polymeric fluids: a review of recent findings. *J. Phys. Condens. Matter*, **15**, S909.

Willbourn, A.H. (1958) The glass transition in polymers with the $(CH_2)_n$ group. *Trans. Faraday Soc.*, **54**, 717.

Wolf, K. (1951) Beziehungen zwischen mechanischem und elektrischem Verhalten von Hochpolymeren. *Kunststoffe*, **41**, 89.

11

Non-linear Viscoelastic Behaviour

In Chapter 5, we introduced linear viscoelasticity. In this scheme, the observed creep or stress relaxation behaviour can be viewed as the defining characteristic of the material. The creep compliance function – the ratio of creep strain $e(t)$ to the constant stress σ – is a function of time only and is denoted as $J(t)$. Similarly and necessarily, the stress relaxation modulus, the ratio of stress to the constant strain, is the function $G(t)$. Any system in which these two conditions do not apply is *non-linear*. Then, the many useful and elegant properties associated with the linear theory, notably the Boltzmann superposition principle, no longer apply and theories to predict stress or strain are approximations that must be supported by experiment.

The non-linearity may arise for a variety of reasons. First, the linear theory has been developed for small strains,[1] and to generalize it to large strain requires decisions on the appropriate definitions of both strain and stress, in effect making it necessary to create a new theory. Typical polymer applications may require the material to operate at strains in excess of 10%, and for elastomers the strains may be up to several hundred percent. Secondly, even at small strains linear behaviour may not be obtained. The behaviour may be quite rich, with the possibility of the polymer being initially linear but becoming non-linear at large times.

Linear viscoelasticity is based on linearity of both elastic strains and viscous flow, and so excludes any considerations of underlying physics that are specific to polymers. Many of the nonlinear theories are developments of the linear theory and so share this characteristic.

Methods of tackling problems of nonlinear viscoelasticity vary from highly pragmatic engineering solutions to rigorous approaches based in sound science. We may categorize them broadly as follows:

1) *The engineering approach*. The design engineer requires the ability to predict behaviour exactly for a proposed situation in terms of as few initial experiments as possible. Empirical relations that describe the performance are adequate, and these need not have any physical significance.
2) *The rheological approach*. There has been a rich variety of attempts to create formal descriptions of non-linear behaviour beginning in the early 1950s. Some have taken

1 By 'small strains', we mean that only linear terms in the strain–displacement equations are required (Section 3.1.5).

Mechanical Properties of Solid Polymers, Fourth Edition. John Sweeney and Peter Hine.

the form of generalizations of linear theory, such as those that resemble the Boltzmann integral – the so-called single-integral theories. Some of these models have been developed to encompass issues such as large deformations, anisotropy, and physical aging. Multiple integral models were proposed in the late 1950s and studied throughout the 1960s, but have now been largely abandoned as their need to be supported by implausibly extensive experimental programmes has been realized. A rich field of single integral models remains in active use.

There is a limited coverage of this field by textbooks. A theoretical survey, using sophisticated mathematical methods, is given in the text by Lockett (1972), and the contrasting approach of the practical engineer is typified in the work of Turner (1966). Ferry's textbook (Ferry 1980) covers both theoretical and practical aspects of polymer viscoelasticity. The more recent work of Lakes (1999) covers some aspects of non-linearity. The textbook on polymer engineering science by Brinson and Brinson (2015) includes accounts of both linear and non-linear viscoelasticity.

Wineman (2009) has reviewed the field of non-linear viscoelasticity, noting its 40-year history, and the limited number of engineering applications hitherto. Drapaca *et al.* (2007) have also produced a review, in which they note the large variety of different and seemingly unconnected theories.

11.1 The Engineering Approach

11.1.1 Isochronous Stress–Strain Curves

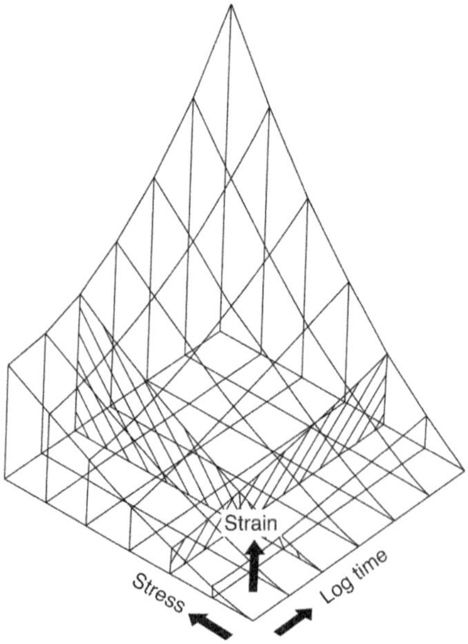

Figure 11.1 The stress–strain time relation obtained from creep. Turner (1966) / John Wiley & Sons.

The aim here is to predict the behaviour for a proposed application, using the minimum of experimental data. A general picture of a non-linear viscoelastic polymer is shown in Figure 11.1. Empirical relations between stress, strain and time can be obtained that will approximately reproduce that part of the surface of Figure 11.1 that is of practical interest. These relations may have no physical significance, and their use may be restricted to very specific stress or strain programmes. In the general case, creep curves need to cover the complete range of stresses over as long a period of time as feasible. However, Turner (1966) has shown that when stress and time dependence are approximately separable, it may be possible to interpolate creep curves at intermediate stresses from a knowledge of two creep curves, combined with a knowledge of the stress–strain curve for a fixed time (say 100 s): so-called isochronal stress–strain curves represented by the vertical lines in Figure 11.2. However, these data cannot give a generally accurate prediction of the non-linear response. When, for instance, the rate of strain applied to a polymer specimen is changed abruptly, a transient 'bump' in stress will be produced in the stress–strain curve; this will not be reproduced in the corresponding path across the smooth surface of Figure 11.1.

Isochronous stress–strain curves remain a useful tool. Yang *et al.* (2006) have used them to assess the viscoelasticity of polyamides and various polyamide composites and to

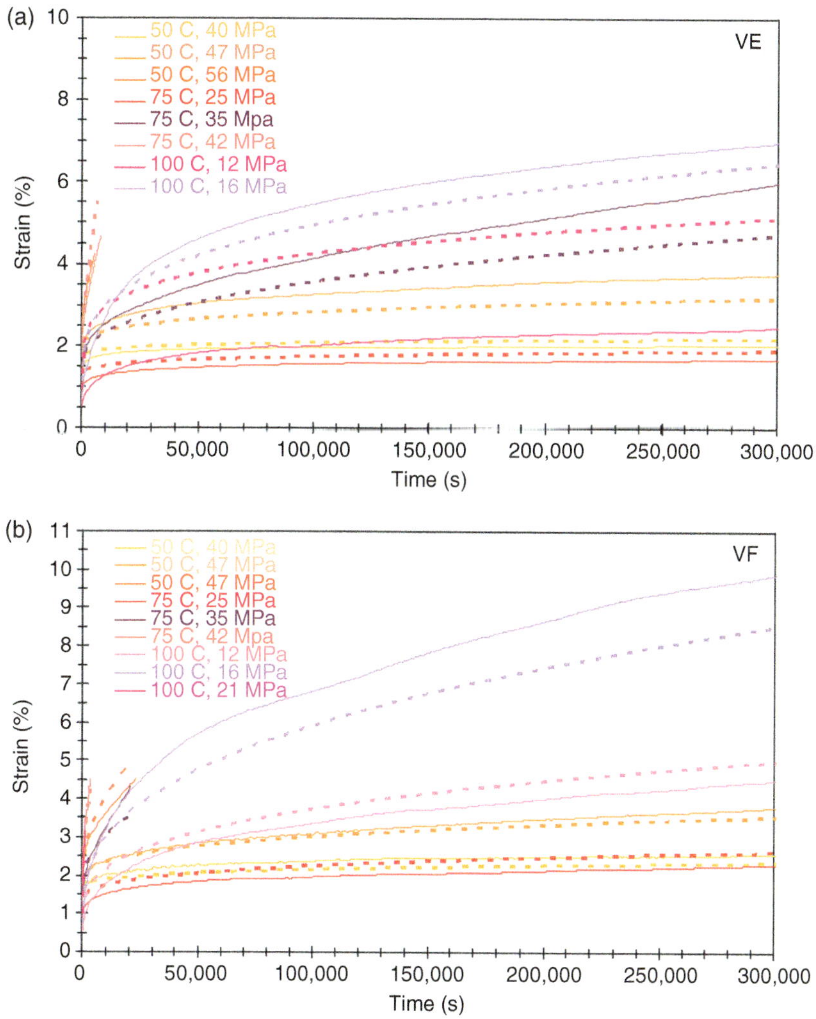

Figure 11.2 Comparison between experimental and predicted results of creep strain vs creep time for (a) vinyl ester and (b) nanocomposite with 0.5 wt% functionalized carbon nanofiber at different stresses and temperatures based on power-law model (–, experiment and ---, prediction). Plaseied and Fatemi (2009) / with permission of SAGE Publications.

establish their linear viscoelastic ranges. Lyons (2003) has used them to analyse the behaviour of glass reinforced polymer composites. However, Hirakawa *et al.* (2003) have questioned their relevance in predicting the performance of polymer geogrids.

11.1.2 Power Laws

To describe the creep behaviour of glassy or tough polymers where the creep strains involved are small (~5% say), as distinct from elastomers where the deformations are large (~100% say), separable stress and time functions have been proposed. This follows the approach adopted for metals over many years, a useful review of which has been given by Abdullah *et al.* (2014).

Pao and Marin (1952, 1953) followed the approach originally suggested by Marin (1937) for metals, where the total creep strain e is considered to consist of three independent components, an elastic strain e_1, a transient recoverable viscoelastic strain e_2 and a permanent non-recoverable plastic strain e_3. At constant stress, the elastic strain is given by $e_1 = \sigma/E$, where E is Young's modulus. The viscoelastic strain is defined by integrating the condition that the transient creep is a function of the stress σ and the transient creep strain, that is $de_2/dt = f(e_2, \sigma)$. The plastic strain is found by integrating the condition that the plastic strain rate is a function of stress only. For simplicity, the functions of stress for both the viscoelastic and the plastic strains are assumed to be simple power laws of stress, and most usually the same power law is adopted.

The total creep strain for loading under a constant stress σ is then proposed to be

$$e = \frac{\sigma}{E} + K\sigma^n(1 - e^{-qt}) + B\sigma^n t, \tag{11.1}$$

where K, n, q and B are material constants.

Findley and Khosla (1955) have attempted to fit the creep of many plastics and plastic laminates to analytical relationships similar to those suggested for metals (Nutting 1943). It was found that the creep strain e_c at time t could be related by an equation of the form

$$e_c(t) = e_0 + mt^n \tag{11.2}$$

where e_0 and m are functions of stress for a given material and n is a material constant. Further work at constant temperature revealed possible forms of e_0 and m as functions of stress:

$$e_c(\sigma, t) = e'_0 \sinh\frac{\sigma}{\sigma_e} + m't^n \sinh\frac{\sigma}{\sigma_m},$$

where m', σ_e and σ_m are material constants. This equation was a good fit to single-step creep data, and so represents Turner's stress–strain time surface of Figure 11.1. It has also been shown to fit accurately the creep data of graphite–epoxy composites (Dillard *et al.* (1987).

Findley's equation (11.2) continues to be used, with various forms for the functions e_0 and m which vary with both stress and temperature. Plaseied and Fatemi (2009) and Amjadi and Fatemi (2021) assumed that m and n have the same functional forms, given by

$$\log(m) = a_m + b_m T + c_m\sigma + d_m\sigma T + e_m T^2$$
$$\log(n) = a_n + b_n T + c_n\sigma + d_n\sigma T + e_n T^2$$

for constants a_m, b_m, a_n, b_n, etc. The effectiveness of predictions for this scheme is shown in Figure 11.2 over a range of temperatures for both vinyl ester and its nanocomposite.

Findley's empirical equations are very useful to design engineers for constant stress loading conditions as they can predict the creep of a given material once the required material parameters are known.

The empirical approaches suggested so far have two principal limitations:

1) They do not provide a general representation for creep, recovery and behaviour under complicated loading conditions.
2) Creep data in these formulations cannot be simply related to stress relaxation and dynamic mechanical data.

11.2 The Rheological Approach

11.2.1 Quasi-linear Viscoelasticity

In the course of the extensive studies of the creep and recovery behaviour of textile fibres already referred to, Leaderman (1943) became one of the first to appreciate that the simple assumptions of linear viscoelasticity might not hold even at small strains. For nylon and cellulosic fibres, he discovered that although the creep and recovery curves may be coincident at a given level of stress – a phenomenon associated with linear viscoelasticity – the creep compliance plots indicated a softening of the material as stress increased, except at the shortest times. Thus, the creep compliance function is a function of both time and stress.

Non-linear viscoelastic theories can be created by generalizing the Boltzmann superposition principle (see Chapter 5). Leaderman (1943, 1962) introduced non-linearity into the Boltzmann integral by including stress dependence into the integrand. Leaderman's integral takes the form

$$e(t) = \int_{-\infty}^{t} J(t-\tau)\frac{\mathrm{d}}{\mathrm{d}\tau}f\{\sigma(\tau)\}\,\mathrm{d}\tau \qquad (11.3)$$

for the strain in terms of the stress history. Here, f is a function of stress relating to the material's instantaneous response that is to be determined experimentally for each polymer. Equivalently we may write

$$e(t) = \int_{-\infty}^{t} J(t-\tau)\frac{\mathrm{d}e(\sigma)}{\mathrm{d}\sigma}\frac{\mathrm{d}\sigma}{\mathrm{d}\tau}\,\mathrm{d}\tau \qquad (11.4)$$

where we can recognize the creep compliance $J(t)$ of the linear theory of Chapter 5 (Equation (5.1)) has been replaced by the nonlinear creep compliance function $J(t)\dfrac{\mathrm{d}e}{\mathrm{d}\sigma}$. Findley and Lai (1967) have adopted a similar approach to investigate experiments in which levels of constant stress were applied in steps to specimens of unplasticized PVC, and referred to it as the 'modified superposition principle'.

For the stress in terms of the strain history, the equivalent to Equation (11.4) is

$$\sigma(t) = \int_{-\infty}^{t} G(t-\tau) \frac{\mathrm{d}\sigma(e)}{\mathrm{d}e} \frac{\mathrm{d}e}{\mathrm{d}\tau} \mathrm{d}\tau \tag{11.5}$$

This form has been proposed by Fung (1972, 1993) working on biological materials, and developed further by workers in the biological field (Pryse *et al.* 2003; Nekouzadeh *et al.* 2007; De Pascalis *et al.* 2014).

More general forms of Equations (11.4) and (11.5) can be created by abandoning the assumption of separability of the relaxation or creep functions. The generalization of Equation (11.5) thus becomes

$$\sigma(t) = \int_{-\infty}^{t} G(t-\tau, e(\tau)) \frac{\mathrm{d}e}{\mathrm{d}\tau} \mathrm{d}\tau \tag{11.6}$$

where the relaxation modulus G is defined in differential form as

$$G(t, e) = \frac{\partial \sigma(t, e)}{\partial e}.$$

This is equivalent to the form proposed by Pipkin and Rogers (1968) and is the first term of a multiple-integral series, discussed below. It has been applied by Pryse *et al.* (2003) who were working on collagen gels. For the strain in terms of the stress history, the analogous relation is

$$e(t) = \int_{-\infty}^{t} J(t-\tau, \sigma(\tau)) \frac{\mathrm{d}\sigma}{\mathrm{d}\tau} \mathrm{d}\tau \tag{11.7}$$

where J is defined as

$$J(t, \sigma) = \frac{\partial e(t, \sigma)}{\partial \sigma}.$$

Pipkin and Rogers tested their model (11.7) using published data on poly(vinyl chloride) under multi-step creep conditions. This kind of testing programme consists of a constant stress being applied for a pre-determined period, and then abruptly changed to a different value for a second time interval, and so on (see Figure 5.6); this testing regime gives a more severe test of the theory than, for instance, constant strain rates. As an illustration, we show their results for the five-step stress history in Figure 11.3(a) and (b).

In summary, we note that all the other approaches to quasi-linear viscoelasticity are special cases of that of Pipkin and Rogers.

11.2.2 More Single-Integral Representations

11.2.2.1 Reduced Time

Nonlinearity can be introduced by retaining the Boltzmann integral form and replacing time by a function of time known as the *reduced time*. In studies of creep in polypropylene,

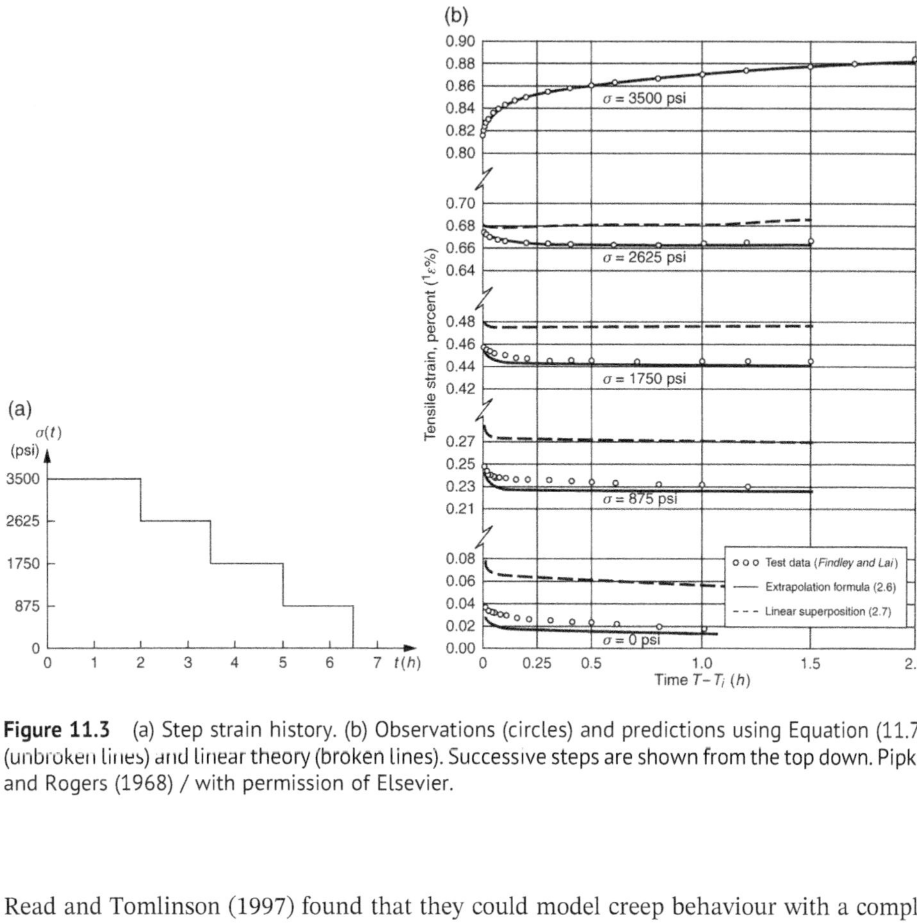

Figure 11.3 (a) Step strain history. (b) Observations (circles) and predictions using Equation (11.7) (unbroken lines) and linear theory (broken lines). Successive steps are shown from the top down. Pipkin and Rogers (1968) / with permission of Elsevier.

Read and Tomlinson (1997) found that they could model creep behaviour with a compliance function J:

$$J(t) = D_0 + \Delta D(1 - \exp(-s^n))$$

where s is the reduced time

$$s = \int_0^t \frac{du}{t_0(\sigma(u))}$$

so that the parameter t_0 is a function of stress. With this representation they were able to apply superposition to successfully model two-step creep. In a later publication Dean *et al.* (2009, 2011) generalized this approach to a single-integral model

$$e(t) = \int_0^t \frac{d\sigma}{d\tau}\left[D_0 + \Delta D\left(1 - \exp\left\{-\left(\int_\tau^t \frac{du}{t_0(\sigma(u))}\right)^n\right\}\right)\right]d\tau \tag{11.8}$$

Dean *et al.* (2011) developed a three-dimensional model from this approach and were able to implement it as a two-dimensional finite element (FE) analysis in the ABAQUS

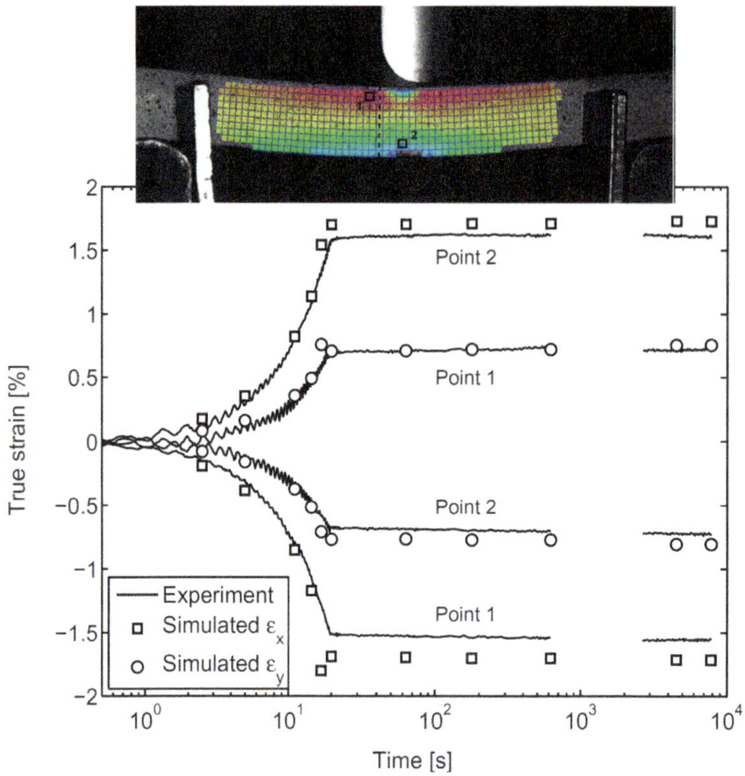

Figure 11.4 Predicted and observed strains at points 1 and 2 on the surface of the bend specimen in the upper image. x and y are horizontal and vertical directions, respectively. Specimen is loaded and then held at constant deflection in three-point bend. Tscharnuter et al. (2012). / with permission of Elsevier.

where $\mathbf{X}$ represents a point in the undeformed reference state. By contrast, Bernstein *et al.* adopted a relative strain measure $\mathbf{F_r}$ appropriate for a fluid:

$$\mathbf{F_r}(t,\tau) = \frac{\partial \mathbf{x}(t)}{\partial \mathbf{x}(\tau)}$$

where $\mathbf{x}$(t) is a point at the 'present' time t and $\mathbf{x}(\tau)$ its position at a previous time τ. Other relative strain measures analogous to those of Chapter 3 can be generated, such as the left Cauchy–Green measure $\mathbf{B_r}$:

$$\mathbf{B}_r = \mathbf{F}_r\mathbf{F}_r^T$$

and invariants

$$J_1 = \mathrm{tr}(\mathbf{B}_r)$$
$$J_2 = \frac{1}{2}\left[(\mathrm{tr}(\mathbf{B}_r))^2 - \mathrm{tr}\left(\mathbf{B}_r^2\right)\right].$$

Noting that

$$\frac{\partial \mathbf{x}(t)}{\partial \mathbf{x}(\tau)} = \frac{\partial \mathbf{x}(t)}{\partial \mathbf{X}(t)} \Big/ \frac{\partial \mathbf{x}(\tau)}{\partial \mathbf{X}(\tau)}$$

in principal directions $\mathbf{B_r}$ is

$$\mathbf{B}_r = \begin{pmatrix} (\lambda_\mathrm{I}(t)/\lambda_\mathrm{I}(\tau))^2 & 0 & 0 \\ 0 & (\lambda_\mathrm{II}(t)/\lambda_\mathrm{II}(\tau))^2 & 0 \\ 0 & 0 & (\lambda_\mathrm{III}(t)/\lambda_\mathrm{III}(\tau))^2 \end{pmatrix}$$

with accompanying invariant expressions

$$J_1(t,\tau) = (\lambda_\mathrm{I}(t)/\lambda_\mathrm{I}(\tau))^2 + (\lambda_\mathrm{II}(t)/\lambda_\mathrm{II}(\tau))^2 + (\lambda_\mathrm{III}(t)/\lambda_\mathrm{III}(\tau))^2 \tag{11.14}$$

$$\begin{aligned} J_2(t,\tau) = {} & (\lambda_\mathrm{I}(t)/\lambda_\mathrm{I}(\tau))^2(\lambda_\mathrm{II}(t)/\lambda_\mathrm{II}(\tau))^2 + (\lambda_\mathrm{II}(t)/\lambda_\mathrm{II}(\tau))^2(\lambda_\mathrm{III}(t)/\lambda_\mathrm{III}(\tau))^2 \\ & + (\lambda_\mathrm{III}(t)/\lambda_\mathrm{III}(\tau))^2(\lambda_\mathrm{I}(t)/\lambda_\mathrm{I}(\tau))^2 \end{aligned} \tag{11.15}$$

and

$$J_3(t,\tau) = (\lambda_\mathrm{I}(t)/\lambda_\mathrm{I}(\tau))^2(\lambda_\mathrm{II}(t)/\lambda_\mathrm{II}(\tau))^2(\lambda_\mathrm{III}(t)/\lambda_\mathrm{III}(\tau))^2 \tag{11.16}$$

analogous to Equations (3.56), (3.57) and (3.58).

Central to this theory is a strain energy function U which is dependent on both time and strain. It is dependent on strain through invariants, ensuring material objectivity (see Section 3.4.2). Since the theory is for incompressible material, U does not depend on the volumetric invariant $J_3 = 1$ and takes the form

$$U = U(J_1(t,\tau), J_2(t,\tau), t-\tau)$$

The theory is of the following form, with the integrand generated by differentiation of U in a manner similar to Equation (3.55) to give the stress tensor $\mathbf{\Sigma}$:

$$\mathbf{\Sigma} = -p\mathbf{I} + 2\int_{-\infty}^{t} \left[\frac{\partial U}{\partial J_1}(J_1(t,\tau),J_2(t,\tau),t-\tau)\mathbf{B}_r(t,\tau) - \frac{\partial U}{\partial J_2}(J_1(t,\tau),J_2(t,\tau),t-\tau)\mathbf{B}_r^{-1}(t,\tau)\right] \mathrm{d}\tau \tag{11.17}$$

$\mathbf{I}$ is the identity tensor and p is the unknown hydrostatic pressure associated with incompressibility as in Equation (3.55). Bernstein (1966) has set out the principal stresses for various homogeneous states of strain. For uniaxial stretching along I, there is zero stress along II and III and we have along I

$$\sigma_\mathrm{I} = -p + 2\int_{-\infty}^{t} \left[\frac{\partial U}{\partial J_1}(J_1(t,\tau),J_2(t,\tau),t-\tau)\frac{\lambda_\mathrm{I}^2(t)}{\lambda_\mathrm{I}^2(\tau)} - \frac{\partial U}{\partial J_2}(J_1(t,\tau),J_2(t,\tau),t-\tau)\frac{\lambda_\mathrm{I}^2(\tau)}{\lambda_\mathrm{I}^2(t)}\right] \mathrm{d}\tau \tag{11.18}$$

For equal strains and zero stress along II and III, $\lambda_{II} = \lambda_{III} = \lambda_I^{-1/2}$ and with $\lambda_I = \lambda$, $\sigma_I = \sigma$

$$0 = -p + 2\int_{-\infty}^{t}\left[\frac{\partial U}{\partial J_1}(J_1(t,\tau),J_2(t,\tau),t-\tau)\frac{\lambda(\tau)}{\lambda(t)} - \frac{\partial U}{\partial J_2}(J_1(t,\tau),J_2(t,\tau),t-\tau)\frac{\lambda(t)}{\lambda(\tau)}\right]\mathrm{d}\tau \tag{11.19}$$

Eliminating p from Equation (11.18) now gives

$$\sigma=2\int_{-\infty}^{t}\left[\left(\frac{\lambda^2(t)}{\lambda^2(\tau)}-\frac{\lambda(\tau)}{\lambda(t)}\right)\frac{\partial U}{\partial J_1}(J_1(t,\tau),J_2(t,\tau),t-\tau)-\left(\frac{\lambda^2(\tau)}{\lambda^2(t)}-\frac{\lambda(t)}{\lambda(\tau)}\right)\frac{\partial U}{\partial J_2}(J_1(t,\tau),J_2(t,\tau),t-\tau)\right]\mathrm{d}\tau \tag{11.20}$$

Zapas and Craft (1965) produced a model for uniaxial stretching by rewriting (11.20) as

$$\sigma(t) = \int_{-\infty}^{t}\left(\frac{\lambda^2(t)}{\lambda^2(\tau)} - \frac{\lambda(\tau)}{\lambda(t)}\right) h\left(\frac{\lambda(t)}{\lambda(\tau)}, t-\tau\right)\mathrm{d}\tau. \tag{11.21}$$

where

$$h\left(\frac{\lambda(t)}{\lambda(\tau)}, t-\tau\right) = 2\left(\frac{\partial U}{\partial J_1} + \frac{\lambda(\tau)}{\lambda(t)}\frac{\partial U}{\partial J_2}\right)$$

This version can be used to illustrate a general characteristic of the K-BKZ model. Since τ runs through all values previous to t, the stress depends on the strain at time t as measured relative to all previous strain states. For a specimen that is unstrained prior to time $t = 0$, there are contributions to the stress at positive times from the part of the strain history occurring at times less than zero. In this respect, the theory differs from those mentioned so far. Splitting the integral in Equation (11.21) into parts before and after zero time, and setting $\lambda(\tau) = 1$ for $\tau < 0$, we obtain

$$\sigma(t) = \left(\lambda^2(t) - 1/\lambda(t)\right)\int_{-\infty}^{0} h(\lambda(t), t-\tau)\,\mathrm{d}\tau + \int_{0}^{t}\left(\frac{\lambda^2(t)}{\lambda^2(\tau)} - \frac{\lambda(\tau)}{\lambda(t)}\right) h\left(\frac{\lambda(t)}{\lambda(\tau)}, t-\tau\right)\mathrm{d}\tau. \tag{11.22}$$

In particular, for stress relaxation starting at zero time the second term is zero. We may then write

$$\sigma(t) = \left(\lambda^2 - 1/\lambda\right) H(\lambda, t), \tag{11.23}$$

where

$$h(\lambda, t) = \frac{\partial H}{\partial t}(\lambda, t). \tag{11.24}$$

Here, the assumption has been made that the function h is zero at large times. This is appropriate for a fluid and in any case causes no significant loss of generality. It is clear from Equations (11.23) and (11.24) that the function h is entirely determined by single-step stress relaxation measurements. Therefore, the stress for a general strain history can be calculated once stress relaxation data have been gathered over the appropriate range of strain. There is no parallel formulation for the strain in terms of the stress history.

In the original paper (Bernstein *et al.* 1963), the authors reported work on the uniaxial tension of plasticized poly(vinyl chloride), sulphur vulcanizates of butyl rubber and polyisobutylene. Very successful predictions were made at extension ratios up to approximately five. Zapas and Craft (1965) applied their formulation to multi-step stress relaxation and creep and recovery of both plasticized poly(vinyl chloride) and polyisobutylene. McKenna and Zapas (1979) applied a modified form of the model to the torsional deformation of PMMA. McKenna and Zapas (1981) have used the model in the analysis of the tensile behaviour of carbon-black-filled butyl rubbers.

The K-BKZ model has been implemented numerically for a range of applications, many of which take advantage of its large strain capability. Erchiqui used the theory within an FE scheme to model thermoforming of HDPE (Erchiqui 2005) and to model thermoforming and blow moulding (Erchiqui 2006). Cha *et al.* (2019) modelled thermoforming of ABS and successfully predicted the wall thickness of the product. The stages of the simulation are shown in Figure 11.5.

11.2.2.4 Other Models of Finite Viscoelasticity

Simo (1987) has developed a highly influential large deformation model of viscoelasticity that is based on thermodynamics and can incorporate damage. As an example of its

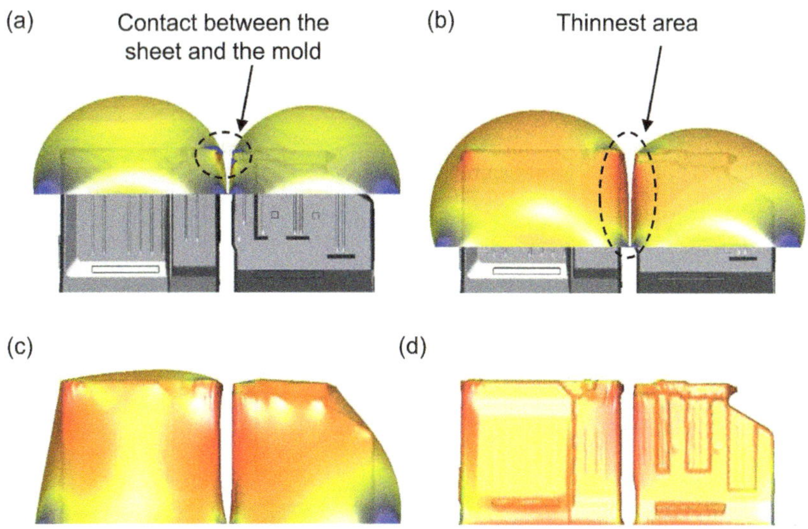

Figure 11.5 Stages in the simulation of a forming process. Initially there is a blowing stage when the sheet is formed into a bubble (a). The mould moves upwards in (b) and it stretches the bubble. Vacuum is applied in stages (c) and (d) to complete the process. Cha et al., 2019 / Springer Nature / CC BY 4.0.

application, we consider the work of Jridi *et al.* (2019) who applied it to experimental results in tension and shear of filled rubbers in both dynamic and static tests. The uniaxial response can be written as (Jridi *et al.* 2019):

$$\begin{aligned}\sigma(t) &= \frac{2}{g_\infty}\left(\lambda^2(t)-1/\lambda(t)\right)\left(\frac{\partial W}{\partial I_1}+\frac{1}{\lambda(t)}\frac{\partial W}{\partial I_2}\right)\\ &+\frac{2}{g_\infty}\left(\lambda^2(t)-1/\lambda(t)\right)\int_0^t\frac{\partial g_1(s)}{\partial s}\left(\frac{\partial W}{\partial I_1}+\frac{1}{\lambda(t-s)}\frac{\partial W}{\partial I_2}\right)ds\\ &-\frac{2}{\lambda(t)g_\infty}\int_0^t\frac{\partial g_1(s)}{\partial s}\frac{\partial W}{\partial I_2}\lambda^2(t-s)ds\end{aligned} \tag{11.25}$$

where W is a strain energy function, g_∞ a material parameter and g_1 defines the time-dependent behaviour. Jridi *et al.* compared the performance of the Simo model with a variant of the K-BKZ model and two other models, namely those of Fosdick and Yu (1998) and Christensen (1980).

Fosdick and Yu (1998) adopted a thermodynamics-based approach that resulted in a model of quasi-linear form. Its response in uniaxial strain has been given as (Jridi *et al.* 2019):

$$\begin{aligned}\sigma(t) &= 2\left(\lambda^2(t)-1/\lambda(t)\right)\left(\frac{\partial W}{\partial I_1}+\frac{1}{\lambda(t)}\frac{\partial W}{\partial I_2}\right)\\ &+\frac{G_0}{2\lambda^2(t)}\int_0^t g_1(t-s)\frac{\partial\lambda^2(s)}{\partial s}ds\\ &-\frac{G_0\lambda(t)}{2}\int_0^t g_1(t-s)\frac{\partial(1/\lambda(s))}{\partial s}ds\end{aligned} \tag{11.26}$$

Christensen (1980) has created a model aimed at elastomers, in the form of a generalization of the kinetic theory of rubber elasticity (see Chapter 4). The uniaxial response is (Jridi *et al.* 2019):

$$\begin{aligned}\sigma(t) &= 2\left(\lambda^2(t)-1/\lambda(t)\right)\left(\frac{\partial W}{\partial I_1}+\frac{1}{\lambda(t)}\frac{\partial W}{\partial I_2}\right)\\ &+\frac{G_0}{2}\lambda^2(t)\int_0^t g_1(t-s)\frac{\partial\lambda^2(s)}{\partial s}ds\\ &-\frac{G_0}{2\lambda(t)}\int_0^t g_1(t-s)\frac{\partial(1/\lambda(s))}{\partial s}ds\end{aligned} \tag{11.27}$$

The variant of the K-BKZ model was created by adding a hyperelastic term to the K-BKZ integral.

Jridi *et al.* compared all four models. For uniaxial tension, predictions were highly influenced by the choice of the strain energy function W, but the K-BKZ, Simo and Fosdick models gave similar and reasonably accurate predictions, generally better than that of Christensen (see Figure 11.6(a)). For dynamic testing, the main issue is the poor performance of K-BKZ in predicting storage modulus, with the best prediction from the Simo model (see Figure 11.6(b)).

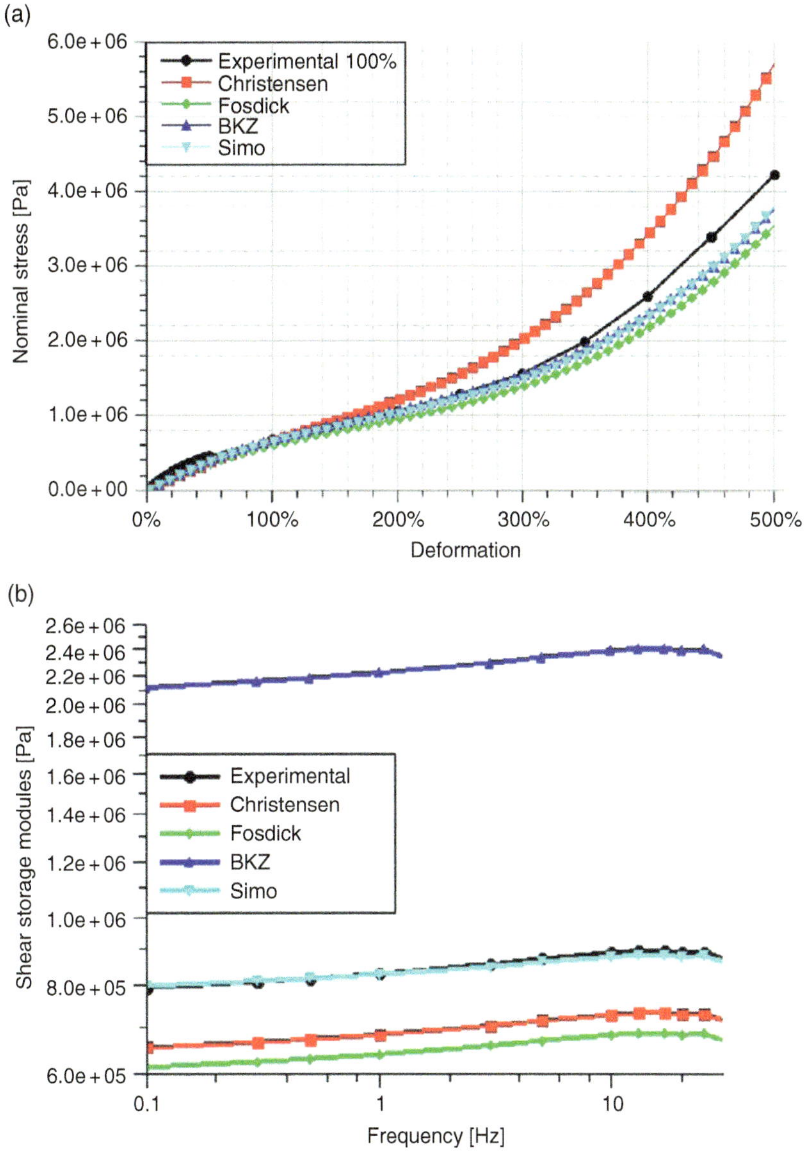

Figure 11.6 Typical results for filled natural rubber (a) Tensile tests. (b) Storage modulus from dynamic testing. Jridi et al. (2019) / Springer Nature.

11.2.3 Multiple Integral Models

The multiple integral approach received a lot of attention in the 1960s and 1970s. In principle, an exact theory is possible by this means, but this is not necessarily relevant as it could require a theory with infinitely many terms. As each term requires additional experimentation, there is potential for a correspondingly unwieldy experimental programme. The question to be answered is whether an acceptably accurate prediction can be achieved with a feasible computational and experimental effort. The history of the application of this method shows that this question has been answered broadly in the negative. The discussion below will set out the main aspects of the theories and the conclusions reached as to their utility.

11.2.3.1 Green and Rivlin

Green and Rivlin (1957) were the pioneers of the multiple integral approach. They took a continuum approach with no significant physical assumptions other than material objectivity (see Section 3.4.2).

This model makes use of the Fréchet series of multiple integrals. In the one-dimensional case, for the strain e at time t in terms of the history of stress σ this takes the form

$$\begin{aligned} e(t) = & \int_{-\infty}^{t} J_1(t-\tau_1)\dot{\sigma}(\tau_1)\mathrm{d}\tau_1 + \int\int_{-\infty}^{t} J_2(t-\tau_1, t-\tau_2)\dot{\sigma}(\tau_1)\dot{\sigma}(\tau_2)\mathrm{d}\tau_1\mathrm{d}\tau_2 \\ & + \iint\int_{-\infty}^{t} J_3(t-\tau_1, t-\tau_2, t-\tau_3)\dot{\sigma}(\tau_1)\dot{\sigma}(\tau_2)\dot{\sigma}(\tau_3)\mathrm{d}\tau_1\mathrm{d}\tau_2\mathrm{d}\tau_3 + \cdots . \end{aligned} \tag{11.28}$$

There is an analogous form of this infinite series for the stress in terms of the strain history. In both cases, the first term is recognizable as the Boltzmann integral (see Chapter 5).

To use this theory, it is necessary to evaluate a sufficient number of the functions J_i, the 'kernel functions' that are essentially the material parameters, so that the series can be calculated to give an acceptable level of accuracy of the stress or strain prediction. In the years following the proposal of the theory, there were a number of specific applications to polymers. A broader discussion centred around several issues, principally: the size of the experimental programme needed to derive the J_i ; their physical significance; the practicality of determining them with sufficient accuracy and the convergence of the series (11.28). Haddad (1995) has given a brief account of this discussion.

Early applications were to relatively simple stress or strain histories. Ward and Onat (1963), working on the creep of oriented polypropylene, noted the inadequacy of Leaderman's approach in representing their results. One critical aspect was that the initial linear elastic response predicted by Leaderman was not mirrored in the polypropylene results. Furthermore, Leaderman's restrictive assumption for the form of the integral kernel function – its separability into functions of stress and of time – led to the prediction of identical creep and recovery curves at the same stress, another qualitative difference from the polypropylene observations. This was their motivation for exploring the Green–Rivlin theory. They concluded that for their study, only the first and third terms of Equation (11.28) were required for an adequate representation. Later, however, Hadley and Ward (1965) also working on polypropylene, concluded that in general more terms were needed, depending on load level and step duration.

On the subject of practicality of the multiple-integral representation, Turner (1966) has remarked that there are in general 'great difficulties' in defining a suitable experimental programme to determine the kernel functions. Lockett (1972, 1965) has quantified the size of an adequate experimental programme and concluded that it is in general impractically large, even for the one-dimensional case. In three dimensions, having made the simplification of incompressible material, he estimated that a programme of 287 experiments would be required for a satisfactory representation. He remarked that this 'demonstrates the futility of mathematical theories which do not consider the consequent experimental requirements' (Lockett 1972, p. 95).

Another aspect that relates to practicality is the degree of accuracy required for the experiments. Gradowczyk (1969) has shown that the kernel functions are very sensitive to experimental error, and that as the number of terms in the representation is increased, so do the errors in the higher order terms. Thus, a point is reached at which the benefit in accuracy of adding a higher order term is outweighed by the accompanying uncertainty in its value.

11.2.3.2 Pipkin and Rogers

Some discussion has taken place on the physical significance of the terms in the Green–Rivlin expansion. The first term corresponds to linear viscoelasticity, and so can be viewed as having physical significance without being related to specific mechanisms. Otherwise, since there are no physical mechanisms implicit in the theory, there would seem to be no reason to expect any direct physical interpretation. The situation seems to be analogous to the use of a polynomial function to fit to a non-linear elastic stress–strain curve. The linear term (if fitted with care) would correspond to the linear elastic regime, and any correlation of the higher-order terms with physical mechanisms would be entirely coincidental. This is essentially the conclusion reached for the Green–Rivlin theory by Yannas and Haskell (1971) who concluded that, apart from for the first term, the kernels have a correspondence to observable quantities that was 'either nil or else cannot be ascertained conveniently'.

To address some of the objections and difficulties of the Green–Rivlin theory outlined above, Pipkin and Rogers (1968) introduced an alternative multiple-integral expansion. This takes the form

$$\begin{aligned}\sigma(t) = & \int_{-\infty}^{t} \mathrm{d}_{e(\tau_1)} R_1[e(\tau_1), t-\tau_1] + \frac{1}{2!}\int\int_{-\infty}^{t} \mathrm{d}_{e(\tau_1)}\mathrm{d}_{e(\tau_2)} R_2[e(\tau_1), t-\tau_1; e(\tau_2), t-\tau_2] \\ & + \frac{1}{3!}\int\int\int_{-\infty}^{t} \mathrm{d}_{e(\tau_1)}\mathrm{d}_{e(\tau_2)}\mathrm{d}_{e(\tau_3)} R_3[e(\tau_1), t-\tau_1; e(\tau_2), t-\tau_2; e(\tau_3), t-\tau_3] + \cdots\end{aligned} \tag{11.29}$$

for the stress in terms of the strain history. The kernel functions can be related to experimental stress relaxation results. In the first term, the quantity $R_1(e, t)$ is the stress in stress relaxation at strain e. This term includes non-linear behaviour and is an extension of the Leaderman integral mentioned above; it is a generalization, in that the kernel function is no longer assumed to be separable into functions of strain and time. The first term represents non-linear superposition and reproduces exactly the experimental response to single-step stress relaxation tests. It will also give approximate predictions for two-step stress relaxation tests. The second term provides the correction needed to produce the exact result for two-step tests. Thus, as the series is built up each term added suggests the experimental programme required for its evaluation. In this way, a physical interpretation is built into the theory and, since each term is non-linear, it is to be expected that fewer terms are necessary than with the Green–Rivlin approach. Most applications of the Pipkin–Rogers theory have involved the evaluation of the first term only, but Mittal and Singh (1986) have applied the theory up to second order for nylon-6.

There have, however, been criticisms of the practicality of this approach. Stafford (1969) has pointed out that the second term requires the evaluation of a function of four variables. He argued that this would make it less experimentally tractable than the two-variable kernel in Green and Rivlin's second term. Nevertheless Pipkin and Rogers demonstrated that the first term of their expansion was more effective than a three-term Green–Rivlin model. The Pipkin–Rogers model was and remains a significant theoretical development since, as noted by Drapaca *et al.* (2007), many of the alternative approaches can be shown to be special cases of it.

11.2.4 Differential Models

The differential approach to linear viscoelasticity can be modified to accommodate nonlinearity. There have been two principal attempts using this approach that have had significant impact. In the first, Smith (1962) generalized linear theory to large strain, using an analysis that can be viewed as a generalization of the Maxwell model. In the contrasting approach of Kitagawa *et al.* (1989), the differential equation of the standard linear solid formed the starting point. Both approaches continue to make significant contributions and will be discussed further below.

11.2.4.1 Large-Strain Behaviour of Elastomers

It is possible to take the equations of linear spring and dashpot models and adapt them to nonlinear conditions. Thus, Smith (1962) has described the large-strain behaviour of elastomers by taking as his starting-point the (linear) Maxwell element. Rewriting Equation (5.15), we have

$$\frac{\mathrm{d}e}{\mathrm{d}t} = \frac{\sigma}{\eta} + \frac{1}{E}\frac{\mathrm{d}\sigma}{\mathrm{d}t}.$$

Imposing a constant strain rate $\mathrm{d}e/\mathrm{d}t = R$, it can be readily shown (see Section 5.25 above) that

$$\sigma = R\,\tau\left(1 - e^{-t/\tau}\right),$$

where $\tau = \eta/E$, and equivalently

$$\sigma = R\,\tau\,E\left(1 - e^{-t/\tau}\right).$$

For a continuous distribution of relaxation times $H(\tau)$, equivalent to an infinite parallel array of Maxwell elements, the total stress is summed by the integral to give

$$\sigma = R\int_{-\infty}^{\infty} \tau H(\tau)\left(1 - e^{-t/\tau}\right)\mathrm{d}\ln\tau.$$

With $R = e/t$, this becomes

$$\frac{\sigma}{e} = \frac{1}{t}\int_{-\infty}^{\infty} \tau H(\tau)\left(1 - e^{-t/\tau}\right)\mathrm{d}\ln\tau + E_e,$$

where the term E_e, equivalent to an additional parallel elastic element, denotes the equilibrium modulus. The quantity

$$\frac{\sigma}{e} = \frac{\sigma\,(e, t)}{e}$$

is clearly a function of time only and is known as the *constant strain-rate modulus* $F(t)$. When generalized to large strain, the theory will become non-linear and F will be a function of both strain and time. Smith assumes that F is separable into a function of strain and a function of time, such that

$$F(t) = \frac{g(e)\sigma(e, t)}{e}.$$

This is also expressible as

$$\log F(t) = \log\left(\frac{g(e)}{e}\right) + \log\sigma(e, t). \tag{11.30}$$

To return linear viscoelasticity, it is required that $g(e)$ approaches unity for small strain. The stress–strain data for Smith's SBR vulcanizate rubber material are plotted in Figure 11.7(a). Log stress against log time plots was obtained for fixed strains and, as shown in Figure 11.7(b), forms parallel linear relationships. This suggests via Equation (11.30) that the quantity $g(e)/e$ is independent of time. It was found that for extension ratios up to 2, $g(e) \cong 1$ provided that σ is understood to denote the true stress. At higher strains, the empirical function

$$g(e) = \lambda\ \exp\left(\frac{1}{\lambda} - \lambda\right)$$

was used, equivalent to that proposed by Martin *et al.* (1956).

In this approach, the non-linearity is introduced as a consequence of large deformation. We would not expect it to be generally successful for polymers as non-linearity is sometimes observed at small strains. However, similar approaches have been successful for rubbers by Guth *et al.* (1946) and Tobolsky and Andrews (1945).

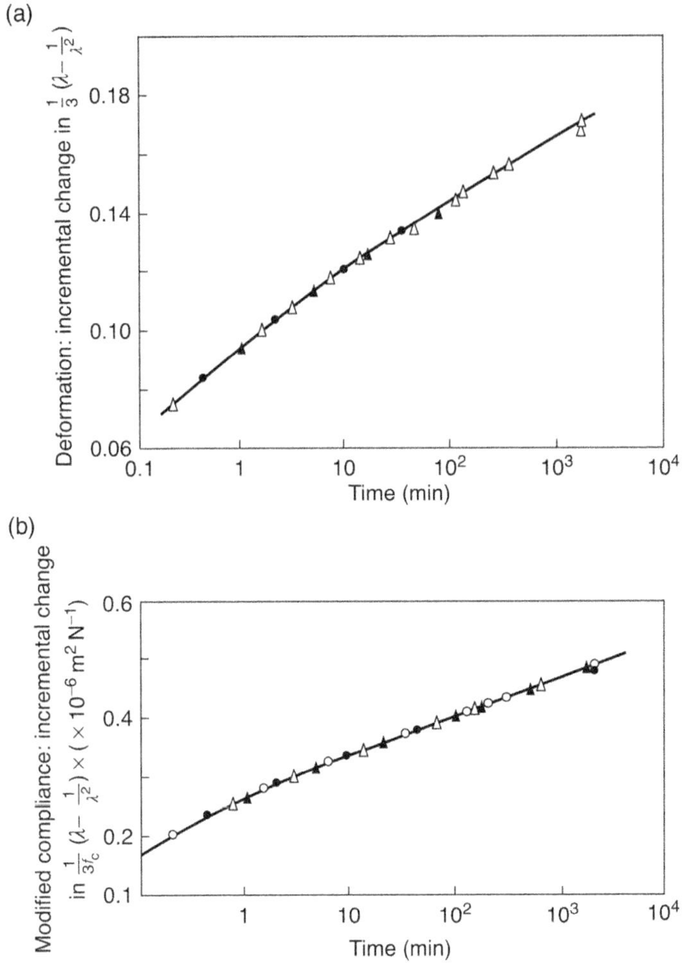

Figure 11.9 (a) Comparison of creep (Δ) and recovery (▲) of plasticized polyvinyl chloride under a constant nominal stress of 0.355 MPa. (b) Creep of plasticized polyvinyl chloride under constant nominal stress: ○, f_0 = 0.444 MPa; ●, f_0 = 0.355 MPa; Δ, f_0 = 0.267 MPa; ▲, f_0 = 0.178 MPa. Adapted from Leaderman, 1962.

11.2.4.3 Overstress Theories

While Smith's analysis above took the Maxwell model as its starting point, a useful alternative is to take the standard linear solid as the basis. This is the case in the study of polypropylene by Kitagawa *et al.* (1989) and of polyethylene by Kitagawa and Takagi (1990). The differential equation of the standard linear solid is given by Equation (11.31):

$$\sigma + \tau\frac{d\sigma}{dt} = E_a e + (E_m + E_a)\tau\frac{de}{dt}. \tag{11.31}$$

Recall that the first term on the right is equal to the stress after a long time, or equivalently the stress produced by loading at an infinitely slow rate. The equation can be generalized to

$$\sigma + K\frac{d\sigma}{dt} = f(e) + M\frac{de}{dt}, \tag{11.32}$$

where $f(e)$ is simply the stress response of the material at infinitely slow strain rate, which we no longer require to be linear. M and K are in general functions of stress, strain and their time derivatives, so that Equation (11.32) defines a non-linear material. It can be rearranged as

$$\sigma - f(e) = M\frac{de}{dt} - K\frac{d\sigma}{dt}. \tag{11.33}$$

The quantity $\sigma - f(e)$ on the left, the excess of stress over that obtaining at equilibrium or very slow strain rate, is termed the *overstress*; theories formulated in this way are termed *overstress theories*. This approach has been applied to metals by Lui and Krempl (1979), who used the term *viscoplastic* to categorize their model. From the form of Equation (11.32), it is clear that the function K could be measured experimentally using stress relaxation tests $\left(\frac{de}{dt} = 0\right)$, or that M could be measured using creep tests $\left(\frac{d\sigma}{dt} = 0\right)$. Additional information can be gained from small strain, high strain rate experiments, where the response of the model can be assumed linear elastic. In these conditions, we may revert to Equation (11.31) and notice that, for fast loading, the time derivative terms become dominant, so that the elastic response of the standard linear solid is characterized by the elastic modulus $E_a + E_m$. Then, inspection of Equation (11.32) reveals that the instantaneous modulus E is given by

$$E = \frac{M}{K} \tag{11.34}$$

for small strains and fast loading. The model is simplified if Equation (11.34) is assumed to apply under all conditions, so that K and M have the same functional form; this approach was adopted by Lui and Krempl (1979), Kitagawa *et al.* (1989) and Kitagawa and Takagi (1990).

Kitagawa *et al.* in their work on polypropylene, and Kitagawa and Takagi, working on polyethylene, used torsion testing. Their equivalent of Equation (11.33) in terms of shear stress and shear strain takes the form

$$\tau - f(\gamma) = M\frac{d\gamma}{dt} - K\frac{d\tau}{dt} \tag{11.35}$$

and Equation (11.10) becomes

$$G = \frac{M}{K}, \tag{11.36}$$

where G is the shear modulus. In both cases, K was found to depend on the overstress and on the strain, and took the form

$$K = K_0 \exp\left(-K_1(\gamma)(\tau - f(\gamma))\right), \tag{11.37}$$

where K_0 was found to be constant and

$$K_1(\gamma) = p_0 + \frac{p_1}{p_2 + \gamma} \tag{11.38}$$

with p_0, p_1 and p_2 material constants.

Stress relaxation tests were used to evaluate K. According to Equation (11.37),

$$\ln K = \ln K_0 - K_1(\gamma)(\tau - f(\gamma)) \tag{11.39}$$

suggesting the form required for graphs to show linear relationships at constant shear strain. Such a plot is shown for the polypropylene data (Kitagawa *et al.*(1989) in Figure 11.10, taken at a series of strain levels. There is a good approximation to linearity, with the common intercept corresponding to the constancy of the parameter K_0. The efficacy of the model in predicting torsional stress–strain curves is shown in Figure 11.11. Here, strain rates are initially constant, but are changed abruptly to different constant rates so that transient stresses are generated. This provides a severe test for the model, which performs well.

Similar work with polyethylene (Kitagawa and Takagi 1990) revealed a comparable level of accuracy. The differential approach has also been applied to high-density polyethylene by Zhang and Moore (1997a, 1997b). Brusselle-Dupend *et al.* (2003) have extended the approach to a more complex model that encompasses unloading as well as loading behaviour in polypropylene.

We can rearrange Equation (11.33) to give an expression for strain rate:

$$\frac{\mathrm{d}e}{\mathrm{d}t} = \frac{K}{M}\frac{\mathrm{d}\sigma}{\mathrm{d}t} + \frac{\sigma - f(e)}{M}$$

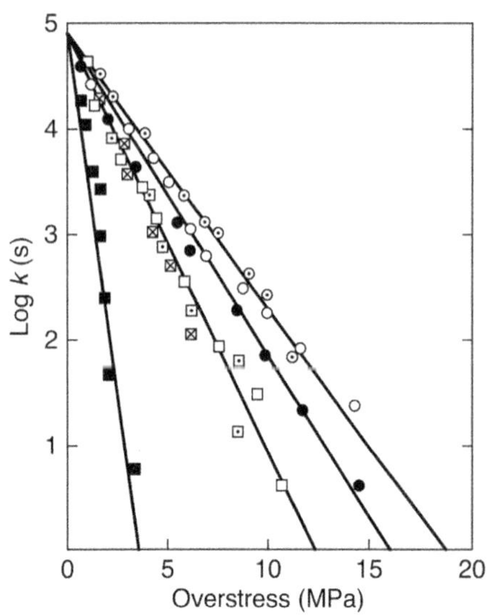

Figure 11.10 The function K of Equations (11.11) and (11.13) for polypropylene obtained from stress relaxation at various strains, with strain applied at varying rates. Symbols correspond to + strain 0.01, strain rate 1.4×10^{-3} s^{-1}; ∗ 0.045, 1.4×10^{-3}; 0.055, 1.4×10^{-3}; □ 0.053, 1.4×10^{-2}; ● 0.108, 1.4×10^{-3}; ⊙ 0.253, 1.4×10^{-4}; ○ 0.267, 1.4×10^{-3}. Kitagawa et al. (1989) / John Wiley & Sons.

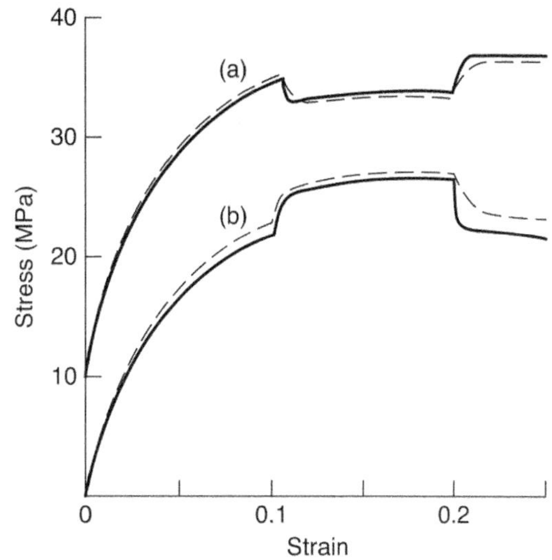

Figure 11.11 Modelling of stress–strain curves with step changes in strain rate. (a) $1.1 \times 10^{-2} \rightarrow 1.4 \times 10^{-3} \rightarrow 1.4 \times 10^{-2}\ s^{-1}$. (b) $1.0 \times 10^{-3} \rightarrow 1.4 \times 10^{-2} \rightarrow 6 \times 10^{-4}\ s^{-1}$. Curve (b) is shifted by 10 MPa along the vertical axis. (Reproduced from Kitagawa, M., Mori, T. and Matsutani, T. (1989) Rate-dependent nonlinear constitutive equation of polypropylene. J. Polym. Sci. B, **27**, 85. Copyright (1989) John Wiley & Sons, Ltd.)

This resembles the expression of Ho and Krempl (2002)

$$\frac{\mathrm{d}e}{\mathrm{d}t} = \frac{1}{E}\frac{\mathrm{d}\sigma}{\mathrm{d}t} + \frac{\sigma - G}{Ek}$$

for uniaxial deformation, where E is elastic modulus and G is now the equilibrium stress. The first term on the right corresponds to elastic strain and the second to inelastic strain. Ho and Krempl generalized this approach to three dimensions:

$$\frac{\mathrm{d}\mathbf{e}}{\mathrm{d}t} = \frac{1+\nu}{E}\frac{\mathrm{d}\mathbf{s}}{\mathrm{d}t} + \frac{3}{2}F\left(\frac{\Gamma}{D}\right)\frac{\mathbf{s}-\mathbf{g}}{\Gamma} \tag{11.40}$$

Here **s** and **g** represent the stress deviator and equilibrium stress deviator, respectively, and **e** is the strain deviator (see Chapter 3). Γ is a scalar invariant of the overstress deviator tensor **s** − **g**, D a constant known as a drag stress, F a viscosity function, and ν Poisson's ratio.

Equation (11.40) relates the inelastic deviatoric strain rate tensor to a deviatoric stress tensor. As such it resembles a flow rule, an essential component of the theory of plasticity, to be discussed in the next chapter. The model of Equation (11.40) along with other developments are known as theories of viscoplasticity based on overstress (VBOs). They are better classified as theories of viscoplasticity rather than viscoelasticity and will be discussed in the next chapter. However, we note that they can be developed from the starting point of the viscoelastic standard linear solid of Equation (11.31).

11.2.5 Current Usage

It is now possible to include quite complex non-linear theories into FE analysis schemes for engineering applications. Multiple-integral theories have made no impact in this area, most probably because of the experimental difficulties that they impose and also the complexity of the numerical schemes required in their implementation. The other forms of constitutive equation discussed in this introduction have provided more fruitful areas for numerical exploitation. The numerical implementation of integral models is a somewhat more challenging computing problem than that for differential or rate theories. In an incremental analysis, an integral model must compute over all of the history of strains in previous increments, whereas in a rate model only the immediately preceding time increment need be considered. However, with current computer technology, this consideration has diminished in importance. This is evidenced by practically realized FE computations with single integral models at their core (Dean *et al.* 2011).

References

Abdullah, Z., Gray, V., Whittaker, M. and Perkins, K. (2014) A critical analysis of the conventionally employed creep lifing methods. *Materials*, **7**, 3371–3398.

Amjadi, M. and Fatemi, A. (2021) Creep behavior and modeling of high-density polyethylene (HDPE). *Polym. Test.*, **94**, 107031.

Beijer, J.G.J. and Spoormakerb, J.L. (2002) Solution strategies for FEM analysis with nonlinear viscoelastic polymers. *Comput. Struct.*, **80**, 1213–1229.

Bernstein, B. (1966) Time-dependent behavior of an incompressible elastic fluid. Some homogeneous deformation histories. *Acta Mech.*, **2**, 329–354.

Bernstein, B., Kearsley, B.A. and Zapas, L.J. (1963) A study of stress relaxation with finite strain. *Trans. Soc. Rheol.*, **7**, 391.

Brinson, H.F. and Brinson, L.C. (2015) *Polymer Engineering Science and Viscoelasticity: An Introduction*, 2nd edn, Springer, New York.

Brusselle-Dupend, N., Lai, D., Feaugas, X. *et al.* (2003) Mechanical behavior of a semicrystalline polymer before necking. Part II: modeling of uniaxial behavior. *Polym. Eng. Sci.*, **43**, 501–518.

Cha, J., Kim, M., Park, D. and Go, J.S. (2019) Experimental determination of the viscoelastic parameters of K-BKZ model and the influence of temperature field on the thickness distribution of ABS thermoforming. *Int. J. Adv. Manuf. Technol.*, **103**, 985–995.

Chailleux, E. and Davies, P. (2005) A non-linear viscoelastic viscoplastic model for the behaviour of polyester fibres. *Mech. Time-Depend. Mater.*, **9**, 147–160.

Christensen, R.N. (1980) A nonlimear theory for viscoelasticity for application to elastomers. *J. Appl. Mech.*, **47**, 762–768.

Crochon, T., Schönherr, T. and Li, C. (2010) On finite-element implementation strategies of Schapery-type constitutive theories. *Mech. Time-Depend. Mater.*, **14**, 359–387.

Crook, R.A. (1993) Damage and the nonlinear viscoelastic response of glassy polycarbonate and LaRC-TPI. *Polym. Eng. Sci.*, **33**, 56.

De Pascalis, R., Abrahams, I.D. and Parnell, W.J. (2014) On nonlinear viscoelastic deformations: a reappraisal of Fung's viscoelastic model. *Proc. R. Soc. A*, **470**(2014), 0058.

Dean, G., McCartney, L.N., Crocker, L. and Mera, R. (2009) Modelling long term deformation behaviour of polymers for finite element analysis. *Plast., Rubber Compos.*, **38**, 433–443.

Dean, G.D., McCartney, L.N., Mera, R. and Urqhart, J.M. (2011) Modeling nonlinear viscoelasticity in polymers for design using finite element analysis. *Polym. Eng. Sci.*, **51**, 2011–2219.

Dillard, D.A., Straight, M.R. and Brinson, H.F. (1987) The nonlinear viscoelastic characterization of graphite/epoxy composites. *Polym. Eng. Sci.*, **27**, 116–123.

Drapaca, C.S., Sivaloganathan, S. and Tenti, G. (2007) Nonlinear constitutive laws in viscoelasticity. *Math. Mech. Solids*, **12**, 475–501.

Erchiqui, F. (2005) Thermodynamic approach of inflation process of K-BKZ polymer sheet with respect to thermoforming. *Polym. Eng. Sci.*, **45**, 1319–1335.

Erchiqui, F. (2006) A new hybrid approach using the explicit dynamic finite element method and thermodynamic law for the analysis of the thermoforming and blow molding processes for polymer materials. *Polym. Eng. Sci.*, **46**, 1554–1564.

Ferry, J.D. (1980) *Viscoelastic Properties of Solids*, John Wiley & Sons, New York. ISBN 0-471-04894-1

Findley, W.N. and Khosla, G. (1955) Application of the superposition principle and theories of mechanical equation of state, strain, and time hardening to creep of plastics under changing loads. *J. Appl. Phys.*, **26**, 821.

Findley, W.N. and Lai, J.S.Y. (1967) A modified superposition principle applied to creep of nonlinear viscoelastic material under abrupt changes in state of combined stress. *Trans. Soc. Rheol.*, **11**, 361–381.

Fosdick, R. and Yu, J.-H. (1998) Thermodynamics, stability and non-linear oscillations of viscoelastic solids – II. History type solids. *Int. J. Non-Linear Mech.*, **35**, 165–188.

Fung, Y.C. (1972) Stress-strain-history relations of soft tissues in simple elongation, in *Biomechanics: Its Foundations and Objectives*, (eds Y.C. Fung, N. Perrone and M. Anliker), Prentice-Hall, Englewood Cliffs, NJ, pp. 181–218.

Fung, Y.C. (1993) *Biomechanics: Mechanical Properties of Living Tissues*, 2nd edn, Springer, New York, NY.

Gradowczyk, M.H. (1969) On the accuracy of the Green–Rivlin representation for viscoelastic materials. *Int. J. Solids Struct.*, **5**, 873–877.

Green, A.E. and Rivlin, R.S. (1957) The mechanics of non-linear materials with memory – part I. *Arch. Ration. Mech. Anal.*, **1**, 1–21.

Guth, E., Wak, P.E. and Anthony, R.L. (1946) Significance of the equation of state for rubber. *J. Appl. Phys.*, **17**, 347.

Haddad, Y.M. (1995) *Viscoelasticity of Engineering Materials, Chapter 6*, Chapman and Hall, London.

Hadley, D.W. and Ward, I.M. (1965) Non-linear creep and recovery behaviour of polypropylene fibres. *J. Mech. Phys. Solids*, **13**, 397–411.

Haj-Ali, R.M. and Muliana, A.H. (2004) Numerical finite element formulation of the Schapery non-linear viscoelastic material model. *Int. J. Numer. Methods Eng.*, **59**, 25–45.

Hirakawa, D., Kongkitkul, W., Tatsuoka, F. and Uchimura, T. (2003) Time-dependent stress–strain behaviour due to viscous properties of geogrid reinforcement. *Geosynth. Int.*, **10**, 176–199.

Ho, K. and Krempl, E. (2002) Extension of the viscoplasticity theory based on overstress (VBO) to capture non-standard rate dependence in solids. *Int. J. Plast.*, **18**, 851–872.

Jridi, N., Arfaoui, M., Hamdi, A., Salvia, M., Bareille, O., Ichchou, M. and Abdallah, J.B. (2019) Separable finite viscoelasticity: integral-based models vs. experimenrts. *Mech. Time-Dependent Mater*, **23**, 295–325.

Kaye, A. (1962) Non-Newtonian Flow in Incompressible Fluids. Note No. 134, College of Aeronautics, Cranfield, UK.

12

Yielding and Instability in Polymers

As pointed out by Wineman and Waldron (1993), there are two approaches to the modelling of yield in polymers: the use of non-linear viscoelasticity and the direct application of metal plasticity. Relatively simple theories of plasticity are available from the metals field. These theories, that embody the classical concepts of plasticity, may still be applied usefully to polymers, for instance in cases where changes in strain rate are small. This restriction arises since, while stress maxima that resemble yield points may be observed in polymers, their values do not denote a unique property as they depend on the rate of strain.

Allied to the subject of yielding is that of instability. Necking in a tensile test specimen is an example of instability and is caused by the underlying yield properties of the material. Yielding may lead to a maximum in the applied force, which may then allow the strain in the specimen to increase with no increase in force – the unstable condition. We should note that yielding and instability are qualitatively different phenomena, in that yielding is an intrinsic material characteristic, whereas instability is a function of the geometry and loading conditions of the loaded body.

A number of different factors contribute to the present interest in the yield behaviour. First, it has been recognized that the classical concepts of plasticity are relevant to forming, rolling and drawing processes in polymers. Secondly, there has been a number of striking experimental studies of 'slip bands' and 'kink bands' in polymers, which suggest that deformation processes in polymers might be similar to those in crystalline materials such as metals and ceramics. Finally, it is evident that distinct yield points are observed, and there is much interest in understanding these in the context of other ideas in polymer science.

Our first task in this chapter is to discuss the relevance of classical ideas of plasticity to the yielding of polymers. Although the yield behaviour is temperature and strain rate dependent, it will be shown that provided that the test conditions are chosen suitably, yield stresses can be measured which satisfy conventional yield criteria. The temperature and time dependence often obscure some generalities of the yield behaviour. For example, it might be concluded that some polymers show necking and cold-drawing, whereas others are brittle and fail catastrophically. Yet another type of polymer (a rubber) extends homogeneously to rupture. A salient point to recognize is that polymers in general can show all these types of behaviour depending on the exact conditions of test (Figure 13.1), quite irrespective of their chemical nature and physical structure. Thus explanations of yield behaviour, which involve, for example, cleavage of crystallites or lamellar slip or amorphous mobility are only

Mechanical Properties of Solid Polymers, Fourth Edition. John Sweeney and Peter Hine.

relevant to specific cases. As in the case of linear viscoelastic behaviour or rubber elasticity, we must first seek an understanding of the relevant phenomenological features, decide on suitable measurable quantities and then provide a molecular interpretation of the subsequent constitutive relations.

12.1 Discussion of the Load–Elongation Curves in Tensile Testing

The most dramatic consequence of yield is seen in a tensile test when a neck or deformation band occurs, as in Figure 12.1, with the plastic deformation concentrated either entirely or primarily in a small region of the specimen. The precise nature of the plastic deformation depends both on the geometry of the specimen and on the form of the applied stresses, and will be discussed more fully later.

The characteristic necking and cold-drawing behaviour is as follows. On the initial elongation of the specimen, homogeneous deformation occurs and the conventional load–extension curve shows a steady increase in load with increasing elongation (AB in Figure 12.2). At the point B, the specimen thins to a smaller cross-section at some point, i.e. a neck is formed. Further elongation brings a fall in load. Continuing extension is achieved by causing the shoulders of the neck to travel along the specimen as it thins from the initial cross-section to the drawn cross-section. The existence of a finite or *natural draw ratio* is an important aspect of polymer deformation and is discussed in Section 12.6 below. Ductile behaviour in polymers does not always give a stabilized neck, so that the requirements for necking and cold-drawing must now be considered in some detail.

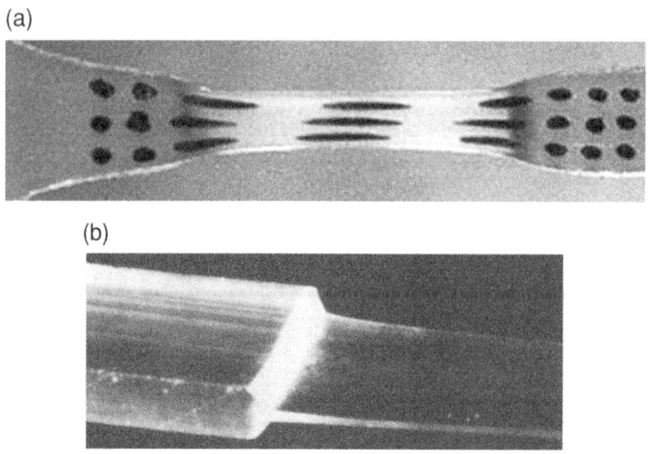

Figure 12.1 Photograph of necks formed in the drawing of (a) isotropic polypropylene and (b) oriented polyethylene.

12.1.1 Necking and the Ultimate Stress

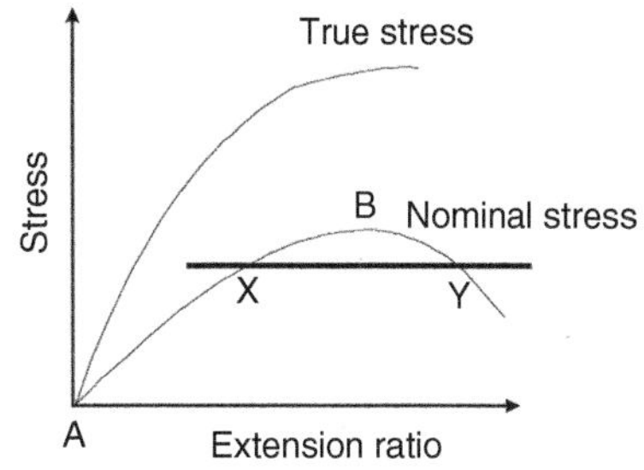

Figure 12.2 Comparison of nominal stress–elongation curve (load–elongation curve) and true stress–elongation curve.

It is important to distinguish between the nominal stress, which is the load at any time during deformation divided by the initial cross-sectional area, and the true stress, which is the load divided by the actual cross-section at any time. The cross-section of the sample decreases with increasing extension, so that the true stress may be increasing when the apparent or conventional stress or load may be remaining constant or even decreasing. This has been very well discussed by Nadai (1963) and Orowan (1949).

Consider the conventional stress–strain curve or the load–elongation curve for a ductile material (Figure 12.2). The ordinate is equal to the nominal stress obtained by dividing the load P by the original cross-sectional area A_0:

$$\sigma_a = P/A_0.$$

This gives a stress–strain curve of the form shown. The load reaches its maximum value at the instant the extension of the sample ceases to be uniform. At this elongation, the specimen begins to neck and consequently the load falls as shown by the last part of the stress–strain curve. Finally, the sample fractures at the narrowest point of the neck.

It is instructive to plot the true tensile stress at any elongation rather than the nominal stress σ_a. This is given by $\sigma = P/A$, where A is the actual cross-section at any time. We now assume, as is usual for plastic deformation, that the deformation takes place at constant volume. Then $Al = A_0 l_0$, and if we put $l/l_0 = \lambda$ where λ is the extension ratio,

$$A = \frac{A_0 l_0}{l} = \frac{A_0}{\lambda}.$$

The true stress is given by

$$\sigma = \frac{P}{A} = \frac{\lambda P}{A_0} = \lambda \sigma_a. \tag{12.1}$$

Thus if we know σ_a, the true stress as a function of λ, i.e. the true stress–strain curve, can be computed. The nominal and true stress–strain curves are compared in Figure 12.2.

Consideration of the nominal stress can lead to insight into the mechanical instability associated with necking. For a tensile specimen of initially uniform cross-section, equilibrium of forces ensures that the nominal stress σ_a is the same all along its length. Therefore, when the curve of σ_a against λ possesses a maximum such as that in Figure 12.2, a small strain (point X) can coexist with a large strain (point Y). Point X corresponds to the unnecked region of the specimen, and point Y to a region that is beginning to neck. If the specimen were stretched to the same elongation without necking, the strain would be uniform at a level somewhere between that of X and that of Y. It is clear that such a state of strain corresponds to a nominal stress higher than the line XY, and therefore to a strain energy higher than that in the necked specimen. On this basis, we would expect the necked state, corresponding to the lower strain energy, to be preferred.

The argument above is based on the assumption that the stress–strain curve completely defines the material behaviour. In reality, with polymers stress depends on strain rate, and since necking is associated with a local increase in strain rate, the issue is more complex. A strong dependence of stress on strain rate can inhibit necking even when the nominal stress reaches a maximum; the existence of the maximum is a necessary condition for necking, but not a sufficient one. Necking in rate-dependent materials has been discussed by Sweeney *et al.* (1999).

Another aspect of the energy/equilibrium argument, that at first seems problematic, is the question of how the material moves from the unnecked state at X to the necked state at Y without violating equilibrium. Intermediate strain states are clearly (Figure 12.2) at a higher nominal stress than that pertaining at X and Y. However, as the neck develops and material moves from one state to the other, the cross-sectional area is non-uniform, and so the true stress varies along the neck. Equilibrium equations then ensure that shear stresses are acting, and so the material is not in a state of uniaxial stress or strain. Therefore, the stress–strain state of the transitional material cannot be plotted on the curve of Figure 12.2. Its combined state of normal and shear stress is such as to maintain equilibrium. This has been explained by Vincent (1960), and corresponding work on metals has revealed the complexity of the stress field in the neck (Bridgeman 1964).

In mathematical terms, the existence of a stress maximum as the condition necessary for necking is

$$\frac{d\sigma_a}{d\lambda} = 0. \tag{12.2}$$

This can be re-expressed in terms of the true stress σ. From Equations (12.1) and (12.2) it is equivalent to

$$\frac{d}{d\lambda}\left(\frac{\sigma}{\lambda}\right) = 0,$$

which becomes

$$\frac{1}{\lambda}\frac{d\sigma}{d\lambda} - \frac{\sigma}{\lambda^2} = 0,$$

i.e.

$$\frac{d\sigma}{d\lambda} = \frac{\sigma}{\lambda}. \tag{12.3}$$

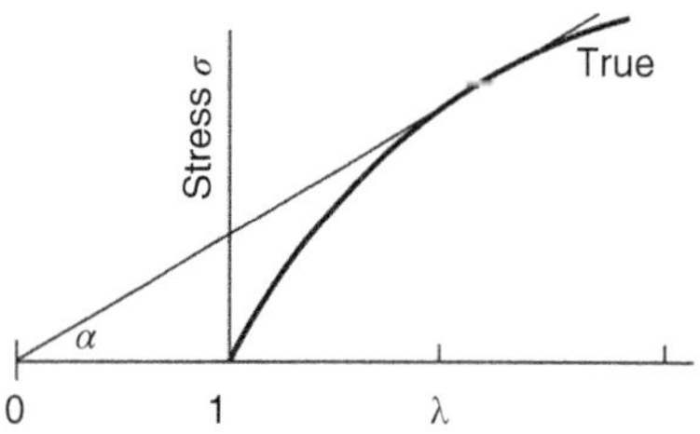

Figure 12.3 The Considère construction.

Equation (12.3) defines a geometric condition for the true stress–strain curve, corresponding to the simple construction due to Considère shown in Figure 12.3. The ultimate stress is obtained when the tangent to the true stress–strain curve $d\sigma/d\lambda$ is given by the line from the point $\lambda = 0$ on the extension axis. The angle α in Figure 12.3 is defined by

$$\tan\alpha = \frac{d\sigma}{d\lambda}.$$

The ultimate stress has a much greater significance than is the case for metals as it is a determining factor in deciding whether a polymer will neck and cold-draw.

The significance of the argument at this stage relates to the failure of plastics in the ductile state. Orowan (1949) first pointed out that for ductile materials the ultimate stress is entirely determined by the stress–strain curve, i.e. by the plastic behaviour of the material, without any reference to its fracture properties, provided that fracture does not occur before the load maximum corresponding to $d\sigma/d\lambda = \sigma/\lambda$ is reached. Yield stress is thus an important property in many plastics and defines the practical limit of behaviour much more than the ultimate fracture, unless the plastic fails by brittle fracture.

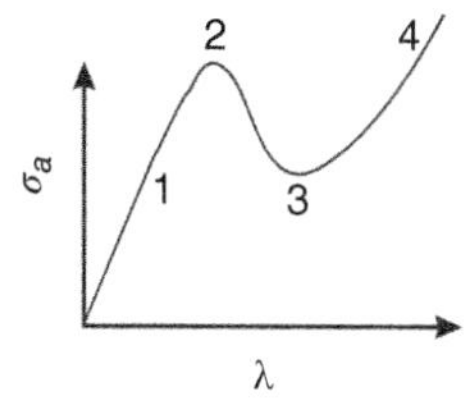

Figure 12.4 Nominal stress–strain behaviour of a necking tensile specimen.

12.1.2 Necking and Cold-Drawing: A Phenomenological Discussion

Figure 12.4 shows that there are four distinct regions on the nominal stress–strain curve of a typical cold-drawing polymer:

1) Initially the stress rises in an approximately linear manner as the applied strain increases.
2) The nominal stress reaches a maximum. During the subsequent fall in stress, the neck shape develops. From this point on, the strain in Figure 12.4 should be understood to be that at the centre of the neck.
3) The nominal stress reaches a minimum. The strain at this point corresponds to the natural draw ratio. In tensile stretching, the strain stays at this approximately constant level for some considerable time as the neck propagates through the specimen and the specimen continues to be elongated – the phenomenon termed 'stable necking'.
4) The specimen is of finite dimensions, so that at some stage the propagating neck occupies the whole of its length – it reaches the grips at both end of the tensile specimen. Now, further elongation causes an increase in strain and a corresponding increase in nominal stress as molecular chains are stretched – phenomenologically, this process may be termed 'strain-hardening'.

In some materials, notably metals, there is no minimum region 3 and the stress continues to decrease. Then, there is no stable necking and the neck continues to stretch, becoming continuously thinner until fracture.

The maximum mentioned above as occurring in region 2 corresponds to the condition (12.2). This condition involves the nominal stress only, and the existence or otherwise of a maximum in the true stress is unspecified by the physical arguments employed so far. Observations show that a maximum in true stress may arise or may not, depending on the polymer and test conditions. This is well illustrated in the work of Amoedo and Lee (1992). In Figure 12.5, two sets of tensile true stress–strain curves are shown, one set for polycarbonate and the other for polypropylene. The stress maximum is clearly present in the case of polycarbonate and, equally clearly, absent in the case of polypropylene. It is

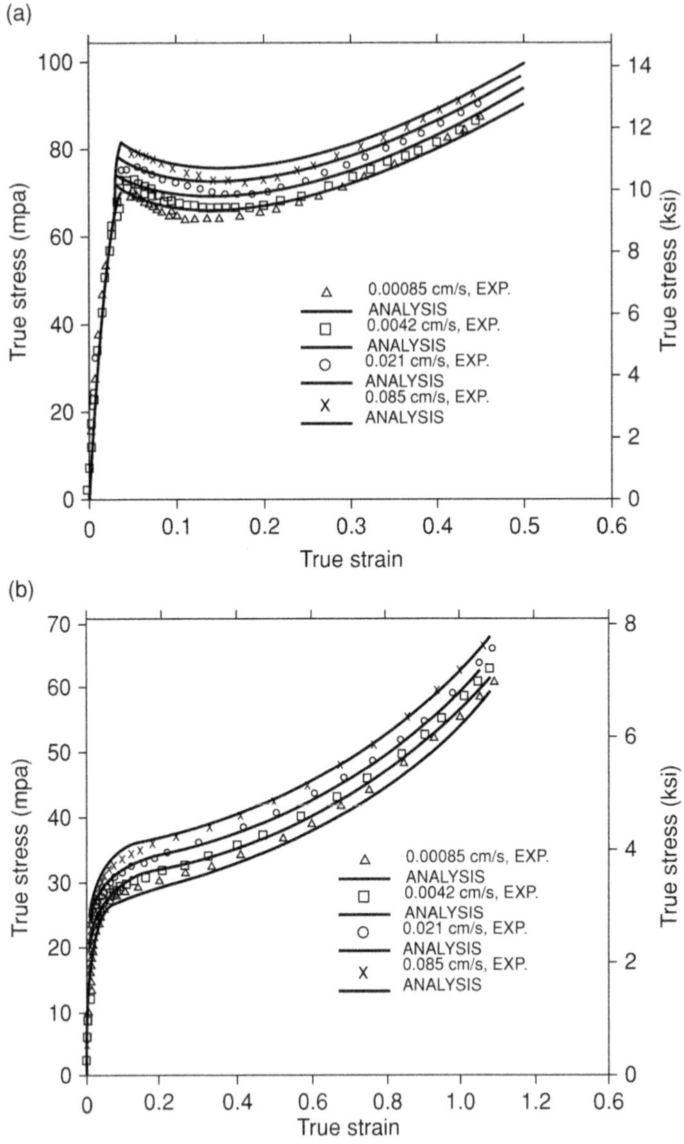

Figure 12.5 True stress–strain curves at room temperature for (a) polycarbonate (b) polypropylene. Amoedo and Lee (1992) / John Wiley & Sons.

of significance here that the comparison is between an amorphous polymer (polycarbonate) and a semi-crystalline one (polypropylene).

There are two ways in which a neck may be initiated. First, if the cross-section of the sample is not uniform, perhaps as a result of a flaw, the element with the smallest effective cross-section will be subjected to the highest true stress, and so will reach the yield point before any other element in the sample. Secondly, a fluctuation in material properties may cause a localized reduction of the yield stress in a given element so that this element reaches the yield point at a lower applied load. When a particular element has reached its yield point, it is easier to continue deformation entirely within this element because it has a lower flow stress stiffness than the surrounding material. Hence, further deformation of the sample is accompanied by straining in only one region and a neck is formed.

12.1.3 Use of the Considère Construction

Corresponding with the above description of the four regions of the nominal stress–strain curve, a parallel description is possible using true stress. The maximum and minimum in regions 2 and 3 are replaced by tangents to the stress–strain curve corresponding to Equation (12.3). In Figure 12.6, two such tangent lines have been drawn to the true stress–strain curve from the point $\lambda = 0$. Inspection of the figure reveals that the slopes of the tangents $\frac{d\sigma}{d\lambda}$ are as given by Equation (12.3). The regions 1–4 correspond to the same physical stages as those in Figure 12.4. When the curve is such that it is possible to construct the tangent to region 2, the polymer has the potential to neck; the existence of the tangent at region 3 implies a potential for stable necking and cold-drawing. Similar deductions can be made from the presence or otherwise of maxima and minima in the nominal stress–strain relations of Figure 12.4.

An example of a stress–strain relation for which a Considère tangent is not possible is the upwardly curving line shown in Figure 12.7(a). In Figure 12.7(b), one tangent is possible that corresponds to necking, but the second one associated with stable necking is not. The typical polymer behaviour of necking followed by strain hardening and stable necking is only possible for curves resembling those of Figure 12.6.

12.1.4 Definition of Yield Stress

Yield stress may most simply be regarded as the minimum stress at which permanent strain is produced when the stress is subsequently removed. Although this definition is satisfactory for metals, where there is a clear distinction between elastic recoverable deformation

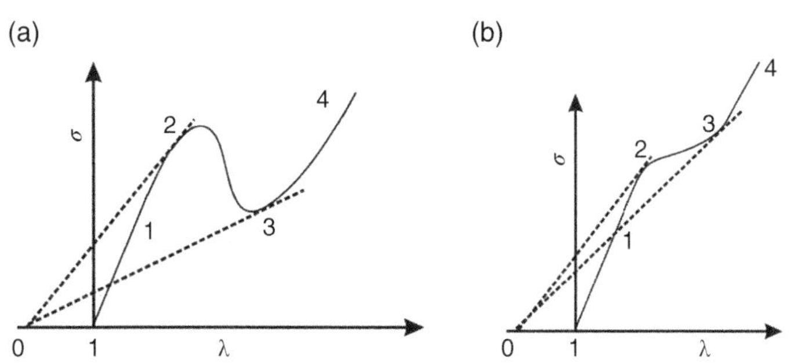

Figure 12.6 Considère tangents. In (a), the true stress has reached a maximum, whereas in (b) there is no maximum.

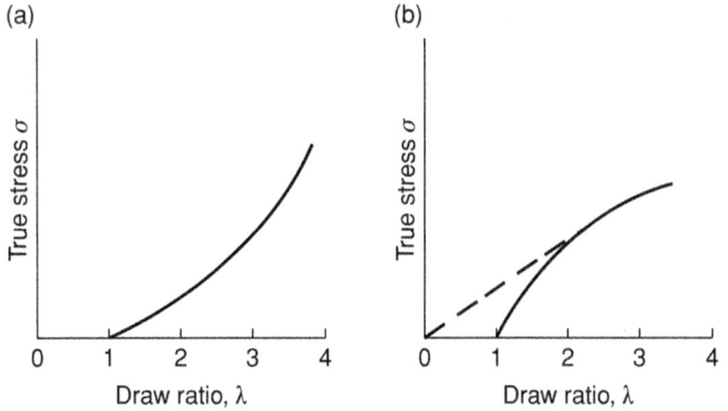

Figure 12.7 Stress–strain curves for which (a) no Considère tangent is possible and (b) only one tangent, corresponding to the onset of necking, has been drawn.

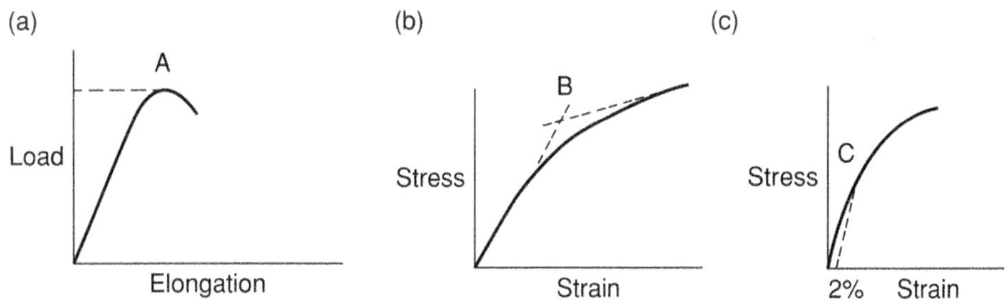

Figure 12.8 (a) The yield stress is defined as the load divided by the cross-sectional area at the point A. (b) The yield stress is defined as the stress at the point B. (c) The yield stress is defined as the stress at the point C.

and plastic irrecoverable deformation, in polymers the distinction is not so straightforward. In many cases, such as the tensile tests discussed above, yield coincides with the observation of a maximum load in the load–elongation curve. The yield stress can then be defined as the true stress at the maximum observed load (Figure 12.8(a)). Because this stress is achieved at a comparatively low elongation of the sample, it is often adequate to use the engineering definition of the yield stress as the maximum observed load divided by the initial cross-sectional area.

In some cases, there is no observed load drop and another definition of yield stress is required. One approach is to determine the stress where the two tangents to the initial and final parts of the load–elongation curve intersect (Figure 12.8(b)).

An alternative is to attempt to define an initial linear slope on the stress–strain curve and then to draw a line parallel to this, which is offset by a specified strain, say 2%. The interception of this line with the stress–strain curve then defines the offset or proof stress, which is considered to be the yield stress (Figure 12.8(c)).

12.2 Ideal Plastic Behaviour

12.2.1 The Yield Criterion: General Considerations

The simplest theories of plasticity exclude time as a variable and ignore any feature of the behaviour, which takes place below the yield point. In other words, we assume a rigid-plastic material whose stress–strain relationship in tension is shown in Figure 12.9. For stresses below the yield stress there is no deformation. Yield can be produced by a wide range of stress states, not just simple tension. In general, it must therefore be assumed that the yield condition depends on a function of the three-dimensional stress field. In a Cartesian axis set, this is defined by the six components of stress, σ_{11}, σ_{22}, σ_{33}, σ_{12}, σ_{23} and σ_{31}. However, the numerical values of these components depend on the orientation of the axis set, and it is crucial that the yield criterion be independent of the observer's chosen viewpoint; the yield criterion must be invariant with respect to changes in the axis set. It is often convenient to make use of the principal stresses. If the material itself is such that its tendency to yield is independent of the direction of the stresses – that is if it is isotropic – then the yield criterion is a function of the principal stresses only

$$f(\sigma_{\mathrm{I}}, \sigma_{\mathrm{II}}, \sigma_{\mathrm{III}}) = \text{constant}.$$

12.2.2 The Tresca Yield Criterion

The earliest yield criterion to be suggested for metals was Tresca's proposal that yield occurs when the maximum shear stress reaches a critical value (Tresca 1864),

$$\sigma_{\mathrm{I}} - \sigma_{\mathrm{III}} = \text{constant}$$

with $\sigma_{\mathrm{I}} > \sigma_{\mathrm{II}} > \sigma_{\mathrm{III}}$. The uniaxial case defines the constant as the uniaxial yield stress σ_Y. to give

$$\sigma_{\mathrm{I}} - \sigma_{\mathrm{III}} = \sigma_Y$$

or more generally

$$\sigma_{\max} - \sigma_{\min} = \sigma_Y \tag{12.4}$$

The situation in this case is illustrated using Mohr circles for the I–III plane (see Chapter 3) in Figure 12.10, where σ denotes normal stress and τ shear stress.

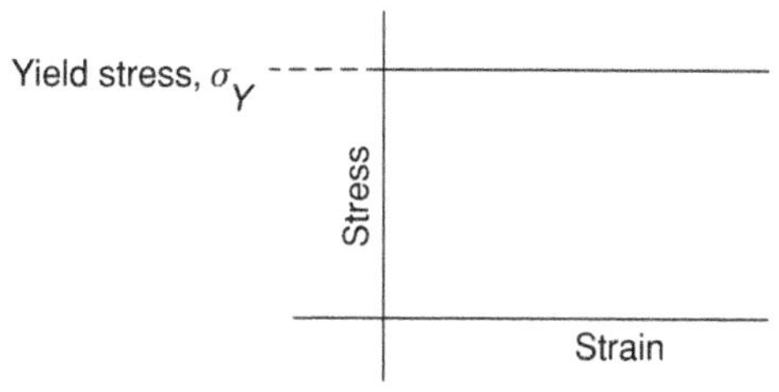

Figure 12.9 Stress–strain relationship for an ideal rigid-plastic material.

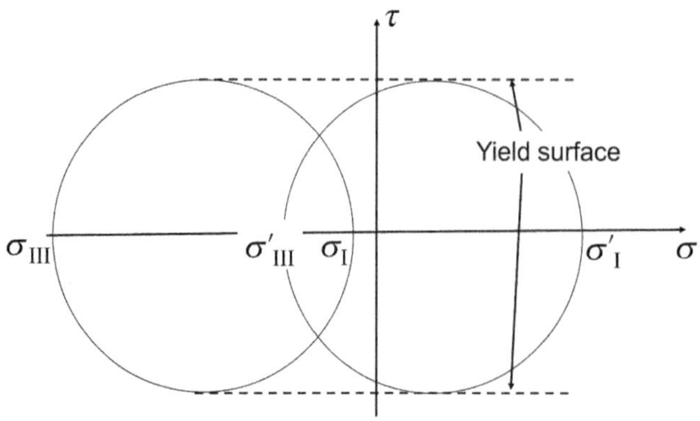

Figure 12.10 Mohr circles showing two states of stress meeting the Tresca failure criterion and yield surface.

Of a similar nature is the Smith critical resolved shear–stress law for the yield of metal single crystals (Smith 1996).

12.2.3 The Coulomb Yield Criterion

The Tresca yield criterion assumes that the critical shear stress is independent of the normal pressure on the plane on which yield is occurring. Although this assumption is valid for metals, it is more appropriate in polymers to consider the possible applicability of the Coulomb yield criterion (Coulomb 1776; Labuz and Zang 2012) which states that the critical shear stress τ for yielding to occur in any plane varies linearly with the stress normal to this plane, that is

$$\tau = \tau_c - \mu\sigma_N. \tag{12.5}$$

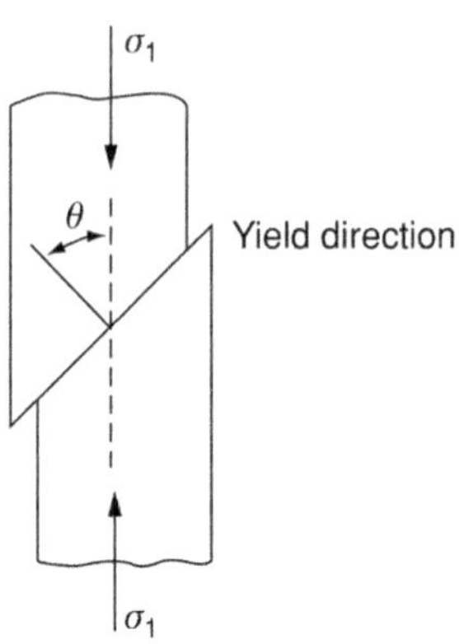

Figure 12.11 The yield direction under a compressive stress σ_1 for a material obeying the Coulomb criterion.

The Coulomb criterion was originally conceived for the failure of soils and τ_c was termed the 'cohesion' and μ the coefficient of internal friction. For a compressive stress, σ_N has a negative sign so that the critical shear stress τ for yielding to occur on any plane increases linearly with the pressure applied normal to this plane.

The Coulomb criterion is often written as

$$\tau = \tau_c - \tan\varphi\sigma_N, \tag{12.6}$$

where μ has been written as $\tan\varphi$, for reasons which will now become apparent.

The example of uniaxial compression under a compressive stress σ_1 where yield occurs on a plane whose normal makes an angle θ with the direction of σ_1 is shown in Figure 12.11. For more general stress states, as with the Tresca criterion yield is determined by the largest and smallest principal

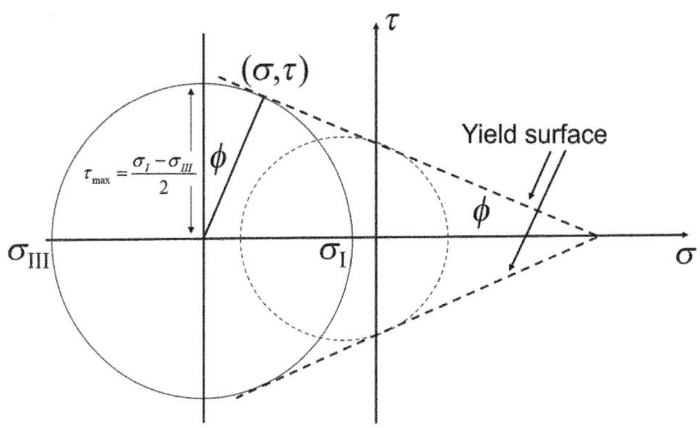

Figure 12.12 Mohr circles showing the operation of the Coulomb criterion and its yield surface.

stresses and takes place in the plane in which they lie. Now we consider a two-dimensional state of stress in the I–III plane, which is the relevant plane when $\sigma_{\text{I}} > \sigma_{\text{II}} > \sigma_{\text{III}}$. In Figure 12.12 we show Mohr circles (see Chapter 3). From the diagram, the stress state (σ, τ) defined by the angle φ is given by

$$\sigma = \frac{\sigma_{\text{I}} + \sigma_{\text{III}}}{2} + \frac{\sigma_{\text{I}} - \sigma_{\text{III}}}{2} \sin\varphi$$

$$\tau = \frac{\sigma_{\text{I}} - \sigma_{\text{III}}}{2} \cos\varphi$$

Writing $\sigma_N = \sigma$ in Equation (12.5), we construct the yield criterion in terms of principal stresses:

$$\frac{\sigma_{\text{I}} - \sigma_{\text{III}}}{2} \cos\varphi = \tau_c - \left(\frac{\sigma_{\text{I}} + \sigma_{\text{III}}}{2} + \frac{\sigma_{\text{I}} - \sigma_{\text{III}}}{2} \sin\varphi\right) \tan\varphi.$$

This reduces to

$$\sigma_{\text{I}} - \sigma_{\text{III}} = 2\tau_c \cos\varphi - (\sigma_{\text{I}} + \sigma_{\text{III}}) \sin\varphi \tag{12.7}$$

Figure 12.12 shows a compressive stress field, with the angle θ between the plane of failure at (σ, τ) and the major (most compressive) stress σ_{III} given by $2\theta = \pi/2 + \varphi$, i.e.

$$\theta = \pi/4 + \varphi/2 \tag{12.8}$$

In the case of a tensile stress field, the circles are shifted to the right of the τ axis and are otherwise unchanged. Then, the major principal stress is σ_{I} which makes the angle with the failure plane:

$$\theta = \pi/4 - \varphi/2 \tag{12.9}$$

Thus, tanφ determines the direction of yield and conversely the direction of yielding can be used to define φ, where tanφ is the coefficient of friction.

We see that the Coulomb yield criterion defines both the stress condition required for yielding to occur and the directions in which the material will deform. Where a deformation band forms, its direction is that which is neither rotated nor distorted by the plastic deformation, because its orientation marks that direction which establishes material continuity between the deformed material in the deformation band and the undistorted material in the rest of the specimen. If volume is conserved, the band direction therefore denotes the direction of shear in a simple shear (by the definition of a shear strain). Thus for a Coulomb yield criterion, the band direction is defined by Equations (12.8) and (12.9).

We may generalize from Equation (12.7) to produce equations for each principal plane to give a three-dimensional yield criterion. The left-hand sides are ambiguous in sign, as the yield condition can be met by the point (σ, τ) in Figure 12.12 or by its reflection in the σ axis, corresponding to a negative value of τ_{max}:

$$\begin{aligned} \pm(\sigma_I - \sigma_{II}) &= 2\tau_c \cos\varphi - (\sigma_I + \sigma_{II}) \sin\varphi \\ \pm(\sigma_{II} - \sigma_{III}) &= 2\tau_c \cos\varphi - (\sigma_{II} + \sigma_{III}) \sin\varphi \\ \pm(\sigma_{III} - \sigma_I) &= 2\tau_c \cos\varphi - (\sigma_{III} + \sigma_I) \sin\varphi \end{aligned} \tag{12.10}$$

Here, the relevant equation is the one that excludes the intermediate principal stress. Note that for $\varphi = 0$ the criterion reduces to Tresca's.

12.2.4 The von Mises Yield Criterion

The von Mises (1913) yield criterion assumes that the yield behaviour is independent of hydrostatic pressure and that the yield stresses in simple tension and compression are equal. It is expressed most simply in terms of the principal components of stress so that

$$(\sigma_I - \sigma_{II})^2 + (\sigma_{II} - \sigma_{III})^2 + (\sigma_{III} - \sigma_I)^2 = \text{constant}.$$

The constant term can be easily expressed in terms of the yield stress σ_Y in uniaxial testing. Then we can assign the values $\sigma_I = \sigma_Y$, $\sigma_{II} = \sigma_{III} = 0$ and the constant on the right is found to be $2\sigma_Y{}^2$:

$$(\sigma_I - \sigma_{II})^2 + (\sigma_{II} - \sigma_{III})^2 + (\sigma_{III} - \sigma_I)^2 = 2\sigma_Y^2 \tag{12.11}$$

In rather more sophisticated terms, the von Mises yield criterion assumes that the yield criterion depends only on the components of the deviatoric stress tensor obtained by subtracting the hydrostatic components of stress from the total stress tensor. In terms of principal components of stress, the deviatoric stress tensor is

$$\begin{bmatrix} \sigma'_I & 0 & 0 \\ 0 & \sigma'_{II} & 0 \\ 0 & 0 & \sigma'_{III} \end{bmatrix} = \begin{bmatrix} \sigma_I + p & 0 & 0 \\ 0 & \sigma_{II} + p & 0 \\ 0 & 0 & \sigma_{III} + p \end{bmatrix},$$

where $p = -\frac{1}{3}(\sigma_I + \sigma_{II} + \sigma_{III}) = -\bar{\sigma}$ is the hydrostatic pressure. The von Mises yield criterion can then be written as

$$\sigma'^2_I + \sigma'^2{}_{II} + \sigma'^2{}_{III} = \text{constant}. \tag{12.12}$$

This illustrates the physical basis of the criterion: yield takes place when the strain energy associated with the deviatoric stress reaches a critical value.

The von Mises yield criterion is often written in terms of the so-called octahedral shear stress τ_{oct}, where

$$\tau_{\text{oct}} = \frac{1}{3}\left\{(\sigma_{\text{I}} - \sigma_{\text{II}})^2 + (\sigma_{\text{II}} - \sigma_{\text{III}})^2 + (\sigma_{\text{III}} - \sigma_{\text{I}})^2\right\}^{\frac{1}{2}} \tag{12.13}$$

giving the yield criterion as τ_{oct} = constant.

In an arbitrary 1-2-3 axis set, the criterion is in the form of the invariant expression

$$(\sigma_{11} - \sigma_{22})^2 + (\sigma_{22} - \sigma_{33})^2 + (\sigma_{33} - \sigma_{11})^2 + 2\left(\sigma_{12}^2 + \sigma_{23}^2 + \sigma_{31}^2\right) = 2\sigma_Y^2 \tag{12.14}$$

We have seen that the Coulomb yield criterion defines both the stresses required for yield and also the directions in which the material deforms. In the case of the von Mises yield criterion, we require a further development of the theory to predict the directions in which plastic deformation starts.

It is important to appreciate that plasticity is different in kind from elasticity, where there is a unique relationship between stress and strain defined by a modulus or stiffness constant. Once we achieve the combination of stresses required to produce yield in an idealized rigid-plastic material, deformation can proceed without altering stresses, and is determined by the movements of the external constraints, for example the displacement of the jaws of the tensometer in a tensile test. This means that there is no unique relationship between the stresses and the total plastic deformation. Instead, the relationships that do exist relate the stresses and the incremental plastic deformation, as was first recognized by St Venant, who proposed that for an isotropic material the principal axes of the strain increment are parallel to the principal axes of stress.

If the material is assumed to remain isotropic after yield there is no dependence on the deformation or stress history. Furthermore, if we assume that the yield behaviour is independent of the hydrostatic component of stress then the principal axes of the strain increment are parallel to the principal axes of the deviatoric stress tensor.

Levy (1870) and von Mises (1913) independently proposed that the principal components of the strain-increment tensor

$$\begin{bmatrix} \mathrm{d}e_{\text{I}} & 0 & 0 \\ 0 & \mathrm{d}e_{\text{II}} & 0 \\ 0 & 0 & \mathrm{d}e_{\text{III}} \end{bmatrix}$$

and the deviatoric stress tensor

$$\begin{bmatrix} \sigma'_{\text{I}} & 0 & 0 \\ 0 & \sigma'_{\text{II}} & 0 \\ 0 & 0 & \sigma'_{\text{III}} \end{bmatrix}$$

are proportional, that is

$$\frac{\mathrm{d}e_{\text{I}}}{\sigma'_{\text{I}}} = \frac{\mathrm{d}e_{\text{II}}}{\sigma'_{\text{II}}} = \frac{\mathrm{d}e_{\text{III}}}{\sigma'_{\text{III}}} = \mathrm{d}\lambda, \tag{12.15}$$

where $\mathrm{d}\lambda$ is not a material constant but is determined by our choice of the extent of deformation of the material, for example by the displacement of the jaws of the tensometer.

Rewriting the individual deviatoric stresses

$$\sigma'_{\mathrm{I}} = \sigma_{\mathrm{I}} + p$$
$$\sigma'_{\mathrm{II}} = \sigma_{\mathrm{II}} + p$$
$$\sigma'_{\mathrm{III}} = \sigma_{\mathrm{III}} + p$$

and adding them gives

$$\sigma'_{\mathrm{I}} + \sigma'_{\mathrm{II}} + \sigma'_{\mathrm{III}} = \sigma_{\mathrm{I}} + \sigma_{\mathrm{II}} + \sigma_{\mathrm{III}} - 3 \times \frac{1}{3}(\sigma_{\mathrm{I}} + \sigma_{\mathrm{II}} + \sigma_{\mathrm{III}}) = 0.$$

From Equation (12.14), we have three relations

$$\sigma'_{\mathrm{I}} = \frac{\mathrm{d}e_{\mathrm{I}}}{\mathrm{d}\lambda}, \sigma'_{\mathrm{II}} = \frac{\mathrm{d}e_{\mathrm{II}}}{\mathrm{d}\lambda}, \sigma'_{\mathrm{III}} = \frac{\mathrm{d}e_{\mathrm{III}}}{\mathrm{d}\lambda}$$

that when added give

$$\sigma'_{\mathrm{I}} + \sigma'_{\mathrm{II}} + \sigma'_{\mathrm{III}} = \frac{\mathrm{d}e_{\mathrm{I}} + \mathrm{d}e_{\mathrm{II}} + \mathrm{d}e_{\mathrm{III}}}{\mathrm{d}\lambda}.$$

It follows that $\mathrm{d}e_{\mathrm{I}} + \mathrm{d}e_{\mathrm{II}} + \mathrm{d}e_{\mathrm{III}} = 0$, that is that deformation takes place at constant volume.

If the stress–strain relations are referred to other than principal axes we have

$$\mathrm{d}e_{ij} = \sigma'_{ij}\mathrm{d}\lambda(i,j = 1,2,3),$$

that is

$$\frac{\mathrm{d}e_{11}}{\mathrm{d}\sigma'_{11}} = \frac{\mathrm{d}e_{22}}{\mathrm{d}\sigma'_{22}} = \frac{\mathrm{d}e_{33}}{\mathrm{d}e'_{33}} = \frac{\mathrm{d}e_{23}}{\mathrm{d}\sigma'_{23}} = \frac{\mathrm{d}e_{31}}{\mathrm{d}\sigma'_{31}} = \frac{\mathrm{d}e_{12}}{\mathrm{d}\sigma'_{12}} = \mathrm{d}\lambda. \tag{12.16}$$

These equations are called the Levy–Mises equations.

12.2.5 Geometrical Representations of the Tresca, von Mises and Coulomb Yield Criteria

The assumption of material isotropy which implies that σ_{I}, σ_{II} and σ_{III} are interchangeable, means that the Tresca and von Mises yield criteria take very simple analytical forms when expressed in terms of the principal stresses. Thus the yield criteria form surfaces in principal stress space, that is that space where the three rectangular Cartesian axes are parallel to the principal stress directions. Points lying closer to the origin than the yield surface represent combinations of stress where yield does not occur; points on or outside the surface represent combinations of stress where yield does occur.

Putting $\sigma_{\mathrm{III}} = 0$ reveals the failure envelope in the σ_{I}–σ_{II} plane in Figure 12.13(a). For the Tresca criterion, the minimum stress is either zero or negative; in the latter case, it is equal to either σ_{I} or σ_{II}. Thus, in the positive quadrant when $\sigma_{\mathrm{I}} > 0$, $\sigma_{\mathrm{II}} > 0$ the smallest stress is zero and criterion (12.4) gives

$$\sigma_{\mathrm{I}} = \sigma_Y \; \sigma_{\mathrm{I}} > \sigma_{\mathrm{II}}$$
$$\sigma_{\mathrm{II}} = \sigma_Y \; \sigma_{\mathrm{II}} > \sigma_{\mathrm{I}}$$

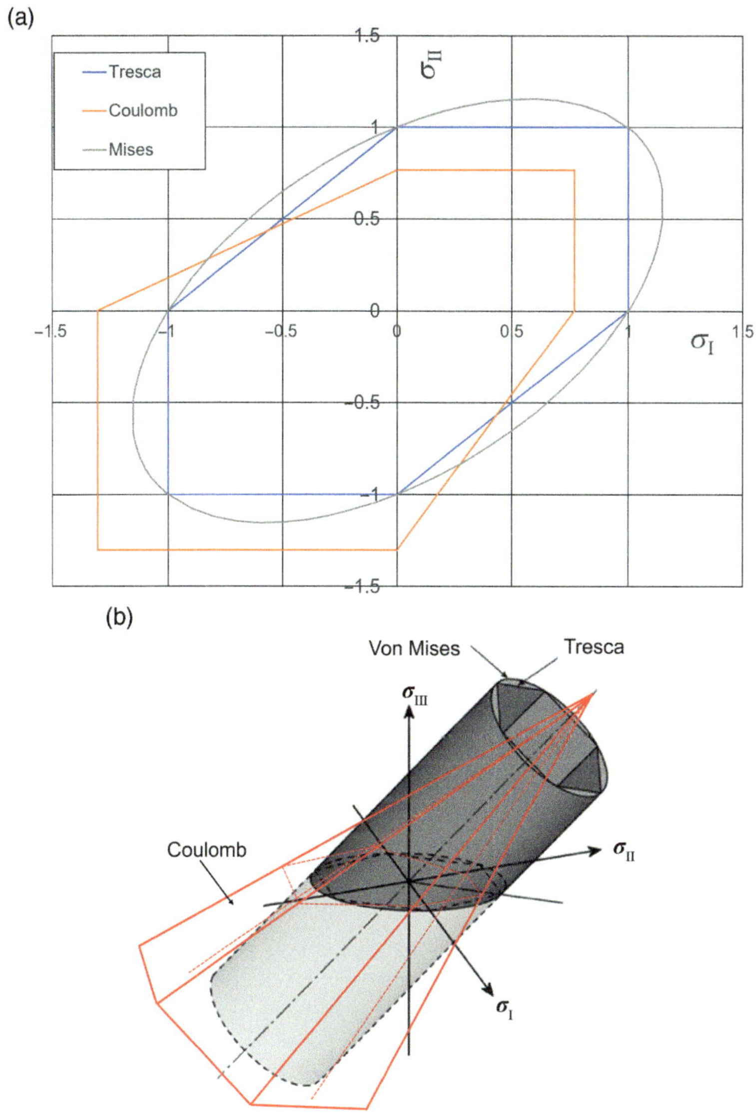

Figure 12.13 (a) Yield surfaces on the plane σ_{III} = 0. Stress is in units of σ_Y. For the Coulomb criterion, τ_c = $\sigma_Y/2$ and φ = 15°. (b) Three-dimensional yield surfaces.

that defines a square of side σ_Y in the σ_I–σ_{II} plane. For the negative quadrant $\sigma_I < 0$, $\sigma_{II} < 0$, the maximum principal stress is $\sigma_{III} = 0$ and the yield surface is defined by

$$\sigma_I = -\sigma_Y \; \sigma_I < \sigma_{II}$$

$$\sigma_{II} = -\sigma_Y \; \sigma_{II} < \sigma_I$$

to give another square of side σ_Y. For the remaining quadrants the intermediate principal stress is $\sigma_{III} = 0$. For the quadrant $\sigma_I < 0$, $\sigma_{II} > 0$ criterion (12.4) gives

$$\sigma_{II} - \sigma_I = \sigma_Y$$

that defines a linear function with intercepts σ_Y on the II axis an $-\sigma_Y$ on the I axis. A similar function applies on the remaining quadrant. The surface is shown in Figure 12.13(a).

A similar process reveals the shape of the Coulomb surface, which is identical to that of the Tresca criterion when $\varphi = 0$. On the plane $\sigma_{III} = 0$, in the positive quadrant $\sigma_I > 0, \sigma_{II} > 0$ the yield surface is defined by the square bounded by the sides

$$\sigma_I = 2\tau_c \cos\varphi/(1 + \sin\varphi)$$
$$\sigma_{II} = 2\tau_c \cos\varphi/(1 + \sin\varphi).$$

In the third quadrant $\sigma_I < 0$, $\sigma_{II} < 0$, the square is defined by sides

$$\sigma_I = -2\tau_c \cos\varphi/(1 - \sin\varphi)$$
$$\sigma_{II} = -2\tau_c \cos\varphi/(1 - \sin\varphi).$$

We can now see the asymmetry in the tensile and compressive behaviour, with a higher hydrostatic pressure giving a yield stress of increased magnitude.

For the remaining quadrants, when $\sigma_I < 0$, $\sigma_{II} > 0$ the stresses are related by the linear function

$$\sigma_{II} = \left(\frac{1 - \sin\varphi}{1 + \sin\varphi}\right)\sigma_I + \frac{2\tau_c \cos\varphi}{1 + \sin\varphi}$$

and for $\sigma_I > 0$, $\sigma_{II} < 0$

$$\sigma_{II} = \left(\frac{1 + \sin\varphi}{1 - \sin\varphi}\right)\sigma_I - \frac{2\tau_c \cos\varphi}{1 - \sin\varphi}.$$

The surface is shown in Figure 12.13(a) for the case $\varphi = 15°$.

The von Mises criterion for $\sigma_{III} = 0$ is, from Equation (12.11)

$$\sigma_I^2 + \sigma_{II}^2 - \sigma_I\sigma_{II} = \sigma_Y^2 \qquad (12.17)$$

This is the equation for an ellipse with major axis at 45° to the σ_I axis as shown in Figure 12.13(a). The von Mises and Tresca criteria coincide for uniaxial and equibiaxial stress, but generally the von Mises condition predicts yielding at a higher stress.

In Figure 12.13(b) we show the surfaces for all values of stress. The Tresca and von Mises yield criteria are independent of the hydrostatic component of stress, so we can replace σ_I, σ_{II} and σ_{III} by $\sigma_I + p$, $\sigma_{II} + p$ and $\sigma_{III} + p$, respectively, without affecting the material's state with regard to yield. Thus, if the point in principal stress space (σ_I, σ_{II}, σ_{III}) lies on the yield surface, so does the point ($\sigma_I + p$, $\sigma_{II} + p$, $\sigma_{III} + p$). This shows that the yield surface must be parallel to the {111} direction, and has the appearance as sketched in Figure 12.13(b). The material isotropy implies equivalence between σ_I, σ_{II} and σ_{III} and hence that the section has a threefold symmetry about the {111} axis. The assumption that the behaviour is the same in tension and compression implies an equivalence between σ_1 and $-\sigma_1$ and so on, and hence we have finally sixfold symmetry about the {111} direction. This is most clearly shown by the Tresca yield surface in Figure 12.13(b).

The Coulomb criterion is not independent of the hydrostatic pressure and so the yield surface is not a cylinder with axis {111}. However it retains the sixfold symmetry of the Tresca criterion and is a cone with hexagonal cross-section. Since for polymers yield stress tends to increase with hydrostatic pressure, the cone in Figure 12.13 is directed so that its cross-section increases as it travels down the {111} axis in the direction of increasingly negative stresses.

12.2.6 Yield Criteria for Anisotropic Materials

A very simple yield criterion for anisotropic materials is the critical resolved shear stress of Schmid (Gottstein 2004). This is concerned with crystal slip. The law states that yield occurs when the resolved shear stress in the slip direction in the slip plane reaches a critical value. Although this law is extensively used in metal plasticity, it is of restricted application in polymers.

Hill (1985) has developed a generalization of the von Mises criterion for anisotropic materials. Anisotropy is defined with respect to specific axes fixed within the material which, in the case of orthotropic materials, are mutually perpendicular. Then, a 1–2–3 axes set can be chosen to align with the directions of orthotropy and the yield criterion defined with respect to the stresses in this axis set. This precludes the use of principal stresses as the principal directions do not in general coincide with the directions of orthotropy. Therefore, Hill's criterion is a generalization of Equation (12.14):

$$F(\sigma_{22}-\sigma_{33})^2+G(\sigma_{33}-\sigma_{11})^2+H(\sigma_{11}-\sigma_{22})^2+2\left(L\sigma_{23}^2+M\sigma_{31}^2+N\sigma_{12}^2\right)=1. \tag{12.18}$$

Here F, G, H, L, M and N are material parameters that define the anisotropy. When they are all equal, the criterion reduces to that of von Mises. Clearly the observed tensile yield stress will depend on the direction of stress according to this criterion. It has recently been applied to particle-filled polymers by Van Dommelen and Meijer (2005), who made use of the direction-dependent yield stress to fit their model.

12.2.6.1 Back Stress and Bauschinger Effect

When a polymer is oriented predominantly in one direction, it is observed that the tensile yield stress along the direction of drawing is greater than in other directions, and also greater than the compressive yield stress. This latter phenomenon is known as the *Bauschinger effect* and has been observed for both glassy and semi-crystalline-oriented polymers for many years, as studied by Brown *et al.* (1968) and Duckett *et al.* (1972). In order to represent this effect and to improve the treatment of the anisotropy of yielding, Brown *et al.* (1968) introduced an *internal stress* or *back stress* σ_i. This can be thought of as the compressive stress in the material that is required to balance the tension resulting from chains extended in the draw direction, which must be overcome before tensile yield processes can begin to act. The internal stress was introduced as a modification to the Hill criterion, which for a material oriented in the 1 direction becomes

$$\begin{aligned}&F(\sigma_{22}-\sigma_{33})^2+G(\sigma_{33}-\sigma_{11}+\sigma_i)^2+H(\sigma_{11}-\sigma_i-\sigma_{22})^2\\&\quad+2\left(L\sigma_{23}^2+M\sigma_{31}^2+N\sigma_{12}^2\right)=1.\end{aligned} \tag{12.19}$$

Brown *et al.* showed that this modified version gave a better fit to the yield in simple shear of oriented polycarbonate than Equation (12.18). For experiments in which the

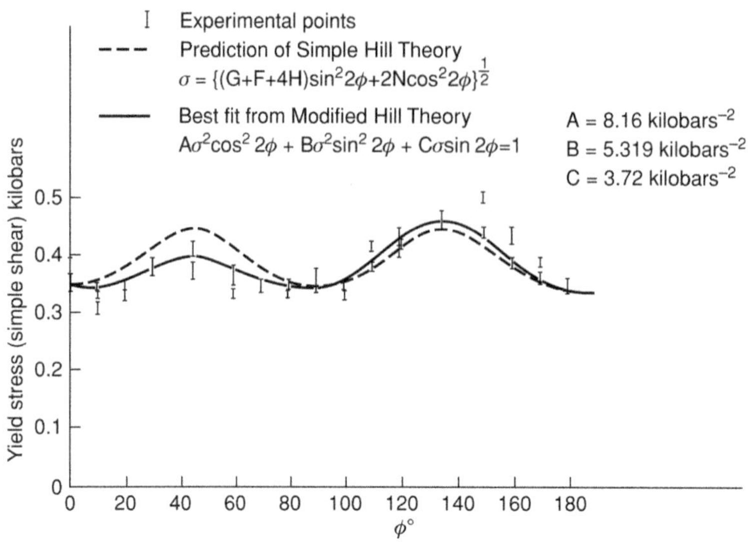

Figure 12.14 Comparison of the effectiveness of the Hill criterion and its modified form with internal stress. Brown et al. (1968) / Taylor & Francis.

direction of shearing was varied with respect to the orientation axis, the results in terms of shear yield stress are plotted in Figure 12.14. The presence of the back stress enables the theoretical fit to show maxima of different heights, in line with the experimental findings.

We have so far not addressed the physical origin of the internal stress. The modified Hill criterion is not a complete constitutive material model. It would ideally arise as a consequence of such a model, in which case it would be possible to explore its other implications and its evolution during the drawing process. The question of, for instance, whether it is entropic or energetic is not addressed. These issues will be discussed further below in Section 12.11.

12.2.7 The Plastic Potential

The Levy–Mises equations define one of a number of possible flow rules that can be derived via an argument that depends upon a concept known as the plastic potential. This idea has been discussed by Hill (1985). It is assumed that the components of the plastic strain increment tensor are proportional to the partial derivatives of the plastic potential, which is a scalar function of stress. The flow rule can thus be generated by this differentiation process. We may choose to assume, for a particular form of yield criterion, that the plastic potential has the same functional form as the yield criterion; then, the derived flow rule is described as being *associated* with the yield criterion (or as an *associative* flow rule). However, this assumption is not obligatory and when it is not true we will be applying a yield criterion together with a *non-associated* flow rule. This is discussed further by de Souza Neto *et al.* (2008).

As an example, suppose that we have a plastic potential f of the same functional form as the von Mises criterion. Then, in principal directions we assume that

$$f = \frac{d\lambda}{6}\left((\sigma_{I} - \sigma_{II})^2 + (\sigma_{II} - \sigma_{III})^2 + (\sigma_{III} - \sigma_{I})^2\right),$$

where $d\lambda$ is a constant. For each principal direction, we may obtain an expression for the plastic strain increment. Taking the 1 direction as an example,

$$\begin{aligned} de_{I} &= \frac{\partial f}{\partial \sigma_{I}} = \frac{d\lambda}{6}(2(\sigma_{I} - \sigma_{II}) - 2(\sigma_{III} - \sigma_{I})) \\ &= d\lambda\left(\sigma_{I} - \frac{1}{3}(\sigma_{I} + \sigma_{II} + \sigma_{III})\right) = d\lambda(\sigma_{I} - \overline{\sigma}) = d\lambda\sigma_{I}'. \end{aligned}$$

This reproduces the first of the Levy–Mises Equation (12.15). The Levy–Mises flow rule is thus associated with the von Mises criterion.

Similarly, let us on the basis of Equation (12.17) construct the plastic potential for the Hill criterion

$$f = \frac{d\lambda}{2}\left[F(\sigma_{22} - \sigma_{33})^2 + G(\sigma_{33} - \sigma_{11})^2 + H(\sigma_{11} - \sigma_{22})^2 + 2\left(L\sigma_{23}^2 + M\sigma_{31}^2 + N\sigma_{12}^2\right)\right].$$

Then, we can construct Hill's flow rule (Hill 1985) from the expressions given by

$$de_{ij} = \frac{\partial f}{\partial \sigma_{ij}} (i,j = 1,2,3),$$

which results in

$$\begin{aligned} de_{11} &= d\lambda[H(\sigma_{11} - \sigma_{22}) + G(\sigma_{11} - \sigma_{33})] \\ de_{22} &= d\lambda[F(\sigma_{22} - \sigma_{33}) + H(\sigma_{22} - \sigma_{11})] \\ de_{33} &= d\lambda[G(\sigma_{33} - \sigma_{11}) + F(\sigma_{33} - \sigma_{22})] \\ de_{23} &= d\lambda L\sigma_{23} \\ de_{13} &= d\lambda M\sigma_{13} \\ de_{12} &= d\lambda N\sigma_{12}. \end{aligned} \tag{12.20}$$

12.3 Historical Development of Understanding of the Yield Process

We have seen that yield is often associated with a load drop on the load–extension curve, and always involves a change in slope on the true stress–strain curve. This load drop has sometimes been attributed either to adiabatic heating of the specimen or to the geometrical reduction in cross-sectional area on the formation of a neck. Present knowledge leads to the conclusion that yielding is an intrinsic property of the material, and that temperature rises and necking are secondary consequences. This is supported by the observations in Section 12.5.4, to the effect that the Eyring process gives a consistent model of yield in polymers. Localized or geometrical effects can have no relevance to this molecular-based model. However, temperature and geometrical effects are present during yielding and must be understood if the phenomenon is to be interpreted correctly.

12.3.1 Adiabatic Heating

Under conventional conditions of cold-drawing, where the specimen is extended at strain rates of the order of 10^{-2} s^{-1} or higher, a considerable rise of temperature occurs in the region of the neck. Marshall and Thompson (1954), following Müller (1949), proposed that cold-drawing involves a local temperature rise and that necking occurs because of strain softening produced by the consequent fall in flow stress with rising temperature. The stability of the drawing process was then attributed to the stability of a localized process of heat transfer through the shoulders of the neck, with extension taking place at constant tension throughout the neck.

Hookway (1958) later attempted to explain the cold-drawing of nylon 6.6 on somewhat similar grounds, suggesting that there is a possibility of local melting in the neck due to a combination of hydrostatic tension and temperature.

There is no doubt that an appreciable rise in temperature does occur at conventional drawing speeds, and the ideas of Marshall and Thompson are very relevant to an understanding of the complex process of fibre drawing. Calorimetric measurements by Brauer and Müller (1954) have, however, shown that at slow rates of extension the increase in temperature is quite small (~10°C) and not sufficient to give an explanation for necking and cold-drawing in terms of adiabatic heating. In addition, Lazurkin (1958) demonstrated that necking could still take place under quasi-static conditions, for elastomers below their glass transition temperature, cold-drawn at very low speeds. A comparable result was shown by Vincent (1960) for (semi-crystalline) polyethylene which cold-draws at very slow extension rates at room temperature.

The adiabatic heating explanation arose at least in part because the initial yield process was not regarded as distinct from the drawing process. It is now recognized that up to the yield point the deformation of the sample is homogeneous and generally quite small strains are involved, whereas once a neck forms, the deformation is inhomogeneous and large strains are involved in the neck. The work of plastic deformation can then lead to a large rise in temperature in the neck. For example, thermal imaging has been used during tensile testing (Hadriche *et al.* 2010) to give simultaneous measurements of stress and temperature for polycarbonate. Here it is clear that the temperature begins to rise after the yield point, as shown in Figure 12.15. It is argued that provided the drawing is carried out at a low strain rate, any heat which is generated will be conducted away from the neck sufficiently rapidly for no significant temperature rise to occur. As the strain rate is increased and the process becomes more nearly adiabatic, the effective temperature at which the drawing is taking place is increased. In particular, heat is conducted into the unyielded portion of the sample, and so lowers the yield stress of the undeformed material and reduces the force necessary to propagate the neck.

The observed temperature rise in the neck has been found to be in approximate agreement with that calculated from the work done in drawing, assuming that no heat is generated due to crystallization. Sweeney *et al.* (2018) measured temperature rises in tensile specimens of PET by thermal imaging, and concluded that 70% of the strain energy had been converted to heat. Testing temperatures were below or just within the glass transition range, so no crystallization would be expected for PET according to Bazhenov and Kechek'yan (2013).

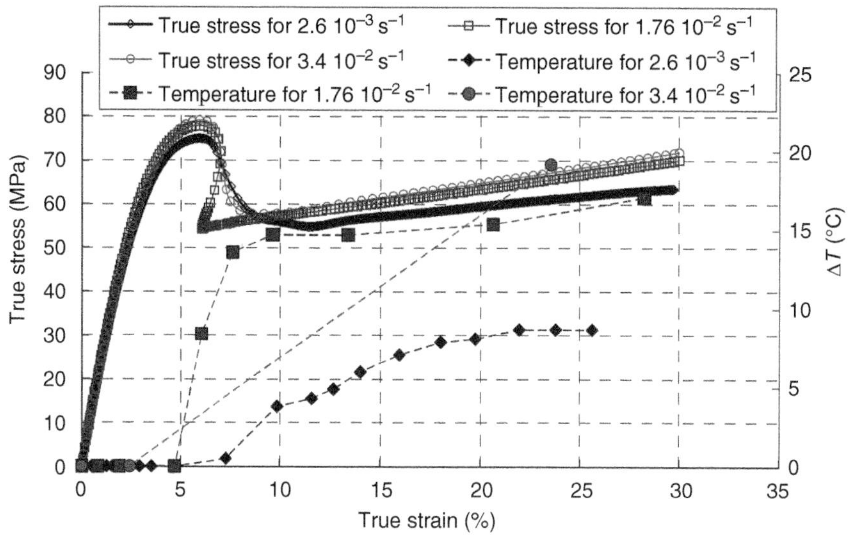

Figure 12.15 Stresses and temperatures during tensile testing of PC at room temperature at two strain rates. Hadriche et al. 2010 / Trans Tech Publications.

12.3.2 The Isothermal Yield Process: The Nature of the Load Drop

There is no doubt that a temperature rise does occur in cold-drawing under many conditions of test. We have shown, however, that there is very good evidence to support the view that necking can still take place under quasi-static conditions where there is no appreciable temperature rise. Vincent (1960) therefore proposed that the observed fall in load is a geometrical effect because the fall in cross-sectional area during stretching is not compensated for by an adequate degree of strain hardening. This effect, called strain softening, was attributed to the reduction in the slope of the stress–strain curve with increasing strain.

Contrary to this latter explanation of the load drop in terms of geometric softening, results reported by Whitney and Andrews (1967) showed a yield drop in compression for polystyrene and PMMA where there are no geometrical complications. Brown and Ward (1968) then made a detailed investigation of yield drops in PET, studying isotropic and oriented specimens, in tension, shear and compression. They concluded that in most cases there is clear evidence for the existence of an intrinsic yield drop, that is that a fall in true stress can occur in polymers, as in metals. This is reflected in the work of Amoedo and Lee (1992), shown above in Figure 12.5(a). We have seen from the energetic argument in Section 12.1.1 that an intrinsic stress drop can be physically associated with necking.

There is, however, a significant difference between polymers and many metals with regard to yield behaviour. For a polymer, as shown in Figure 12.5, at most one maximum is observed on the load–extension curve, in contrast with metals (illustrated by mild steel in Figure 12.16), where often two maxima are observed on a typical load–extension curve. The first maximum (point A in Figure 12.16) called the upper yield point, represents a fall in true stress, an intrinsic load drop, and corresponds to a sudden increase in the amount of plastic strain which relaxes the stress. From B to C, Lüders bands propagate throughout the

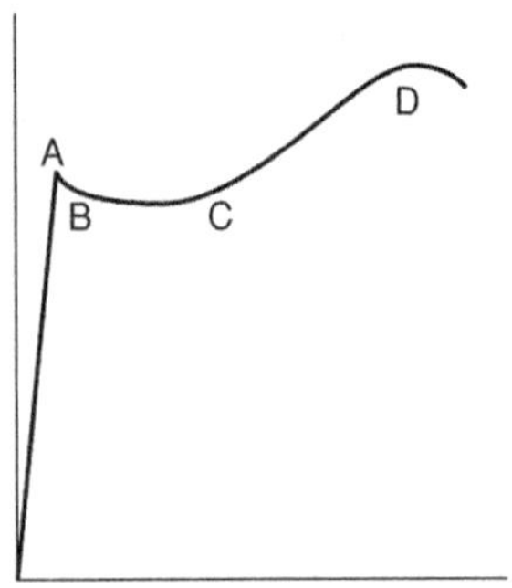

Figure 12.16 Load–extension curve in tension of mild steel.

specimen. Lüders bands have also been observed in polymers (Allison and Ward 1967). At C, the specimen is homogeneously strained and the stress begins to rise as the material work hardens uniformly. A second maximum is observed at point D and is always associated with the beginning of necking in the specimen. Necking occurs when the effects of strain hardening of the metal are overwhelmed by the geometrical softening due to the reduction in the cross-sectional area of the specimen as it is strained, that is the Orowan–Vincent explanation, discussed in Section 12.1.1.

The second maximum, as we have seen previously, is not observed if the true stress–strain curve is plotted instead of the load–extension curve. The first maximum, on the other hand, would exist on the true stress–strain curve. It is called an intrinsic yield point, because it relates to the intrinsic behaviour of the material.

In polymers, as we have emphasized, no more than one maximum is observed in the load–extension curve. The investigations of Whitney and Andrews (1967) and Brown and Ward (1968) show that this maximum combines the effect of the geometrical changes and an intrinsic load drop, and cannot be attributed to the geometrical changes alone. In particular, the cold-drawing results are not accounted for by a decrease in the slope of the true stress–strain curve, as suggested in the explanation of Vincent. It is important to note that not every element of the material follows the same true stress–strain curve, since the stress for initiation is greater than for propagation of yielding, so confirming (as has already been noted in Chapter 11, Section 11.1.3) that it will not be possible to give a complete explanation of necking and cold-drawing in terms of the Considère construction on a true stress–strain curve.

12.4 Experimental Evidence for Yield Criteria in Polymers

Many studies of the yield behaviour of polymers have bypassed the question of strain rate and temperature and sought to establish a yield criterion as discussed in Section 12.2. In very general terms, such studies divide into two categories: (1) those which attempt to define a yield criterion on the basis of determining yield for different stress states and (2) those which confine the experimental studies to an examination of the influence of hydrostatic pressure on the yield behaviour.

12.4.1 Yield for Multiaxial States of Stress

One of the earliest examples of the use of multiaxial stress states is that of Bowden and Jukes (1968), who carried out plane-strain compression tests on PMMA. The experimental set-up is shown in Figure 12.17. A particular advantage of this technique is that yield behaviour can be observed in compression for materials, which normally fracture in a tensile test. In this case, PMMA was studied at room temperature, i.e. below its brittle–ductile transition temperature in tension. The yield point in compression σ_1 was measured for various values

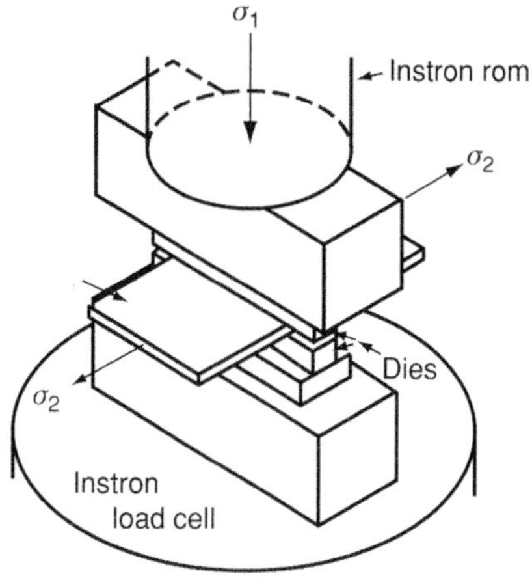

Figure 12.17 The plane-strain compression test. Bowden and Jukes (1968) / Springer Nature.

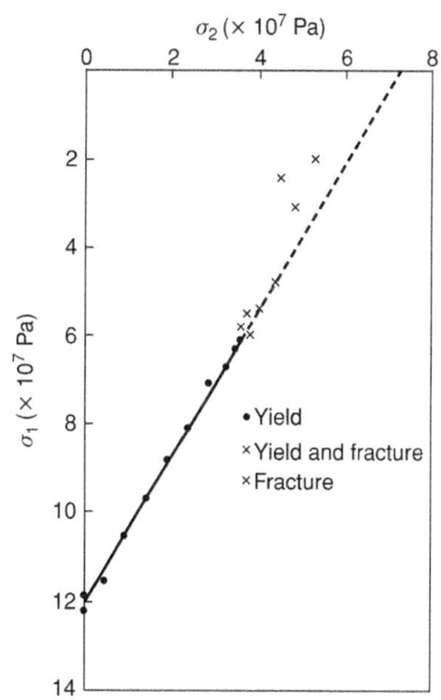

Figure 12.18 Measured values of the compressive yield stress σ_1 (true stress) plotted against applied tensile stress σ_2 (nominal stress). The full circles denote ductile yield, the crosses, brittle fracture, and the combined points, tests where ductile yielding occurred, followed immediately by brittle fracture. Bowden and Jukes. (1968) / Springer Nature.

of applied tensile stress σ_2. The results, shown in Figure 12.18, give $\sigma_1 = -110.0 + 13.65\sigma_2$, where both σ_1 and σ_2 are expressed as true stresses in units of MPa. The results therefore clearly do not fit the Tresca criterion, where $\sigma_1 - \sigma_2 =$ constant at yield; neither do they fit a von Mises yield criterion. They are, however, consistent with a Coulomb yield criterion with $\tau = 47.4 - 1.58\sigma_N$ (see Equation (12.5).

Another early study is that of Bauwens (1970) who studied tension, compression and tension-torsion of PVC. The observed difference in tensile and compressive yield stresses was sufficient to disprove the Tresca and von Mises theories, and to suggest the significance of hydrostatic pressure. Bauwens incorporated pressure into the von Mises yield criterion, which can be written $\tau_{oct} =$ constant (see 12.2.4 above), by writing

$$\tau_{oct} + Ap = \text{constant}$$

where A is a constant and $p = -(\sigma_{\text{I}} + \sigma_{\text{II}} + \sigma_{\text{III}})/3$ is the hydrostatic pressure. This embodies the concept that yielding occurs when a critical value of a combination of deviatoric and volumetric strain energies is reached. For tensile and compressive yield stress σ_{Yt} and σ_{Yc}, respectively, this is equivalent to (Hachour *et al.* 2014)

$$\frac{3\sqrt{2}(\sigma_{yt} - \sigma_{yc})p}{\sigma_{yt} + \sigma_{yc}} + \left((\sigma_{\text{I}} - \sigma_{\text{II}})^2 + (\sigma_{\text{II}} - \sigma_{\text{III}})^2 + (\sigma_{\text{III}} - \sigma_{\text{I}})^2\right)^{1/2} = \frac{2\sqrt{2}\sigma_{yt}\sigma_{yc}}{\sigma_{yt} + \sigma_{yc}} \tag{12.21}$$

Raghava *et al.* (1973) also worked on PVC with a programme of uniaxial tension, compression, and biaxial tension in the form of tensile stretching of thin-walled tubes subject to internal pressure. They investigated an alternative modification to the von Mises criterion:

$$6(\sigma_{yt} - \sigma_{yc})p + \left((\sigma_{\text{I}} - \sigma_{\text{II}})^2 + (\sigma_{\text{II}} - \sigma_{\text{III}})^2 + (\sigma_{\text{III}} - \sigma_{\text{I}})^2\right) = 2\sigma_{yt}\sigma_{yc} \tag{12.22}$$

which they attributed ultimately to early work of Schleicher (1926).

Bowden and Jukes (1972) extended their 1968 work beyond the Coulomb criterion and introduced pressure dependence into both the von Mises and Tresca theories. Their von Mises model is essentially that of Equation (12.22), while their Tresca-based theory takes a form adapted from Equation (12.4) so that $\sigma_{\text{max}} - \sigma_{\text{min}}$ becomes a function of pressure rather than a constant. This can be written as (Hachour *et al.* 2014)

$$\sigma_{\text{max}} - \sigma_{\text{min}} = 2\frac{\sigma_{yt}\sigma_{yc}}{\sigma_{yt} + \sigma_{yc}} - 3\frac{\sigma_{yt} - \sigma_{yc}}{\sigma_{yt} + \sigma_{yc}}p \tag{12.23}$$

Hachour *et al.* (2014) have conducted a comprehensive experimental programme on high-density polyethylene to assess the performance of the various pressure-dependent yield criteria. In addition to uniaxial tension and compression tests, they performed tests with combined shear-tension and shear compression stress fields. For these non-uniaxial tests they used 'butterfly' specimens that can be stretched or compressed along different axes to vary the proportion of shear to normal stress. The results for Bauwens (1970) of Equation (12.21), Raghava *et al.* (1973) of Equation (12.22) and the modified Tresca of Equation (12.23) are compared in Figure 12.19. The Bauwens and Raghava models cannot be distinguished by the available data and are generally more effective than the modified Tresca criterion, but with relatively poor predictions of shear-tension.

12.4.2 Direct Evidence for the Influence of Hydrostatic Pressure on Yield Behaviour

There have been a number of detailed investigations of the influence of hydrostatic pressure on the yield behaviour of polymers (Ainbinder *et al.* 1965; Pae *et al.* 1968; Biglione *et al.* 1969; Maers *et al.* 1969; Rabinowitz *et al.* 1970; Christiansen *et al.* 1971; Matashugi *et al.* 1975). Because it illustrates clearly the relationship between a yield criterion which depends on hydrostatic pressure and the Coulomb yield criterion, an experiment will be discussed where Rabinowitz *et al.* (1970) determined the torsional stress–strain behaviour of isotropic PMMA under hydrostatic pressures up to 700 MPa. The results are shown in Figure 12.20.

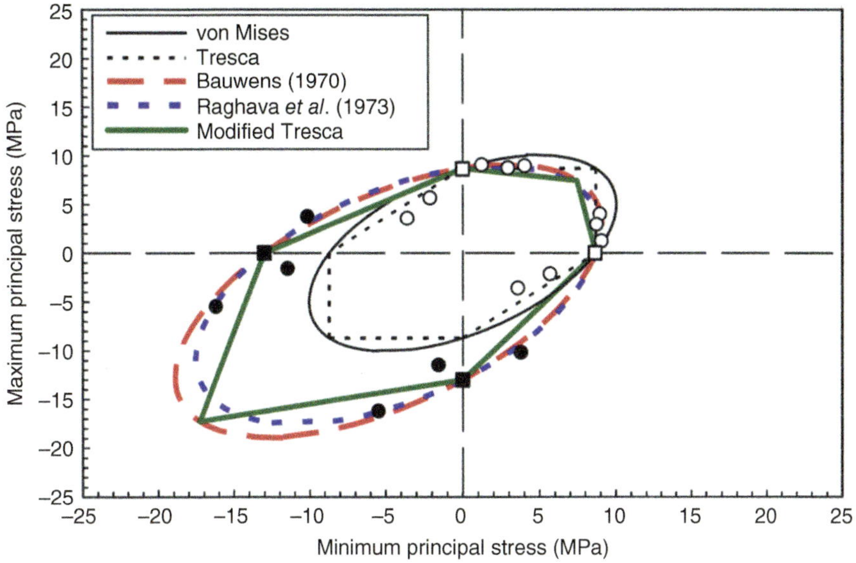

Figure 12.19 Yield predictions for conventional Tresca and von Mises and pressure-dependent yield criteria of equations of Equations (12.21–12.23) against experimental points. Open circles – combined shear-tension; solid circles – combined shear-compression; open squares – uniaxial tension; solid squares – uniaxial compression. Hachour et al. 2014.

Figure 12.20 Shear stress–strain curves for PMMA showing fracture envelope. Adapted from Rabinowitz et al., 1970.

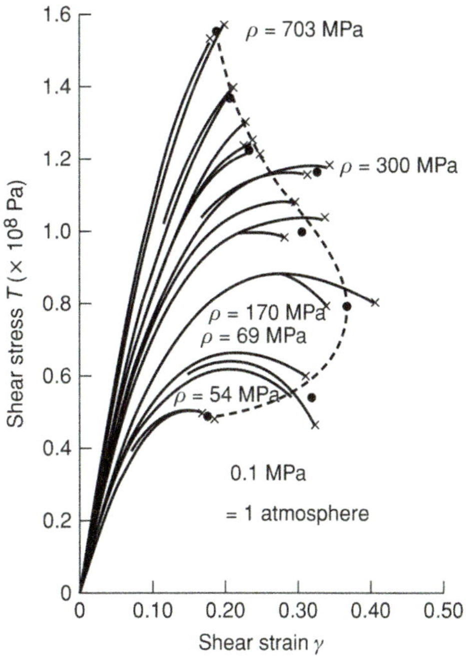

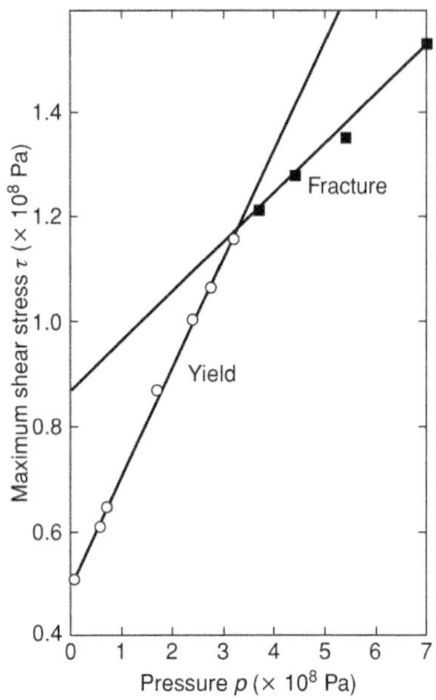

Figure 12.21 Maximum shear stress τ as a function of hydrostatic pressure p for PMMA. (○) Yield; (■) fracture. Adapted from Rabinowitz et al., 1970.

There is a substantial increase in the shear yield stress up to a hydrostatic pressure of about 300 MPa. After this pressure brittle failure occurs, unless prevented by protecting the specimens from the hydraulic fluid (Harris *et al.* 1971) (e.g. by coating with a layer of solidified rubber solution). A study of polyethylene under conditions of combined pressure and tension has shown that the yield stress of polyethylene increases approximately linearly up to pressures of 850 MPa (Sweeney *et al.* 1986). The strain at which yield occurs also increases with increasing pressure, similar to the results of other workers for tensile tests under pressure. The shear yield stress increases linearly with pressure to an excellent approximation (Figure 12.21).

There are two other ways in which these results can be presented. First, recalling Section 12.2.3 and Figures 12.10 and 12.12, the Mohr circle diagram can be constructed from the data, as shown in Figure 12.22 where Bowden and Jukes's results appear as crossed points. This diagram leads naturally to a Coulomb yield criterion (see Figure 12.12)

However, it is equally reasonable to interpret Figure 12.21 directly in terms of the equation

$$\tau_Y = \tau_{Y0} + \alpha p, \tag{12.24}$$

where τ_Y is the shear yield stress at pressure p, τ_{Y0} is the shear yield stress at atmospheric pressure and α is the coefficient of increase of shear yield stress with hydrostatic pressure.

We will see that this simple form of pressure-dependent yield criterion is more satisfactory than the Coulomb criterion when a representation is developed which includes the effects of temperature and strain rate on the yield behaviour. In physical terms, the hydrostatic pressure can be seen as changing the state of the polymer by compressing the polymer significantly, unlike the situation in metals where the bulk moduli are much larger (~100 GPa compared with ~5 GPa). Although such experimental evidence as exists is not

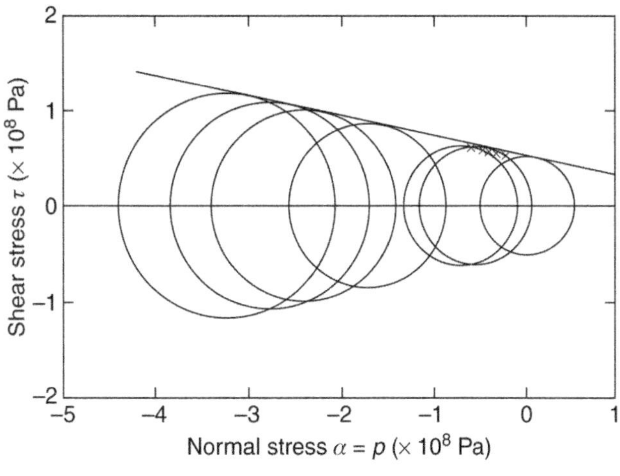

Figure 12.22 Mohr circles for yield behaviour of polymethyl methacrylate obtained from results of Rabinowitz, Ward and Parry. The crosses are the results of Bowden and Jukes. Adapted from Rabinowitz et al., 1970.

unequivocal in this respect, it seems likely that the yield criterion for the polymer subjected to hydrostatic pressure is still given by Equation (12.14), i.e. pressure has the sole effect of increasing the magnitude of the yield stresses.

Recent studies of yield behaviour, using a variety of multi-axial stressing experiments, can all be adequately described by a generalization of Equation (12.24), i.e. a generalized von Mises equation where τ is replaced by the octahedral shear stress (see Section 12.5.1 for a fuller development).

Finally, it can be noted that the coefficient α in Equation (12.24) depends on the temperature of measurement and increases markedly near a viscoelastic transition. Briscoe and Tabor (1978) have pointed out that α is equivalent to the coefficient of friction μ in sliding friction, and show that there is good numerical agreement between values of μ and the values of α obtained from yield stress/pressure measurements.

12.5 The Molecular Interpretations of Yield

Two principal approaches have been used to model the yield behaviour of polymers. The first approach addresses the temperature and strain-rate dependence of the yield stress in terms of the Eyring equation for thermally activated processes (Halsey *et al.* 1945). This approach has been applied to many amorphous and crystalline polymers (see Section 12.5.1), and links have been established with molecular relaxation processes determined by dynamic mechanical and dielectric measurements and with non-linear viscoelastic behaviour determined by creep and stress relaxation. The Eyring approach assumes that the yield process is velocity controlled, i.e. the yield process relates to existing thermally activated processes that are accelerated by the application of the yield stress to the point where the rate of plastic deformation reaches the applied macroscopic strain rate. This approach has been very successful in modelling the yield behaviour of polymers at comparatively high temperatures (Brereton *et al.* 1974; Truss *et al.* 1984).

The second approach is based on classical ideas of crystal plasticity. It is considered that yield involves applying a critical stress to cause the movement of dislocations or disclinations. For this reason, yield is considered to be nucleation controlled. This approach was originally proposed by Bowden and Raha (1974) and Argon (1973) and has been applied by Young (1974), Argon (1973) and others to the yield behaviour of both amorphous and crystalline polymers. Young developed this crystal plasticity approach by assuming that the yield stress relates to the energy required to nucleate screw dislocations within the crystalline lamellae of a crystalline polymer. Research by Ward and co-workers (1971) and Crist *et al.* (1989) have confirmed that the Young approach appears to be valid for yield at low strain rates in polyethylene at −60°C and temperatures below this. Detailed structural studies do suggest that at higher temperatures interlamellar shear occurs, although Young (1974), O'Kane and Young (1995) and Darras and Seguela (1993) have successfully used the crystal plasticity approach to model the yield behaviour of bulk crystalline and annealed semi-crystalline polymers at much higher temperatures.

12.5.1 Thermally Activated Processes and the Eyring Equation

We have shown (Section 5.2.7) that the standard linear solid, a three-component spring and dashpot model, provides to a first approximation a description of linear viscoelastic behaviour. Eyring and his colleagues (Halsey *et al.* 1945) assumed that the deformation of a polymer was a thermally activated rate process involving the motion of segments of chain molecules over potential barriers, and modified the standard linear solid so that the movement of the dashpot was governed by the activated process. The model, which now represents non-linear viscoelastic behaviour, is useful because its parameters include an activation energy and an activation volume that may give an indication of the underlying molecular mechanisms. The activated rate process may also provide a common basis for the discussion of creep and yield behaviour.

Macroscopic deformation is assumed to be the result of basic processes that are either intermolecular (e.g. chain-sliding) or intramolecular (e.g. a change in the conformation of the chain), whose frequency ν depends on the ease with which a chain segment can surmount a potential energy barrier of height ΔH. In the absence of stress, dynamic equilibrium exists, so that an equal number of chain segments move in each direction over the potential barrier at a frequency given by

$$\nu = \nu_0 \exp\left(-\frac{\Delta H}{kT}\right). \tag{12.25}$$

The equation above is equivalent to Equation (7.5) describing the frequency of a molecular event, but here we use Boltzmann's constant k rather than the gas constant R as we are interested in the absolute number of molecular events rather than the overall effect of a mole of material.

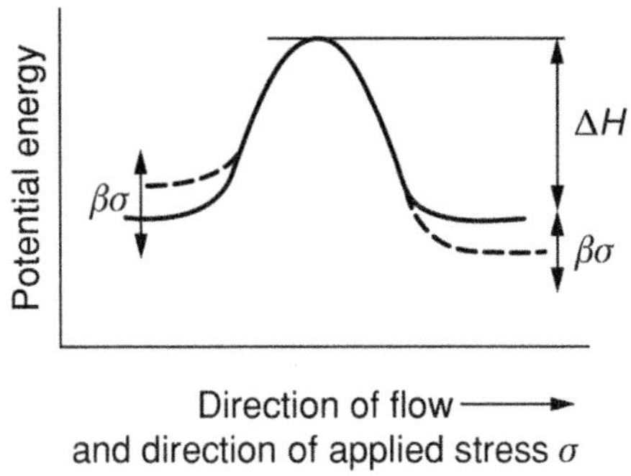

Figure 12.23 The Eyring model for an activated process.

An applied uniaxial stress σ is assumed to produce linear shifts $\beta\sigma$ of the energy barriers in a symmetrical way (Figure 12.23), where β has the dimensions of volume. The flow in the direction of the applied stress is then given by

$$\nu_1 = \nu_0 \exp\left[-\frac{(\Delta H - \beta\sigma)}{kT}\right]$$

compared with a smaller flow in the backward direction of

$$\nu_2 = \nu_0 \exp\left[-\frac{(\Delta H + \beta\sigma)}{kT}\right].$$

The net flow in the forward direction is then

$$\nu' = \nu_1 - \nu_2 = \nu_0 \exp\left(-\frac{\Delta H}{kT}\right)\left[\exp\left(\frac{\beta\sigma}{kT}\right) - \exp\left(-\frac{\beta\sigma}{kT}\right)\right]. \tag{12.26}$$

The resemblance of the large bracket to the sinh function should be noted.

While we have developed the argument using a normal stress σ, the molecular events involved are driven by shear stress. In the uniaxial case the shear stress τ is simply related to the normal stress. Assuming that the net flow in the forward direction is directly related to the rate of change of strain, we then have

$$\frac{\mathrm{d}e}{\mathrm{d}t} = \dot{e} = \dot{e}_0 \exp\left(-\frac{\Delta H}{kT}\right) \sin\mathrm{h}\left(\frac{V\tau}{kT}\right), \tag{12.27}$$

where $\dot{e}_0$ is a constant pre-exponential factor, τ is the octahedral shear stress of Equation (12.13) and V, simply related to β, is termed the *activation volume*. This quantity is related to the volume swept out by a molecular-scale entity during a plastic deformation process.

The rate of strain Equation (12.27) defines an 'activated' viscosity, in effect defining strain rate as a nonlinear function of stress. It can be viewed as defining a nonlinear dashpot which could, for instance, be substituted for the dashpot in a Maxwell model, to create a nonlinear viscoelastic model. Such a model could also be termed viscoplastic, acknowledging that (12.27) defines plastic strain. In the following sections we discuss uniaxial deformations to demonstrate the Eyring process in some fundamental cases and its applicability in practice.

12.5.2 Applications of the Eyring Equation to Creep

We may derive an alternative to Equation (12.27) by making the approximation $\sin hx \approx \frac{1}{2} \exp x$ to arrive at

$$\dot{e} = \frac{1}{2}\dot{e}_0 \exp\left(\frac{V\tau - \Delta H}{kT}\right) \tag{12.28}$$

Strain rate depends on stress and temperature; we also know that over the course of a creep experiment it will vary as strain develops. Following Mindel and Brown (1973), this can be incorporated by including a strain dependence *F(e)* and adapting the equation so that

$$\dot{e} = F(e) \exp\left(\frac{V\tau - \Delta H}{kT}\right)$$

which implies

$$\ln(\dot{e}) = \ln(F(e)) + \frac{V\tau - \Delta H}{kT} \tag{12.29}$$

There is no specific justification in incorporating strain dependence in this way, but the usefulness of this approach can be explored using experiments. For a material that conforms with Equation (12.29), plotting the logarithm of creep strain rate against strain will produce curves that can be shifted up and down the vertical $\ln(\dot{e})$ axis as determined by the value of τ or equivalently the uniaxial tensile stress σ. This approach was introduced by Sherby and Dorn (1958) who investigated the creep under constant stress of glassy PMMA at different temperatures by applying step temperature changes. They constructed plots of creep rate versus total creep strain for varying stress levels which could be shifted to superpose onto a single curve. More recent examples of Sherby–Dorn plots are given by Erartsin *et al.* (2021), an example of which is given in Figure 12.24 for glass-reinforced polypropylene. Typically, there is a downward trend of strain rate with strain. It appears that the curves could be approximately superposed by vertical shifting.

However, Sherby–Dorn plots are used to illustrate creep behaviour when superposition is not applicable. An example is the work of Ward and Wilding (1984) on oriented

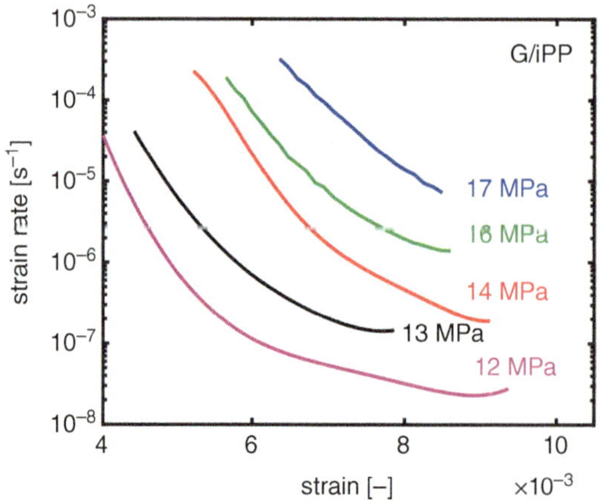

Figure 12.24 Creep rates as a function of total creep strain for glass-reinforced polypropylene at indicated stress levels. Erartsin et al., 2021 / SAGE Publications / CC BY 4.0.

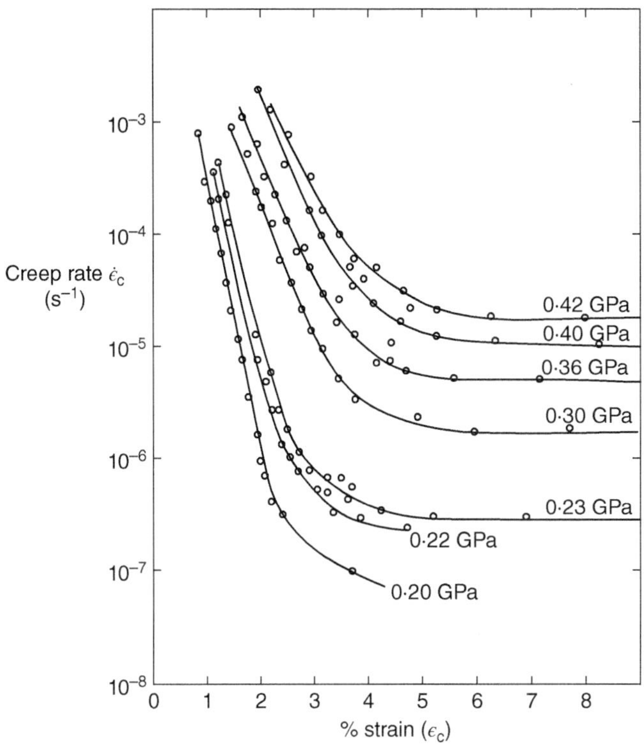

Figure 12.25 Sherby–Dorn plots of creep of ultrahigh modulus polyethylene at different temperatures. Ward and Wilding (1984) / John Wiley & Sons.

polyethylene fibres shown in Figure 12.25. Here, given sufficient time, the curves reach constant 'plateau' strain rates and do not appear to be superposable. This suggests that the creep behaviour cannot be modelled by a single Eyring process as in Equation (12.28), and indeed models that include two Eyring processes have been developed in a series of publications on oriented polyethylenes (Wilding and Ward 1981, 1984; Ward and Wilding 1984). This work followed similar approaches to represent the strain rate dependence of yield stress (Robertson 1963; Roetling 1965; Bauwens-Crowet *et al.* 1969a). Initially Wilding and Ward (1981) used the Eyring rate process to model creep and showed that at high strains, which correspond to long creep times, the creep rate reached a constant value called the plateau (or equilibrium) creep rate (such as in Figure 12.25). For polymers of low relative molecular mass, the stress and temperature dependence of the final creep rate could be modelled by a single activated process. For polymers of a higher molecular mass, and for copolymers, irreversible creep was observed only at high stress levels, which suggested that there are two Eyring processes coupled in parallel (Figure 12.26). Process A has the smaller tensile activation volume (~0.05 nm^3) and larger pre-exponential factor, and is activated only at high stress levels. Process B has the larger tensile activation volume (~1 nm^3) and a smaller pre-exponential factor, and is operative at low stress levels. At low stresses, there will be

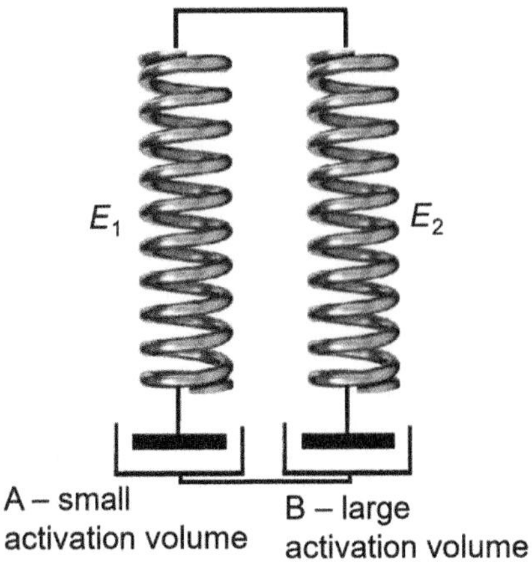

Figure 12.26 The two-process model for permanent flow creep.

little permanent flow because B carries almost the whole load. Although the essentials of the creep and recovery behaviour can be represented by a model containing two activated dashpots, a spectrum of relaxation times would be required to give an accurate fit to experimental data over the complete time and strain scales.

We generate expressions for the stresses in the two-process model by noting that the total stress is the sum of the stress in each arm and the total strain is the same in each arm. For an individual arm, the strain e is the sum of the elastic and Eyring or plastic trains e_E e_p, respectively:

$$e = e_E + e_p \tag{12.30}$$

Differentiating with respect to time gives

$$\dot{e} = \dot{e}_E + \dot{e}_p \tag{12.31}$$

We now replace the plastic strain rate $\dot{e}_p$ with the Eyring equation of Equation (12.27), which is

$$\dot{e}_p = \dot{e}_0 \exp\left(-\frac{\Delta H}{kT}\right) \sinh\left(\frac{V\tau}{kT}\right) = C \sinh(V^*\sigma) \tag{12.32}$$

where the constants C and V^* have been introduced for brevity, and we have made use of the simple relation between shear and normal stress in uniaxial conditions. The stress τ in Equation (12.32) is the octahedral shear stress of Equation (12.13), so we have $\tau = \frac{\sqrt{2}}{3}\sigma$ and

$$V^* = \frac{\sqrt{2}}{3}\frac{V}{kT} \tag{12.33}$$

Also,

$$C = \dot{e}_0 \exp\left(-\frac{\Delta H}{kT}\right) \tag{12.34}$$

We now generate an expression for the stress in a single arm using Equation (12.32):

$$\sigma = \frac{1}{V^*} \sinh^{-1}\left(\frac{\dot{e}_p}{C}\right) \tag{12.35}$$

The total stress in the two-process model of Figure 12.26 is the sum of the stresses in each arm. Adopting obvious notations for the parameters and strain rates in arms A and B, the total stress is

$$\sigma = \sigma_A + \sigma_B = \frac{1}{V_A^*} \sinh^{-1}\left(\frac{\dot{e}_{pA}}{C_A}\right) + \frac{1}{V_B^*} \sinh^{-1}\left(\frac{\dot{e}_{pB}}{C_B}\right) \tag{12.36}$$

In general the viscous strain rates $\dot{e}_{pA}$ and $\dot{e}_{pB}$ are different from each other. However, when the stress has reached a plateau level and has become constant with time in each arm, the stress in each arm is constant, and so the springs are at constant strain and the elastic strain rates are zero. Then, since the total strain rates in the two arms are equal, Equation (12.31) gives

$$\dot{e} = \dot{e}_{EA} + \dot{e}_{pA} = \dot{e}_{EB} + \dot{e}_{pB} \Rightarrow \dot{e}_{pA} = \dot{e}_{pB} = \dot{e}_p$$

so that there is a common plastic strain rate and we may write (12.36) as

$$\sigma = \sigma_A + \sigma_B = \frac{1}{V_A^*} \sinh^{-1}\left(\frac{\dot{e}_p}{C_A}\right) + \frac{1}{V_B^*} \sinh^{-1}\left(\frac{\dot{e}_p}{C_B}\right)$$

This is essentially the form used by Wilding and Ward (1981), Ward and Wilding (1984) and Liu and Truss (1994). For the general case when the plastic strain rates are different in the two arms, the solution for stresses and strains in the model can be generated numerically, by enforcing the equilibrium between spring and Eyring dashpot in each arm to solve for the plastic strain at successive time increments. This approach has been used by Sweeney *et al.* (2019), (2014), where in both cases the model was generalized to large elastic strains by replacing the linear springs with neo-Hookean elements.

12.5.3 Applications of the Eyring Equation to Stress Relaxation

Guiu and Pratt (1964) have shown how a model consisting of an Eyring dashpot in series with an elastic element leads to a simple equation to describe stress relaxation curves in tension. Suppose that there is a total uniaxial strain e on the system of Figure 12.27, consisting of an elastic component e_E and an Eyring or plastic component e_p. We have already the solution for the plastic strain rate in Equation (12.31). We approximate this by using $\sinh x \approx \frac{1}{2} \exp x$ to give

$$\dot{e}_p = \dot{e}_0 \exp\left(-\frac{\Delta H}{kT}\right) \frac{1}{2} \exp\left(\frac{V\tau}{kT}\right) = D \exp(V^* \sigma) \tag{12.37}$$

where Equation (12.33) still applies and

$$D = \frac{1}{2} \dot{e}_0 \exp\left(-\frac{\Delta H}{kT}\right) \tag{12.38}$$

Figure 12.27 The single process model for stress relaxation.

Assuming a linear relation for the elastic component, with modulus E, Equation (12.31) can now be rewritten so that only stresses appear on the right-hand side:

$$\dot{e} = \frac{\dot{\sigma}}{E} + D\exp(V^*\sigma).$$

Under conditions of stress relaxation, the total strain rate is zero, and the stress decays in a manner governed by the relation

$$0 = \frac{\dot{\sigma}}{E} + D\exp(V^*\sigma). \tag{12.39}$$

This can be solved by separation of variables to give the stress as

$$\sigma_0 - \sigma = \frac{1}{V^*}\ln(1 + DEV^*\exp(V^*\sigma_0)t) \tag{12.40}$$

where σ_0 is the stress at time $t = 0$.

Equation (12.40), the Guiu and Pratt expression, has been shown to be remarkably effective in representing stress relaxation curves for polymers. Escaig (1982) has discussed its general utility. Sweeney *et al.* (2019) have used numerical fitting of the equation to estimate Eyring parameters for PLA. Sweeney and Ward (1990) applied the expression successfully to the stress relaxation behaviour of highly oriented polyethylene fibres at small strains. However, emerging from both these studies is the major drawback of the Guiu–Pratt approach in that, as a Maxwell-type model, it has no capability of representing stress or strain recovery behaviour (see 11.2.5.2). A parallel model such as that in Figure 12.26, with the potential for stress to be exchanged between the two arms, has the capability for recovery. Since stress and strain recovery are essentially universal phenomena in polymers, a two-arm model is a minimum requirement for realistic behaviour. The Guiu–Pratt approach can however give useful first estimates of Eyring parameters and help in the definition of a two-process model, as in Sweeney *et al.* (2019).

12.5.4 Applications of the Eyring Equation to Yield

Suppose the system of Figure 12.27 is subjected to a constant total strain rate $\dot{e}$, starting at zero load. Initially, the stress is low and causes only a small rate of strain in the Eyring dashpot, and the elastic spring is stretched. Continued stretching of the spring increases the stress and thus the rate of strain $\dot{e}_V$ in the Eyring dashpot, until eventually it becomes equal to the total applied strain rate $\dot{e}$. At this point, the spring ceases to extend, and so the stress in it, which is equal to the total stress on the system, becomes constant. Thus, behaviour resembling yield is predicted. It is easily shown that this yield stress depends on strain rate. Adopting the notation of the previous section, once the spring reaches its state of constant strain, $\dot{e}_E = 0$. Then, using Equation (12.31)

$$\dot{e} = \dot{e}_p$$

and from Equation (12.37)

$$\dot{e} = D\exp(V^{*}\sigma_Y) \tag{12.41}$$

where σ_Y denotes the yield stress. Rearranging gives the relation

$$\sigma_Y = \frac{1}{V^{*}} \ln\left(\frac{\dot{e}}{D}\right) \tag{12.42}$$

or, regaining the expressions for V^{*} and D in Equations (12.33) and (12.38),

$$\sigma_Y = \frac{3kT}{\sqrt{2}V} \ln\left[\frac{2\dot{e}}{\dot{e}_0} \exp\left(\frac{\Delta H}{kT}\right)\right]. \tag{12.43}$$

These equations imply a linear relationship between $\frac{\sigma_Y}{T}$ and $\ln(\dot{e})$ This has been demonstrated for a number of polymers. For example, the work of Bauwens-Crowet *et al.* (1969b) on polycarbonate is illustrated in Figure 12.28. Plots such as these can supply information on the Eyring parameters; in particular, the slopes enable estimates of the activation volume. As established by Bauwens-Crowet *et al.* (1974), yield stress corresponding to a stress maximum at constant rate of strain is equivalent to equilibrium creep stress at the same rate of creep strain, so that the two types of result can be plotted together. Plots of the form in Figure 12.28 are used routinely to establish Eyring parameters, examples being Costa *et al.* (2008) and Pini *et al.* (2021) for PVDF; Sedighiamiri *et al.* (2012) for polyethylene; Engels *et al.* (2010), Söntjens *et al.* (2012), Sweeney *et al.* (2019) for PLA and van Erp *et al.* (2009) for oriented polypropylene.

Since the slopes of the plots such as those in Figure 12.28 are simply related to the activation volume, we would expect a more complex picture to emerge for materials that

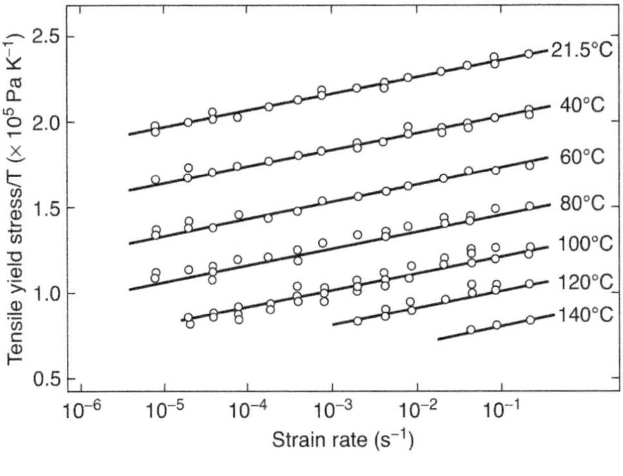

Figure 12.28 Measured ratio of yield stress to temperature as a function of the logarithm of strain rate for polycarbonate. The set of parallel straight lines is calculated from Equation (12.43). Bauwens-Crowet et al. (1969) / John Wiley & Sons.

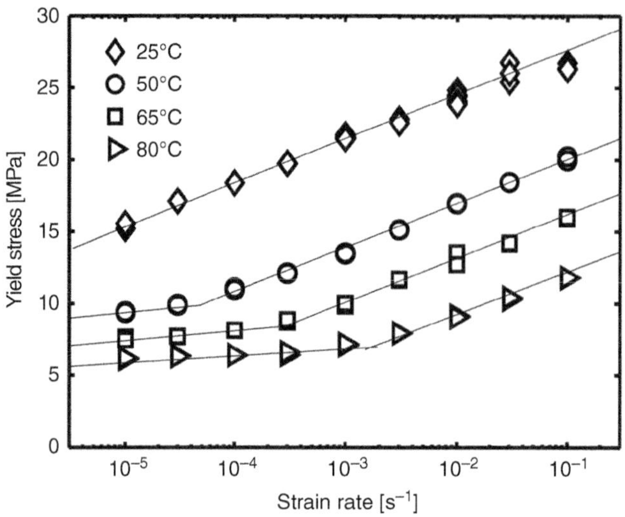

Figure 12.29 Yield of polyethylene. Sedighiamiri et al. (2012) / John Wiley & Sons.

demand to be represented using more than one Eyring process, such as by a model shown in Figure 12.26. This is indeed the case, that is typified by the plots of Figure 12.29 (Sedighiamiri *et al.* 2012). Here the results at the lower temperatures show two distinct slopes, corresponding to two processes. The use of two-process systems

resembling that in Figure 12.26 for modelling yield precedes their use for modelling creep, having begun with Roetling (1965), continued by Truss *et al.* (1984) and with relatively recent applications such as from Sedighiamiri *et al.* (2012),

The relevance of the two-process model has been demonstrated more directly by the observation of double yield points in polyethylene. Brooks *et al.* (1992) performed uniaxial room temperature tensile tests on three polyethylene grades and, by observing both stress maxima and the onset of necking, were able to distinguish two yield points in each material. They concluded that one yield was associated with interlaminar shear and the other with intralaminar shear (or *c*-slip). The work of Seguala and Darras (1994) and Gupta and Rose (1998) supported these conclusions.

Haward and Thackray (1968) made use of the Eyring process in a model that covered both yield and post-yield behaviour to large strains. The large strain capability was achieved by incorporating an inverse Langevin chain element (see Section 4.4.4). The model is shown in Figure 12.30. The Hookean spring provides the initial elastic response. The total stress is shared between the Eyring process and the Langevin spring, with the latter contributing a level of strain hardening. Predictions were made using numerical techniques. The simple three-element model produced realistic stress–strain curves, an example of which is shown in Figure 12.31. Tests were at relatively low strain rates, and the range of strain rates was relatively low at a maximum of around three decades.

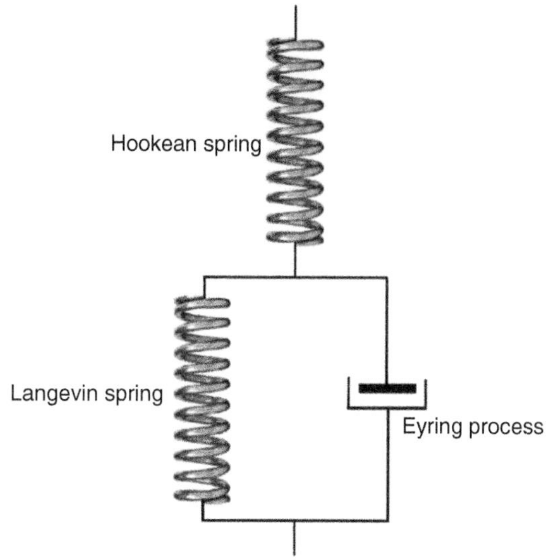

Figure 12.30 Schematic diagram of the Haward–Thackray model.

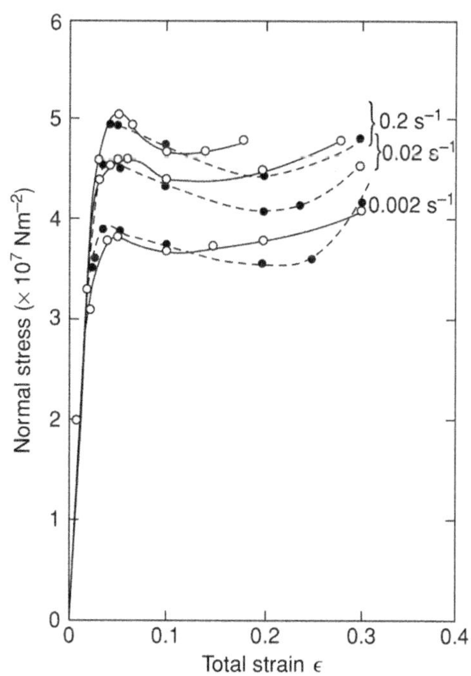

Figure 12.31 Stress–strain curves for cellulose nitrate at 23°C. Experimental curves (0) and calculated curves, Langevin model (e). n 1/2 = 0.30, N = 1.57 × 10^{26} chain m^{-3}●. Haward and Thackray (1968) / John Wiley & Sons.

12.5.5 Physical Interpretation of Eyring Parameters

Haward and Thackray (1968) have compared the Eyring activation volumes obtained from yield stress data with the volume of the 'statistical random link'. The latter was obtained from solution studies, by assuming that the real chain can be represented by an equivalent

Table 12.1 A comparison of the statistical segment volume for a polymer measured in solution with the flow volumes derived from the Eyring theory (after Haward and Thackray (1968).

Polymer	Volume of statistical link in solution (nm^3)	Eyring flow volume V (nm^3)
Polyvinyl chloride	0.38	8.6
Polycarbonate	0.48	6.4
Polymethyl methacrylate	0.91	4.6
Polystyrene	1.22	9.6
Cellulose acetate	2.06	8.8
Cellulose trinitrate	2.62	6.1
Cellulose acetate	2.05	17.4

chain with freely jointed links of a particular length. Table 12.1 is based on data collated by Haward and Thackray and shows that the activation volumes are very large in molecular terms and range from about two to 10 times that of the statistical random link. The result suggests that yield involves the cooperative movement of a larger number of chain segments than would be required for a conformational change in dilute solution.

This view has been supported by the work of Buckley *et al.* (1996), who modelled PET subject to large biaxial deformations. They arrived at an activation volume of around 12 nm^3, equivalent to 50 monomer units. Costa *et al.* (2008) compared the activation volumes obtained from mechanical observations of poled and unpoled PVDF, and found that poling increased the activation volume from 0.49 to 0.80 nm^3; they associated this change with a greater degree of cooperativity in the poled material.

Acknowledging the findings of Haward and Thackray, Fotheringham and Cherry (1978) pursued the concept of co-operative motions by analysing the probability of a number of Eyring events occurring simultaneously in order to enable a deformation. This resulted in a modified Eyring equation equivalent to

$$\dot{e} = \dot{e}_0 \exp\left[-\frac{n\Delta H}{kT}\right] \sinh^n\left(\frac{V\tau}{kT}\right) \tag{12.44}$$

Results for linear polyethylene were fitted to give a value of about three for n and an activation volume of 0.5 nm^3, which is in the same range as the volume swept out by an elementary displacement of a defect moving through the crystal lattice.

This approach has been progressed by Povolo and Hermida (1995), Povolo *et al.* (1996) and Richeton *et al.* (2005), with success particularly with amorphous polymers. In many cases the plot of yield stress against log strain rate is nonlinear, with an increasing slope. Such a plot can be approximated by two linear slopes, but a better fit can sometimes be obtained using the co-operative model which can give a smoothly increasing slope. With the co-operative model, smaller activation volumes result in smaller argumentsof the hyperbolic sine function, so that it can no longer be approximated with $\sinh x \approx \frac{1}{2}\exp x$ and the logarithmic plot is not linear. The co-operative model proved very effective for observations in both tension and compression, with values of n 6.65 for PMMA (Povolo and Hermida

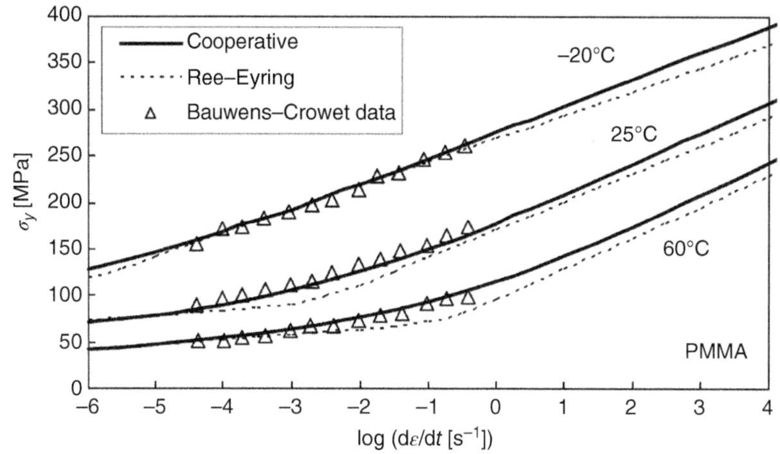

Figure 12.32 Comparison of the co-operative model and the two-process Eyring ('Lee-Eyring') model for polycarbonate in compression. Richeton et al. (2005) / with permission of Elsevier.

1995), 10 for PVC (Povolo *et al.* 1996) and 3.91 for polycarbonate (Richeton *et al.* 2005). The case of polycarbonate in compression is shown in Figure 12.32 for a typical example, where the co-operative model performs better than the two-process model.

We may conclude that, in the case of semi-crystalline polymers, the two-process model may be identified with two specific mechanisms (for instance inter-laminar and intra-laminar shear – see previous section) with distinct slopes as in Figure 12.29. For amorphous polymers, fitting the yield – log strain rate plot with two slopes can be unnaturally restrictive, as shown in Figure 12.32, and the co-operative model gives a superior performance.

12.5.6 The Robertson Model

Robertson (1966) has developed a slightly more elaborate version of the Eyring viscosity theory. For simplicity, it is considered that there are only two rotational conformations, the *trans* low-energy state and the *cis* high-energy state, which Robertson terms the 'flexed state'. Applying a shear stress τ causes the energy difference between the two stable conformational states of each bond to change from ΔU to $(\Delta U - \tau V \cos\theta)$. $\tau V \cos\theta$ represents the work done by the shear stress in the transition between the two states and θ is the angle defining the orientation of a particular element of the structure with respect to the shear stress.

Prior to application of stress the fraction of elements in the high-energy state is

$$\chi_i = \frac{\exp\{-\Delta U/k\theta_g\}}{1 + \exp\{-\Delta U/k\theta_g\}},$$

where $\theta_g = T_g$ if the test temperature $T < T_g$ and $\theta_g = T$ if $T > T_g$, i.e. below T_g the configurational state 'freezes' at that which exists at T_g. For application of a shear stress τ at a temperature T, the fraction of elements in the upper state with orientation θ is given by

$$\chi_f(\theta) = \frac{\exp\{-(\Delta U - \tau V\cos\theta)/kT\}}{1 + \exp\{-(\Delta U - \tau V\cos\theta)/kT\}}.$$

Clearly the fraction of flexed elements increases for orientations such that

$$\frac{\Delta U - \tau V\cos\theta}{kT} \leq \frac{\Delta U}{k\theta_g}.$$

For one part of the distribution of structural elements, applying the stress tends to make for an equilibrium situation where there are more flexed bonds, and this can be regarded as corresponding to a rise in temperature. For the other part of the distribution, the effect of stress can be regarded as tending to lower the temperature. Robertson now argues that the *rate* at which conformational changes occur is very dependent on temperature (cf. WLF equation). Hence, the rate of approach to equilibrium is much faster for these elements that flex under the applied stress, so that changes in the others can be ignored in calculating the maximum flexed-bond fraction which can occur under a given applied stress. This maximum corresponds to a rise in temperature to a temperature θ_1. The strain rate $\dot{e}$ at θ_1 is calculated from the WLF equation

$$\dot{e} = \frac{\tau}{\eta_g}\exp\left\{-2.303\left[\left(\frac{C_1^g C_2^g}{\theta_1 - T_g + C_2^g}\right)\frac{\theta_1}{T} - C_1^g\right]\right\},$$

where C_1^g, C_2^g are the universal WLF parameters (see Section 7.4.1) and η_g is the 'universal' viscosity of a glass at T_g.

Duckett *et al.* (1970) have modified the Robertson model to include the effect of the hydrostatic component of pressure p. It was proposed that p also does work during the activation event and that the energy difference between the two states should therefore be

$$\Delta U - \tau V\cos\theta + pV_p$$

where V_p is the pressure activation volume. Figure 12.33 shows that in this modified form the Robertson model can bring consistency to yield data in tension and compression for polymethyl methacrylate, together with the measured effect of hydrostatic pressure.

12.6 Yield Considered to Relate to the Movement of Dislocations or Disclinations

It is well known (Cottrell 1953) that plastic deformation in crystals can occur when the applied shear stress causes one plane of atoms to slip over another plane because there is an imperfect match between these adjacent planes at a particular point in the crystal lattice. These points of imperfection are called dislocations (Taylor 1934) and were identified by electron diffraction techniques to relate to specific crystal defects. Dislocations were observed in polyethylene single crystals by Peterman and Gleiter (1972) and give credence to the idea that yield in crystalline polymers can be understood in similar terms to those used by metallurgists for crystalline solids.

A starting point for the discussion is the Frenkel argument for predicting the maximum theoretical shear strength of a crystal (Taylor 1934; Kelly and Macmillan 1987). For a simple

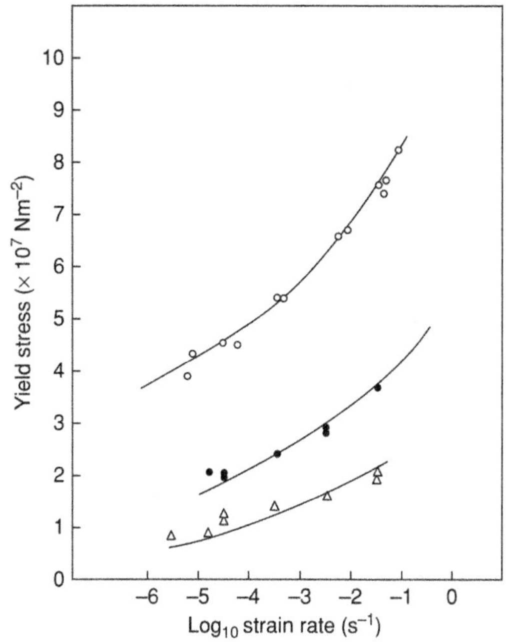

Figure 12.33 Yield stress of polymethyl methacrylate as a function of strain rate. ○, compression at 23°C; Δ, tension at 90°C; ●, tension at 60°C. Curves represent the best theoretical fit (Duckett et al. 1970).

lattice of identical atoms with a repeat distance b in the direction of shear on planes separated by a distance h it can most simply be considered that the shear stress τ follows a sine curve with shear displacement x to give

$$\tau = k \sin 2\pi \frac{x}{b}. \tag{12.45}$$

The shear stress is zero when each atom is situated half way between one equilibrium position and the next, that is $x = \frac{b}{2}$ and the shear strain is 0.5. In the simplest case, the slip of one plane will involve a movement where $x = b$. This characteristic slip distance b is called the Burgers vector.

For small strains

$$\tau = 2\pi k \frac{x}{b} \tag{12.46}$$

and the shear strain is given by

$$\gamma = \frac{x}{h}.$$

The shear modulus G is then

$$\frac{\tau}{\gamma} = 2\pi k \frac{h}{b} \tag{12.47}$$

and the maximum shear stress

$$\tau_{\max} = \frac{\tau}{\gamma} = k = \frac{Gb}{2\pi h}. \tag{12.48}$$

Since $b \sim h$ it follows that the maximum shear stress is

$$\tau_{\max} \sim \frac{G}{2\pi}$$

and occurs when $x = \frac{b}{4}$, where the shear strain $\frac{x}{h} = \frac{x}{b} = 0.25$.

On this approach, yield can be considered to be nucleation controlled as distinct from the viscoelastic approach that can be considered to be velocity controlled. It implies a direct link between the shear modulus and the shear yield stress. Brown (1971a,b) proposed that there is good empirical evidence for this supposition, which had been suggested by other workers previously (Allison and Ward 1967; Allison *et al.* 1966; Vincent 1967), and it is an essential ingredient of nucleation controlled yield behaviour as developed formally by Bowden and Raha (1974), Argon (1973) and others.

The discussion so far has considered a regular lattice, but Bowden and Argon have proposed that in an amorphous polymer in the glassy state similar ideas can be developed. Following Li and Gilman (1970), the analogy in amorphous polymers to dislocations in crystalline polymers has been called *disclinations.*

12.6.1 The Young Theory

Following treatments given by Kelly and Macmillan (1987), Bowden and Raha (1974), Young (1974) and others, the increase in energy U associated with forming a dislocation loop of Burgers vector b and radius R in a solid with a shear modulus G under an applied shear stress τ is given approximately by

$$U = 2\pi R \frac{Gb^2}{4\pi} \ln \frac{2R}{r_0} - \pi R^2 \tau b, \tag{12.49}$$

where r_0 is the core radius of the dislocation.

The energy of the loop increases as R increases until it reaches a maximum value U_c at R_c. This is found by differentiating Equation (12.49) and is given by

$$U_c = \frac{Gb^2 R_c}{4} \left[\ln \frac{2R_c}{r_0} - 1 \right] \tag{12.50}$$

at

$$R_c = \frac{Gb}{4\pi\,\tau} \left[\ln \frac{2R_c}{r_0} + 1 \right]. \tag{12.51}$$

The core radius r_0 can be calculated by the argument developed in Equations (12.45–12.48) above where $\tau \sim \frac{G}{2\pi}$ and leads to a value for $r_0 \sim b$.

The discussion so far considers the theoretical shear stress of a crystal in the absence of thermal fluctuations. Frank (1950) considered that there are always local thermal fluctuations, which must be taken into account. At any temperature T, there is a significant chance of thermal fluctuations in the timescale of the experiments supplying an energy up to 50 kT. Furthermore, this discussion only relates the yield stress to the elastic energy, whereas U is strictly the activation enthalpy. Analogous to the site model theory (see Section 7.3), we should discuss the Gibbs free energy ΔG, where $\Delta G = T\Delta S$ and the shear strain rate is

$$\dot{\gamma} = \exp\left(-\frac{\Delta G}{kT}\right) = \exp\left(-\frac{U}{kT}\right)\exp\left(\frac{\Delta S}{k}\right). \tag{12.52}$$

Frank's assumption is equivalent to putting $\frac{\Delta S}{k} = 1$ and $U = 50\,\text{kT}$ at the applied strain rate, which for typical experiments is $10^{-3}\,\text{s}^{-1}$.

Young has followed these ideas to relate the yield behaviour in crystalline polymers to the lamellar thickness. Following Shadrake and Guiu (1976), Young showed that the change in the Gibbs free energy ΔG_a associated with the nucleation of a screw dislocation in a lamella of thickness *d*, having a Burgers vector of magnitude *b* in the chain direction gives a shear yield stress

$$\tau_y = \frac{K}{4\pi}\exp\left[-\left(\frac{2\pi\Delta G_a}{dKb^2} + 1\right)\right], \tag{12.53}$$

where $K = (c_{44}c_{55})^{1/2}$ with c_{44}, c_{55} the shear moduli.

d should more correctly be interpreted as the stem length (i.e. the length traversed by the polymer chains within the crystalline lamella) and not the lamellar thickness. The yield stress then depends on the stem length, and on temperature and strain rate through the shear modulus term *K*. ΔG_a is assumed to be 50 kT in accordance with Frank's assumption.

12.6.2 The Argon Theory

Argon (1973) has proposed a theory of yielding for glassy polymers based on the concept that deformation at a molecular level consists in the formation of a pair of molecular kinks. The unit process of deformation is shown in Figure 12.34. The resistance to double kink

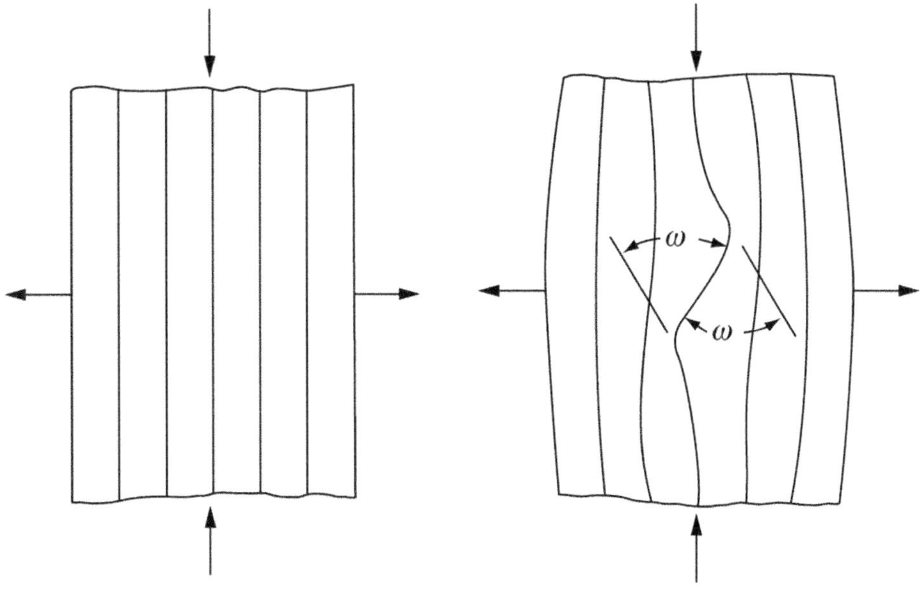

Figure 12.34 Schematic representation of the unit process of deformation consisting of unbending and bending a molecular segment.

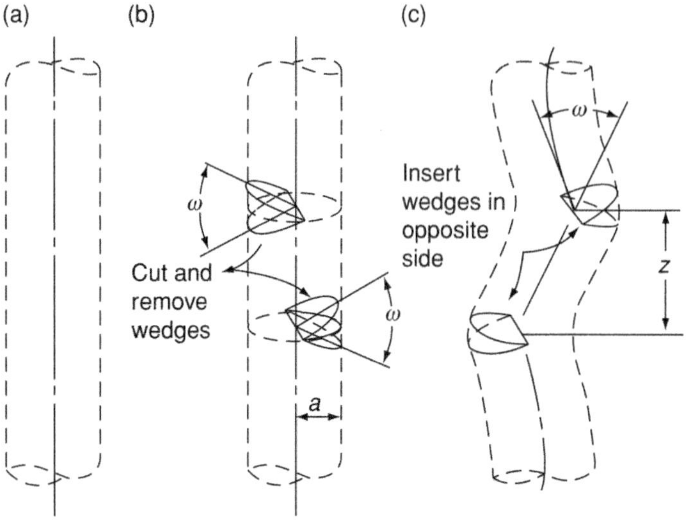

Figure 12.35 Modelling of a molecular kink pair by a pair of wedge disinclination loops: (a) outline of polymer molecule in an elastic surrounding made up of other neighbouring molecules, (b) make the circular cuts of radius a at distance 2a apart and cut and remove wedges of angle ω and (c) insert cut wedges into opposite side and join all parts together. Argon (1973) / Taylor & Francis.

formation is considered to arise from the elastic interactions between a chain molecule and its neighbour, i.e. from intermolecular forces in contrast to the Robertson theory where intramolecular forces are the primary consideration. The intermolecular energy change associated with a double kink is then calculated by modelling these as two wedge disclination loops as proposed by Li and Gilman (1970) (Figures 12.35 and 12.36).

The activation energy (strictly enthalpy) for the formation of a pair of molecular links under an applied shear stress τ is given by

$$\Delta H^* = \frac{3\pi G\omega^2 a^2}{16(1-\nu)}\left[1 - 6.75(1-\nu)^{5/6}\left(\frac{\tau}{G}\right)^{5/6}\right], \tag{12.54}$$

where G, ν are the shear modulus and Poisson's ratio, respectively, a is the molecular radius and ω the angle of rotation of the molecular segment (Figure 12.28). The shear strain rate is then

$$\dot{\gamma} = \gamma_0 \Omega C \nu_a \exp\left[-\frac{\Delta H^*}{kT}\right], \tag{12.55}$$

where γ_0 is the shear strain in the local volume $\Omega = \pi a^2 z_{eq}$ (z_{eq} is the equilibrium molecular segment length), C is the total volume density of potentially rotatable segments in the polymer, and ν_a is a frequency factor of the order of (but somewhat smaller than) the atomic frequency.

It can be noted that when Equation (12.54) is substituted into Equation (12.55) the resultant equation is quite similar in form to the Eyring equation

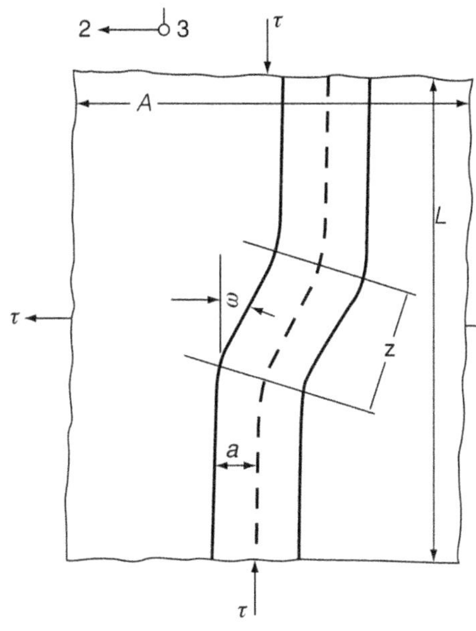

Figure 12.36 Plastic strain increment by formation of a pair of kinks in a polymer molecule. Argon (1973) / Taylor & Francis.

$$\dot{\gamma} = \dot{\gamma}_0 \exp\left[-\left(\frac{\Delta H - \tau v}{kT}\right)\right].$$

The shear yield stress τ is given from Equation (12.55) as

$$\tau = \frac{0.102G}{1-\nu}\left[1 - \frac{16(1-\nu)}{3\pi G\omega^2 a^3}kT\ln\frac{\dot{\gamma}_0}{\dot{\gamma}}\right]^{6/5}, \tag{12.56}$$

where $\dot{\gamma}_0 = \gamma_0 \nu_a \Omega C$.

12.6.3 Experimental Studies Based on the Nucleation Approach

The first experimental studies based on the nucleation approach were for glassy polymers. Bowden and Raha (1974) assumed that the dislocation loop theory set out above for crystalline solids could be regarded as modelling the behaviour of a small volume of amorphous solid. The two key variables in Equation (12.56) are the shear modulus and the Burgers vector b. Bowden and Raha used literature values for the shear modulus and assumed values of b, which gave reasonable fits to the temperature dependence of the shear stress. Argon (1973) developed Li and Gilman's ideas to explain the concept of disclinations and fitted results of Ward and co-workers on glassy polyethylene terephthalate.

Figure 12.37 shows Argon's fit to very extensive data for polyethylene terephthalate. The fit is good, but it may be noted that if we replace the factor 6/5 in Equation (12.36) by unity, which makes a comparatively small difference numerically, then this equation reduces to

$$\tau = \frac{0.102G}{1-\nu} - \frac{16 \times 0.102kT}{3\pi\omega^2 a^2}\ln\frac{\dot{\gamma}_0}{\dot{\gamma}}, \tag{12.57}$$

which is of similar form to the Eyring equation where

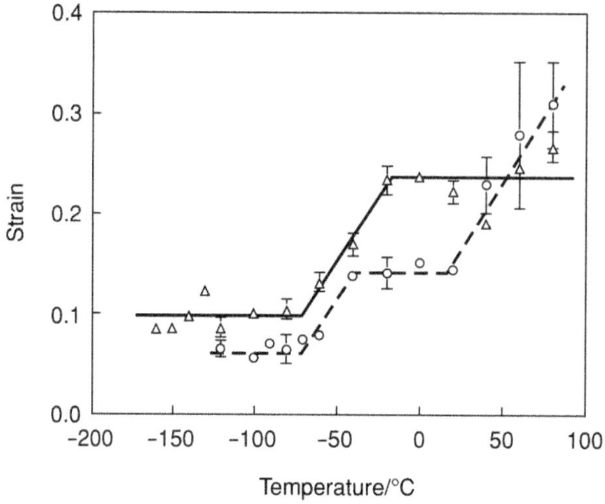

Figure 12.39 Yield strain for polyethylene at strain rates Δ2.08×10^{-3} s^{-1} and ○ 8.3 s^{-1}. Brooks et al. (1997) / John Wiley & Sons.

to the behaviour at very low temperatures. In this respect, it is interesting to recall that as we approach absolute zero the ratio of yield stress to modulus approaches a limiting value, which is consistent with classical theoretical shear strength arguments.

As clearly discussed, the mechanical behaviour of polymers changes rapidly as the temperature is reduced or the strain rate increased. Brooks *et al.* have shown that for polyethylene there is a sudden transition in the yield strain at temperatures below ambient, the exact temperature depending on the sample morphology. Figure 12.39 shows results for linear PE. It was also found that this temperature marks the change from classical elastic–plastic behaviour to time-dependent viscoelastic behaviour where the yielded samples show evidence of interlamellar shear. Below the transition temperature, the yield behaviour is consistent with nucleation-controlled yield as proposed by Bowden, Young and others, and an excellent correlation is obtained between the yield stress and the lamellar stem thickness.

12.7 The Billon Model

Billon (2012) has developed a material model for deformations at or near to the glass transition that takes an approach completely different from those discussed so far. It takes as its basis the entropic model of Edwards and Vilgis (1986) as described in Chapter 4 (Equations (4.36) and (4.37)). It incorporates time-dependent viscoplastic behaviour by allowing the parameters that define the model – cross-link number N_c, sliplink number N_s, slipperiness factor η and finite strain extensibility factor α – to vary in response to deformation in a manner consistent with thermodynamics. Various phenomena can be modelled

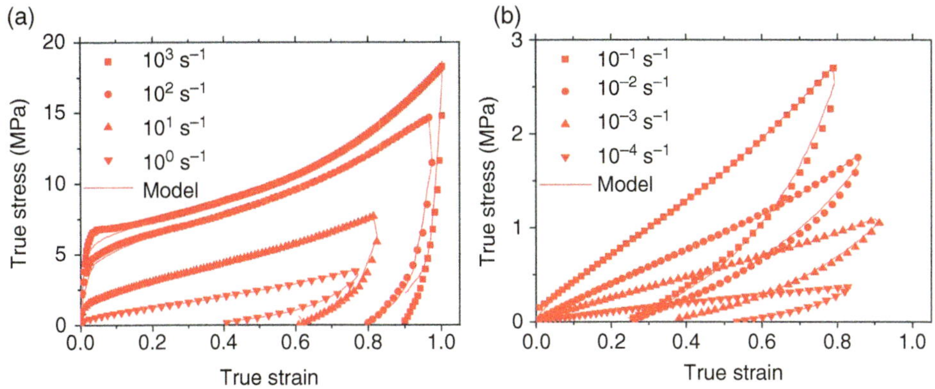

Figure 12.40 Comparing tensile loading/unloading experiments (markers)with model (lines) for a PMMA material at 130°C. (a) and (b) show separate ranges of strain rate for clarity. (Federico et al. 2020).

by the evolution of the parameters. For instance, an increase in η could correspond to chain disentanglement; or an increase in N_c and/or α to an increase in crystallinity. The parameter changes represent microstructural changes driven by energy that is assumed to originate in the partial recovery of elastic network strain. Some of this elastic energy is dissipated. There may be relationships between the parameters; an example discussed is that $N_c + N_s$ is constant. Temperature dependence of the parameters has been included using the WLF relation (see Chapter 7). The model has been applied to PMMA successfully at large deformations in both loading and unloading (Billon 2012). A complete three-dimensional presentation of the model has been given by Maurel-Pantel *et al.* (2015), who applied it to semi-crystalline polyamide. A further variation of the model has been applied to PMMA (Federico *et al.* 2020). The performance of this model in tensile loading and unloading is shown in Figure 12.40.

12.8 Multi-axial Deformation: Three-Dimensional Plasticity

The discussion so far has been dominated by one-dimensional behaviour, reflecting the most convenient and customary materials testing methods. However, any engineering application will be for a three-dimensional body, subject to multi-axial stresses. It is now feasible to implement plasticity models in numerical schemes to perform analyses of structures making a three-dimensional approach essential.

The value of splitting the stress into hydrostatic and deviatoric components, and the strain into volumetric and deviatoric components, has been demonstrated in the analysis of compressible material in Section 3.4.1. On the basis that creep is caused by shearing of molecules past one another, we would expect creep to be only associated with deviatoric stress. This expectation is in line with the adoption of shear stress as the major driver for plastic deformation in equations such as (12.27); shear stress, in its many forms including the octahedral shear stress of Equation (12.13), is independent of the hydrostatic stress and therefore dependent only on the deviatoric stress. However, we know that pressure can influence plastic flow in polymers as it controls the yield stress as in Equation (12.24). When considering plasticity in three-dimensional stress fields, it makes sense to start with the effect of pressure.

12.8.1 Pressure Dependence and the Eyring Model

We have seen that the effect of pressure on the shear yield stress of a polymer can be very well represented by Equation (12.24)

$$\tau_Y = \tau_{Y0} + \alpha p \tag{12.24}$$

At yield,

$$\tau_Y - \alpha p = \tau_{Y0} \tag{12.59}$$

With this in mind we may generalize the Eyring model to include pressure dependence. On the right-hand side of Equation (12.59) the quantity τ_{Y0} depends on strain rate and temperature, defined by the Eyring equation for yield stress (12.43) with $\tau_{Y0} = \frac{\sqrt{2}}{3}\sigma_{Y0} = \frac{kT}{V}\ln\left[\frac{2\dot{e}}{\dot{e}_0}\exp\left(\frac{\Delta H}{kT}\right)\right]$,

With this expression Equation (12.59) gives

$$\frac{kT}{V}\ln\left(\frac{2\dot{e}}{\dot{e}_0}\exp\left(\frac{\Delta H}{kT}\right)\right) = \tau_Y - \alpha p$$

from which

$$\dot{e} = \frac{\dot{e}_0}{2}\exp\left(-\frac{\Delta H}{kT}\right)\exp\left(\frac{V}{kT}(\tau_Y - \alpha p)\right) \tag{12.60}$$

We now introduce the shear activation volume $V_s = V$ and the pressure activation volume $V_p = \alpha V_s$. Also, since we are moving towards a three-dimensional formulation, we now replace the linear strain rates with octahedral strain rates defined by

$$\dot{\gamma}_{\text{oct}} = \frac{2}{3}\left[(\dot{e}_{\text{I}} - \dot{e}_{\text{II}})^2 + (\dot{e}_{\text{II}} - \dot{e}_{\text{III}})^2 + (\dot{e}_{\text{III}} - \dot{e}_{\text{I}})^2\right]^{1/2}. \tag{12.61}$$

Now Equation (12.45) becomes

$$\dot{\gamma} = \frac{\dot{\gamma}_0}{2}\exp\left(-\frac{\Delta H}{kT}\right)\exp\left(-\frac{V_p p}{kT}\right)\exp\left(\frac{V_s \tau}{kT}\right) \tag{12.62}$$

This is now the general flow equation with τ as the octahedral shear stress. Finally we can recover the hyperbolic sine function from the original Eyring equation:

$$\dot{\gamma} = \dot{\gamma}_0 \exp\left(-\frac{\Delta H}{kT}\right)\exp\left(-\frac{V_p p}{kT}\right)\sinh\left(\frac{V_s \tau}{kT}\right) \tag{12.63}$$

A positive hydrostatic pressure has the effect of decreasing the available free volume, so will slow down the event rates relating to all directions. The hyperbolic sine function is therefore inappropriate for the hydrostatic pressure, and so we retain an exponential function for this term.

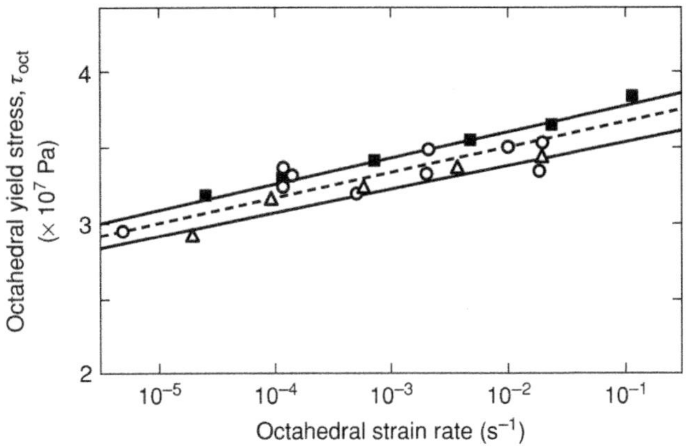

Figure 12.41 The strain rate dependence of the octahedral shear stress at atmospheric pressure using data from torsion (○), tension (Δ) and compression (■). Duckett et al. (1978) / John Wiley & Sons.

Bauwens (1967) arrived at an identical form by considering separately the action of deviatoric stress to produce strain, and that of hydrostatic stress to contribute to the formation of a hole to accommodate the chain segment motion.

Figure 12.41 shows results for polycarbonate at atmospheric pressure (Duckett *et al.* 1978) using data from torsion, tension and compression. It can be seen that on average the values of τ_{oct} lie in the order compression > torsion > tension. The differences are therefore consistent with the observed linear dependence of τ_{oct} on pressure shown by direct measurement of the yield stress in torsion over a range of hydrostatic pressures (see Section 12.4.2), and there is good numerical agreement between the two sets of measurements.

12.8.2 Numerical Implementation

To make use of the models here for practical purposes such as to analyse stresses in components or, at large deformation, to simulate processes, a numerical implementation is needed. This is usually in the form of the finite element method (FEM), where some commercial codes offer user-defined material behaviour.

The Eyring model is the most frequently used plasticity mechanism in this context. It is combined in various ways with elastic and hyperelastic elements. To give examples, large deformation FEM has been applied by Jarrah *et al.* (2021) to shape memory foams, Olley and Sweeney (2011) and Sweeney *et al.* (2009) to polypropylene sheet, Sweeney *et al.* (2012) and O'Conner *et al.* (2013) for polypropylene under thermoforming conditions, Johnsen *et al.* (2019) to low-density polyethylene, Naz *et al.* (2010) to model fracture of ultra-high-molecular-weight polyethylene, Tamrakar *et al.* (2020) to epoxy resin, Li and Buckley (2009) to strain localization in glassy polymers, Fleischhauer *et al.* (2012) to epoxy, PMMA and polycarbonate and Kershah *et al.* (2020) to anisotropic polypropylene. These examples, along with the vast majority of Eyring implementations, use the conventional

Eyring model such as that embodied in Equation (12.63). An exception to this is the work of Bouvard *et al.* (2013) who used the co-operative model of Fotheringham and Cherry (1978) of Equation (12.44) to model amorphous glassy polymers.

The Argon model has been applied in a number of FEM analyses, most frequently with glassy polymers. Examples are the analysis of localization of plastic flow in glassy polymer by Kweon and Benzerga (2013), the work of Holopainen (2014) on a damage model for glassy polymer, the study of the fracture of notched polycarbonate by Faye *et al.* (2015), the study of scratching in PMMA by Jiang *et al.* (2017), analysis of the fracture of epoxy by Torres *et al.* (2018) and the work by Tang *et al.* (2022) on shear banding and strain softening in glassy polymer.

12.9 Cold-Drawing, Strain Hardening and the True Stress–Strain Curve

12.9.1 General Considerations

We have seen that strain hardening is a necessary prerequisite for cold-drawing (see Section 12.1.2). There are two possible sources of strain hardening:

1) Drawing causes molecular alignment so that the drawing stress (often called the flow stress) is increased. This is a general phenomenon, true for both crystalline and amorphous polymers. (Note that the theories of mechanical anisotropy developed in Sections 8.6 and 8.7 apply to the final drawn material and do not relate directly to the strain-hardening effect.)
2) Strain-induced crystallization may occur at the high degree of extension occurring in cold-drawing. This may be similar to the crystallization occurring in rubber at high degrees of stretching (see Section 4.4.6). It may involve at a morphological level extended chain crystallization or the formation of shish–kebab structures.

12.9.2 Cold-Drawing and the Natural Draw Ratio

We have seen in Section 12.1 above that cold-drawing through a neck leads to the movement of the neck to accommodate the stretching of more material to what is termed the natural draw ratio. Cold-drawing occurs at temperatures below the glass transition, sometimes as much as ~150°C below. The yield process and subsequent cold-drawing do not involve long-range molecular flow but are associated with molecular re-arrangements between points of entanglement and/or cross-linkage. This view is consistent with the observation of yield, necking and cold-drawing, in highly cross-linked rubbers at temperatures below their glass transition. It is evident that cross-linking does not prevent the required molecular rearrangements.

The natural draw ratio for amorphous polymers is very sensitive to the degree of pre-orientation, i.e. the molecular orientation in the polymer before cold-drawing. This was reported for polyethylene terephthalate by Marshall and Thompson (1954) and for PMMA and polystyrene by Whitney and Andrews (1967).

It has been proposed (Allison *et al.* 1966) that the sensitivity of natural draw ratio to pre-orientation arises as follows. The extension of an amorphous polymer to its natural draw ratio is regarded as equivalent to the extension of a network to a limiting extensibility. This limiting extensibility is then a function of the original geometry of the network and the nature of the links of which it is comprised.

During fibre spinning, the network forms immediately below the point of extrusion from the small holes, and the fibre is subsequently stretched in the rubber-like state before cooling further and being collected as a frozen stretched rubber. Quantitative stress-optical measurements have confirmed this part of the hypothesis (Pinnock and Ward 1966). Cold-drawing then extends the network to its limiting extensibility. The ratio of the extended to unextended lengths of the network is a constant independent of the division of the extension between the spinning, hot-drawing and cold-drawing processes, provided that the junction points holding the network together are not ruptured nor the links in the chain broken.

The dimensions of the unstrained network can be measured by shrinking the pre-oriented fibres back to the state of zero strain, that is isotropy (Allison *et al.* 1966). These results can then be combined with measurements of the natural draw ratio to give the maximum extensibility for the network.

Consider the cold-drawing of a sample of length l_1 (Figure 12.42). If the fibre were allowed to shrink back to its isotropic state, length l_0, the shrinkage s would be defined by

$$s = \frac{l_1 - l_0}{l_1}. \tag{12.64}$$

Drawing to a length l_2 gives a natural draw ratio

$$N = \frac{l_2}{l_1}. \tag{12.65}$$

Combining Equations (12.64) and (12.65), we have

$$\frac{l_2}{l_0} = \frac{N}{1 - s}. \tag{12.66}$$

Table 12.2 shows collected results for a series of PET filaments. It can be seen that N varied from 4.25 to 2.58 and s from 0.042 to 0.378, but the ratio l_2/l_0 calculated from Equation (12.66) remained constant at a value of about 4.0.

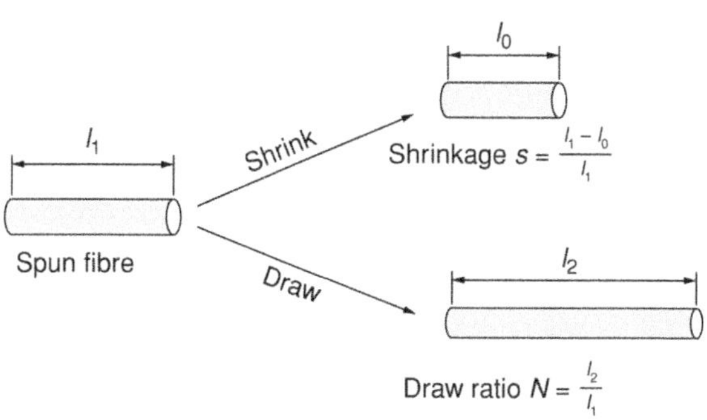

Figure 12.42 A representation of the shrinkage and drawing processes.

Table 12.2 Value of $l_2/l_0 = N/(1 - s)$ for samples of differing amounts of pre-orientation (polymer: polyethylene terephthalate, see Vincent (1967).

Initial birefringence (× 10_3)	Natural draw ratio, *N*	Shrinkage, *s*	1 − *s*	*N*/(1 − *s*)
0.65	4.25	0.042	0.958	4.44
1.6	3.70	0.094	0.906	4.08
2.85	3.32	0.160	0.840	3.96
4.2	3.05	0.202	0.798	3.83
7.2	2.72	0.320	0.680	4.01
9.2	2.58	0.378	0.622	4.14

It is, of course, possible that the natural draw ratio is determined directly by the strain-hardening requirements. This does not invalidate the hypothesis that cold-drawing involves the extension of a molecular network, but suggests that strain hardening increases very rapidly as the network reaches its limiting extensibility.

12.9.3 The Concept of the True Stress–True Strain Curve and the Network Draw Ratio

Vincent (1960) proposed the concept of a true stress–true strain curve for the plastic deformation of a ductile polymer, based on the following experiment. First, the polymer is subjected to extension beyond the yield point less than that required to produce failure. The polymer sample is then allowed to relax by removing the load, after which the loading regime is repeated, again stopping the extension before ultimate failure. This procedure is repeated, each successive loading taking the sample closer to failure. As shown in Figure 12.43, it is found that on each reloading the new true stress–true strain curve could be made to coincide with the curve obtained by taking the sample to failure in a single experiment. It is vital to plot true stress (load divided by current cross-sectional area) versus true strain (natural logarithm of length divided by initial length).

Another very important idea is that drawing involves the stretching of a polymer network in which the junction points of the network are formed by physical entanglements. Even if crystallization occurs, the deformation of the polymer network determines the overall constraints on the macroscopic deformation, which can occur during the fibre or film extrusion process and in subsequent tensile drawing or die drawing processes. The development of molecular orientation is a major factor in the development of physical properties such as tensile modulus and tensile strength and hence relates to the deformation of the molecular network.

From the viewpoint of practical polymer processing, it is well known that fibre and film properties can be empirically correlated with draw ratio, but it is important to recognize that the true correlation is with the total network draw ratio which occurs from the point where the first network forms, that is it is important to relate the structure and properties of the drawn material to the network draw ratio.

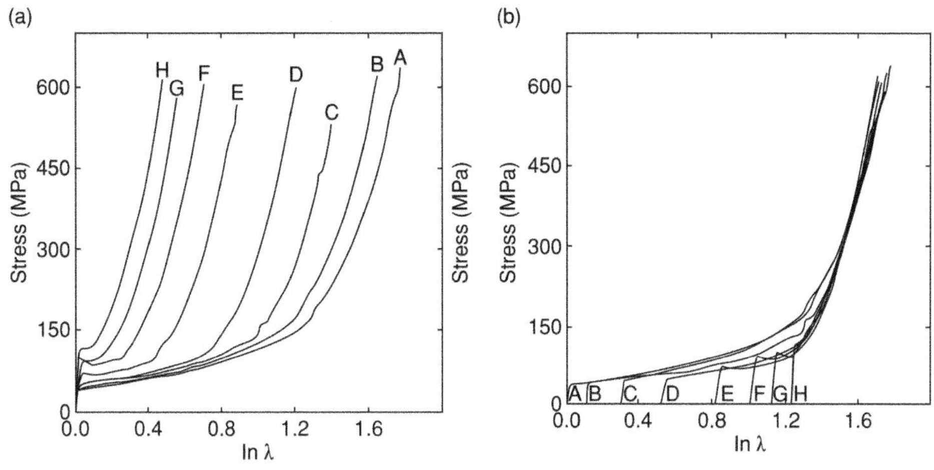

Figure 12.43 Curve-matching procedure to obtain the network draw ratios for polyester fibres. Curves in plot (a) are shifted horizontally to coincide in the plot (b). Long and Ward (1991) / John Wiley & Sons.

Long and Ward (1991) showed how the concept of a true stress–true strain curve could be combined with the concept of a molecular network to provide a better understanding of the relationship of the properties of oriented polymers to the processing route. It is assumed that these properties will relate to the network draw ratio for multi-stage processing even if there are dramatic changes in morphology. The additional ingredient is to determine the network draw ratio at each stage by superimposing true stress–true strain curves. This procedure of curve matching was first proposed by Brody (1983) and has been used by fibre technologists (Bessey and Jaffe 2000). Long and Ward studied the conventional two-stage process for producing oriented polyester fibres where melt spinning is followed by stretching above the glass transition temperature (not by cold-drawing as described in Section 12.1). The network draw ratio in the melt spinning stage was determined by curve matching (Figure 12.43) and the total network draw ratio λ_{net} calculated as $\lambda_{net} = \lambda_s \lambda_{ha}$ where λ_{ha} is the draw ratio imposed by stretching the spun yarn in the solid phase at a temperature above T_g. As shown in Figures 12.44 and 12.45 (Long and Ward 1991), the birefringence and modulus of the oriented fibres relate very well to the network draw ratio.

The assumption that plastic deformation of polymers involves a network draw ratio has been developed further to predict the properties of the oriented polymer. It was recognized that for fibres and films the magnitude of the birefringence could be related to the draw ratio. For cold-drawing, as discussed in Section 8.6.3, the molecular orientation follows the pseudo-affine deformation scheme, and it is possible to calculate the birefringence quantitatively. Assuming an aggregate model the anisotropic mechanical behaviour can also be predicted. For drawing above T_g, which is more usual in commercial processes, it can be assumed that this is akin to stretching a rubber network. It has been customary to follow the Kuhn and Grün (1942) model where the actual network is replaced by an equivalent network of identical chains each containing freely jointed links. This is called

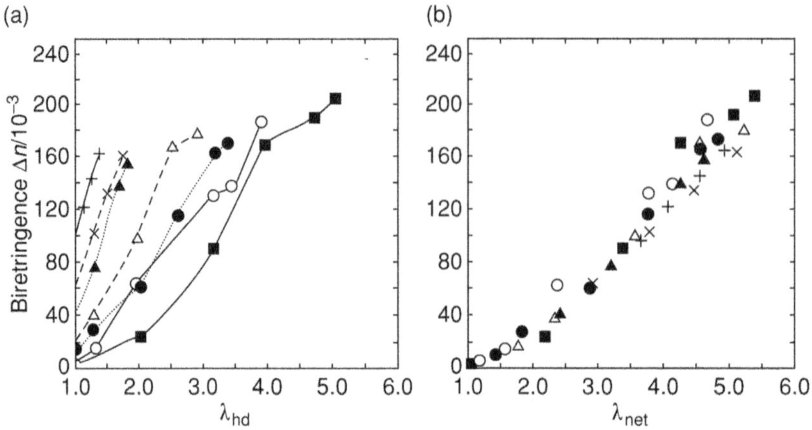

Figure 12.44 Birefringence versus total (plot a, λ_{hd}) and network (plot b, λ_{net}) draw ratio for polyester fibres. Long and Ward (1991) / John Wiley & Sons.

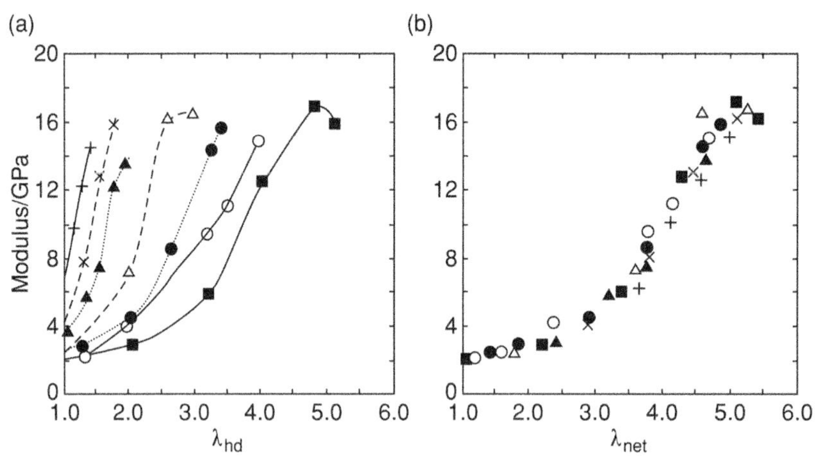

Figure 12.45 Modulus versus total (plot a, λ_{hd}) and network (plot b, λ_{net}) draw ratio for polyester fibres. Long and Ward (1991) / John Wiley & Sons.

the affine deformation scheme, because the network junction points undergo identical displacements to points marked on the macroscopic body (see Section 8.6.3). Recent research has refined this semi-phenomenological approach by more appropriate molecular modelling procedures involving rotational isomeric state calculations (Cail *et al.* 2007).

Recent research has also addressed two issues. First, to what extent does the macroscopic measured draw ratio reflect the network draw ratio, and whether slippage occurs at a molecular level. Although curve matching attempts to resolve this problem it has also been instructive to invite more sophisticated molecular studies, for example by scanning near-field optical microscopy (Ube *et al.* 2007).

Secondly, there is the question of any relaxation of the molecular network, which can occur, during any further heating of the oriented polymer. Early work by Pinnock and Ward (1966) showed excellent correlation between shrinkage force and birefringence for spun PET fibres, with the rate of recovery to isotropy correlating well with time temperature equivalence according to the WLF equation. Work by Hine *et al.* (2007) suggests that in oriented polystyrene it is possible to identify relaxation processes relating to the different stress relaxation modes described for entangled networks (see Section 7.6). The concept of a true stress/true strain/strain rate surface can also be used to predict the creep to failure of oriented polyethylene, as discussed in Section 13.6.3.

12.9.4 Strain Hardening and Strain Rate Sensitivity

The discussions of the true stress–true strain curve and the network draw ratio are clearly consistent with the concept of a molecular network, which provides a physical understanding of what is known to engineers as strain hardening and has also been shown to be relevant to slow crack propagation (see Section 13.6.3).

However, it is also relevant to consider that plastic flow in polymers is determined by a thermally activated process following Eyring (see Section 12.5.1).

Therefore, the flow stress can be represented by a viscosity stress acting in parallel with a network stress, akin to the Voigt or Kelvin model of Section 5.2.5. The effect of strain (total network deformation) and strain rate on the flow stress σ can thus be represented mathematically by

$$\mathrm{d}\sigma = \left(\frac{\partial \sigma}{\partial e}\right)_{\dot{e}} \mathrm{d}e + \left(\frac{\partial \sigma}{\partial \dot{e}}\right)_{e} \mathrm{d}\dot{e}, \tag{12.67}$$

where $\left(\frac{\partial \sigma}{\partial e}\right)_{\dot{e}}$ represents the strain-hardening spring E_v and $\left(\frac{\partial \sigma}{\partial \dot{e}}\right)_{e}$ represents the strain rate sensitivity dashpot η_v.

In terms of the strain-hardening modulus, this has been developed by the use of Kuhn and Grün models and Kratky models to relate the development of molecular orientation and mechanical anisotropy (see Section 8.6.3). With regard to the strain rate sensitivity the strain rate-dependent viscosity has been developed by studies of creep and yield behaviour (see Section 12.5.1).

12.9.5 Process Flow Stress Paths

The concept of a true stress–true strain curve has been extended to include the effect of strain rate to provide a baseline for quantitative modelling of polymer processing, including tensile drawing, hydrostatic extrusion and die drawing. It is necessary to take into account the effect of the rate of deformation that was explicitly neglected in the analyses described above. The key assumption is that the current flow stress depends only on the total plastic strain (the strain hardening relating to the deformation of the molecular network) and the current strain rate (the relevant strain rate sensitivity). As before, properties relate to the network draw ratio, sometimes called the effective draw ratio. It is therefore possible to

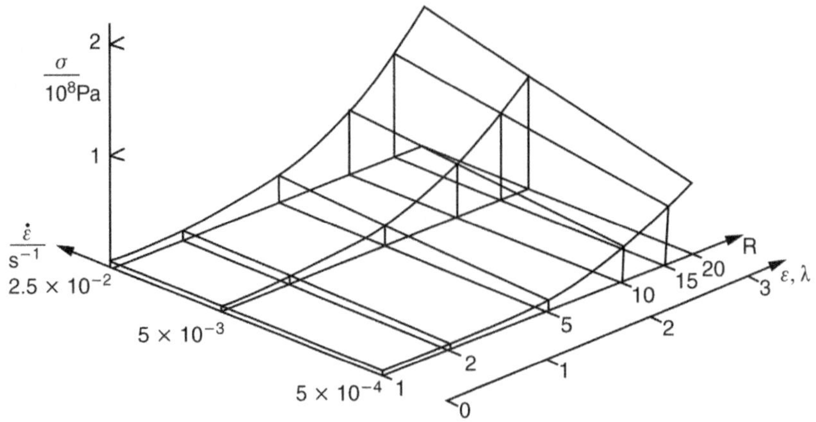

Figure 12.46 True stress–true strain–strain rate surface for high-density polyethylene stretched at 100°C. Adapted Hope and Ward, 1981.

consider that any engineering process involves taking the polymer on a chosen route across a flow stress/strain/strain rate surface. Figure 12.46 shows Ward and Coates' illustration of such a surface for a polyethylene polymer at 100°C, from which the major increases in flow stress with strain and strain rate can be seen (Ward and Coates 2000). Figure 12.47 shows their image of two typical flow process paths for tensile drawing and hydrostatic extrusion. It can be seen that in tensile drawing where a neck is formed (see Section 12.1.2) there is a peak in flow stress at the high strain rates in the neck, whereas in hydrostatic extrusion both strain and strain rate increase monotonically to a maximum value at the exit of the conical die.

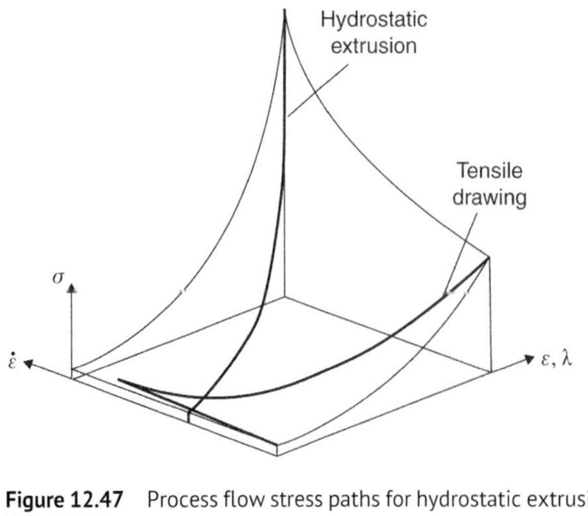

Figure 12.47 Process flow stress paths for hydrostatic extrusion and die drawing. Adapted Hope and Ward, 1981.

These ideas have been developed extensively by Ward and co-workers to provide quantitative analyses of the mechanics of hydrostatic extrusion and die drawing for a range of polymers (Coates and Ward 1980; Coates *et al.* 1980; Hope and Ward 1981; Mohanraj *et al.* 2007).

12.9.6 Neck Profiles

Coates and Ward showed that neck profiles in tensile drawing which are determined by the strain rate field in the neck can be related to the strain hardening and strain rate sensitivity by Equation (12.67). This equation can be rewritten as

$$\frac{\partial \dot{e}}{\partial \lambda} = \frac{\frac{\mathrm{d}\sigma}{\mathrm{d}\lambda} - \left(\frac{\partial \sigma}{\partial \lambda}\right)_{\dot{e}}}{\left(\frac{\partial \sigma}{\partial \dot{e}}\right)_{\lambda}},$$

where we have chosen to represent strain by the draw ratio λ to emphasize the relationship with the tensile drawing process.

A sharp neck implies a high value of $\left(\frac{\partial \dot{e}}{\partial \lambda}\right)$, and $\left(\frac{\partial \dot{e}}{\partial \lambda}\right)$ represents a specific path across the true stress, strain (λ) and strain rate surface.

There will be a very large positive change in the strain rate in an element of material crossing the process surface if both the strain hardening is small compared with $\frac{\mathrm{d}\sigma}{\mathrm{d}\lambda}$ (which is essentially the Considère line) and the strain rate sensitivity $\left(\frac{\partial \sigma}{\partial \dot{e}}\right)_{\lambda}$ is low. This will give a sharp neck.

These ideas were developed further in several publications (Coates *et al.* 1980; Hope and Ward 1981; Mohanraj *et al.* 2007).

12.9.7 Crystalline Polymers

The plastic deformation of crystalline polymers, in particular polyethylene, has been studied intensively from the viewpoint of changes in morphology. This is discussed in Chapter 9. It is now evident that very drastic reorganization occurs at the morphological level, with the structure changing from a spherulitic to a fibrillar type as the degree of plastic deformation increases. The molecular reorientation processes are very far from being affine or pseudo-affine and can also involve mechanical twinning in the crystallites. It is surprising that some of the continuum ideas for mechanical anisotropy are nevertheless still relevant, although they must be appropriately modified.

In a few highly crystalline polymers, notably high-density polyethylene, extremely large draw ratios, ~30 or more, have been achieved by optimizing the chemical composition of the polymers and the drawing conditions (Capaccio and Ward 1973, 1974). These high draw ratios lead to oriented polymers with very high Young's moduli as discussed in Chapter 9, Section 9.4. In spite of the much more complex deformation processes in a crystalline polymer, it has been concluded (Capaccio *et al.* 1973) that the molecular topology and the deformation of a molecular network are still the overriding considerations in determining the strain-hardening behaviour and the ultimate draw ratio achievable. For high-molecular-weight, high-density polyethylene, the key network junction points are physical entanglements, as in amorphous polymers. For low-molecular-weight, high-density polyethylene, both physical entanglements and crystallites where more than one molecular chain is incorporated, can provide the network junction points. Junction points associated with the crystallites will be of a temporary nature. Very high draw ratios involve the breakdown of the crystalline structure and the unfolding of molecules, so that the simple ideas of a molecular network suggested for amorphous polymers have to be extended and modified.

12.10 Shear Bands

As we have seen above in Section 12.1, sometimes the tensile stretching of a polymer results in strain localization. So far it has been assumed that the localization takes the form of a neck, but an alternative geometric form is possible – the shear band. In uniaxial straining, localization of strain occurs in a narrow band at an oblique angle to the straining direction, as illustrated in Figure 12.48. Shear bands have been observed in many ductile materials. Nadai (1963), for instance, gives an account of their occurrence in mild steel. Bowden (1973) has described the phenomenon for polymers.

Given that a shear band has formed in isotropic material under uniaxial conditions, a simple analysis is available to predict the angle at which it occurs with respect to the direction of straining. It is assumcd (Nadai (1963) that the material in the shear band is plastic and that the material outside it remains elastic. For an incompressible and isotropic material, an axial strain e is accompanied by a transverse strain $-e/2$; this is consistent with the Levy–Mises flow rule. At some oblique angle to the axis, the normal strain must be zero. It is argued that the shear band forms at this angle, since this corresponds to there being no additional constraint on the band material and therefore the least stress. The strains within the band are generally much larger than those in the surrounding elastic material, but at this angle the zero plastic strain in the band is essentially compatible with the small elastic strain in the adjacent unyielded material. For the angle θ shown, the normal strain e_n is given by

$$e_n(\theta) = e\cos^2\theta - (e/2)\sin^2 \theta SD$$

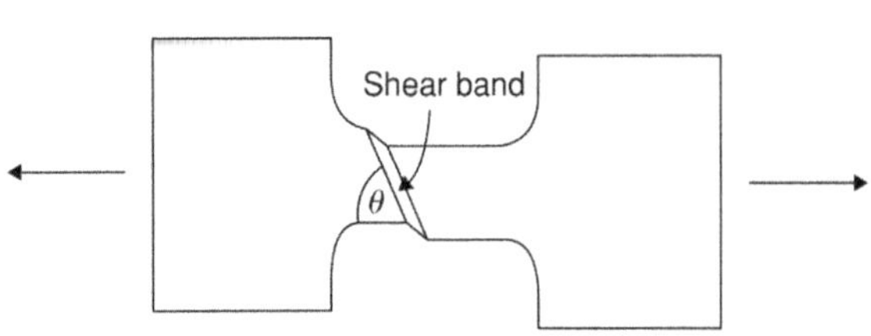

Figure 12.48 Schematic diagram of shear band in tensile specimen.

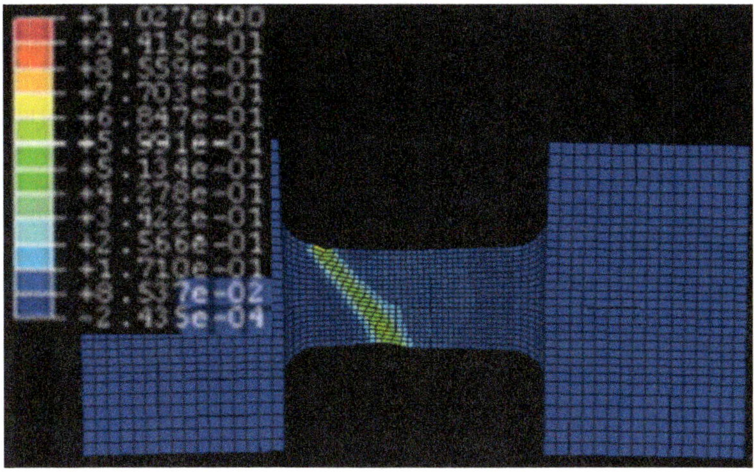

Figure 12.49 Finite element model of tensile specimen of polycarbonate showing shear banding. Analysis is using ABAQUS 6.8, with contours of maximum principal true strain. (Reproduced with permission from Sweeney, J., Caton-Rose, P., Spares, R. et al. (2007) A unified model of necking and shearbanding in amorphous and semicrystalline polymers. J. Appl. Polym. Sci., **106**, 1095. Copyright (2007) John Wiley & Sons, Inc.)

so that when $e_n = 0$, $\tan\theta = \sqrt{2}$, giving an angle of 54.7°. This was shown to be consistent with experiments on steel (Nadai (1963). Early work on polyvinyl chloride by Bauwens (1967) confirmed the result.

Given an appropriate constitutive equation that includes a flow rule, numerical methods should provide a means of modelling shear banding. The FEM has been used in this way by Lu and Ravi-Chandar (1999), who devised a simplified constitutive model with no strain rate dependence. A more realistic constitutive model, combining the plasticity model of Argon and an entropic network, was used by Wu and van der Giessen (1994, 1996) in finite element modelling of shear bands in polycarbonate and polystyrene. Sweeney *et al.* (2007) found that a Maxwell-type series model, incorporating a Gaussian elastic element and an Eyring process operating with a Levy–Mises flow rule, was sufficient to capture shear banding behaviour in polycarbonate. Results for a tensile specimen are shown in Figure 12.49. The angle of the band is consistent with the simple analysis outlined above.

When discussing the issue of instabilities, the question arises as to why in some instances they take the form of a neck and in other cases that of a band. The solution appears to lie in the nature of the deformation mechanism that is dominant. We generally have a combination of elastic and plastic mechanisms, with the latter governed by a flow rule that will tend to favour shear banding. On this basis our expectation would be that, if most of the total strain was due to plasticity, a shear band would be observed, whereas otherwise the instability would take the form of a neck. This question was explored by Sweeney *et al.* 2007 by changing the relative strengths of the Eyring and Gaussian mechanisms. With material parameters appropriate for polycarbonate and the Eyring process producing most of the strain, a shear band was predicted as shown. With parameters appropriate for polypropylene and most of the deformation due to the elastic mechanism, a neck was predicted. This contrasting behaviour was in line with experimental observation.

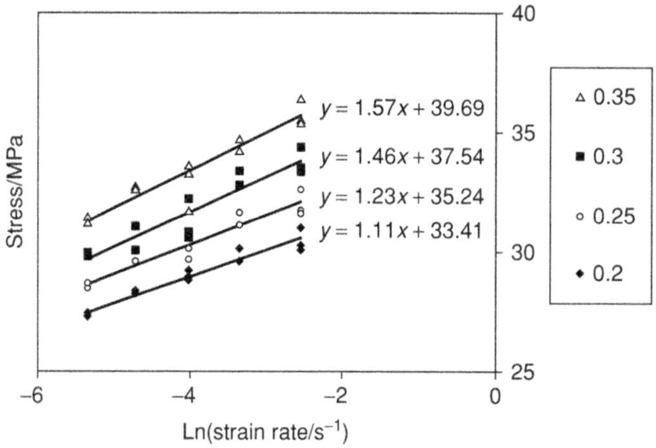

Figure 12.51 Plots of tensile stress at a range of levels of true strain against log strain rate. Sweeney et al. (2011) / John Wiley & Sons.

Senden *et al.* 2010). In this case, it was concluded that strain hardening could be viewed as originating entirely from a strain-dependent Eyring process.

12.11.1 The Bauschinger Effect

The Bauschinger effect is the term for asymmetry in the yield response of a material between tension in compression. For isotropic polymers the effect is small (the yield stress in compression being slightly higher than that in tension) and can be seen as a consequence of the differing levels of hydrostatic pressure. It can thus be adequately modelled by the inclusion of the pressure activation volume in the Eyring process. For oriented polymer, however, the asymmetry is much greater (see the early results for oriented polypropylene of Duckett *et al.* (1972), where a draw ratio of 5 increased the yield stress by a factor of 8).

As shown by Senden *et al.* (2010), conventional material models that incorporate entropic strain hardening give a qualitatively incorrect prediction of the Bauschinger effect. This can be illustrated by investigating the effects of cyclic loading. For a two-arm model such as that of Haward and Thackray, when loaded in tension and then unloaded, during unloading the stress in the arm containing the Eyring process becomes negative while the total stress is still positive. The Eyring process then yields, so that the total stress shows yield behaviour while still tensile. This is shown in Figure 12.52(a), and is clearly unrealistic when compared with the observed pattern of 12.52(b).

The discussion above has introduced an alternative model of strain hardening, the strain-dependent Eyring process. With a model incorporating a strain-hardening arm of this nature, a reversal of the direction of straining would rapidly change the sign of the stress in this arm. This is in contrast to the behaviour of the entropic mechanism where large strains need to be removed to reverse the stress. The revised ideas of strain hardening thus provide a probable means of gaining a realistic prediction of the Bauschinger effect. The

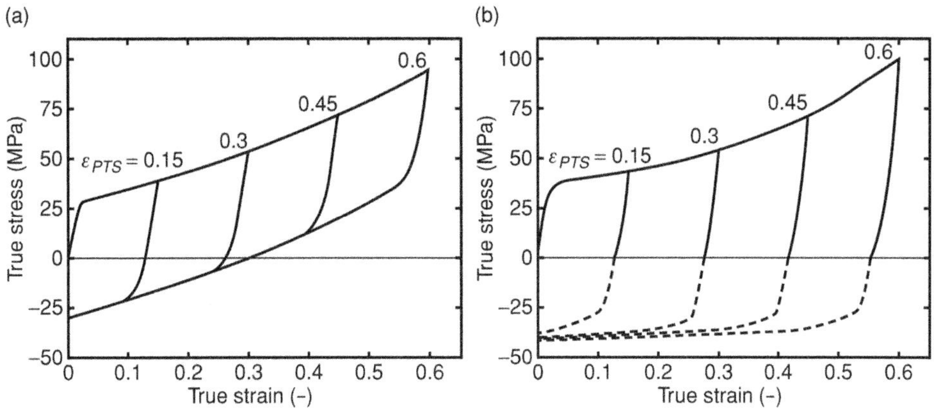

Figure 12.52 (a) Predictions of the response of polycarbonate to cyclic loading, based on a constitutive model incorporating entropic strain hardening. (b) Realistic unloading behaviour. Senden et al. (2010) / John Wiley & Sons.

work in this area is, however, still incomplete. According to Senden *et al.* (2010), for polycarbonate the strain hardening would be best modelled by a combination of entropic and strain-dependent Eyring processes.

References

Ainbinder, S.B., Laka, M.G. and Maiors, I.Y. (1965) Effect of hydrostatic pressure on mechanical properties of plastics. *Mech. Compos. Mater.*, **1**, 50.

Allison, S.W. and Ward, I.M. (1967) The cold drawing of polyethylene terephthalate. *Brit. J. Appl. Phys.*, **18**, 1151.

Allison, S.W., Pinnock, P.R. and Ward, I.M. (1966) The cold drawing of polyethylene terephthalate. *Polymer*, **7**, 66.

Amoedo, J. and Lee, D. (1992) Modeling the uniaxial rate and temperature dependent behavior of amorphous and semi-crystalline polymers. *Polym. Eng. Sci.*, **32**, 1055–1065.

Argon, A.S. (1973) A theory for the low-temperature plastic deformation of glassy polymers. *Philos. Mag.*, **28**, 839.

Arruda, E.M. and Boyce, M.C. (1993) Evolution of plastic anisotropy in amorphous polymers during finite straining. *Int. J. Plast.*, **9**, 697.

Bauwens, J.C. (1967) Déformation plastique des hauts polymères vitreux soumis á un système de contraintes quelconque. *J. Polym. Sci. A2*, (5), 1145.

Bauwens, J.C. (1970) Yield condition and propagation of Lüders' lines in tension-torsion experiments on poly (vinyl Chloride). *J. Polym. Sci.: Part A-2*, **8**, 893–901.

Bauwens-Crowet, C., Bauwens, J.A. and Homès, G. (1969a) Tensile yield-stress behavior of glassy polymers. *J. Polym. Sci. A2*, **7**, 735.

Bauwens-Crowet, C., Bauwens, J.A. and Homès, G. (1969b) Tensile yield-stress behavior of poly(vinyl chloride) and polycarbonate in the glass transition region. *J. Polym. Sci. A2*, **7**, 1745–1754.

Bauwens-Crowet, C., Ots, J.-M. and Bauwens, J.-C. (1974) The strain-rate and temperature dependence of yield of polycarbonate in tension, tensile creep and impact tests. *J. Mater. Sci. Lett.*, **9**, 1197–2101.

Bazhenov, S.L. and Kechek'yan, A.S. (2013) Heating of polymers during neck propagation. *Polym. Sci., Ser. A*, **55**, 404–414.

Bessey, W. and Jaffe, M. (2000), in *Solid Phase Processing of Polymers*, (eds I.M. Ward, P.D. Coates and M.M. Dumoulin), Hanser, Munich. Chap. 4.

Biglione, G., Baer, E. and Radcliffe, S.V. (1969), in *Fracture 1969*, (ed. P.L. Pratt), Chapman and Hall, London, p. 520.

Billon, N. (2012) New constitutive modeling for time-dependent mechanical behavior of polymers close to glass transition: fundamentals and experimental validation. *J. Appl. Polym. Sci.*, **125**, 4390–4401.

Bouvard, J.L., Francis, D.K., Tschopp, M.A., Marin, E.B., Bammann, D.J. and Horstemeyer, M.F. (2013) An internal state variable material model for predicting the time, thermomechanical, and stress state dependence of amorphous glassy polymers under large deformation. *Int. J. Plast.*, **42**, 168–193.

Bowden, P.B. (1973), in *The Physics of Glassy Polymers*, (ed. R.N. Haward), Applied Science, London.

Bowden, P.B. and Jukes, J.A. (1968) The plastic yield behaviour of polymethyl-methacrylate. *J. Mater. Sci.*, **3**, 183.

Bowden, P.B. and Jukes, J.A. (1972) The plastic flow of isotropic polymers. *J. Mater. Sci.*, **7**, 52–63.

Bowden, P.B. and Raha, S. (1974) Amolecular model for yield and flow in amorphous glassy polymers making use of a dislocation analogue. *Philos. Mag.*, **29**, 149.

Boyce, M.C., Parks, D.M. and Argon, A.S. (1988) Large inelastic deformation of glassy polymers. Part I: rate dependent constitutive model. *Mech. Mater.*, **7**, 15.

Boyce, M.C., Parks, D.M. and Argon, A.S. (1989) Plastic flow in oriented glassy polymers. *Int. J. Plast.*, **5**, 593.

Boyce, M.C., Socrate, M. and Llana, P.G. (2000) Constitutive model for the finite deformation stress–strain behavior of poly(ethylene terephthalate) above the glass transition. *Polymer*, **41**, 2183.

Brauer, P. and Müller, F.H. (1954) Über die Temperaturüberhöhung in der Fließzone während der Kaltverstreckung Anwendung von Leuchtstoffen zur Temperaturmes-sung. *Kolloidzeitschrift*, **135**, 65.

Brereton, M.S., Croll, S.G., Duckett, R.A. *et al.* (1974) Nonlinear viscoelastic behavior of polymers – an implicit equation approach. *J. Mech. Phys. Solids*, **22**, 97.

Bridgeman, P.W. (1964) *Studies in Large Plastic Flow and Fracture*, Harvard University Press, Cambridge, New York.

Briscoe, B.J. and Tabor, D. (1978), in *Polymer Surfaces*, (eds D.T. Clark and W.J. Feast), John Wiley & Sons, New York. Chap. 1

Brody, H. (1983) The extensibility of polyethylene terephthalate fibers spun at high wind-up speeds. *J. Macromol. Sci. Phys.*, **B22**, 19.

Brooks, N.W.J. and Mukhtar, M. (2000) Temperature and stem length dependence of the yield stress of polyethylene. *Polymer*, **41**, 1475.

Brooks, N.W., Duckett, R.A. and Ward, I.M. (1992) Investigation into double yield points in polyethylene. *Polymer*, **33**, 1872.

Brooks, N.W., Unwin, A.P., Duckett, R.A. *et al.* (1997a) *Deformation, yield and fracture of polymers*. Conference Papers. Cambridge, UK, pp. 53–56.

Brooks, N.W., Unwin, A.P., Duckett, R.A. *et al.* (1997b) Temperature and strain rate dependence of yield strain and deformation behavior in polyethylene. *J. Polym. Sci. B: Phys. Edn.*, **35**, 545.

Brooks, N.W.J., Duckett, R.A. and Ward, I.M. (1998) Temperature and strain-rate dependence of yield stress of polyethylene. *J. Polym. Sci. B Polym. Phys.*, **36**, 2177.

Brooks, N.W., Ghazali, M., Duckett, R.A. *et al.* (1999a) Effects of morphology on the yield stress of polyethylene. *Polymer*, **40**, 821.

Brooks, N.W.J., Duckett, R.A. and Ward, I.M. (1999b) Effects of crystallinity and stress state on the yield strain of polyethylene. *Polymer*, **40**, 7367.

Brown, N. (1971a) Fundamental tensile stress-strain relationship for yielding of linear polymers. *Bull. Am. Phys. Soc.*, **16**, 428.

Brown, N. (1971b) The relationship between yield point and modulus for glassy polymers. *Mater. Sci. Eng.*, **8**, 69.

Brown, N. and Ward, I.M. (1968) Load drop at upper yield point of a polymer. *J. Polym. Sci. A2*, **6**, 607.

Brown, N., Duckett, R.A. and Ward, I.M. (1968) The yield behaviour of oriented polyethylene terephthalate. *Philos. Mag.*, **18**, 483.

Buckley, C.P., Jones, D.C. and Jones, D.P. (1996) Hot drawing of poly(ethylene terephthalate) under biaxial stress: application of a three-dimensional glass-rubber constitutive model. *Polymer*, **37**, 2403–2414.

Cail, J.J., Stepto, R.F.T. and Ward, I.M. (2007) Experimental studies and molecular modelling of the stress–optical and stress–strain behaviour of poly(ethylene terephthalate). Part III: measurement and quantitative modelling of birefringence–strain, stress–strain and stress–optical properties. *Polymer*, **48**, 1379.

Capaccio, G. and Ward, I.M. (1973) Properties of ultra-high modulus linear polyethylenes. *Nature Phys. Sci.*, **243**, 43.

Capaccio, G. and Ward, I.M. (1974) Preparation of ultra-high modulus linear polyethylenes; effect of molecular weight and molecular weight distribution on drawing behaviour and mechanical properties. *Polymer*, **15**, 233.

Christiansen, A.W., Baer, E. and Radcliffe, S.V. (1971) The mechanical behaviour of polymers under high pressure. *Philos. Mag.*, **24**, 451.

Coates, P.D. and Ward, I.M. (1980) Neck profiles in drawn linear polyethylene. *J. Mater. Sci.*, **15**, 2597.

Coates, P.D., Gibson, A.G. and Ward, I.M. (1980) An analysis of the mechanics of solid phase extrusion of polymers. *J. Mater. Sci.*, **15**, 359.

Costa, C.C.M., Sencadas, V., Pelicano, I., Martins, F., Rocha, J.G. and Lanceros-Mendez, S. (2008) Microscopic origin of the high-strain mechanical response of poled and non-poled poly(vinylidene fluoride) in the b-phase. *J. Non-Cryst. Solids*, **354**, 3871–3876.

Cottrell, A.H. (1953) *Dislocations and Plastic Flow in Crystals*, Clarendon Press, Oxford.

Coulomb, C.A. (1776) Essai sur une application des règles des maximis et minimis à quelques problèmes de statique relatifs à l'architecture. *Mem. Math. Phys.*, **7**, 343.

Crist, B., Fisher, C.J. and Howard, P.R. (1989) Mechanical properties of model polyethylenes: tensile elastic modulus and yield stress. *Macromolecules*, **22**, 1709.

Darras, O. and Seguala, R. (1993) Tensile yield of polyethylene in relation to crystal thickness. *J. Polym. Sci. B Polym. Phys.*, **31**, 759.

De Souza Neto, E.A., Peric, D. and Owen, D.R.J. (2008) *Computational Methods for Plasticity*, John Wiley & Sons, Chichester.

Duckett, R.A., Rabinowitz, S. and Ward, I.M. (1970) The strain-rate, temperature and pressure dependence of yield of isotropic poly(methylmethacrylate) and poly(ethylene terephthalate). *J. Mater. Sci.*, **5**, 909.

Duckett, R.A., Ward, I.M. and Zihlif, A.M. (1972) Direct measurements of the reverse stress asymmetry in the yielding of anisotropic polypropylene. *J. Mater. Sci.*, **7**(Letters), 480–482.

Duckett, R.A., Goswami, B.C., Smith, L.S.A. *et al.* (1978) Yielding and crazing behavior of polycarbonate in torsion under superposed hydrostatic-pressure. *Brit. Polym. J.*, **10**, 11.

Edwards, S.F. and Vilgis, T.H. (1986) The effect of entanglements in rubber elasticity. *Polymer*, **27**, 483.

Engels, T.A.P., Sönjens, S.H.M., Smit, T.H. and Govaert, L.E. (2010) Time-dependent failure of amorphous polylactides in static loading conditions. *J. Mater. Sci. Mater. Med.*, **21**, 89–97.

Erartsin, O., van Drongelen, M. and Govaert, L.E. (2021) Identification of plasticity-controlled creep and fatigue failure mechanisms in transversely loaded unidirectional thermoplastic composites. *J. Compos. Mater.*, **55**, 1947–1965.

Escaig, B. (1982) Kinetics and thermodynamics of plastic flow in polymeric glasses, in *Plastic Deformation of Amorphous and Semi-crystalline Materials*, (eds B. Escaig and C. G'Sell), Les Éditions de Physique, Les Ulis, pp. 187–225.

Faye, A., Parameswaran, V. and Basu, S. (2015) Mechanics of dynamic fracture in notched polycarbonate. *J. Mech. Phys. Solids*, **77**, 43–60.

Federico, C.E., Bouvard, J.L., Combeaud, C. and Billon, N. (2020) Modelling strain rate and temperature dependent mechanical response of PMMAs at large deformation from below to above Tg. *Polymer*, **202**, 122710.

Fleischhauer, R., Dal, H., Kaliske, M. and Scheider, K. (2012) A constitutive model for finite deformation of amorphous polymers. *Int. J. Mech. Sci.*, **65**, 48–63.

Fotheringham, D.G. and Cherry, B.W. (1978) The role of recovery forces in the deformation of linear polyethylene. *J. Mater. Sci.*, **13**, 951.

Frank, F.C. (1950) *Symposium on Plastic Deformation of Crystalline Solids*, Carnegie Institute of Technology, Pittsburgh.

Gottstein, G. (2004) *Physical Foundations of Materials Science*, Springer, Berlin.

Guiu, F. and Pratt, P.L. (1964) Stress relaxation and the plastic deformation of solids. *Phys. Status Solidi*, **6**, 111.

Gupta, V.B. and Rana, S.K. (1998) Double yield in tensile deformation of high-density polyethylene fiber. *J. Macromolecular Sci. Phys.*, **B37**, 783.

Hachour, K., Zaïri, F., Naït-Abdelaziz, M., Gloaguen, J.M., Aberkane, M. and Lefebvre, J.M. (2014) Experiments and modeling of high-crystalline polyethylene yielding under different stress states. *Int. J. Plast.*, **54**, 1–18.

Hadriche, I., Ghorbel, E., Masmoudi, N. and El Halouani, F. (2010) Influence of strain rate on the yielding behavior and on the self heating of thermoplastic polymers loaded under tension. *Key Eng. Mater.*, **446**, 63–72.

Halsey, G., White, H.J. and Eyring, H. (1945) Mechanical properties of textiles. *I. Text. Res. J.*, **15**, 295.

Harris, J.S., Ward, I.M. and Parry, J.S.C. (1971) Shear strength of polymers under hydrostatic pressure: surface coatings prevent premature fracture. *J. Mater. Sci.*, **6**, 110.

Haward, R.N. (1993) Strain hardening of thermoplastics. *Macromolecules*, **26**, 5860.

Haward, R.N. and Thackray, G. (1968) The use of a mathematical model to describe isothermal stress-strain curves in glassy thermoplastics. *Proc. Roy. Soc. A*, **302**, 453.

Hill, R. (1985) *The Mathematical Theory of Plasticity*, Oxford University Press, New York.

Hine, P.J., Duckett, R.A. and Read, D.J. (2007) Influence of molecular orientation and melt relaxation processes on glassy stress–strain behavior in polystyrene. *Macromolecules*, **40**, 2782.

Holopainen, S. (2014) Influence of damage on inhomogeneous deformation behavior of amorphous glassy polymers. Modeling and algorithmic implementation in a finite element setting. *Eng. Fract. Mech.*, **117**, 28–50.

Hookway, D.C. (1958) The cold-drawing of nylon 6.6. *J. Text. Inst.*, **49**, 292.

Hope, P.S. and Ward, I.M. (1981) An activated rate theory approach to the hydrostatic extrusion of polymers. *J. Mater. Sci.*, **16**, 1511.

Hope, P.S., Ward, I.M. and Gibson, A.G. (1980) The hydrostatic extrusion of poly-methylmethacrylate. *J. Mater. Sci.*, **15**, 2207.

Hoy, R.S. and Robbins, M.O. (2008) Strain hardening of polymer glasses: entanglements, energetics, and plasticity. *Phys. Rev. E*, **77**, 031801.

Jarrah, H.R., Zolfagharian, A., Hedayati, R., Serjouei, A. and Bodaghi, M. (2021) Nonlinear finite element modelling of thermo-visco-plastic styrene and polyurethane shape memory polymer foams. *Actuators*, **10**, 46.

Jiang, H., Zhang, J.W., Yang, Z.R., Jiang, C.K. and Kang, G.Z. (2017) Modeling of competition between shear yielding and crazing in amorphous polymers' scratch. *Int. J. Solids Struct.*, **124**, 215–228.

Johnsen, J., Clausen, A.H., Gryttenb, F., Benallal, A. and Hopperstad, O.S. (2019) A thermo-elasto-viscoplastic constitutive model for polymers. *J. Mech. Phys. Solids*, **124**, 681–701.

Kelly, A. and Macmillan, N.H. (1987) *Strong Solids*, 3rd edn, Clarendon Press, Oxford.

Kershah, T., Anderson, P.D. and van Breemen, L.C.A. (2020) Uniaxial and biaxial response of anisotropic polypropylene. *Macromol. Theory Simul.*, **29**, 2000018.

Kuhn, W. and Grun, F. (1942) Relationships between elastic constants and stretching double refraction of highly elastic substances. *Kolloid. Z*, **101**, 248.

Kweon, S. and Benzerga, A.A. (2013) On the localization of plastic flow in glassy polymers. *Eur. J. Mech. A. Solids*, **39**, 251–267.

Labuz, J.F. and Zang, A. (2012) Mohr-Coulomb failure criterion. *Rock Mech. Rock. Eng.*, **45**, 975–979.

Lazurkin, Y.S. (1958) Cold-drawing of glass-like and crystalline polymers. *J. Polym. Sci.*, **30**, 595.

Levy, M. (1870) Mémoire sur les équations générales des mouvements intérieurs des corps ductiles au delà des limites en élasticité pourrait les ramener à leur premier état. *C. R. Acad. Sci. (Paris)*, **70**, 1323.

Li, H.X. and Buckley, C.P. (2009) Evolution of strain localization in glassy polymers: a numerical study. *Int. J. Solids Struct.*, **46**, 1607–1623.

Li, J.C.M. and Gilman, J.J. (1970) Disclination loops in polymers. *J. Appl. Phys.*, **41**, 4248.

Liu, Y. and Truss, R.W. (1994) A study of tensile yielding of Isotactic polypropylene. *J. Polym. Sci. B Polym. Phys.*, **32**, 2037–2047.

Long, S.D. and Ward, I.M. (1991) Shrinkage force studies of oriented polyethylene terephthalate. *J. Appl. Polym. Sci.*, **42**, 1921.

Lu, J. and Ravi-Chandar, K. (1999) Inelastic deformation and localization in polycarbonate under tension. *Int. J. Solids Struct.*, **36**, 391.

Maers, D.R., Pae, K.D. and Sauer, J.A. (1969) Effects of hydrostatic pressure on the mechanical behavior of polyethylene and polypropylene. *J. Appl. Phys.*, **40**, 4229.

Marshall, I. and Thompson, A.B. (1954) The cold drawing of high polymers. *Proc. Roy. Soc.*, **A221**, 541.

Matsushigi, K., Radcliffe, S.V. and Baer, E. (1975) The mechanical behaviour of polystyrene under pressure. *J. Mater. Sci.*, **10**, 833.

Maurel-Pantel, A., Baquet, E., Bikard, J., Bouvard, J.L. and Billin, N. (2015) A thermo-mechanical large deformation constitutive model for polymers based on material network description: application to a semi-crystalline polyamide 66. *Int. J. Plast.*, **67**, 102–126.

van Melick, H.G.H., Govaert, L.E. and Meijer, H.E.H. (2003) On the origin of strain hardening in glassy polymers. *Polymer*, **44**, 2493.

Mindel, M.J. and Brown, N. (1973) Creep and recovery of polycarbonate. *J. Mater. Sci.*, **8**, 863.

von Mises, R. (1913) Mechanik der festen Korper im plastisch deformablen Zustand. *Gottinger Nach. Math. -Phys. Kl.*, **1913**, 582.

Mohanraj, J., Bonner, M.J., Barton, D.C. *et al.* (2007) Analysis and design of profiled dies for the polymer wire die-drawing process. *Proc. Inst. Mech. Eng., Part E: J. Process Mech. Eng.*, **221**, 47.

Müller, F.H. (1949) Zum Problem der Kaltverstreckung hochpolymerer Substanzen. *Kolloidzeitschrift*, **114**, 59; **115**, 118 (1949); **126**, 65 (1952)

Nadai, A. (1963) *Theory of Flow and Fracture of Solids*, McGraw-Hill, New York.

Naz, S., Sweeney, J. and Coates, P.D. (2010) Analysis of the essential work of fracture method as applied to UHMWPE. *J. Mater. Sci.*, **45**, 448–459.

Nikolov, S. and Raabe, D. (2006) Yielding of polyethylene through propagation of chain twist defects: temperature, stem length and strain-rate dependence. *Polymer*, **47**, 1696.

O'Connor, C.P.J., Martin, P.J., Sweeney, J., Menary, G., Caton-Rose, P. and Spencer, P.E. (2013) Simulation of the plug-assisted thermoforming of polypropylene using a large strain thermally coupled constitutive model. *J. Mater. Process. Technol.*, **213**, 1588–1600.

O'Kane, W.J. and Young, R.J. (1995) The role of dislocations in the yield of polypropylene. *J. Mater. Sci. Lett.*, **14**, 433.

Olley, P. and Sweeney, J. (2011) A multiprocess eyring model for large strain plastic deformation. *J. Appl. Polym. Sci.*, **119**, 2246–2260.

Orowan, B. (1949) Fracture and strength of solids. *Rep. Prog. Phys.*, **12**, 185.

Pae, K.D., Maers, D.R. and Sauer, J.A. (1968) Stress–strain behavior of polypropylene under high pressure. *J. Polym. Sci. Part B: Polymer Letters*, **6**, 773.

Petermann, J. and Gleiter, H. (1972) Direct observation of dislocations in polyethylene crystals. *Philos. Mag.*, **25**, 813.

Pini, T., van Drongelen, M., Remmers, J.J.C., Geers, M.G.D. and Govaert, L.E. (2021) Deformation and failure kinetics of polyvinylidene fluoride: influence of crystallinity. *J. Polym. Sci.*, **59**, 1209–1220.

Pinnock, P.R. and Ward, I.M. (1966) Stress-optical properties of amorphous polyethylene terephthalate fibres. *Trans. Faraday Soc.*, **62**, 1308.

Povolo, F. and Hermida, E.B. (1995) Phenomelogical description of strain rate and temperature-dependent yield stress of PMMA. *J. Appl. Polym. Sci.*, **58**, 55–68.

Povolo, F., Schwarz, G. and Hermida, E.B. (1996) Temperature and strain rate dependence of the tensile yield stress of PVC. *J. Appl. Polym. Sci.*, **61**, 109–117.

Rabinowitz, S., Ward, I.M. and Parry, J.S.C. (1970) The effect of hydrostatic pressure on the shear yield behaviour of polymers. *J. Mater. Sci.*, **5**, 29.

Raghava, R.S., Caddell, R.M. and Yeh, G.S.Y. (1973) The macroscopic yield behaviour of polymers. *J. Mater. Sci.*, **8**, 225–232.

Richeton, J., Ahzi, S., Daridon, L. and Rémond, Y. (2005) A formulation of the cooperative model for the yield stress of amorphous polymers for a wide of strain rates and temperatures. *Polymer*, **46**, 6035–6043.

Robertson, R.E. (1963) On the cold-drawing of plastics. *J. Appl. Polym. Sci.*, **7**, 443.

Robertson, R.E. (1966) Theory for the plasticity of glassy polymers. *J. Chem. Phys.*, **44**, 3950.

Roetling, J.A. (1965) Yield stress behaviour of polymethylmethacrylate. *Polymer*, **6**, 311.

Schleicher, F. (1926) Die energiegrenze der elastizität. Z.Agnew.Math. *Mech.*, **6**, 199–216.

Sedighiamiri, A., Govaert, L.E., Kanters, M.J.W. and van Dommelen, J.A.W. (2012) Micromechanics of semicrystalline polymers: yield kinetics and long-term failure. *J. Polym. Sci.: B Polym. Phys. Ed.*, **50**, 1664–1679.

Seguela, R. and Darras, D. (1994) Phenomenological aspects of the double yield of polyethylene and related copolymers under tensile loading. *J. Mater. Sci.*, **29**, 5342.

Senden, D.J.A., van Dommelen, J.A.W. and Govaert, L.E. (2010) Strain hardening and its relation to Bauschinger effects in oriented polymers. *J. Polym. Sci. B Polym. Phys.*, **48**, 1483.

Shadrake, L.G. and Guiu, F. (1976) Dislocations in polyethylene crystals: line energies and deformation modes. *Philos. Mag.*, **34**, 565.

Sherby, O.D. and Dorn, J.B. (1958) Anelastic creep of polymethyl methacrylate. *J. Mech. Phys. Solids*, **6**, 145.

Smith, W.F. (1996) *Principles of Materials Science and Engineering*, 3rd edn, McGraw-Hill, New York.

Söntjens, S.H.M., Engels, T.A.P., Smit, T.H. and Govaert, L.E. (2012) Time-dependent failure of amorphous poly-D,L-lactide: influence of molecular weight. *J. Mech. Behav. Biomed. Mater.*, **13**, 69–77.

Sweeney, J. and Ward, I.M. (1990) A unified model of stress relaxation and creep applied to oriented polyethylene. *J. Mater. Sci.*, **25**, 697–705.

Sweeney, J., Duckett, R.A. and Ward, I.M. (1986) The fracture toughness of a tough polyethylene using tension testing in a high-pressure environment. *J. Mater. Sci. Lett.*, **5**, 1109.

Sweeney, J., Shirataki, H., Unwin, A.P. *et al.* (1999) Application of a necking criterion to PET fibers in tension. *J. Appl. Polym. Sci.*, **74**, 3331–3341.

Sweeney, J., Caton-Rose, P., Spares, R. *et al.* (2007) A unified model of necking and shearbanding in amorphous and semicrystalline polymers. *J. Appl. Polym. Sci.*, **106**, 1095.

Sweeney, J., Spares, R. and Woodhead, M. (2009) A constitutive model for large multiaxial deformations of solid polypropylene at high temperature. *Polym. Eng. Sci.*, **49**, 1902–1908.

Sweeney, J., Naz, S. and Coates, P.D. (2011) Modeling the tensile behavior of ultrahigh-molecular-weight polyethylene with a novel flow rule. *J. Appl. Polym. Sci.*, **121**, 2936.

Sweeney, J., O'Connor, C.P.J., Spencer, P.E., Pua, H., Caton-Rose, P. and Martin, P.J. (2012) A material model for multiaxial stretching and stress relaxation of polypropylene under process conditions. *Mech. Mater.*, **54**, 55–69.

Sweeney, J., Bonner, M. and Ward, I.M. (2014) Modelling of loading, stress relaxation and stress recovery in a shape memory polymer. *J. Mech. Behav. Biomed. Mater.*, **37**, 12–23.

Sweeney, J., Spencer, P.E., Vgenopoulos, D., Babenko, M., Boutenel, F., Caton-Rose, P. and Coates, P.D. (2018) Application of activated barrier hopping theory to viscoplastic modeling of glassy polymers. *Mech. Time-Depend. Mater.*, **22**, 145–165.

Sweeney, J., Spencer, P.E., Nair, K. and Coates, P.D. (2019) Modelling the mechanical and strain recovery behaviour of partially crystalline PLA. *Polymers*, **11**, 1342.

Tamrakar, S., Ganesh, R., Sockalingam, S., Haque, B.Z. and Gillespie, J.W. (2020) Strain rate-dependent large deformation inelastic behavior of an epoxy resin. *Compos. Mater.*, **54**, 71–87.

Tang, W., Li, D., Peng, Y. and Wu, P. (2022) Shear band development under simple shear and the intrinsicality of strain softening of amorphous glassy polymers. *Polymer*, **255**, 125115.

Taylor, G.I. (1934) The mechanism of plastic deformation of crystals. Part I. Theoretical. *Proc. Roy. Soc.*, **A145**, 362.

Torres, D., Guo, S., Villar, M.P., Araujo, D. and Estevez, R. (2018) Calibration of a cohesive model for fracture in low cross-linked epoxy resins. *Polymers*, **10**, 1321.

Tresca, H. (1864) Mémoire sur l'ecoulement des corps solides soumis a de fortes pressions. *C. R. Acad. Sci. (Paris)*, **59**, 754; **64**, 809 (1867)

Truss, R.W., Clarke, P.L., Duckett, R.A. *et al.* (1984) The dependence of yield behavior on temperature, pressure, and strain rate for linear polyethylenes of different molecular weight and morphology. *Polym. Sci. B: Polym. Phys.*, **22**, 191.

Ube, T., Aoki, H., Ito, S. *et al.* (2007) Conformation of single PMMA chain in uniaxially stretched film studied by scanning near-field optical microscopy. *Polymer*, **48**, 6221.

Van Dommelen, J.A.W. and Meijer, H.E.H. (2005) Micromechanics of particle-modified semi-crystalline polymers: influence of anisotropy due to transcrystallinity and/or flow, in *Mechanical Properties of Polymers Based on Nanostructure and Morphology*, (eds G.H. Michler and F.J. Baltá-Calleja), Taylor and Francis, Florida.

Van Erp, T.B., Reynolds, C.T., Peijs, T., Van Dommelen, J.A.W. and Govaert, L.E. (2009) Prediction of yield and long-term failure of oriented polypropylene: kinetics and anisotropy. *J. Polym. Sci. B Polym. Phys.*, **47**, 2026–2035.

Vincent, P.I. (1960) The necking and cold-drawing of rigid plastics. *Polymer*, **1**, 7.

Vincent, P.I. (1967) *Encyclopedia of Polymer Science & Technology*, John Wiley & Sons, New York.

Ward, I.M. (1971) The yield behaviour of polymers: review. *J. Mater. Sci.*, **6**, 1397.

Ward, I.M. (1984) The role of molecular networks and thermally activated processes in the deformation-behavior of polymers. *Polym. Eng. Sci.*, **24**, 724.

Ward, I.M. and Coates, P.D. (2000), in *Solid Phase Processing of Polymers*, (eds I.M. Ward, P.D. Coates and M.M. Dumoulin), Hanser, Munich. Chap. 1.

Ward, I.M. and Wilding, M.A. (1984) Creep behavior of ultrahigh-modulus polyethylene: influence of draw ratio polymer composition. *J. Polym. Sci. Polym. Phys. Ed.*, **22**, 561–575.

Whitney, W. and Andrews, R.D. (1967) Yielding of glassy polymers: volume effects. *J. Polym. Sci. C*, **16**, 2981.

Wilding, M.A. and Ward, I.M. (1981) Creep and recovery of ultra high modulus polyethylene. *Polymer*, **22**, 870–876.

Wilding, M.A. and Ward, I.M. (1984) Creep and stress-relaxation in ultra high modulus polyethylene. *J. Mater. Sci.*, **19**, 629–636.

Wineman, A.S. and Waldron, W.K. (1993) Interaction of nonhomogeneous shear, nonlinear viscoelasticity, and yield of a solid polymer. *Polym. Eng. Sci.*, **33**, 1217–1228.

Wu, P.D. and van der Giessen, E. (1994) Analysis of shear band propagation in amorphous glassy polymers. *Int. J. Solids Struct.*, **31**, 1493.
Wu, P.D. and van der Giessen, E. (1996) Computational aspects of localized deformations in amorphous glassy polymers. *Eur. J. Mech. A/Solids*, **15**, 799.
Young, R.J. (1974) A dislocation model for yield in polyethylene. *Philos. Mag.*, **30**, 85.
Young, R.J. (1988) Screw dislocation model for yield in polyethylene. *Mater. Forum*, **11**, 210–216.

Further Reading

Bergström, J.S. (2015) *Mechanics of Solid Polymers*, Elsevier, Amsterdam.
Halary, J.L., Lauprêtre, F. and Monnerie, L. (2011) *Polymer Materials: Macroscopic Properties and Molecular Interpretations*, John Wiley & Sons Ltd, Hoboken, NJ.

13

Fracture

13.1 Definition of Tough and Brittle Behaviour in Polymers

The mechanical properties of polymers are greatly affected by temperature and strain rate, and the load–elongation curve at a constant strain rate changes with increasing temperature as shown schematically (not necessarily to scale) in Figure 13.1. At low temperatures, the load rises approximately linearly with increasing elongation up to the breaking point, when the polymer fractures in a brittle manner. At higher temperatures, a yield point is observed and the load falls before failure, sometimes with the appearance of a neck, that is ductile failure, but still at quite low strains (typically 10–20%). At still higher temperatures, under certain conditions, strain hardening occurs, the neck stabilizes and cold-drawing ensues. The extensions in this case are generally very large, up to 1000%. Finally, at even higher temperatures, homogeneous deformation is observed, with a very large extension at break. In an amorphous polymer, this rubber-like behaviour occurs above the glass transition temperature, so the stress levels are very low.

For polymers, the situation is clearly more complicated than that for the brittle–ductile transition in metals, as there are in general four regions of behaviour and not two. It is of considerable value to discuss the factors that influence the brittle–ductile transition, and then to consider further factors that are involved in the observation of necking and cold-drawing.

Ductile and brittle behaviour are most simply defined from the stress–strain curve. Brittle behaviour is designated when the specimen fails at its maximum load, at comparatively low strains (say <10%), whereas ductile behaviour shows a peak load followed by failure at a lower stress (Figure 13.1(a) and (b)).

The distinction between brittle and ductile failure is also manifested in two other ways: (1) the energy dissipated in fracture and (2) the nature of the fracture surface. The energy dissipated is an important consideration for practical applications and forms the basis of the Charpy and Izod impact tests (discussed in Section 13.8). At the testing speeds under which the practical impact tests are conducted it is difficult to determine the stress–strain curve, so impact strengths are customarily quoted in terms of the fracture energy for a standard specimen.

Mechanical Properties of Solid Polymers, Fourth Edition. John Sweeney and Peter Hine.

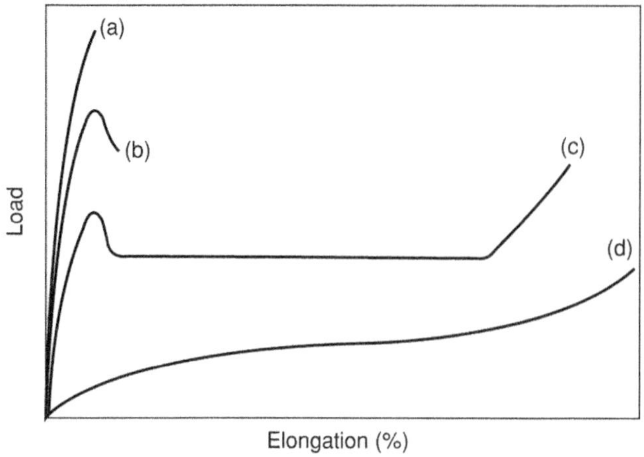

Figure 13.1 Load–extension curves for a typical polymer tested at four temperatures showing different regions of mechanical behaviour: (a) brittle fracture; (b) ductile failure; (c) necking and cold-drawing; (d) homogeneous deformation (quasi-rubber-like behaviour).

The appearance of the fracture surface also can be an indication of the distinction between brittle and ductile failure, although the present state of knowledge concerning the crack propagation is not sufficiently extensive to make this distinction more than empirical.

13.2 Principles of Brittle Fracture of Polymers

Modern understanding of the fracture behaviour of brittle materials stems from the seminal research of Griffith (1921) on the brittle fracture of glass. The Griffith theory of fracture, which is the earliest statement of linear elastic fracture mechanics, has been applied extensively to the fracture of glass and metals, and more recently to polymers. Although it was conceived initially to describe the propagation of a crack in a perfectly elastic material at small elastic strains (hence linear elastic), subsequent work has shown that it is still applicable for situations including localized plastic deformation at the crack tip, which does not lead to general yielding in the specimen.

13.2.1 Griffith Fracture Theory

First, Griffith considered that fracture produces a new surface area and postulated that for fracture to occur the increase in energy required to produce the new surface must be balanced by a decrease in elastically stored energy.

Secondly, to explain the large discrepancy between the measured strength of materials and those based on theoretical considerations, he proposed that the elastically stored energy is not distributed uniformly throughout the specimen but is concentrated in the neighbourhood of small cracks. Fracture thus occurs due to the spreading of cracks that originate in pre-existing flaws.

In general, the growth of a crack will be associated with an amount of work $\mathrm{d}W$ being done on the system by external forces and a change $\mathrm{d}U$ in the elastically stored energy U. The difference between these quantities, $\mathrm{d}W - \mathrm{d}U$, is the energy available for the formation of the new surface. The condition for growth of a crack by a length $\mathrm{d}c$ is then

$$\frac{\mathrm{d}W}{\mathrm{d}c} - \frac{\mathrm{d}U}{\mathrm{d}c} \geq 2\gamma \frac{\mathrm{d}A}{\mathrm{d}c}, \tag{13.1}$$

where γ is the surface free energy per unit area of surface and $\mathrm{d}A$ is the associated increment of surface. The factor 2 arises from the fact that two surfaces are created when a crack grows. If there is no change in the overall extension Δ when the crack propagates, $\mathrm{d}W = 0$ and

$$-\left(\frac{\mathrm{d}U}{\mathrm{d}c}\right)_{\Delta} \geq 2\gamma \frac{\mathrm{d}A}{\mathrm{d}c}. \tag{13.2}$$

In this condition the crack propagates without additional external work; it is self-sustaining, or catastrophic, crack growth. The elastically stored energy decreases to drive the crack and so $-(\mathrm{d}U/\mathrm{d}c)_{\Delta}$ is essentially a positive quantity.

Griffith calculated the change in elastically stored energy using a solution obtained by Inglis (1913) for the problem of a plate, pierced by a small elliptical crack that is stressed at right angles to the major axis of the crack. Equation (13.1) then allows the fracture stress σ_B of the material to be defined in terms of the crack dimension c (see Figure 13.2) by the relationship

$$\sigma_B = (2\gamma E^*/\pi c)^{1/2}, \tag{13.3}$$

where E^* is either equal to Young's modulus E for a thin sheet in plane stress or equal to the *reduced modulus* $E/(1-\nu^2)$,= for a thick sheet in plane strain where ν is equal to Poisson's ratio. This presupposes isotropic elastic material with a single Young's modulus.

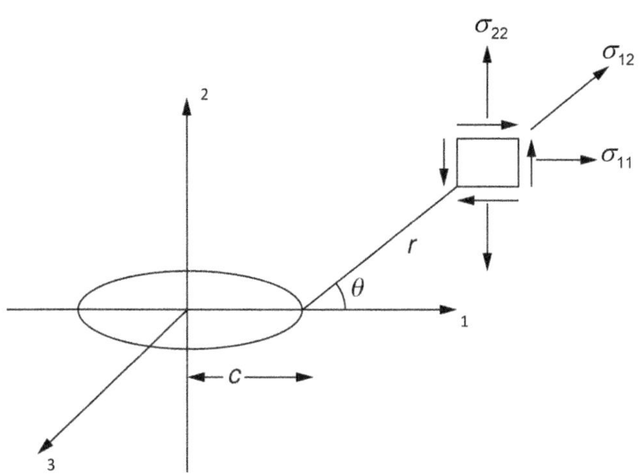

Figure 13.2 The stress field near an idealized crack of length 2c.

13.2.2 The Irwin Model

An alternative formulation of the problem due to Irwin (1957) considers the stress field near an idealized crack length $2c$ (Figure 13.2). In two-dimensional polar coordinates with the 1-axis as the crack axis and $r \ll c$,

$$\begin{aligned}
\sigma_{11} &= \frac{K_{\mathrm{I}}}{(2\pi r)^{1/2}} \cos(\theta/2)[1 - \sin(\theta/2)\sin(3\theta/2)] \\
\sigma_{22} &= \frac{K_{\mathrm{I}}}{(2\pi r)^{1/2}} \cos(\theta/2)[1 + \sin(\theta/2)\sin(3\theta/2)] \\
\sigma_{33} &= \upsilon(\sigma_{11} + \sigma_{22}) \quad \text{for plane strain} \\
\sigma_{33} &= 0 \quad \text{for plane stress} \\
\sigma_{12} &= \frac{K_{\mathrm{I}}}{(2\pi r)^{1/2}} \cos(\theta/2)\sin(\theta/2)\sin(3\theta/2) \\
\sigma_{23} &= \sigma_{31} = 0.
\end{aligned} \tag{13.4}$$

In these equations, θ is the angle between the axis of the crack and the radius vector.

The value of Irwin's approach is that the stress field around the crack is identical in form for all types of loading situation normal to the crack, with the magnitude of the stresses (i.e. their intensity) determined by K_{I}, which is constant for given loads and geometry; K_{I} is called the stress intensity factor, the subscript I indicating loading normal to the crack. This crack opening mode I is distinct from modes II and III, which are defined in the next section. As we approach the crack tip, σ_{11}, σ_{12} and σ_{22} clearly become infinite in magnitude as r tends to zero, but the products $\sigma_{11}\sqrt{r}$, etc. and hence K_{I} remain finite.

For an infinite sheet with a central crack of length $2c$ subjected to a uniform stress σ, it was shown by Irwin that

$$K_{\mathrm{I}} = \sigma(\pi c)^{1/2}. \tag{13.5}$$

He postulated that, when σ reaches the fracture stress σ_B, K_{I} has a critical value given by

$$K_{\mathrm{IC}} = \sigma_B(\pi c)^{1/2}. \tag{13.6}$$

The fracture toughness of the material then can be defined by the value of K_{IC}, termed the critical stress intensity factor, which defines the stress field at fracture.

There is clearly a link with the earlier Griffith formulation in that Equation (13.6) can be written as

$$\sigma_B = \left(K_{\mathrm{IC}}^2/\pi c\right)^{1/2}, \tag{13.7}$$

which is comparable in form to Equation (13.3) and implies a relationship between K_{IC} and γ.

13.2.3 Fracture Modes

In this analysis we have considered cases of cracks being opened directly by tension; this form of fracture is known as opening, or mode I. The subscript I on the stress intensity

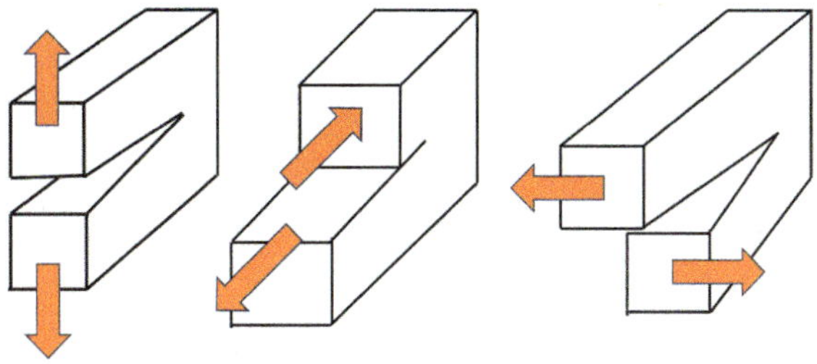

Figure 13.3 The three modes of fracture.

K_I refers to this mode. There are two more elementary modes of fracture, known as sliding mode II and tearing mode III, shown in Figure 13.3. In many practical instances of loaded flaws, the loading will be a combination of two or three modes; this known as *mixed-mode* fracture. For example, a linear flaw at an oblique angle to a uniaxial stress field will be loaded in both modes I and II. To predict failure in the presence of combined modes of stress intensity, a fracture criterion – a function of K_I, K_{II} and K_{III} similar in nature to a yield criterion – is necessary. A review of fracture criteria for mixed modes I and II has recently been published (Hua *et al.* 2023). The field of fracture mechanics is however dominated by measurements and applications of mode I, which is of primary practical importance.

13.2.4 The Strain Energy Release Rate

In linear elastic fracture mechanics, it is useful also to consider the energy G available for unit increase in crack length, which is called the 'strain energy release rate'. Following Equation (13.1) above, G is

$$G = \frac{dW}{dA} - \frac{dU}{dA} = \frac{1}{B}\left[\frac{dW}{dc} - \frac{dU}{dc}\right], \tag{13.8}$$

where B is the thickness of the specimen. It is assumed that fracture occurs when G reaches a critical value, G_c. The equivalent Equation to (13.1) is then

$$G \geq G_c \tag{13.9}$$

and G_c corresponds to 2γ in the Griffith formulation. In current practice G_c is generalized to include all work involved in the creation of fracture, not just the surface energy, and is termed the *fracture energy* or *the critical strain energy release rate*. On this basis Equation (13.3) is generalized to

$$\sigma_B = (E^* G_c/\pi c)^{1/2} \tag{13.10}$$

Work additional to the surface energy includes, for instance, that which is needed to produce plastic deformation of the surfaces during the fracture process. The distinction between surface energy and fracture energy was made at an early stage by Rivlin and Thomas (1953) who were investigating the tearing of rubber.

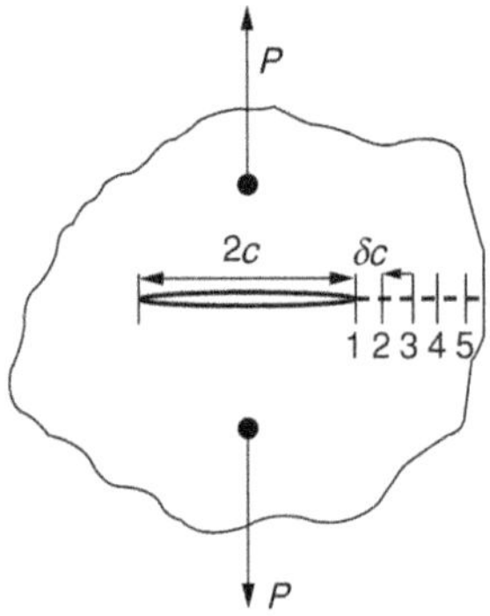

Figure 13.4 Schematic diagram of a specimen with a centre crack of length 2c. The loading is assumed to be sufficiently distant from the crack for it to be equivalent to a uniform remote stress, so we have effectively an infinite body.

Comparison of Equations (13.7) and (13.10) shows

$$G_c = K_{IC}^2/E^* \tag{13.11}$$

and below the fracture stress

$$G = K_I^2/E^* \tag{13.12}$$

Although the Griffith and Irwin formulations of the fracture problems are equivalent, most recent studies of polymers have followed Irwin. Before discussing results for polymers, it is useful to show how G_c can be calculated.

Consider a sheet of polymer with a crack of length $2c$ (Figure 13.4). We now define a quantity termed the compliance of the cracked sheet, C, which is the reciprocal of the slope of the linear load–extension curve from zero load up to the point at which crack propagation begins. At the latter point, the load is P and the extension is Δ, so $C = \Delta/P$.

This quantity C is not to be confused with an elastic stiffness constant as defined in Section 8.1. The work done in an elemental step of crack propagation is illustrated by Figure 13.5. As the crack moves from 4 to 5, for example, the energy available for the formation of a new crack surface is the difference between the work done (45XY) and the increase in elastic-stored energy (triangle 05Y – triangle 04X). This energy corresponds to the area of the shaded triangle in Figure 13.5, and for an increase of crack length dc is given by $\frac{1}{2}P^2 \mathrm{d}C$. Hence

$$G = \frac{P^2}{2B}\frac{\mathrm{d}C}{\mathrm{d}c}, \tag{13.13}$$

which is generally known as the Irwin–Kies relationship (Irwin and Kies 1954).

Here G_c can be determined directly by combining a load–extension plot from a tensile testing machine with determination of the movement of the crack across the specimen,

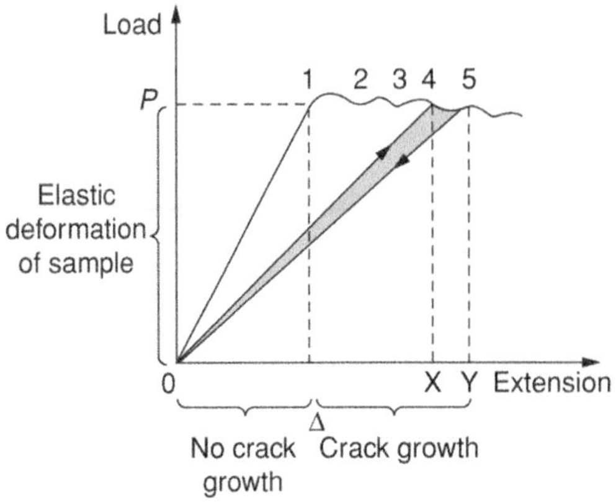

Figure 13.5 The load–extension curve for the specimen shown in Figure 13.3.

noting the load P for given crack lengths (points 1, 2, 3, 4, 5 in Figure 13.5). Alternatively, the strain energy release rate can be determined by computation for a given geometry, using a method such as the J-integral below.

13.3 Finite Geometries

Equations (13.6–13.8) were developed for an isolated crack within an infinite body, so that the stresses are not affected by conditions at a boundary. Equations (13.4) still give a valid representation of the form of the stresses for the region close to the crack tip even when a boundary is at a finite distance from it; however, the value of K_{I} can be affected. This effect can be represented by the finite width correction factor Y, which appears in a corrected version of Equation (13.5):

$$K_{\mathrm{I}} = Y\sigma(\pi c)^{1/2} \tag{13.14}$$

For particular geometries, Y can be calculated numerically. For instance, the crack can be at an edge as in Figure 13.6. The edge is free to deform away from its initial straight line shape, so the crack faces at the edge are not constrained to be parallel (unlike the faces at the centre of the crack in Figure 3.3). As a result there is a higher stress concentration, reflected in a value of Y greater than unity. Koiter (1965) has given the value for a shallow edge notch (Figure 13.6(a)) as $Y = 1.1215$. In general, Y will depend on the depth of the crack. This is so for an edge notch in a body of finite width (Figure 13.6(b)) where Y depends on the ratio c/b. For this case, Bowie and Neal (1965) produced numerical results for bodies of both finite width and length for crack lengths in the range $0.05 \leq c/b \leq 0.7$ and two specimen lengths, corresponding to values of $2L/b$ of 1.0 and 3.0. They solved the problem as posed in Figure 13.6(b), where there is a constant stress σ on the horizontal boundaries. This condition produces displacements of the horizontal boundaries that are not constant, with larger displacements on the right-hand side above and below the crack;

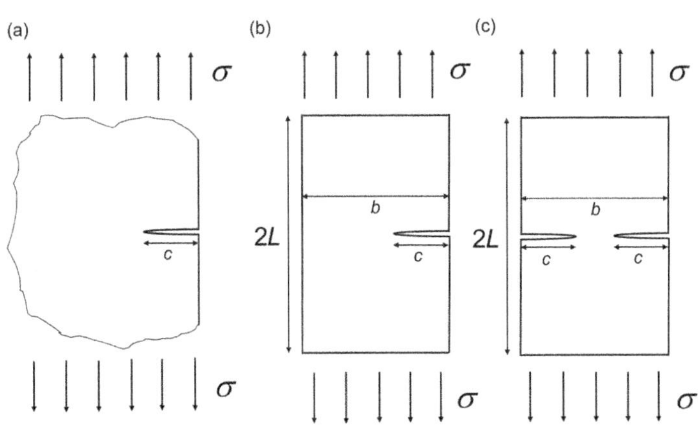

Figure 13.6 (a) Edge notch in a semi-infinite body. (b) Single-edge crack in a body of finite width and length. (c) Double-edge notched body of finite width and length.

the effect becomes more severe with smaller L. This loading can be approximated experimentally by using tensile grips that are free to rotate, such as around a pin joint. Alternatively, the geometry of Figure 13.6(b) can be subject to conditions on the upper and lower boundaries so that they are displaced at constant distance vertically with no horizontal contraction. This will give somewhat lower stress concentration as there is less tendency to open out the crack. It can be implemented experimentally by using tensile grips that allow no rotation. The two loading methods – rotationally free and rotationally constrained – have been compared for this geometry by Hammond and Fawaz (2016), who computed results using a finite element method and also reviewed the work of others. The height-to-width ratios of the geometries for rotationally restrained loading were varied in in the range $0.8 \leq 2L/b \leq 10$, whereas for unrestrained loading results are presented that are considered valid for $2 \leq 2L/b$. Results for finite width correction factors are shown in Figure 13.7 for cracks in the range $0.01 \leq c/b \leq 0.975$.

The complexities created by the single-edge notch geometry's tendency to rotate at its loaded ends are alleviated by the symmetric double-edge notch specimen, shown in Figure 13.6(c). Bowie (1964a) produced numerical results for two values of height to width ratio $2L/b$, 1.0 and 3.0, and crack lengths in the range $0.06 \leq c/b \leq 0.38$ for the stress boundary conditions shown in Figure 13.6(c). However, the applied uniform stress causes some curvature at the boundaries, with regions vertically above and below the cracks being deformed more. Bowie (1964b) explored the importance of this by analysing the problem with constant displacement boundary conditions, using a geometry $2L/b = 1$ and crack lengths in the range $0.07 \leq c/b \leq 0.28$. Bowie's results with stress boundary conditions are compared with those with displacement boundary conditions in Figure 13.8, where they have been recast as values of Y. The boundary curvature that results when stress boundary conditions are applied is accompanied by a greater level of crack opening than

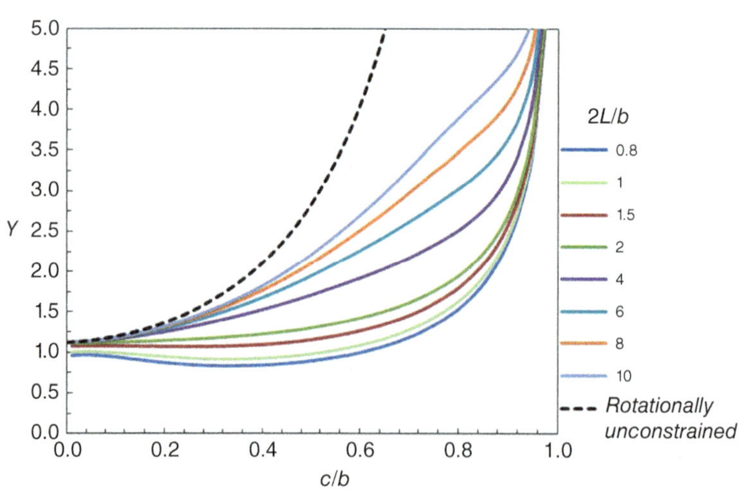

Figure 13.7 Finite width correction factors for single-edge notch specimens as a function of normalized crack length c/b. For rotationally constrained specimens (constant displacement applied to the upper and lower horizontal edges in Figure 13.6(b)) a range of aspect ratios 2L/b is shown. Rotationally unconstrained results are valid for 2L/b>2. Adapted from Hammond and Faraz, 2016.

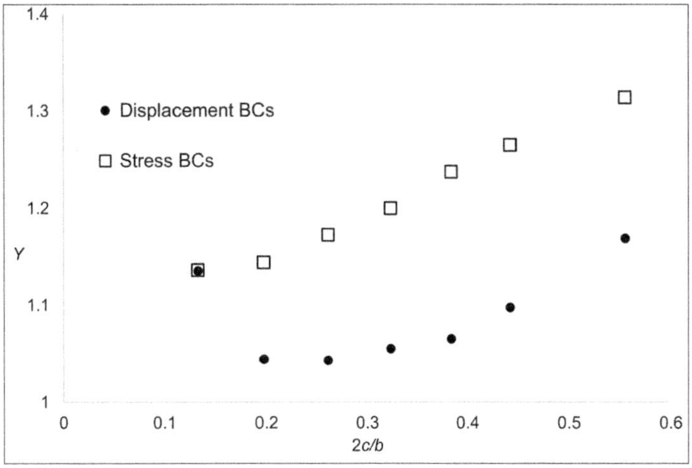

Figure 13.8 Finite width correction factors for double-edge notched specimens for aspect ratio $2L/b = 1$ for differently loaded boundaries. Rooke and Cartwright, 1976; Murakami, 1986; Al Laham, 1999.

with displacement boundary conditions, resulting in higher Y. This effect lessens as the aspect ratio $2L/b$ is increased, as the curvature of the boundaries is less when they are more distant from the cracks.

Fracture testing with any specimen geometry can be interpreted, and K_{IC} measured, provided the appropriate value of Y is known. A number of works have been published that give finite width correction factors for wide ranges of notched fracture geometries, including external cracks, internal cracks and bending tests (Rooke and Cartwright 1976; Murakami 1986; Al Laham 1999).

13.4 Elastic Anisotropy

In many relevant cases, such as with an oriented or a fibre-reinforced polymer, the material is elastically anisotropic. Fracture mechanics of orthotropic elastic material has been investigated by Sih *et al.* (1965). They found that the dependence of stress on distance from the crack tip follows the same inverse square root form as in an isotropic material, so that the intensity K_I can be carried over into orthotropic theory. Fracture energy G_c can be defined physically as the energy needed to produce a unit of crack surface area in the same way as for an isotropic material, but will now in general be dependent on the crack direction. For instance, in an oriented polymer, the crack must sever many more molecular chains per unit area when travelling transverse to the orientation direction than along it. Similarly, the strain energy release rate G will depend on the orientation of the stress field relative to the direction of the planes of elastic orthotropy, the initial orientation of the crack and the direction of its growth. The latter two directions may be different, corresponding to crack growth that is not self-similar. According to Sih *et al.* for a crack aligned with the 1 direction the relationship between K_I and G is given by

$$K_{\mathrm{I}}^{2} = G\left[\left(\frac{s'_{11}s'_{22}}{2}\right)^{1/2}\left(\left(\frac{s'_{22}}{s'_{11}}\right)^{1/2} + \frac{2s'_{12}+s_{66}}{2s'_{11}}\right)^{1/2}\right]^{-1} = GE_{\mathrm{orth}}$$

$$s'_{ij} = \begin{cases} s_{ij} & \text{for plane stress} \\ s_{ij} - s_{i3}s_{j3}/s_{33} & \text{for plane strain} \end{cases} \tag{13.15}$$

where the s_{ij} are the material compliances (see Chapter 8, Section 8.22) and i and j take the values 1 and 2. This is the analogue of Equation (13.12). The function E_{orth} is known as the orthotropic modulus. Since the 1 and 2 directions are fixed within the material, its value depends on the orientation of the crack relative to the material directions of symmetry, as does the value of K_{I}. Finite width correction factors Y can be calculated as defined in Equation (13.14) on the understanding that they will depend on the crack orientation.

For the case of finite bodies, a number of numerical studies have produced finite width corrections for anisotropic materials. An early example is that of Bowie and Freese (1972) for a centrally cracked plate. Snyder and Cruse (1975) provided results for both centre-cracked and edge-notched specimens. It is not practicable to produce sets of Y values for particular test geometries as for isotropic material, as Y depends on the specific levels of anisotropy and must be calculated for each material. This has been facilitated in more recent times by the widespread use of finite element analysis, combined with techniques such as the J-integral (see section below). For example, Sweeney and Ward (1988) used this technique for the analysis of fracture in oriented polyethylene. Since these early developments, the application of finite element analysis in this area has expanded greatly and has recently been reviewed by Ayatollahi *et al.* (2022).

13.5 Controlled Fracture in Brittle Polymers

In its simplest form the Griffith theory, and the linear elastic fracture mechanics (LEFM) that developed from it, ignores any contribution to the energy balance arising from the kinetic energy associated with movement of the crack. A basic study of the brittle fracture of polymers is therefore likely to be most rewarding if the fracture takes place slowly so that a negligible amount of energy is dissipated in this way.

The double cantilever beam specimen of Figure 13.9 is frequently used for this purpose. The compliance C of this specimen can be derived simply using beam theory. The deflection Δ in response to a load P is then given by

$$\Delta = \frac{64c^3}{EBb^3}P \tag{13.16}$$

The compliance is then given by

$$C = \frac{\Delta}{P} = \frac{64c^3}{EBb^3} \text{ and } \frac{\mathrm{d}C}{\mathrm{d}c} = \frac{192c^2}{EBb^3} \tag{13.17}$$

Figure 13.9 The double cantilever beam specimen.

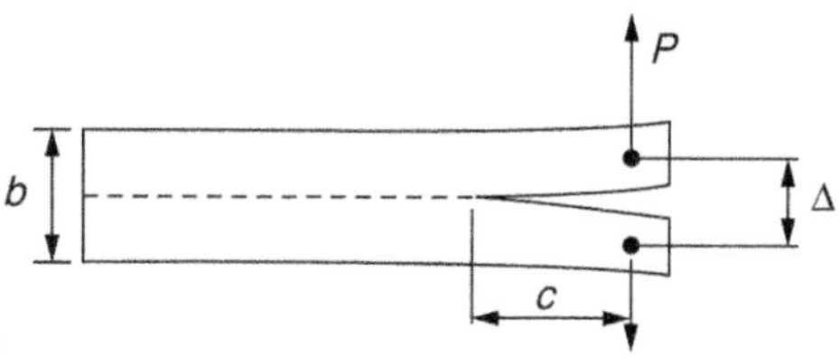

and Equation (13.13) gives

$$G = \frac{P^2}{2B}\frac{dC}{dc} = \frac{96c^2P^2}{B^2Eb^3} \tag{13.18}$$

At crack initiation at critical load P_c,

$$G_c = \frac{96c^2P_c^2}{B^2Eb^3} \tag{13.19}$$

or alternatively in terms of the critical extension Δ_c,

$$G_c = \frac{3\Delta_c^2b^3}{128c^4}E \tag{13.20}$$

The stress intensity K_I can be obtained from Equation (13.18) using Equation (13.12) or (13.15). For instance, for isotropic material in plane stress conditions,

$$K_I = 4\sqrt{6}\frac{P^2}{Bb^{3/2}} \tag{13.21}$$

Under these slow crack growth conditions, studies of brittle fracture in polymers were undertaken by Benbow and Roesler (1957) at ICI in the United Kingdom and by Berry (1963) at GE in the United States. Benbow and Roesler devised a method of fracture in which flat strips of poly(methyl methacrylate) (PMMA) were cleaved by gradually propagating a crack down the middle, as in a cantilever double beam. Essentially, their experiments involved determining the deflection Δ for a given length c (symbols as in Figure 13.9). The results were expressed in terms of the surface energy γ. Following Equation (13.20)

$$\frac{G_c}{E} = \frac{2\gamma}{E} = \frac{3\Delta^2b^3}{128c^4}. \tag{13.22}$$

Knowing a value for Young's modulus, E, the surface energy γ can be found.

Berry adopted a slightly different approach to evaluate γ, and also confirmed the validity of Equation (13.3) for the fracture of PMMA and polystyrene by measuring the tensile strength of small samples containing deliberately introduced cracks of known magnitude.

Katsamanis and Delides (1988) reviewed measurements of fracture surface energy of PMMA. Their summary of results is included in Table 13.1. The very important conclusion to be drawn from these values for fracture surface energy is that they are very much greater than values estimated from the assumption that the energy required to form a new surface originates in the simultaneous breaking of chemical bonds, which might appear to provide an upper theoretical estimate. Take the bond dissociation energy as 400 kJ/mole (McMurry

Table 13.1 Fracture surface energies of PMMA (J m^{-2}).

Method and source	
Tensile; Irwin and Kies (1952)	**400**
Cleavage; Benbow and Roesler (1957)	490
Cleavage; Benbow (1961)	420
Cleavage; Svensson (1961)	450
Tensile: Berry (1961)	300
Cleavage; Berry (1963)	140
Cleavage: Broutman and McGarry (1965)	120
Edge loaded sheet: Van dan Boogaert (1966)	330
Cleavage: Curtis (1970)	220
Tensile, laser induced crack: Dobbs *et al.* (1973)	160 stable 260 catastrophic
Indentation: Puttick (1978)	260
Tensile: Zimmermann *et al.* (1984)	250–300

2023) and the concentration of molecular chains as 1 chain per 0.2 nm^2, giving 5×10^{18} molecular chains m^{-2}. To form 1 m^2 of new surface requires about 1.5 J, which is two orders of magnitude less than that obtained from cleavage and tensile measurements.

It is now generally accepted that more energy is required in the fracture process than that directly associated with the creation of the surfaces. Andrews (1966) suggested that the quantity measured in the fracture of polymers should be described by a 'surface work parameter'; we now understand 'fracture energy' as that required to drive the crack, including any associated plastic work or viscoelastic dissipation. For some polymers, the phenomenon of crazing, where fracture is accompanied by significant plastic work, can account for the discrepancy between the calculated surface energy and the observed fracture energy; this is discussed below. However, polymers that do not craze have fracture energies at levels similar to those that do. Recently, Deng *et al.* (2023) proposed that during polymer fracture, energy must be released or dissipated in molecular chains distant from the crack tip. They showed that a method used for calculating fracture energy in rubbers by Lake and Thomas (1967) persistently underestimated by 1–2 orders of magnitude. They developed a model that includes the nonlinear large strain behaviour of molecular chains, and predicts that molecular chain relaxations of molecules distant from the crack tip contribute to the fracture energy. These predictions of fracture energy are realistic.

13.6 Crazing in Glassy Polymers

As established in previous chapters, polymers are mechanically time-dependent and so it is to be expected that fracture phenomena are not always explicable by the linear elastic fracture mechanics expounded so far in this chapter. Crazing is an example of fracture that does not grow catastrophically on initiation. Early observations by Kambour (1964, 1966) showed that a thin wedge of porous material termed a craze can form at a crack tip in a glassy polymer, as shown schematically in Figure 13.10. A micrograph of a craze is shown in Figure 13.11, showing its complex, irregular and three-dimensional structure. The craze is a crack bridged by polymer tendrils which may become oriented. As the craze propagates, more tendrils are created. It forms under plane strain conditions, so that the polymer is not free to contract laterally and there is a consequent reduction in density. Crazing has been described as proceeding in three stages: initiation;

Figure 13.10 Schematic diagram of a craze.

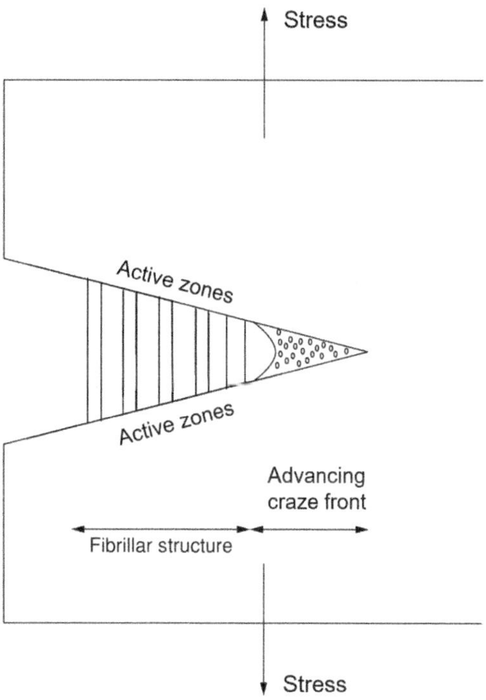

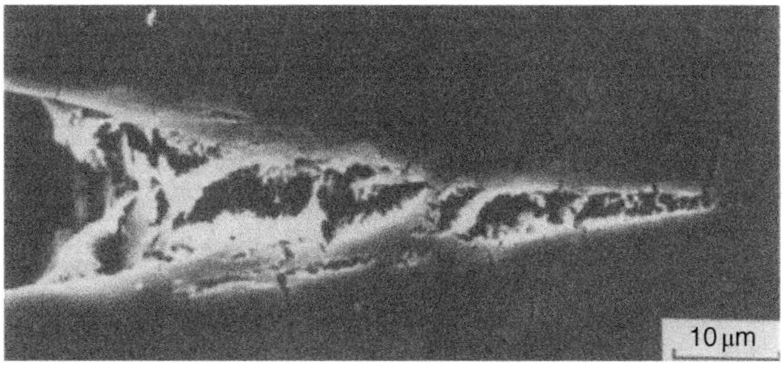

Figure 13.11 Optical image of a craze. Brown et al. (1987) / with permission of Elsevier.

widening, when fibrils are created and stretched; and breakdown, when fibrils break and a crack is formed (Kramer 1983). During widening, individual fibrils lengthen, but above a certain distance from the craze tip TEM observations have revealed (Donald and Kramer 1981; Kramer 1983; Kramer and Berger 1990) that the fibril diameters are essentially equal, so that for any individual fibril its diameter is constant and its volume must be increasing. Therefore, the lengthening of the fibrils cannot be due to creep, but rather to a continuous creation of fibril material as molecules are drawn into the fibrils from the 'active zones' at the interface between the craze and the bulk material (see Figure 13.10). It should be borne in mind that these observations are from work on glassy polymer; work on polyethylene at room temperature has been interpreted as involving orientation and creep of fibrils after their formation (Battachyarya and Brown 1984; O'Connell *et al.* 1995). The historical development of this subject is complex, and reviews at various times (Kambour 1973; Argon 2011) have given useful summaries.

Several workers (Brown and Ward 1973; Doll and Weidmann 1976) attempted to determine the craze profile by examining the crack tip region in PMMA in an optical microscope. It has been found that the craze profile is very similar to the shape of the plastic zone in the model proposed by Dugdale (1960) for metals, which will now be described.

13.6.1 The Dugdale Model

Equations (13.4) imply that there is an infinite stress at the crack tip. In practice this clearly cannot be so, and there are two possibilities. First, there can be a zone where shear yielding of the polymer occurs. In principle, this can occur in both thin sheets where conditions of plane stress pertain and in thick sheets where there is a plane strain. Secondly, for thick specimens under conditions of plane strain, the stress singularity at the crack tip can be released by the formation of a craze, which is a line zone, in contrast to the approximately oval (plane stress) or kidney-shaped (plane strain) shear yield zones. As indicated, its shape approximates very well to the idealized Dugdale plastic zone where the stress singularity at the crack tip is cancelled by the superposition of a second stress field in which the stresses are compressive along the length of the crack (Figure 13.12). A constant compressive stress is assumed and is identified with the craze stress. It is not the yield stress, and crazing and shear yielding are different in nature and respond differently to changes in the structure of the polymer.

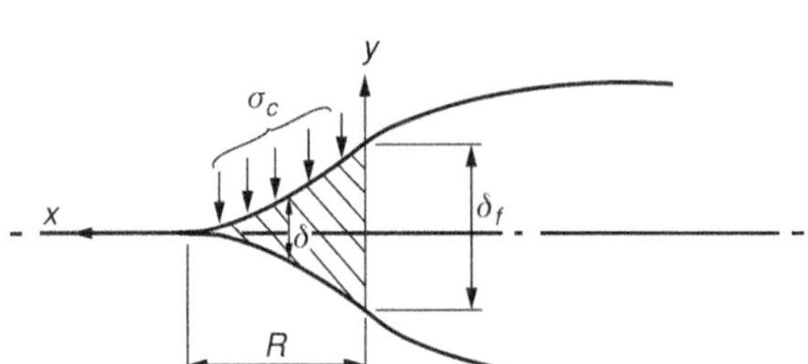

Figure 13.12 The Dugdale plastic zone model for a craze.

Rice (1968a) has shown that the length of the craze for a loaded crack on the point of propagation is

$$R = \frac{\pi}{8}\frac{K_{\mathrm{IC}}^2}{\sigma_{\mathrm{c}}^2} \tag{13.23}$$

and the corresponding separation distance δ between the upper and lower surfaces of the craze is

$$\delta = \frac{8}{\pi E^*}\sigma_{\mathrm{c}}R\left[\zeta - \frac{x}{2R}\log\left(\frac{1+\zeta}{1-\zeta}\right)\right], \tag{13.24}$$

where $\zeta = (1 - x/R)^{1/2}$ and E^* is the reduced modulus as defined in 13.2.1.

The crack opening displacement (COD) δ_t is the value of the separation distance δ at the crack tip, where $x = 0$, and is therefore

$$\delta_t = 8\sigma_{\mathrm{c}}R/\pi E^* = K_{\mathrm{IC}}^2/\sigma_{\mathrm{C}}E^*. \tag{13.25}$$

The fracture toughness of the polymer then relates to two parameters δ_t and σ_{c} (the craze stress), the product of which is equal to G_{IC}, the critical strain energy release rate. Direct measurements of craze shapes for several glassy polymers, including polystyrene, poly(vinyl chloride) and polycarbonate (Doll and Waldman 1976; Pitman and Ward 1979), confirmed the similarity to a Dugdale plastic zone. A result of some physical significance is that the COD is often insensitive to temperature and strain rate for a given polymer, although it has been shown to depend on molecular mass. For constant COD, the true dependence of G_{IC} on strain rate and temperature is determined only by the sensitivity of the craze stress to these parameters. Because as in Equation (13.12) $G_{\mathrm{IC}} = K_{\mathrm{IC}}^2/E^*$, the fracture toughness K_{IC} will in addition be affected by E^*, which is also dependent on strain rate and temperature.

While the Dugdale model conforms with observation that the craze develops well ahead of the crack tip, and achieves some success with the prediction of craze shape, its origins in metal plasticity suggest that it will not help in understanding the underlying physics.

13.6.2 Physical Origins of Craze Initiation

The analysis above involves both fracture toughness and the craze stress, which has a role similar to yield stress. Throughout the history of the study of crazing, there has been discussion of whether craze initiation is governed by yielding or is more closely related to fracture. Some early work sought a stress-based criterion for craze initiation. Sternstein *et al.* (1968) examined the formation of crazes in the vicinity of a small circular hole punched in the centre of PMMA strips pulled in tension. When the solutions for the elastic stress field in the vicinity of the hole were compared with the craze pattern, it was found that the major principal stress acts along the craze plane normal and therefore parallels the molecular orientation axis of the crazed material.

Sternstein and Ongchin (1969) extended this investigation by examining the formation of crazes under biaxial stress conditions, and found that the conditions for crazing involved both the principal stresses σ_1 and σ_2. The most physically acceptable explanation of these results was proposed by Oxborough and Bowden (1973), who suggested that crazing occurs when the extensional strain in any direction reaches a critical value e_1, which depends on the hydrostatic component of stress. This amounts to a stress criterion as the critical strain was related elastically to the stress.

The stress-based approach resembles the development of a yield criterion A radically different approach is one based on the growth of a pre-existing flaw and involves the concepts of fracture mechanics. Bucknall (2007) has introduced a Griffith-style relation between craze initiation stress and an associated fracture energy. Combining Equations (13.11) and (13.14) gives the relation

$$\sigma_B = \frac{1}{Y}\sqrt{\frac{E^* G_c}{\pi c}} \tag{13.26}$$

for a fracture stress σ_B. Bucknall has produced an analogous relation for craze stress σ(craze) and flaw size a_0:

$$\sigma(\text{craze}) = \frac{1}{Y}\sqrt{\frac{E^* G_c}{\pi a_0}} \tag{13.27}$$

There has been evidence of crazes originating from pre-existing flaws for many years. Argon and Hannoosh (1977) performed tension–torsion experiments on polystyrene specimens with well-characterized surface grooves, where crazes were generated in a reproducible manner. They also produced specimens with almost flaw-free surfaces and no internal entrapped dust particles. When tested mechanically, they underwent macroscopic yielding that was associated with internal crazes; they referred to this phenomenon as *intrinsic crazing*, concluding that it was not associated with pre-existing flaws. Bucknall developed his theory by introducing a strain energy release rate for a nascent crack G(nasc) to replace G_c in Equation (13.27). This was justified by observations that the G_c of fracture mechanics is the fracture energy needed to initiate catastrophic, i.e. self-sustaining, crack growth. Craze growth, however, is maintained by a continuous application of force. The stress in the craze may be assumed to be the yield stress σ_Y; then an elementary energy argument gives the strain energy release rate for the craze as

$$G(\text{nasc}) = \sigma_Y \delta_{\text{craze}} \tag{13.28}$$

where δ_{craze} the COD (see previous section) of the craze. This gives a typical value of G(nasc) of 0.6 J m^{-2}, which is very small compared with the fracture energies in Table 13.1 above, which are associated with catastrophic crack propagation. Rewriting Equation (13.27) now gives a new criterion for craze initiation:

$$\sigma(\text{nasc}) = \frac{1}{Y}\sqrt{\frac{E^* G(\text{nasc})}{\pi a_0}} \tag{13.29}$$

This allows for crazing to be initiated from very small flaws, perhaps including ones that are ordinarily undetectable, such as those with length scales at the molecular level.

The approach has been recently applied to the crazing of toughened PLA (McCutcheon *et al.* 2020).

The role of voids at the molecular level – 'nanovoids' – has been discussed in relation to craze initiation by Liu and Yee (2000). A population of nanovoids is to be expected on statistical grounds; their tendency to either grow or collapse is assumed to govern their capability of initiating crazes. In an experimental study of copolymers in which the mobility of chain molecules was varied, Liu and Lee found that increased mobility retarded craze formation. This was interpreted as the result of nanovoids collapsing as a consequence of conformational changes in chains local to them.

This concept has been studied further using molecular dynamics. The statistical nature of this method of analysis results in inhomogeneities in density to develop, which can grow under tension to generate cavitation (Kulmi and Basu 2006). In a study of crazing, Mahajan and Hartmaier (2012) used a model system of 160 molecular chains each consisting of 200 units and found that it gave rise to heterogeneities with potential for craze initiation in sufficient numbers to enable the study. They applied different multiaxial loads to the system, and found that they could produce stress maxima in stress–strain curves, corresponding to mechanical failure. Both shear yielding and crazing were induced, depending on the nature of the loading; some results are shown in Figure 13.13. In general, shear yielding is associated with conformational changes in molecules and crazing is associated with cavitation.

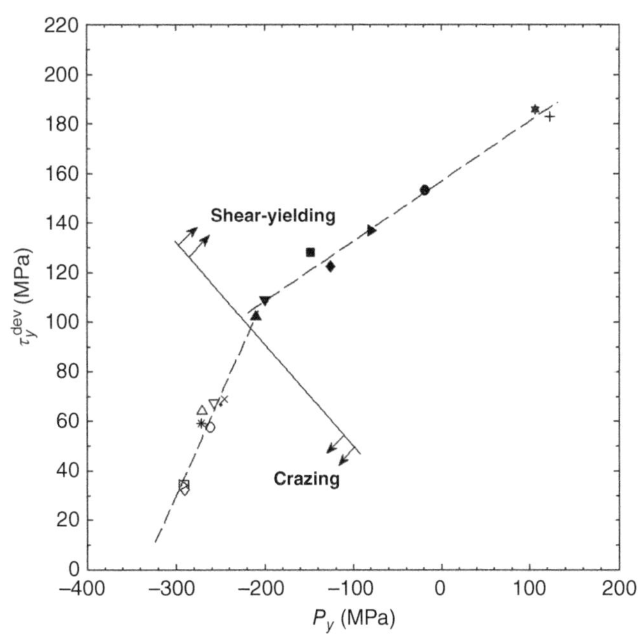

Figure 13.13 Results for an initially identical atomistic model sample subject to different multiaxial stress fields. The points denote deviatoric shear stress maxima, i.e. failure and are plotted against hydrostatic pressure $P_y = -\bar{s}$. The open symbols represent crazing and the filled symbols yield. Mahajam and Hartmaier (2012) / American Physical Society.

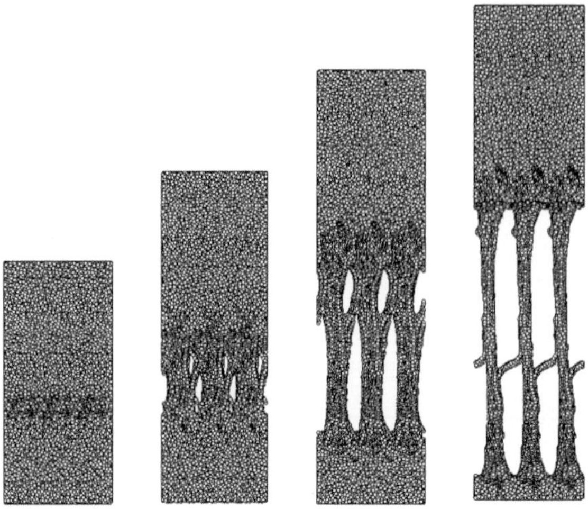

Figure 13.14 Simulations of craze initiation and growth. Model is stretched vertically to extension ratios 1.1, 1.4, 1.6 and 2.1 for images from left to right. Venkatesan et al. (2015) / with permission of Elsevier.

Yield stress increases with pressure, as seen in Chapter 12, whereas hydrostatic tension promotes crazing.

Venkatesan and Basu (2015) also used molecular dynamics to study craze initiation in glassy polymers. They concluded, like Mahajan and Hartmaier, that crazes nucleate in the material as a result of structural fluctuations that result in nanovoids. Their model systems were of size $10 \times 10 \times 47.5$ nm with stretching along the long direction, giving craze dimensions similar to those observed experimentally. The models contained between 32,000 and 100,000 atoms with maximum molecular length of 1000 atoms. They produced successful simulations of craze initiation and subsequent development of fibrils as the specimen was stretched along the directions of the fibril axes (see Figure 13.14). This modelling did not simulate the propagation of the craze through the specimen.

13.6.3 Craze Growth

Argon and Salama (1977) have proposed that the craze front advances by a meniscus instability mechanism in which craze tufts are produced by the repeated break-up of the concave air/polymer interface at the crack tip, as illustrated in Figure 13.15. The process is essentially the continuous nucleation of voids. This approach, further developed by Kramer (1983) and equivalent to a minimum surface energy condition, has become generally accepted (Donald 1997, Bucknall 2007). Estevez *et al.* (2000), who also accepted the meniscus instability mechanism, made a detailed mathematical model of crazing. They made use of a constitutive model of polymer deformation that included the phenomena of rate-dependent yielding and strain hardening, based on work of Wu and van der Giessen (1996). Craze initiation was governed by the criterion of Sternstein and Ongchin (1969).

The craze was modelled as a crack with surfaces forces pulling the crack faces together to represent the action of the tendrils – a 'cohesive surface model'. Craze breakdown was assumed to occur at a critical value of craze width, at which point the crack surface forces were assigned zero values to produce a microcrack. The analysis was implemented numerically using the finite element method. The result is a successful model of the interplay between yielding and crazing.

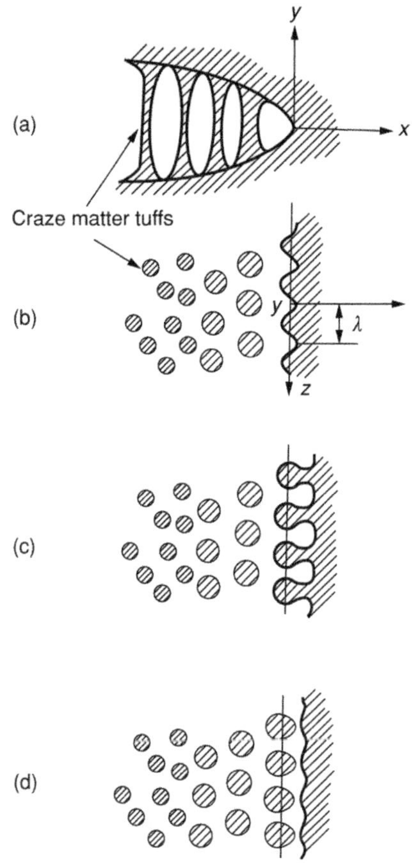

Figure 13.15 Schematic diagram showing craze matter production by the mechanism of meniscus instability: (a) outline of a craze tip; (b) cross-section in the craze plane across craze matter tufts; (c, d) advance of the craze front by a completed period of interface convolution. Hannoosh et al. (1977) / with permission Elsevier.

13.6.4 Crazing in the Presence of Fluids and Gases: Environmental Crazing

The crazing of polymers by environmental agents is of considerable practical importance and has been studied extensively with notable early contributions from Bernier and Kambour (1968), Kambour *et al.* (1973, 1974), Andrews and Bevan (1972), Marshall *et al.* (1970), Williams and Marshall (1975), Brown and Imai (1975), Imai and Brown (1976) and Brown *et al.* (1978). The subject of environmental stress cracking has been reviewed by Robeson (2013). In general environmental agents, which can be fluids or solids, reduce the stress or strain required to initiate both cracking and crazing.

The degree of reduction of the strain required to induce fracture has been shown to depend on the relative values of the solubility parameters of polymer and the environmental agent. The solubility parameter is simply related to the cohesive energy. For a material with volume V and cohesive energy E_{coh}, the solubility parameter δ is given by (Krevelen and Nijenhuis 2009)

$$\delta = \left(\frac{E_{\mathrm{coh}}}{V}\right)^{1/2}$$

Bernier and Kambour (1968), Kambour *et al.* (1973, 1974) showed that the critical strain for crazing reached a minimum as the solubility of the environmental agent was varied, as shown in Figure 13.16, reaching minima around where the solubility of the polymer δ_p was equal to that of the environmental agent δ_e.

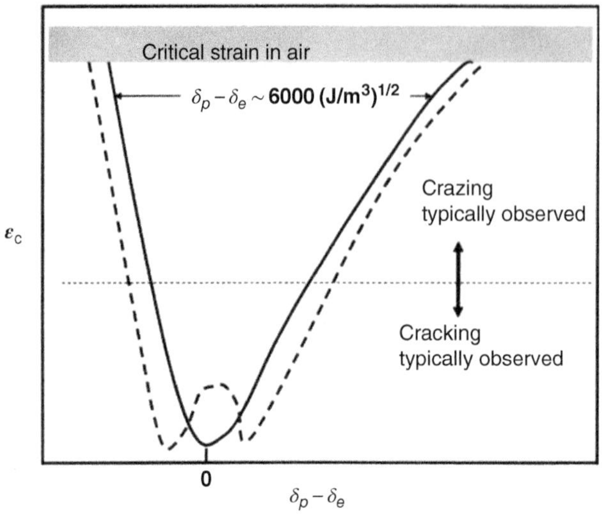

Figure 13.16 Critical strain as a function of solubility. The full line resembles the results of Bernier and Kambour (1968) and the dashed line those of Kambour et al. (1974). Robeson (2013) / John Wiley & Sons.

Andrews and Bevan (1972), adopting a more analytical approach and applying the ideas of fracture mechanics, performed fracture tests on single-edge-notched tensile specimens, where a central edge crack of length c is introduced into a large sheet of polymer that is then loaded in tension. The fracture stress is related to a surface work parameter, essentially the fracture energy G_c by an equation identical in form to Equation (13.3) above. The critical stress for crack and craze propagation σ_c was indeed proportional to $c^{-1/2}$, so that G_c values could be determined. For constant experimental conditions, a range of values was obtained from which a minimum value G_{c0} was estimated. From tests in a given solvent over a range of temperatures, it was found that values of G_{c0} decreased with increasing temperatures up to a characteristic temperature, above which G_{c0} remained constant at value G^*_{c0} The values of G^*_{c0} for the different solvents were shown to be a smooth function of the difference between the solubility parameters of the solvent and the polymer, reaching a minimum when this difference was zero (Figure 13.17). The functional behaviour of fracture energy with respect to solubility is shown to be similar to that of critical strain as shown in Figure 13.16.

These findings were explained using an elastic–plastic model of the swelling of an aspherical cavity under a negative hydrostatic pressure p, which has two terms so that

$$p = \frac{2\gamma_\tau}{r} + \frac{2\sigma_Y}{3}\psi, \tag{13.30}$$

where γ_τ is the surface tension between the solvent in the void and the surrounding polymer, σ_Y is the yield stress and ψ is a factor close to unity. The effect of temperature is to change the yield stress, so that with increasing temperature σ_Y falls eventually to zero at T_c, which is the glass transition temperature of the plasticized polymer. Above T_c, the fracture surface energy G^*_{c0} relates solely to the intermolecular forces represented by the surface tension γ_τ.

Brown has pointed out that gases at sufficiently low temperatures make almost all linear polymers craze (Brown and Imai 1975; Imai and Brown 1976; Brown *et al.* 1978; Brown 1980). Parameters such as the density of the crazes and the craze velocity increase with the pressure of the gas and decrease with increasing temperature. It was concluded that the surface concentration of the absorbed gas was a key factor in determining its effectiveness as a crazing agent.

In a related but somewhat different development, Marshall *et al.* (1970) and Williams and Marshall (1975) studied the rate of craze growth in PMMA in methanol. In all cases, the craze growth depended on the initial stress intensity factor K_0, calculated from the load and the initial notch length. Below a specific value of K_0 termed K_0^* the craze would decelerate and finally arrest. For $K_0 > K_0^*$, the craze would decelerate initially and finally propagatc at constant speed.

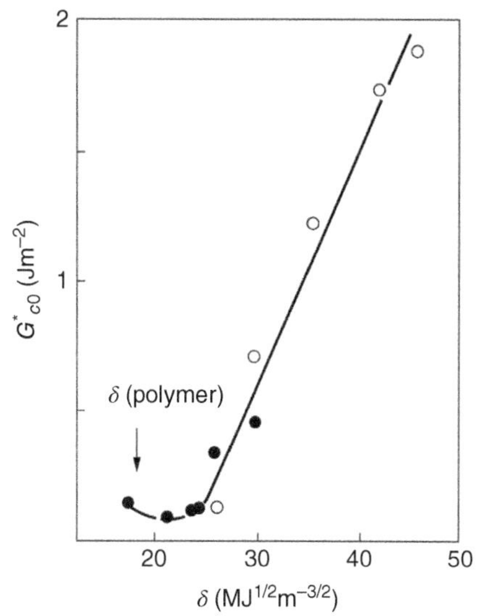

Figure 13.17 Variation of G^*_{c0} for PMMA with the solubility parameter of the solvent: (•) pure solvents; (°) water isopropanol mixtures. Andrews and Bevan (1972) / with permission of Elsevier.

It was argued that the controlling factor determining craze growth was the diffusion of methanol into the craze. Where $K_0 < K_0^*$ the methanol is considered to diffuse along the length of the craze, and it may be shown that the length of the craze x is proportional to the square root of the time of growth (Figure 13.18). In the second type of growth, where $K_0 > K_0^*$, it is considered that the methanol diffuses through the surface of the specimens, maintaining the pressure gradient in the craze and producing craze growth at constant velocity.

More recently environmental crazing has been used as a method for engineering surface modification (Yarysheva *et al.* 2020) and as a tool for producing porous structures (Arzhakova *et al.* 2020; Trofimchuk *et al.* 2020; Yarysheva *et al.* 2021).

13.7 Controlled Fracture in Tough Polymers

The development of brittle fracture, as outlined in Sections 13.2.1–13.2.3, is directly applicable to the failure of many glassy polymers, including PMMA and polystyrene, which have been studied intensively by LEFM. A key criterion for quantitative analysis is that extensive yielding should not occur, either at the crack tip or in the body of the specimen, for which explicit rules have been proposed. In essence, the load–extension curves should be of the form shown in Figure 13.4, i.e. elastic deformation up to the point of initiation of crack growth. In practice, this means that fracture occurs under conditions of plane strain for specimens of minimum

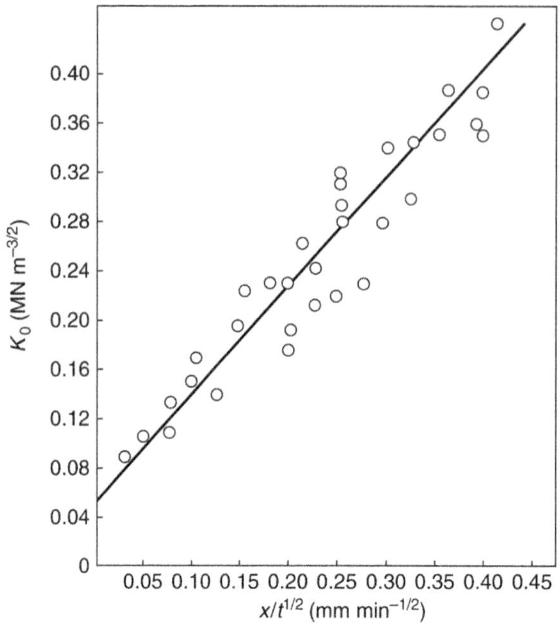

Figure 13.18 Craze growth behaviour for poly(methyl methacrylate) in methanol at 20°C. Adapted from Williams and Marshall, 1975.

thickness, whereas for thin specimens plane stress conditions apply and the stress–strain curve will show a yield point as shown in Figure 13.1, curves (a) and (b).

The fracture of most semi-crystalline polymers, notably polyethylene, polypropylene and nylon, cannot be described by LEFM based on the theory of Griffith and Irwin because large-scale yielding occurs at the crack tip prior to failure. For these materials and for toughened polymers and polymer blends other approaches have been developed, three of which will be discussed in detail:

1) The *J*-integral.
2) Essential work of fracture.
3) Crack opening displacement (COD).

13.7.1 The *J*-Integral

Rice (1968b) defined a quantity termed the *J*-integral, which describes the flow of energy into the crack tip region. It is defined for the two-dimensional problem of a straight crack in the *x* direction (Figure 13.19), and Γ is any contour surrounding the crack tip. In a two-dimensional stress field associated with a crack tip,

$$J = \int_{\Gamma} W \mathrm{d}x_2 - \mathbf{T} \cdot \frac{\mathrm{d}\mathbf{u}}{\mathrm{d}x_1} \mathrm{d}s \tag{13.31}$$

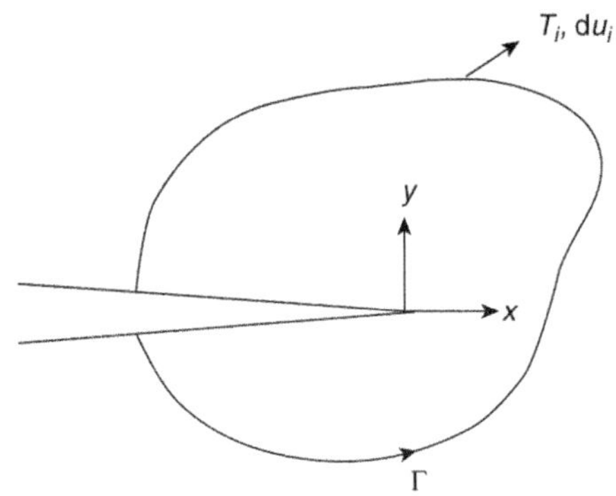

Figure 13.19 Line integral contour.

where $W = \int \sigma_{ij} \mathrm{d}\varepsilon_{ij}$ is the strain energy density relating to the stress and strain components σ_{ij} and ε_{ij} in the crack tip region, $\mathbf{T}$ is the surface traction vector, $\mathbf{u} = (u_1, u_2)$ is the vector of displacements. Rice showed that J is independent of the path Γ chosen for integration of the total energy. In the case of both linear and nonlinear elasticity, of both isotropic and anisotropic material, J equates to the strain energy release rate G. If a stress field associated with a crack is known, the integral can be used to evaluate G. This applies to stress fields in numerical form such as are created using finite element analysis. The J-integral can thus be used to evaluate G for specific geometries of experimental fracture specimens, and, via the use of Equation (13.12) or (13.15), the stress intensity and finite-width correction factors Y.

A more direct approach to its application to fracture experiments has been developed by Begley and Landes (1972). It is most instructive to follow the exposition adopted by Landes and Begley (1979) and later by Chan and Williams (1983).

When the displacements are prescribed, the J-integral is more simply defined as the rate of decrease of potential energy U with crack length. As shown in Figure 13.20, the J-integral is then given by the shaded area and

$$J = -\frac{1}{B}\frac{\mathrm{d}U}{\mathrm{d}a}, \tag{13.32}$$

where B is the specimen thickness and a is the crack length. It can be seen that the difference between constant load and constant displacement is second order and can be neglected.

Sumpter and Turner (1973) have expressed J as the sum of elastic and plastic components J_e and J_p, respectively

$$J = J_e + J_p,$$

where

$$J_e = \frac{\eta_e U_e}{B(W-a)} \quad \text{and} \quad J_p = \frac{\eta_p U_p}{B(W-a)},$$

and η_e and η_p are elastic and plastic work factors dependent on the specimen geometry.

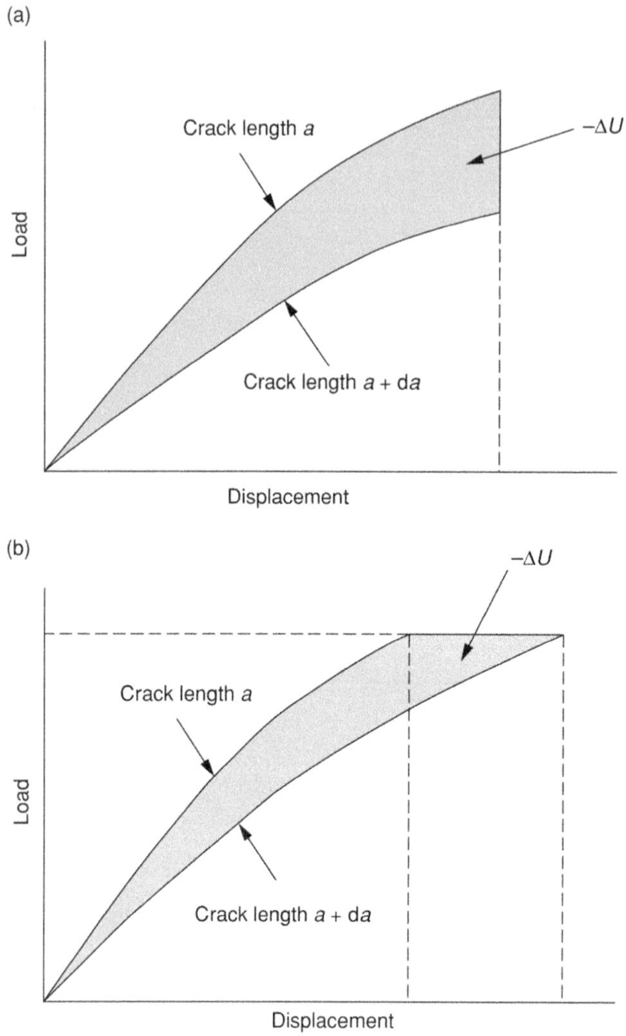

Figure 13.20 Load–displacement curves for crack growth: (a) constant displacement; (b) constant load.

For single-edge-notch blend specimens, when

$$0.4 < \frac{a}{w} < 0.6 \quad \text{and} \quad \eta_e = \eta_p = 2,$$

then

$$J = \frac{2U}{B(W-a)}. \tag{13.33}$$

The accepted procedure is to introduce a sharp crack into a single-edge-notch sample. This is usually done by machining a notch, which is then sharpened by insertion of a razor blade, either by tapping (for brittle polymers) or by slicing (for ductile polymers). After each loading, the specimens (either in tension or bending) are broken up so that the amount of crack extension Δa can be measured. This can be done by cooling the specimen to low temperatures and then breaking it to observe the difference in the fracture surface that occurs at the point of initial crack extension.

The value of J is calculated from Equation (13.33) and plotted against Δa as shown in Figure 13.21(b). Initially the crack extends by blunting the initial sharp crack. If it is assumed that the blunted crack has a semi-circular profile (this is not precisely the case for polymers) so that $\Delta a = \delta/2$, where δ is the COD, the crack blunting line is given by

$$J = \sigma_y \delta = 2\sigma_y \Delta a, \tag{13.34}$$

where σ_y is the yield stress or craze stress.

The J resistance curve shows a point of inflection where true crack growth starts, and this defines an equivalent quantity to the G_{IC} of LEFM for plane strain brittle fracture, which is termed J_{IC}. The sample size requirements for a valid J_{IC} are given by

$$a, W - a \quad \text{and} \quad B \geq 25\left(\frac{J_{IC}}{\sigma_y}\right).$$

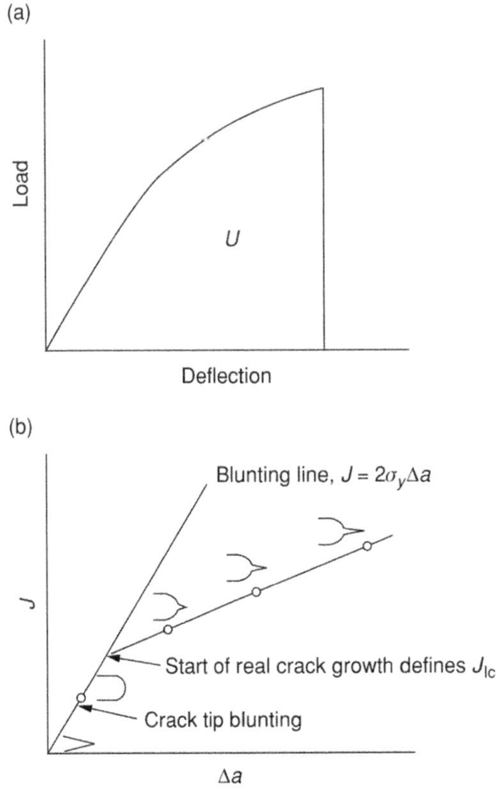

Figure 13.21 Procedure for J_{IC} measurement: (a) load identical specimens to a range of deflections; (b) plot of J against Δa, showing crack tip blunting and crack growth. Adapted from Chan and Williams, 1983.

In very tough polymers, an idealized plot like that of Figure 13.21 is not obtained and a more arbitrary procedure has been proposed (ESIS Technical Committee 1995). The J resistance curve is represented by a power law

$$J = C_1 \Delta a^{C_2}, \tag{13.35}$$

where C_1 and C_2 are fitting parameters, and a crack initiation value $J_{0.2}$ is defined that is determined from the J value for a crack extension of 0.2 mm.

The J-integral can be used to analyse experiments in more sophisticated ways. Digital Image Correlation (DIC) has been available for some years to measure strains on the surface of a stressed body and has been applied to cracked specimens (McNeill *et al.* 1987). Using a speckle pattern it provides strain fields in digital form, which can be associated with a stress field. It is therefore possible to implement a J-integral on such a field to provide a measurement of strain energy release rate based on many data points taken from a large area around a crack. Leitão and Gilabert (2021) have demonstrated this technique using epoxy fracture specimens. They also ran finite element analyses of the specimens, providing numerical stress fields on which they also applied the J-integral, to add validation. Similarly, Kojouri *et al.* (2023) applied a J-integral formulation to DIC fields on PMMA and HDPE specimens, while running parallel finite element analyses; see Figure 13.22.

13.7.2 Essential Work of Fracture

Another approach to the fracture of ductile polymers stems from the recognition that for such materials the crack tip deformation zone has two components, as shown in

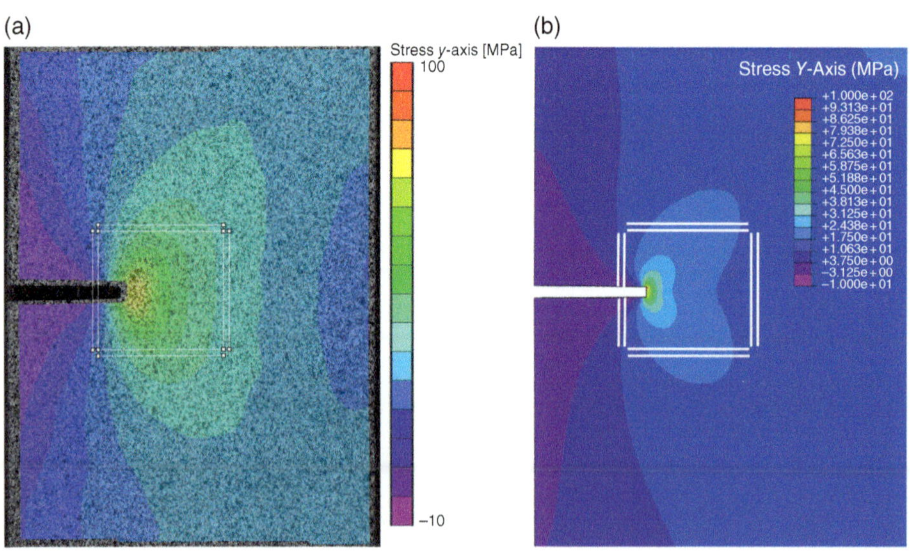

Figure 13.22 Stresses around a crack tip. (a) Experimentally derived stress field using DIC. (b) Contours of stress from a corresponding finite element model. Contour colours in both figures correspond to the same stress levels. (Leitão, N., Gilabert, F.A. (2021). A unified J-Integral-based procedure to investigate at different loading regimes the fracture by FEM simulations and image analysis. J. Mech. Phys. Solids, **149**, 104318.)

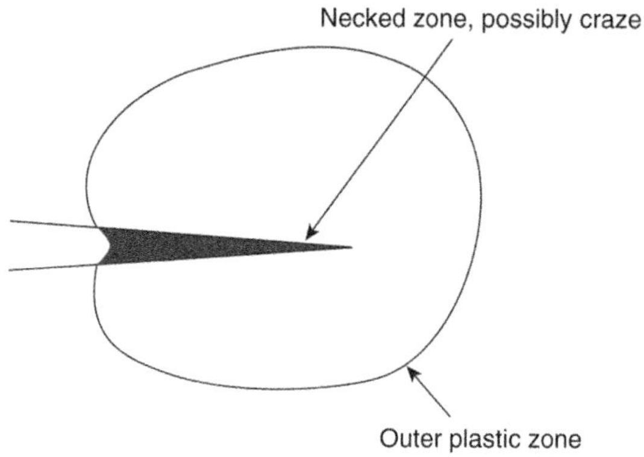

Figure 13.23 Schematic diagram of process zone in ductile fracture specimen. Wu and Mai (1996) / John Wiley & Sons.

Figure 13.23. There is an inner zone where the fracture process occurs – which could involve a combination of shear yielding and crazing – and an outer zone where extensive yielding and plastic deformation occur. This approach was originally proposed by Broberg (1968), and has been developed by Mai and Cotterell (1986), Hashemi and Williams (2000), Wu and Mai (1996) and others.

The measurements are carried out on deeply notched specimens, either single-edge-notched tension (SENT) or double-edge-notched tension (DENT) (see Figure 13.24). The total work of fracture W_f consists of two components. First, there is the work expended in the inner zone W_e, which is called the essential work of fracture. This relates directly to the energy required to fracture the sample and is therefore proportional to the ligament length l. Secondly, there is what is termed the non-essential work of fracture W_p, the energy dissipated in the outer plastic zone, where shear yielding and other forms of plastic deformation can occur. This component is proportional to the second power of the ligament length.

We have

$$W_e + W_p, \tag{13.36}$$

which can be written as

$$W_f = w_e Bl + \beta w_p Bl^2, \tag{13.37}$$

where w_e is the specific essential work of fracture, w_p is the specific non-essential work of fracture, B is the specimen thickness and β is a shape factor for the plastic zone. More conveniently, the specific total work of fracture W_f is given by

$$W_f = \frac{W_f}{lB} = w_e + \beta w_p l. \tag{13.38}$$

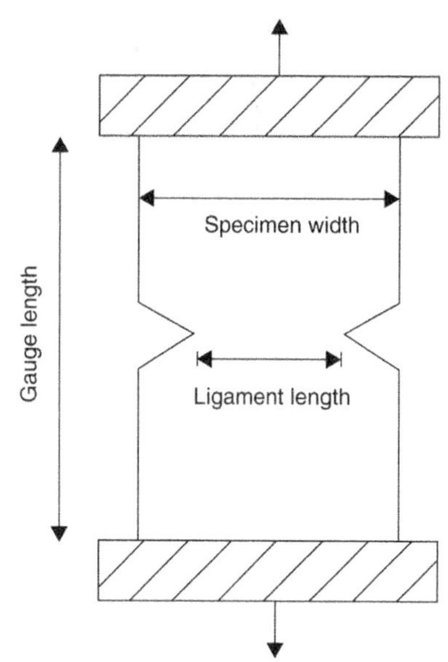

Figure 13.24 Schematic diagram of a DENT test specimen for essential work of fracture test.

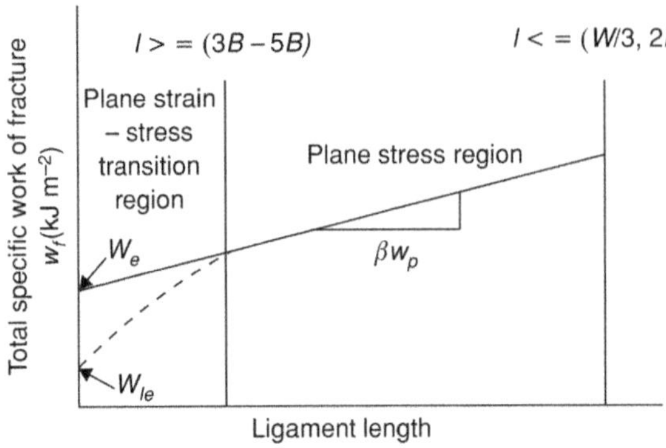

Figure 13.25 Schematic diagram showing specific total work of fracture against ligament length.

It is necessary to consider whether the fracture occurs under conditions of plane stress or of plane strain. Figure 13.25 shows schematically how the total specific work of fracture varies with ligament length. At high ligament length, conditions of plane stress pertain and Equation (13.38) applies to give an extrapolated value of W_e, the essential work of fracture for plane stress fracture; βw_p is the work dissipation in the outer plastic zone. At low ligament length, there is a transition from plane stress to plane strain and extrapolation to zero ligament length gives W_{le}, the plane strain essential work of fracture.

Wu and Mai (1996) have examined the relationship between the essential work of fracture method and the *J*-integral method. Figure 13.26 shows w_f as function of ligament length for

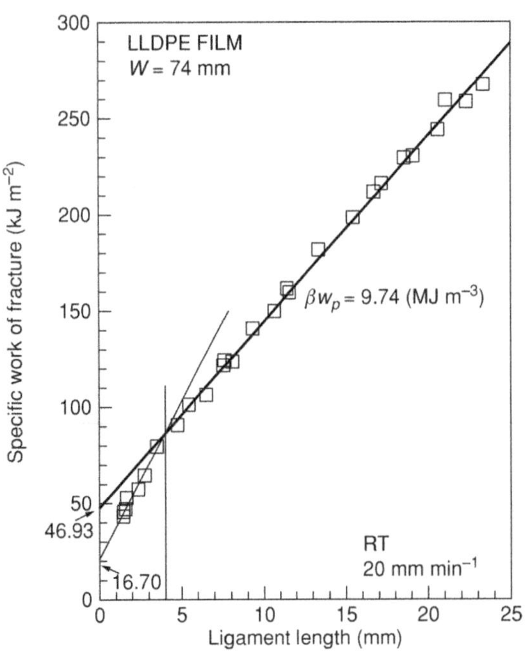

Figure 13.26 Specific work of fracture versus ligament length for linear low-density polyethylene films. Wu and Mai (1996) / John Wiley & Sons.

a DENT specimen of thickness 0.285 mm. At large ligament lengths, failure occurs under conditions of plane stress; at low ligament lengths, failure occurs under plane strain. The corresponding values for w_e are 46.93 and 16.70 kJ/m^2. Wu and Mai concluded that the plane stress w_e obtained by linear extrapolation is equivalent to the plane strain J_{IC}, as proposed earlier by Mai and Cotterell (1986).

The validity of results obtained using this method depends on the specimen dimensions being within specified ranges. Most importantly, the specimen dimensions control whether plane strain or plane stress conditions apply. Williams and Rink (2007), gathering experimental data from a number of laboratories, have produced guidelines for the standardization of the test and the interpretation of results.

The very simple analysis that is enabled by the use of Equation (13.38) requires the assumption that the shape factor β is a constant. This is equivalent to the shape of the plastic zone ahead of the crack tip remaining the same for all the crack lengths studied. Naz *et al.* (2010), working with ultra-high-molecular-weight polyethylene, used finite element modelling to predict the plastic zone shape in double-edge-notched specimens, and concluded that the shape was highly dependent on the notch depth. However, a re-analysis of their fracture results using the model plastic zone shapes did not result in significant changes in the result for w_e. This suggests that the method is robust.

13.7.3 Crack Opening Displacement

For LEFM, there is an explicit relationship between K_{IC} and the COD, which is exemplified by the Dugdale plastic zone model (Equation (13.25)). For tough polymers, it cannot be assumed that this is still the case. Nevertheless, the measurement of COD can still be a valuable tool, and it has been used extensively for assessing the toughness of polyethylene gas pipes, especially by Lu and Brown (1986) and Huang and Brown (1990). The COD is measured under plane strain conditions or nearly plane strain conditions (which cannot always be assumed for very tough samples), where a damage zone forms at the root of the notch, a craze similar to that observed in a glassy polymer (see Section 13.4). These tests are conducted under conditions of slow crack growth and usually under constant stress. Typically single-edge notch specimens such as that shown in Figure 13.5(b) are loaded in tension.

Figure 13.27 shows the COD versus time for a polyethylene copolymer loaded at a constant stress of 3 MPa (O'Connell *et al.* 1995). There is an initial linear portion followed by an accelerating rate. Experimental observations show that the point at which the COD rate starts to accelerate T_p, is associated with the first signs of fracture at the base of the craze, with failure occurring in the mid-rib of the fibrils of the craze, and this gives the failure time for slow crack growth.

Zhou and Brown (1996) developed the Pennsylvania Notch Specimen (PENT) single-edge notch geometry. Their testing method involves the application of a constant load and the continuous monitoring of the COD using a strain gauge. In this work and generally COD development follows the pattern of Figure 13.27, with a linear region followed by acceleration. Other qualitative observations are that the angle of the crack – that between the two crack faces – stays approximately constant (at around 7–10°) during crack growth, so that crack length can be estimated from COD. The time derivative of COD – the COD rate – was investigated, and variations in it were associated with micro-fracture events.

Girard *et al.* (2007) used an adaptation of the PENT specimen to measure COD in PVDF using an optical technique. They found that, above a critical stress level σ_{crit}, COD grew in a manner similar to that in Figure 13.27, as shown in Figure 13.28(a). In contrast, at stresses below σ_{crit}, rather than accelerating the COD rate reaches a steady value, as shown in

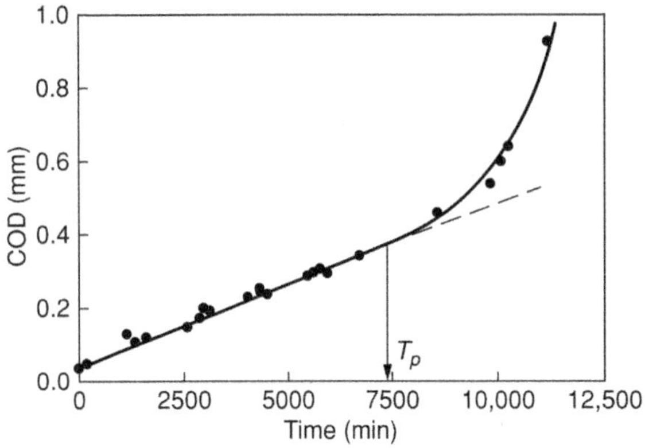

Figure 13.27 Crack opening displacement (COD) for polyethylene copolymer at a bulk stress of 3 MPa. O'Connell et al. (1995) / with permission of Elsevier.

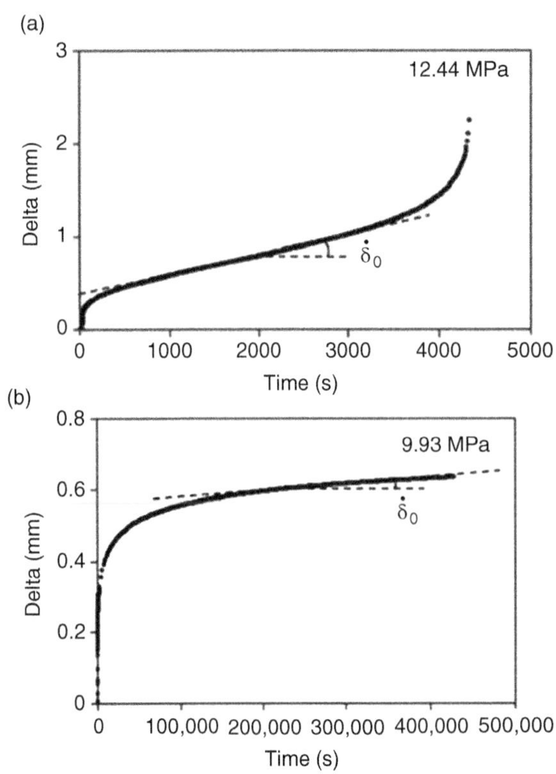

Figure 13.28 COD at constant stress as a function of time for PVDF. (a) Stress above critical value; (b) stress below critical value. Girard et al. (2007) / with permission of Elsevier.

Figure 13.28(b). At the higher stress, the increase in COD rate after around 3000 s corresponds to yielding of the whole specimen section ahead of the notch tip. At the lower stress, damage remains localized ahead of the crack tip.

For the high- and low-stress regimes in Figure 13.28(a) and (b), respectively, steady-state COD rates ($\dot{\delta}_0$) were observed. These rates depend on the stress, and would be expected to have different dependencies in the high- and low-stress regimes. An understanding of the stress dependence of these steady-state rates can help in predicting how COD develops. A power law function of stress has been applied by a number of workers (Lu and Brown 1986; Huang and Brown 1990). Such a law has been proposed by Girard *et al.* (2007), as follows:

$$\dot{\delta}_0 = C\sigma^n \tag{13.39}$$

for an applied constant stress σ. n is an experimentally determined exponent and C depends on the temperature; the COD rate increases with temperature. Girard *et al.* arrived at a form for the temperature dependence of C by using the theory of the activated process. Following this approach, we take an exponential form from Equation (12.27) for the Eyring process. Then,

$$C \propto \exp\left(-\frac{\Delta H}{kT}\right) \tag{13.40}$$

By measuring steady-state COD rates at a constant stress over a range of temperatures, it is possible to generate data to determine ΔH by fitting to the logarithmic equation derived from combining (13.39) and (13.40):

$$\ln\left(\dot{\delta}_0\right) = \ln(A) - \frac{\Delta H}{kT} \tag{13.41}$$

where A is the constant of proportionality in Equation (13.40). The success of this procedure opens up the potential to conduct short-term testing at high temperatures and predict the result of longer term tests at lower temperatures. Thus, programmes of accelerated testing become possible to predict the lifetime performances of loadbearing structures at service temperatures.

Additionally, measurement of COD is used as a tool to assess the effects of environment on slow crack growth. Methods of creep testing of notched specimens have been reviewed by Almomani *et al.* (2023). A recent example is the work of Thuy *et al.* (2022) on high-density polyethylene for a range of environments. This provides an additional way of investigating environmental effects as addressed in Section 13.6.4 above.

13.8 Factors Influencing Brittle–Ductile Behaviour: Brittle–Ductile Transitions

13.8.1 The Ludwik–Davidenkov–Orowan Hypothesis

Many aspects of the brittle–ductile transition in metals, including the effect of notching, which we will discuss separately, have been discussed in terms of the Ludwik–Davidenkov–Orowan (LDO) hypothesis that brittle fracture occurs when the yield stress exceeds

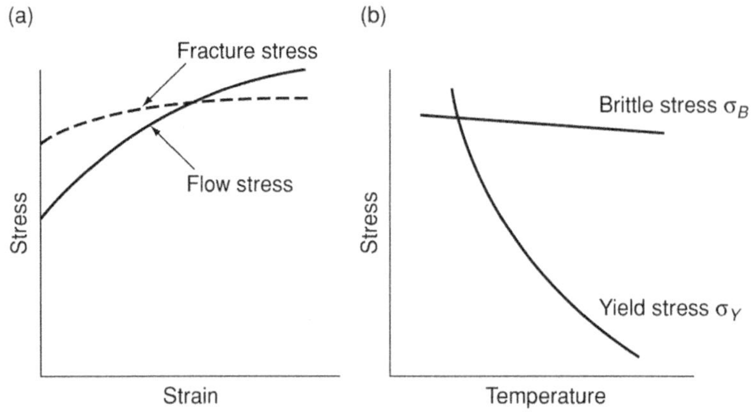

Figure 13.29 Diagrams illustrating the Ludwik–Davidenkov–Orowan theories of brittle–ductile transitions. (a) Fracture occurs when the flow stress exceeds the fracture stress. (b) At constant strain rate and varying temperature, the failure mode corresponding to the lower stress prevails; this pattern may change as strain rate changes.

a critical value (Orowan 1949), as illustrated in Figure 13.29(a). It is assumed that brittle fracture and plastic flow are independent processes, giving separate characteristic curves for the brittle fracture stress σ_B and the yield stress σ_Y as a function of temperature at constant strain rate (as shown in Figure 13.29(b)). Changing strain rate will produce a shift in these curves. It is then argued that whichever process, either fracture or yield that occurs at the lower stress will be the operative one. Thus, the intersection of the σ_B/σ_Y curves defines the brittle–ductile transition, and the material is ductile at all temperatures above this point. This argument presupposes that the fracture stress σ_B is a well-defined material parameter, contradicting Griffith's now well-established view that strength depends on flaw size. It is relevant here that the behaviour of Figure 13.29(a), while recounted by Orowan (1949), was first proposed by Ludwik in 1923, only shortly after the publication of Griffith (1921). Further developments are needed to produce a theory of the brittle–ductile transition in line with current knowledge. However, the use of a value of stress σ_B to characterize crack initiation did serve as a useful starting point.

Vincent (1960) and others (Stearne and Ward 1969; Clarke 1982) have shown that the brittle stress is not much affected by strain and temperature (e.g. by a factor of 2 in the temperature range −180°C to +20°C). The yield stress, on the other hand, is greatly affected by strain rate and temperature, increasing with increasing strain rate and decreasing with increasing temperature. (A typical figure would be a factor of 10 over the temperature range −180°C to +20°C.) These ideas are clearly illustrated by results for PMMA shown in Figure 13.30(a). The brittle–ductile transition will therefore be expected to move to higher temperatures with increasing strain rate (Figure 13.30(b)). The effect can be illustrated by varying the strain rate in a tensile test on a sample of nylon at room temperature: at low strain rates the sample is ductile and cold-draws, whereas at high strain rates it fractures in a brittle manner.

A further complication in varying strain rate occurs at low speeds, where within a certain temperature range cold-drawing occurs. It is possible that at high speeds the heat is not conducted away rapidly enough, so that strain hardening is prevented and the specimen fails in

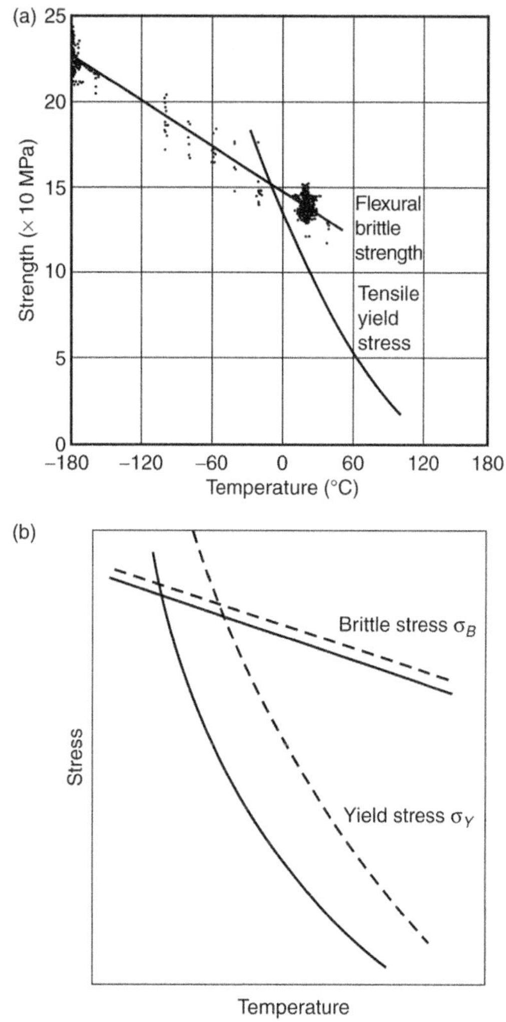

Figure 13.30 (a) Effect of temperature on brittle strength and tensile yield stress of PMMA. Vincent (1961) / John Wiley & Sons. (b) Diagram illustrating the effect of strain rate on the brittle–ductile transition: (—) low strain rate; (- - -) high strain rate.

a ductile manner. Such an isothermal–adiabatic transition does not affect the yield stress and therefore does not affect the brittle–ductile transition, but it does cause a considerable reduction in the energy to break and may be operative in impact tests, even if brittle fracture does not intervene. It has been proposed therefore that there are two critical velocities at which the fracture energy drops sharply as the strain rate is increased: the isothermal–adiabatic transition and at higher strain rates, the brittle–ductile transition. Changes in ambient temperature have very little effect on the position of the isothermal–adiabatic transition but have a large effect on the brittle–ductile transition.

It was thought at first that the brittle–ductile transition was related to mechanical relaxation and in particular to the glass transition, which is true for natural rubber, polyisobutylene and polystyrene but is not the case for most thermoplastics. It was then proposed (Hoff and Turner 1957) that where there is more than one mechanical relaxation the brittle–ductile transition may be associated with a lower temperature relaxation. Although again it appeared that there might be cases where this is correct, it was soon shown that this hypothesis has no general validity. Because the brittle–ductile transition occurs at fairly high strains, whereas the dynamic mechanical behaviour is measured in the linear, low-strain region, it is unreasonable to expect that the two can be linked directly. It is certain that fracture, for example, depends on several other factors such as the presence of flaws, which will not affect the low-strain dynamic mechanical behaviour. The subject was discussed extensively by Boyer (1968) and by Heijboer (1968).

Razavi *et al.* (2020) have reviewed the LDO hypothesis and criticized it on the grounds that it does not address the causes of yielding or explain why fracture does not happen in the ductile regime. They suggest that yielding will occur if conditions are favourable (such as under the control of an activated process), and as a result the stress will be prevented from becoming high enough to cause fracture. This is a credible picture, particularly so when a plastic zone grows across the whole of a loaded section. However, plasticity may result in highly localized behaviour such as in a shear band or neck (see Section 12.1.1).

13.8.2 Plastic Zones Around Cracks

The classic sharp flaw has a singularity in stress at its tip which is strictly non-physical. In reality, an infinitely large stress is avoided on two essential grounds: first that the crack is not ideally sharp; and second that the material will yield so that the stress remains finite. In polymers the second reason is more important than the first, with a plastic zone generally surrounding the crack tip, as analysed in Section 13.7.2 above. The idea inherent in the LDO hypothesis that yielding and fracture are independent is thus undermined. Where yielding is small-scale LEFM can be used, but otherwise more sophisticated methods are required as set out above. For large-scale yielding, the notch is blunted and fracture may be avoided altogether.

The extent of the plastic zone may be estimated by use of the crack tip, Equation (13.4), in combination with a yield criterion such as are set out in Chapter 12. We use the von Mises form for stresses in the 1-2-3 axis set of Equation (12.14). The plane $x_3 = 0$ of the crack is a symmetry plane so that $\sigma_{23} = \sigma_{31} = 0$. First we analyse plane stress conditions with $\sigma_{33} = 0$. Equation (12.14) then reduces to

$$\sigma_{11}^2 + \sigma_{22}^2 - \sigma_{11}\sigma_{22} + \sigma_{12}^2 = \sigma_Y^2 \tag{13.42}$$

Now we rewrite the required Equations (13.4) in the form

$$\begin{aligned} \sigma_{11} &= \frac{K_\mathrm{I}}{(2\pi r)^{1/2}} f_{11}(\theta) \\ \sigma_{22} &= \frac{K_\mathrm{I}}{(2\pi r)^{1/2}} f_{22}(\theta) \\ \sigma_{12} &= \frac{K_\mathrm{I}}{(2\pi r)^{1/2}} f_{12}(\theta) \end{aligned} \tag{13.43}$$

These stresses in Equation (13.42) give rise to the result

$$r = \frac{1}{2\pi}\left(\frac{K_I}{\sigma_Y}\right)^2 (f_{11}^2(\theta) + f_{22}^2(\theta) - f_{11}(\theta) f_{22}(\theta) + f_{12}^2(\theta))$$

At the critical stress intensity, for any value of θ, $r = r_p$ defines the boundary of the plastic zone:

$$r_p = \frac{1}{2\pi}\left(\frac{K_{IC}}{\sigma_Y}\right)^2 (f_{11}^2(\theta) + f_{22}^2(\theta) - f_{11}(\theta) f_{22}(\theta) + f_{12}^2(\theta)) \tag{13.44}$$

Contours of the plastic zones are shown nondimensionally in Figure 13.31. For the plane strain case $\sigma_{33} = -\nu(\sigma_{11} + \sigma_{22})$. With this condition Equation (12.14) becomes

$$\left(1 + \nu^2 - \nu\right)\left(\sigma_{11}^2 + \sigma_{22}^2\right) - \left(1 + 2\nu - \nu^2\right)\sigma_{11}\sigma_{22} + \sigma_{12}^2 = \sigma_Y^2$$

and the plane strain equivalent to Equation (13.44) is

$$r_p = \frac{1}{2\pi}\left(\frac{K_{IC}}{\sigma_Y}\right)^2 \left(\left(1 + \nu^2 - \nu\right)\left(f_{11}^2(\theta) + f_{22}^2(\theta)\right) - \left(1 + 2\nu - \nu^2\right)\left(f_{11}(\theta) f_{22}(\theta)\right) + f_{12}^2(\theta)\right) \tag{13.45}$$

The plastic zones in plane stress and plane strain are compared in Figure 13.31. The smaller plastic zone size in plane strain results in generally more brittle behaviour under these conditions. It should be noted that these zones are based entirely on the elastic stresses; in

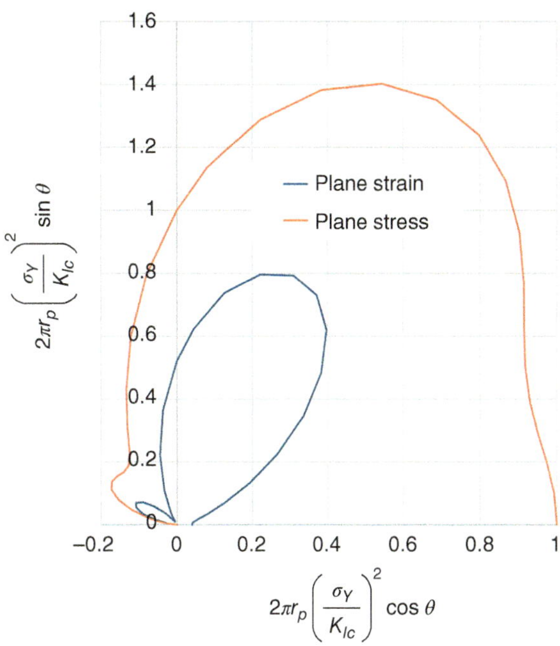

Figure 13.31 Plastic zone boundaries. Crack tip is at the origin with the crack along the horizontal axis, set to grow in the positive direction. Plane strain calculation is made for Poisson's ratio $\nu = 0.4$.

reality, yielding will change the stress field, and a true equilibrium solution can only be obtained using an elastic–plastic analysis. The zones shown provide a useful approximation.

Approximate estimates of plastic zone sizes are available using elementary formula. For plane stress, Figure 13.31 suggests a very approximate dimension of the zone such that

$$2\pi r_p \left(\frac{\sigma_Y}{K_{IC}}\right)^2 \simeq 1$$

leading to an estimate

$$r_p \simeq \frac{1}{2\pi}\left(\frac{K_{IC}}{\sigma_Y}\right)^2 \tag{13.46}$$

For plane strain it is around three times smaller, leading to the approximation

$$r_p \simeq \frac{1}{6\pi}\left(\frac{K_{IC}}{\sigma_Y}\right)^2 \tag{13.47}$$

These forms have been proposed by Irwin (1958). They are useful for estimating the validity of applications of LEFM, by establishing that the plastic zone is small in comparison with the dimensions of the stressed body under investigation. Plastic zones have been the subject of further detailed study recently (Torabi and Shahbazian 2020).

It is clear that a stressed body with large (in comparison with its dimensions) plastic zones will tend to fail in a ductile manner. It follows that if the size of a stressed component is increased and all other factors, including flaw size, are kept constant, then its response to stress may change from ductile to brittle. This shows that the ductile–brittle transition is not a material property, but is rather characteristic to specific geometry and stress conditions. We now have a sufficiently good understanding of both yield and fracture to include both phenomena within numerical models of stressed bodies. It then becomes possible to determine whether a fracture criterion is met, resulting in a brittle response, or whether ductile yielding will result.

13.8.3 Modelling the Ductile–brittle Transition

The ductile or brittle nature of the material response is determined by the outcome of the competition between yielding and fracture. This competition can be modelled explicitly using numerical methods.

Gearing and Anand (2004) studied the behaviour of blunt-notched polycarbonate beams in bending, and cylindrical notched bars in tension, both experimentally and with finite element analysis. In the finite element analyses, a viscoplastic constitutive model was used; brittle failure was defined by the elastic volumetric strain reaching a critical value, and ductile failure by the effective plastic stretch λ^p (a scalar invariant of the plastic strain tensor) reaching a critical level. This enabled the competition between brittle and ductile failure to be represented numerically. Values for the critical quantities in the brittle and ductile criteria were established that enabled quantitative prediction of brittle and ductile behaviour. In the models, ductile failure was represented by removing elements where the criterion had been attained. Figure 13.32 shows experimental observations of the development of

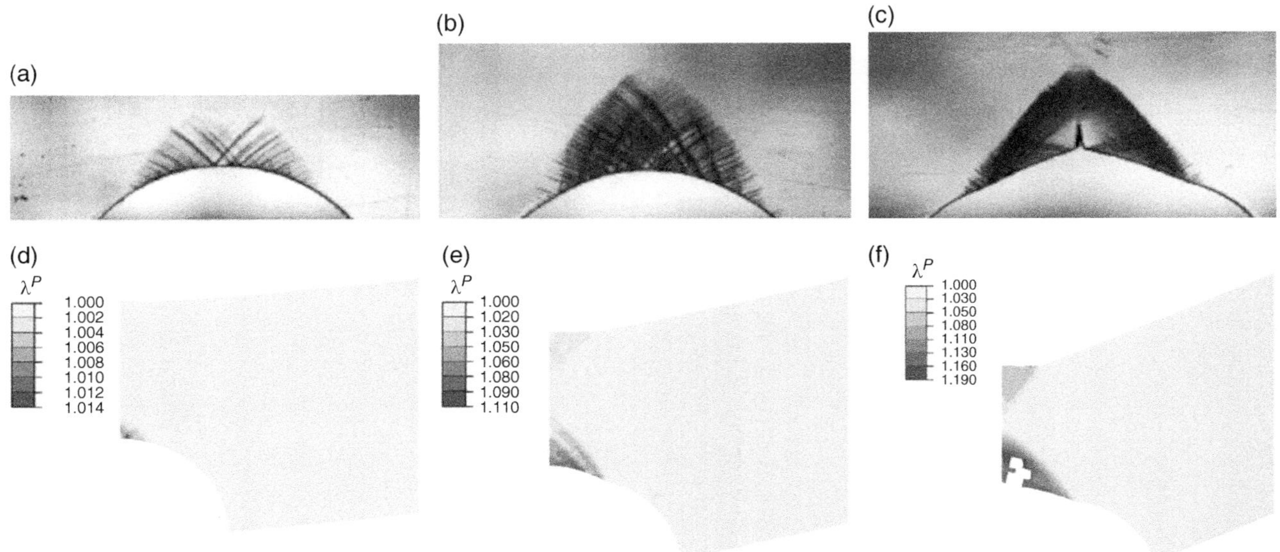

Figure 13.32 (a–c): progressive images of blunt notch in a four-point bend test, showing cracking in (c). (d–f): corresponding model results at the same deformation levels as (a–c), respectively. Reproduced from Gearing et al. (2004) / with permission of Elsevier.

ductile fracture in blunt notched beams in four-point bend together with model results at the same stages of deformation.

For materials that exhibit crazing, the competition between yielding and crazing determines the brittle–ductile transition (Estevez *et al.* 2000). Kattekola *et al.* (2013) have produced such an analysis for polycarbonate. Crazing was assumed to be by void nucleation, governed by a critical value of mean (hydrostatic) stress. Fracture, in the form of ductile tearing, was assumed to begin at a critical value of crack opening tensile stress. Three-dimensional finite element schemes analysed notched bodies with varying thickness and notch radius. Ductile–brittle transitions governed by both notch radius and specimen thickness were modelled successfully.

Van Melick *et al.* (2003) modelled polystyrene containing spherical voids or inclusions. They also used a hydrostatic stress-based criterion for craze nucleation together with a viscoplastic model of polymer deformation. They were able to model brittle-to-ductile transitions resulting from changing geometry. By keeping the void volume constant and increasing the sizes of the voids, the distance between voids was decreased. This led to a brittle-to-ductile transition as the plastic zones occupied a greater proportion of the material between the voids. The predictions agreed well with experimental data on non-adhering core–shell rubber-filled polystyrene.

The modelling of the brittle–ductile transition continues to progress. Jiang *et al.* (2017) have produced a craze initiation criterion with the addition of strain-rate dependence for the analysis of scratch damage in polycarbonate and PMMA. Aranda-Ruiz *et al.* (2020) have studied polycarbonate over a wide range of strain rates, observing the expected ductile-to-brittle transition as rate increases, followed by a further brittle-to-ductile transition at very high rates (40,000 s^{-1}). They have modelled the phenomena using three-dimensional finite element analysis.

13.9 The Impact Strength of Polymers

The ability of a structural part to maintain its integrity and to absorb a sudden impact is often a relevant issue when selecting a suitable material. Impact testing of polymers is thus a subject of some importance and is extensively employed, although many of the results obtained are of an empirical and hence comparative nature. It is useful when fracture information is required of materials that are ductile at low strain rates; experimenters can take advantage of the ductile–brittle transition that accompanies increasing strain rate.

The two major types of impact test are categorized as flexed beam and falling weight.

13.9.1 Flexed-Beam Impact

Examples of flexed-beam impact are the Izod and Charpy impact test, in which a small bar of polymer is struck with a heavy pendulum. In the Izod test, the bar is held vertically by gripping one end in a vice and the other free end is struck by the pendulum. In the Charpy test, the bar is supported near its ends in a horizontal plane and struck either by a single-pronged or two-pronged hammer so as to implement a rapid three-point or four-point bend test, respectively (Figure 13.33(a)). The final height of the hammer after fracture is less than the initial height; the implied loss of energy can be used to estimate the energy of fracture.

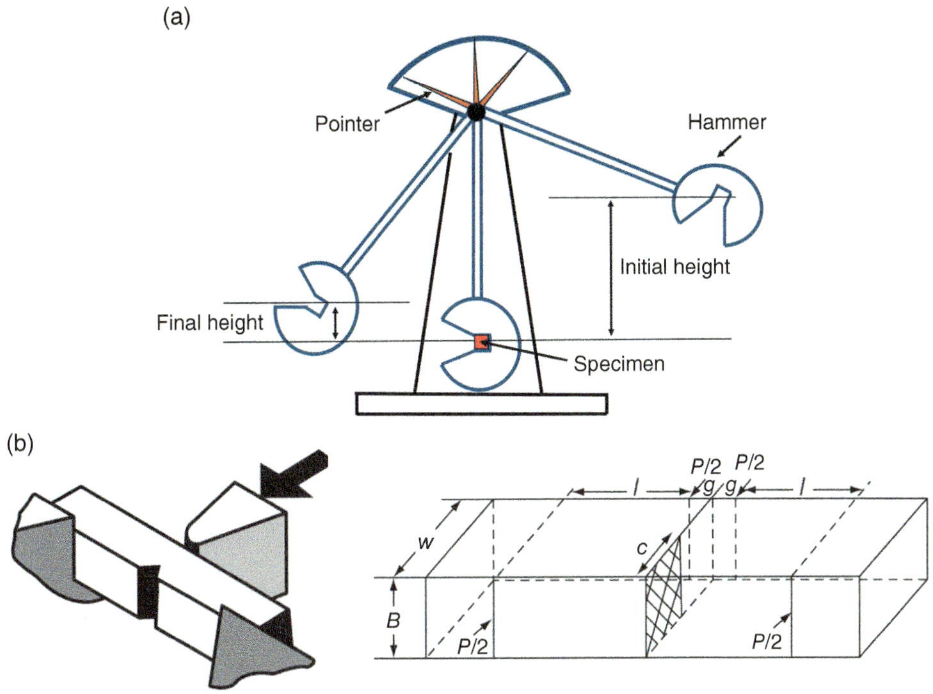

Figure 13.33 (a) Schematic drawing of a Charpy or impact tester. (b) The notched Charpy impact specimen.

However, for many years instrumented impact testers have been used (Radon and Turner 1969; Dear and MacGillivray 1991) that give direct estimates of fracture forces, such as by including a load cell within the hammer. It is customary to introduce a centre notch into the specimen to add to the severity of the test. The standard Charpy impact specimen has a 90° V-notch with a tip radius of 0.25 mm. For polymers, a very much sharper notch is often adopted by tapping a razor blade into a machined crack tip, which has important consequences for interpretation of the subsequent impact test.

The interpretation of impact tests is not straightforward, and it is necessary to consider several alternatives, as follows:

1) It was proposed independently by Brown (1973), and by Marshall et al. (1973), that Charpy impact tests on sharply notched specimens can be analysed quantitatively in terms of linear elastic fracture mechanics. It is assumed that the polymer deforms in a linear elastic fashion up to the point of failure, which occurs when the change in stored elastic energy due to crack growth satisfies the Irwin–Kies relationship (Equation (13.13) above). So that

$$G_c = \frac{K_c^2}{E^*} = \frac{P_0^2}{2B}\frac{dC}{dc},$$

where E^* is the reduced modulus as defined in 13.2.1, P_0 is the load immediately prior to fracture, C and B are the specimen compliance and thickness, respectively, and c is the crack length as in Equation (13.18) in Section 13.5. Because the elastically stored energy in the specimens immediately prior to failure is $U_0 = P_0^2 C/2$,

$$G_c = \frac{U_0}{B}\frac{1}{C}\frac{\mathrm{d}C}{\mathrm{d}c}, \tag{13.48}$$

where U_0 is determined in a commercial impact tester from the potential energy lost due to impact. The total measured impact energy U_I must be reduced by the kinetic energy of the sample U_k to give $U_0 = U_\mathrm{I} - U_k$.

It is conventional to follow Plati and Williams (1975) and express Equation (13.48) as

$$G_c = \frac{U_0}{BW}\frac{1}{C}\frac{\mathrm{d}C}{\mathrm{d}(c/W)}. \tag{13.49}$$

We then have

$$U_1 = BW\phi(c/W)G_c + U_k, \tag{13.50}$$

where $\phi(c/W) = \dfrac{c}{\mathrm{d}C/\mathrm{d}(c/W)}$ can be calculated (see Williams 1980, Chapter 4), and a plot of U_I versus $BW\phi$ produces a straight line with G_0 as slope and U_k as the intercept. This approach has been shown to give values for G_c that are independent of specimen geometry for impact tests on razor-notched samples of several glassy polymers, including PMMA, polycarbonate (Fraser and Ward 1977) and poly(ether sulphone) (Hine 1981). Similar results have been obtained also for razor-notched samples of polyethylene (Truss *et al.* 1983).

2) Vincent (1971) and others have recognized that the impact strength depends on the geometry of the notch, which led Fraser and Ward (1974) to propose that for comparatively blunt notches (i.e. those not introduced by a razor blade or a sharp cutting tool) failure occurs when the stress at the root of the notch reaches a critical value. This stress, which in a glassy polymer marks the stress required to initiate a craze, can be calculated by assuming that the deformation is elastic. On this hypothesis, the Charpy test, as undertaken in the impact tester, can be regarded as a four-point bend test with the maximum bending moment at the specimen centre $M = Pl/2$, where P is the applied load and $2l$ is the specimen span (Figure 13.33). Immediately prior to fracture, $M = M_0$, $P = P_0$ and the elastically stored energy is $U_0 = \frac{1}{2}(2M_0/l)^2 C$, where C is the sample compliance. Hence

$$M_0 = \frac{l}{2}\sqrt{\frac{2U_0}{C}} \tag{13.51}$$

where C is calculable from specimen geometry.

According to the elementary theory of bending, the nominal stress at the root of the notch σ_n is given by $\sigma_n = (M/I)y$, where I is the second moment of area ($= Bt^3/12$ for a rectangular beam) and y is the distance from the root of the notch to the neutral axis.

The actual stress at the root of the notch is the product of the nominal stress derived above and the stress concentration factor α_k. Calculations of α_k for general shapes of notch are available in the literature (Pluvinage 2004), but when the crack length c is much greater than the notch tip radius ρ, α_k reduces to the simple expression $\alpha_k = 2\sqrt{c/\rho}$.

It has been shown that the impact behaviour of blunt-notched specimens of PMMA is consistent with a critical stress at the root of the notch (Fraser and Ward 1974), and similar considerations apply to polycarbonate (Fraser and Ward 1977) and poly(ether sulphone) (Hine 1981) in the absence of shear lips. In these instances, it appears therefore that the maximum local stress is the fracture criterion, independent of specimen geometry.

3) The most unsophisticated interpretation of the flexed bend impact test is that it is a measure of the energy required to propagate the crack across the specimen, irrespective of whether the specimen is notched or unnotched. The loss in energy of the hammer is simply related to the difference between its initial and final heights (see Figure 13.33) as indicated by the pointer's final position. This energy is equated to the energy U_0 required for fracture. Notch sensitivity is ignored and only the energy of propagation is involved. In the elastic case with U_0 defined as above

$$G_c = \frac{U_0}{A} = \frac{U_0}{B(W-c)}, \tag{13.52}$$

where the area of the uncracked cross-section is $A = B(W - c)$.

Plati and Williams (1975) have given a solution for a plastic analysis. When complete yielding of the material ahead of the crack is assumed at the constant yield stress σ_Y, the fracture energy is now denoted by J_c rather than G_c and the energy divided by area is

$$J_c = \frac{\sigma_Y B(W-c) \times u}{B(W-c)} = \sigma_Y u \tag{13.53}$$

If the energy is calculated for an elastic body failing at a maximum stress σ_Y the result is

$$U_0 = \frac{u}{2}\sigma_Y B(W-c)$$

as the stress increases linearly from zero to σ_Y with average value $\sigma_Y/2$. In contrast, for the fully yielded case the stress is at constant σ_Y and the energy is

$$U_{\text{plastic}} = u\sigma_Y B(W-c)$$

The energies differ by a factor of 2. The elastic analysis as embodied in Equation (13.51) can therefore be adapted for plastic conditions by doubling the value of the energy calculated from the peak load. Plati and Williams showed results from high-impact polystyrene (HIPS) and acrylonitrile–butadiene–styrene (ABS) polymers that agreed with values of J_c from impact tests.

Later work has extended the J-integral and essential work of fracture methods described in Section 13.7 to impact tests.

Bramuzzo (1989) used high-speed photography to monitor the crack propagation in a three-point bend test in parallel with determining the force–time curve. In this way, resistance curves were obtained for polypropylene copolymers by plotting the J-integral as fraction of crack length. Martinatti and Riccio (1994) used the multi-specimen technique to determine the J_R curves for rubber-toughened polypropylenes using an instrumented

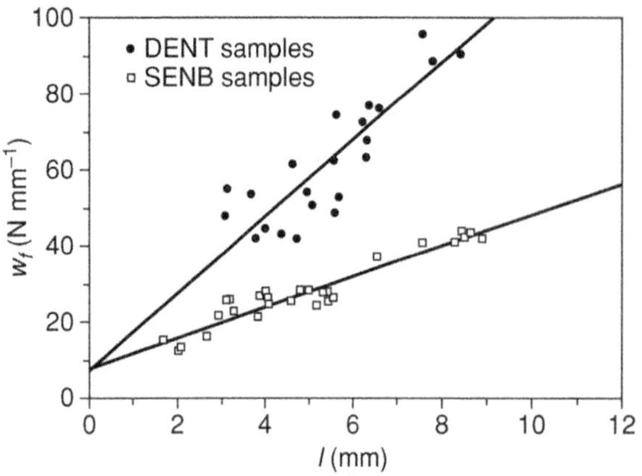

Figure 13.34 Specific total work of fracture w_f versus ligament length l for deeply double-edge-notched tension (DENT) (•) and single-edge-notched bend (SENB) (□) 3 mm thick polypropylene random copolymer samples. Fasce et al. (2001) / John Wiley & Sons.

Charpy test where the hammer of the pendulum could be stopped at different displacements of the specimen. The crack advancement of each loading was measured after successive fractures at low temperatures using an optical microscope.

In further work using the *J*-integral, Ramsteiner (1999) determined $J_{0.2}$ values by constructing the *J* values as a function of crack length, impacting specimens of HDPE with different masses from the same height to give a constant impact velocity of 2 m/s. Finally, Fasce *et al.* (2001) have attempted to apply the essential work of fracture methodology to impact testing of two PP copolymers and ABS using pre-cracked specimens of different notch deeply DENT and single-edged-notched bend (SENB) depth. In both cases, the specific total work of fracture was plotted against ligament length (Figure 13.40).

Fasce *et al.* (2001) used the essential work of fracture to analyse impact fracture of ductile polymers such as polypropylene and ABS; some of their work is summarized in Figure 13.34. Huang and Paul (2006) combined Izod impact testing with the essential work of fracture method of analysis to compare nylon 6 with a toughened polyamide.

13.9.2 Falling-Weight Impact

Another testing geometry widely used to assess the ability of a material to resist high strain rates is the falling weight impact test. Here, a flat specimen is freely supported on an annulus (often 76 mm in diameter) and then impacted by a metal dart with a hemispherical tip (with a typical radius of 6 mm). The metal dart is usually connected to a freely falling weight (typically 25 kg although this is usually adjustable) which can also be dropped from a variety of heights which allows the impact velocity to be varied and explored. In a further enhancement, the sample and its holding fixtures are often surrounded by a temperature controlled oven which can go to sub-zero temperatures, as the combination of low temperature and high strain rates is the most severe test of the impact strength of a polymer.

There are two main modes of operation of the falling weight impact test:

1) In the first configuration known as the staircase method (and as described for instance in ASTM D1709) the drop height of dart is kept at a fixed value while the drop weight can be varied in a series of weight increments. Starting from a low value, the weight is increased in steps until damage is first observed (often by backlighting the underside of the sample). Next the weight is reduced by one increment and tested again. If failure occurs the weight is reduced again and if failure does not occur the weight is increased again. This is continued until 10 fractured samples have been observed. The final result is the weight required for 50% of the sample to fail and gives an insight into the energy required for crack initiation to occur. As the test is carried out at heights designated by the testing standards, the samples can be compared between different laboratories. A similar staircase method can be applied to the quality control of a final moulded polymer component, establishing the height at which 50% of the components will fail.
2) In the second configuration (for example ASTM D5628), a piezoelectric load cell is connected between the dart and falling weight allowing the force/time curve during the impact event to be measured. This second method is what is termed an 'excess energy' method, where the impact energy far exceeds the energy required to initiate failure as found in configuration 1. In a typical arrangement, the sample to be tested is clamped between metal holding plates using air pressure. The opening in the plates is 76 mm and the dart has a diameter of 12.7 mm. The dart is connected to the impact weight of 25 kg and is dropped from a height of 200 mm (giving an impact velocity of 3.33 m/s). The force/time curve is measured during the impact, and this can be analysed to determine the maximum force and displacement at impact, and also integrated to determine the impact energy. By convention, the impact energy is usually integrated up to the point where the measured force drops to 50% of the maximum value.

In the majority of published studies, the second configuration, that is the excess energy method, is used as it allows a more quantitative analysis. Moore and his colleagues have described the application of this configuration to polymers and to polymer composites (Johnson *et al.* 1986; Jones *et al.* 1986; Moore and Prediger 1988). It is emphasized that for any reasonable attempt at interpretation the following must be carried out:

1) Measurement of the force–time curve so that the input energy to maximum force can be determined as well as the total impact energy.
2) Photography of the tension surface during the impact event.

For fibre composites, Moore and colleagues showed that the peak on the force–time curve corresponds well with the energy required to initiate a crack. It was shown also that for both composites and polymers the total fracture energy corresponded quite well with that determined from notched Charpy tests.

Only for the Charpy test, and to a rather lesser extent the Izod test, has a satisfactory theoretical analysis been achieved. Even for these tests, however, there is still a gap between the engineering analysis and any accepted interpretation in physical terms. For example, although it seems likely that the brittle failure of razor-notched impact specimens is associated with the craze at the crack tip, there is no convincing numerical link between craze parameters and the fracture toughness K_{IC}, as exists for the cleavage fracture of compact tension specimens (see Section 13.2). Again, although the mechanics point to a critical stress criterion for some blunt-notched specimens and there is an empirical correlation with the craze stress determined in other ways, the magnitude of the critical stress is very great and suggests that a more sophisticated explanation may be required. For the brittle epoxy resins, which do not show a craze at the crack tip, Kinloch and Williams (1980) have suggested that the fracture of both razor-notched and blunt-notched specimens can be described by a critical stress at a critical distance ($\sim$10 μm) below the root of the notch.

As the temperature and strain rate in a polymer change, the nature of the stress–strain curve can alter remarkably. It is therefore natural to seek correlations between the area beneath the stress–strain curve and the impact strength, and between dynamic mechanical behaviour and the impact strength. Attempts to make such correlations directly have met with mixed success (Evans *et al.* 1960), which is not surprising in view of the complex quantitative interpretations of impact strength suggested above.

Vincent (1974) has examined the statistical significance of a possible inverse correlation between impact strength and dynamic modulus and concluded that, at best, this correlation only accounts for about two-thirds of the variance in impact strength. Factors such as the influence of molecular mass, and details of molecular structure such as the presence of bulky side groups, are not accounted for. He also reported impact tests over a wide temperature range on some polymers, notably polytetrafluorethylene and polysulphone, where peaks in brittle impact strength were observed at temperatures close to dynamic loss peaks, suggesting that in some instances it may be necessary to consider the relevance of a more generalized form of fracture mechanics (Andrews 1974), where the viscoelastic losses occurring during loading and unloading must be taken into account.

Crouch and Huang (1994) produced multi-specimen resistance curves for toughened nylon by impacting SENT three-point bend specimens to different levels of crack growth using a falling-weight impact tower. Force–time curves were determined to obtain the total energy up to maximum deflection.

In common with other macroscopic mechanical tests, impact strength has been found to depend on molecular aspects such as the morphology and the molecular weight distribution. Many authors have shown that the impact strength also correlates well with dynamic properties of a polymer and the value of $\tan\delta$. Although they used an Izod geometry, Wada and Kasahara (1967) showed a strong correlation between impact strength and the loss modulus measured from DMA measurements.

Ippitsu *et al.* (2023) used the first configuration described above and investigated low-energy falling weight tests to study the onset of damage and subsequent mechanical properties of isotactic polypropylene. They linked this to small-angle x-ray scattering measurements of samples exposed to a variety of impact energies and linked this to the crystalline and amorphous fractions.

Karger-Kocis and colleagues (1999, 2002) have used the falling weight geometry to investigate the effect of molecular weight and crystal form on the impact strength of isotactic polypropylene. They found that the falling weight impact strength increased with molecular weight and also improved if the β form could be nucleated during processing.

13.10 The Tensile Strength and Tearing of Polymers in the Rubbery State

13.10.1 The Tearing of Rubbers: Extension of Griffith Theory

The Griffith theory of fracture implies that the quasi-static propagation of a crack is a reversible process. Rivlin and Thomas (1953) and Thomas (1955) recognized, however, that this may be unnecessarily restrictive, and also that the reduction in elastically stored energy due to the crack propagation may be balanced by changes in energy other than that due to an increase in surface energy. Their approach was to define a quantity termed the 'tearing energy', which is the energy expended per unit thickness per unit increase in crack length. The tearing energy includes surface energy, energy dissipated in plastic flow processes and energy dissipated irreversibly in viscoelastic processes. Provided that all these changes in energy are proportional to the increase in crack length and are primarily determined by the state of deformation in the neighbourhood of the tip of the crack, then the total energy will still be independent of the shape of the test piece and the manner in which the deforming forces are applied.

In formal mathematical terms, if the crack increases in length by an amount dc, an amount of work $G_c t dc$ must be done, where G_c is the fracture energy, referred to in rubbers as the tearing energy, and t is the thickness of the sheet. Assuming that no external work is done, this can be equated to the change in elastically stored energy as for Equation (13.2) above, giving (Tsunoda *et al.* 2000)

$$-\left[\frac{\partial U}{\partial c}\right]_l = G_c t \tag{13.54}$$

The suffix l indicates that differentiation is carried out under conditions of constant displacement of the parts of the boundary that are not force-free. Equation (13.54) is similar in form to Equation (13.2) above where G_c has replaced 2γ (see Section 13.2.4).

The so-called trouser tear experiment shown in Figure 13.35 is a particularly simple case where the equation can be evaluated immediately. After making a uniform cut in a rubber sheet, the sample is subjected to tear under the applied forces F. The stress distribution at the tip of the tear is complex, but provided that the legs are long it is independent of the depth of the tear. In contrast to all the testing considered so far, this is testing in mode III (see Section 13.2.3). Given the complexity of the energy contributions to G_c, it is possible that its value may vary between different modes.

Typically, the end of each leg is gripped in the jaws of a tensile testing machine, and the moving jaw travels at constant speed. In a successful test a steady state is reached as a tear progresses at a constant force F. As long as the crack proceeds parallel to the specimen axis, its speed is equal to the speed of the testing machine. If the sample tears a distance Δc under the force F, and changes in extension of the material between the tip of the tear and the legs are ignored, the work done is given by $\Delta W = F\Delta c$. Because the tearing energy $G_c = \Delta W / t\Delta c$, $G_c = F/t$ and can be measured easily.

The tearing energy is the energy required to extend the rubber to its maximum elongation and does not relate directly to tensile strength but depends on the shape of the stress–strain curve together with the viscoelastic nature of the rubber. For example, we may contrast two

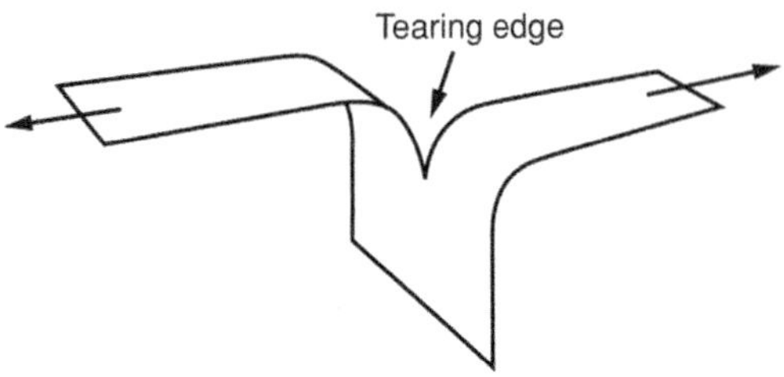

Figure 13.35 The standard 'trouser tear' experiment.

different rubbers, the first possessing a high tensile strength but a very low elongation to fracture and very low viscoelastic losses, and the second possessing a low tensile strength but a high elongation to fracture and high viscoelastic losses. In spite of its comparatively low tensile strength, the second rubber may still possess a high tearing energy.

The steady-state conditions in the trouser tear test make it simple to carry out and to analyse. However, it has been known since the early days of rubber testing that rubber failure in practice often occurs in an unsteady manner, such as in slip-stick cracking where high growth rates are interrupted by stationary or low speed cracks (Greensmith and Thomas 1955; Thomas 1960). These workers established that tearing energy was in many cases an increasing function of the crack tip velocity, though in some cases the function was not monotonic. A major complication in rubber fracture is that some rubbers, in particular natural and polyisoprene rubbers, crystallize in response to a threshold strain level (Osumu *et al.* 2023). Given the large strains at the crack tip, strain-induced crystallization can be expected to have significant effects on rubber fracture, and has been implicated in the complex relationship between tearing energy and testing speed. This phenomenon continues to be investigated, for example by Sakulkaew *et al.* (2011), who have studied tearing energy as a function of the time derivative of the strain energy release rate rather than of crack speed. Using screw-driven, pneumatic and falling weight testing machines, they experimented over a range of rate in excess of six decades. Some results for two natural rubber compounds are shown in Figure 13.36.

The results show stick-lip behaviour in the trouser tear testing, which the authors attribute to strain-induced crystallization. There is little rate dependence in the material NR0-S1. For NR0-S3, with higher crosslink density, the negative gradient is attributed to the suppression of crystallization as the rate increases, and the subsequent positive slope to viscoelastic effects.

13.10.2 Molecular Theories of the Tensile Strength of Rubbers

Most molecular theories of the strength of rubber treat rupture as a critical stress phenomenon, with the strength of the individual polymer chain being the crucial factor controlling the strength of the material. In much of this work stress concentrations arising from flaws are not considered. This approach would enable comparative studies of rubbers if it is assumed that the effect of flaws is the same in the materials that are being compared. Thus, Flory (1946) found results for the strength of butyl to be proportional to the number of chains that are oriented by stretching, and derived relationships between molecular mass and strength. He argued that, in a crosslinked structure, there are 'terminal chains', that

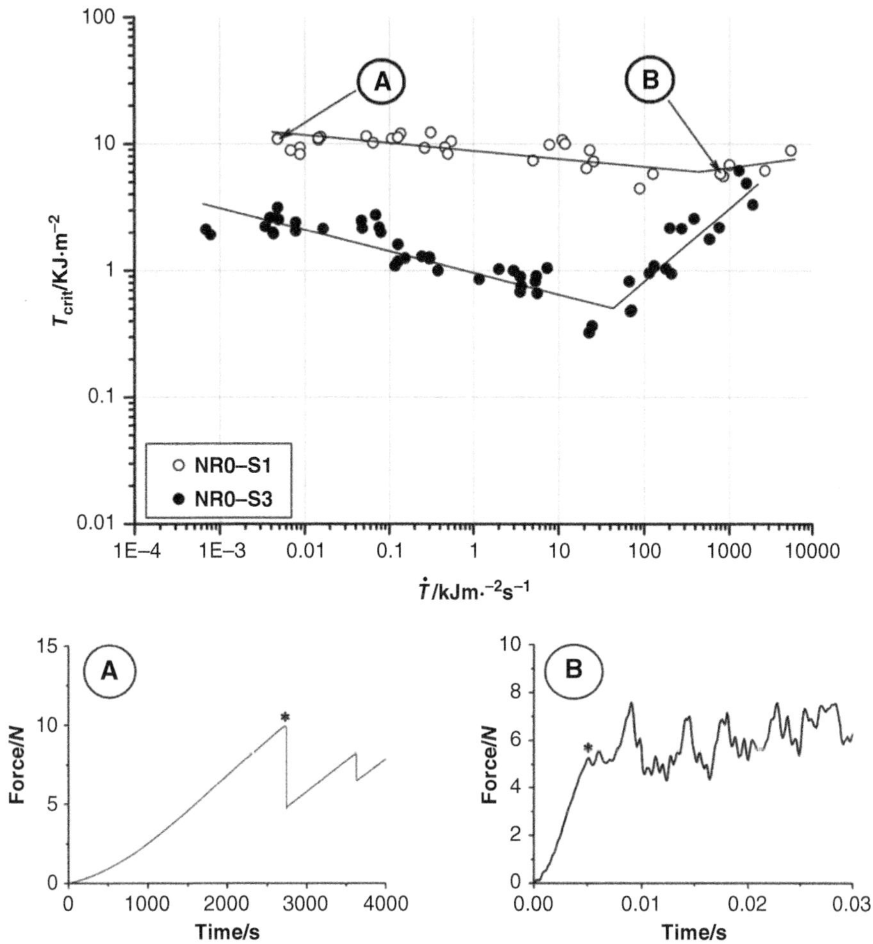

Figure 13.36 Tearing energy T (equivalent to G_c) plotted against its time derivative for two natural rubbers, NR0-S3 having around twice the crosslink density of NR0-S1. Trouser tear plots are shown for the experiments A and B. Sakulkaew et al. (2011) / with permission of Elsevier.

extend from a crosslink to the end of a primary chain (i.e. a chain that pre-existed crosslinking) and that do not participate in the network, and so are not 'effective'. Stress is then determined by the number of effective chains. This in contradiction to the entropic models of Chapter 4, as embodied in Equation (4.30), where stress depends on the number of crosslinks and not on the molecular mass. The proportion of active chains N relative the total number N_{tot} was given by Flory as

$$N = N_{tot}\left(\overline{M}_n - \overline{M}_c\right) / \left(\overline{M}_n + \overline{M}_c\right) \tag{13.55}$$

where $\overline{M}_n$ is the number averaged molecular mass and $\overline{M}_c$ is the average molecular mass between crosslinks. Flory (1946) produced experimental evidence that the tensile strength of some rubbers was proportional to the function N of Equation (13.55), suggesting that strength is a related to the strength of an individual polymer chain. It is accepted that the strength of the rubber is reduced from its theoretical strength in a perfect sample by the presence of flaws. The relevance of equations such as (13.55) becomes feasible if we assume that the strength is reduced from that of a flawless sample by approximately the same factor for different rubbers of the same basic chemical composition.

Bueche (1957) accepted the idea that fracture is controlled by the strength of individual polymer chains, but advanced the argument by envisaging a network of chains that has a distribution of levels of tautness. A highly stressed chain can break and then transfer stress to adjacent chains. Fracture is considered to occur if the load transfer causes other chains to break, causing a 'chain reaction' and catastrophic failure. His experimental results could be fitted to his theory by assuming a strength of a single rubber chain of 4.7 nN. The original chain rupture was assumed to be caused by its increase in tension driving a thermally activated process similar to that of Eyring, exceeding the bonds' activation energies.

Progress on the issue of the mechanism of rupture has been made more recently by single molecule numerical simulations of polyisoprene and polybutadiene by Hanson and Martin (2009). They identified the dissolution of C—C bonds as the mechanism, with *ab initio* quantum mechanical calculations for the strength of single molecules yielding values of 6.8 nN for polyisoprene and 7.2 nN for polybutadiene. The understanding gained of the behaviour of the individual chain has been applied to simulations of three-dimensional networks of polyisoprene by Hansen (2011). He identified initial entropic behaviour of the chain from low strain, proceeding to enthalpic behaviour as the chain straightens until bond rupture. Hansen and Barber (2013) have continued the modelling approach to fracture of three-dimensional networks. Their simulations successfully predict failure without the need for Bueche's 'chain reaction effect'. Comparison with experimental data is difficult as specific material characteristics that are required for the models, such as crosslink density, are not known. However, theoretical strengths are around a factor of two higher than experiments for reasonable estimates of crosslink density. High predictions are to be expected, given that the network models are assumed to be free of flaws and impurities.

13.11 Time and Temperature Effects

13.11.1 Time Dependence in Fracture

Previous chapters have made clear that time dependence is fundamental to polymer mechanical behaviour. This raises the possibility of creating a theoretical framework for time and rate dependence of polymer fracture by replacing elastic constitutive behaviour with viscoelasticity. A theory of linear viscoelastic fracture has been developed by a number of workers, including Knauss (1973) and Schapery (1975), recently reviewed by Knauss (2015). We have seen above in Sections 13.7 and 13.8 that the introduction of viscoplastic analyses can assist in the more realistic modelling of crack blunting and plastic zones. We might expect similar realistic features from the introduction of viscoelasticity, but, as pointed out by Guo *et al.* (2020) this is not the case for linear viscoelasticity. According to the correspondence principle (Anand *et al.* 2022), for a cracked body under stress boundary conditions, the crack tip stresses are the same for linear viscoelasticity as for linear elasticity, while in the former case the strains increase with time. Even when the crack tip singularity has been removed by the introduction of a cohesive zone (Knauss 1973; Schapery 1975), the computed size of dissipative zone is found to be very small (Gent and Lai 1994). Guo *et al.* (2020) developed a nonlinear viscoelastic model that in effect gave rise to rate-dependent yielding resulting in a reduced stress on the approach to the crack tip to give more realistic behaviour. Qi *et al.* (2024) have applied a nonlinear viscoelastic analysis to the fracture of a soft hydrogel.

The effect of time on the growth of cracks and crazes has been analysed by using time-dependent viscoplastic material models combined with a criterion for void growth. The latter criterion may include time dependence via a relation between the time derivative of void growth and the prevailing stress conditions. An example of this is the work of Ricard *et al.* (2014) using finite element analysis. Crack growth is related to the evolution of the material porosity f, the time derivative of which follows a relation (Boisot *et al.* 2011)

$$\dot{f} = (1 - f)\text{trace}\left(\dot{\tilde{\varepsilon}}_p\right)$$

where $\dot{\tilde{\varepsilon}}_p$ is the local strain rate tensor. Fracture is associated with the attainment of a predefined value of f determined by experimental observations of fracture. Thus a damage parameter is defined that controls failure of individual elements, which may then be deleted from the analysis. The progress of a cracking simulation is shown in Figure 13.37.

13.11.2 Strain Rate and Temperature Effects

The influence of strain rate and temperature on the tensile properties of elastomers and amorphous polymers has been studied extensively, particularly by Smith (1958, 1960) and Smith and Stedry (1960), who measured the variation of tensile strength and ultimate stain as a function of strain rate for a number of elastomers. The results for different temperatures could be superimposed, by shifts along the strain rate axis, to give master curves for tensile strength and ultimate strain as a function of strain rate. Results of this nature are shown in Figure 13.38, which summarizes Smith's data for an unfilled styrene–butadiene rubber. Remarkably, the shift factors obtained from superposition of both tensile strength and ultimate strain took the form predicted by the WLF equation (see Section 6.3.2) for the superposition of low-strain linear viscoelastic behaviour of amorphous polymers (Figure 13.39). The actual value for T_g agreed well with that obtained from dilatometric measurements.

This result suggests that, except at very low strain rates and high temperatures where the molecular chains have complete mobility, the fracture process is dominated by viscoelastic effects. Bueche (1955) has treated this problem theoretically and obtained the observed form of the dependence of tensile strength on strain rate and temperature. Later theories have attempted to obtain the time dependence for both tensile strength and ultimate strain, or the time to break at a constant strain rate (Bueche and Halpin 1964; Halpin 1964).

Smith plotted $\log \sigma_B/T$ against $\log e$ for the above and similar data to obtain a unique curve for all strain rates and test temperatures, which he termed the 'failure envelope' for elastomers

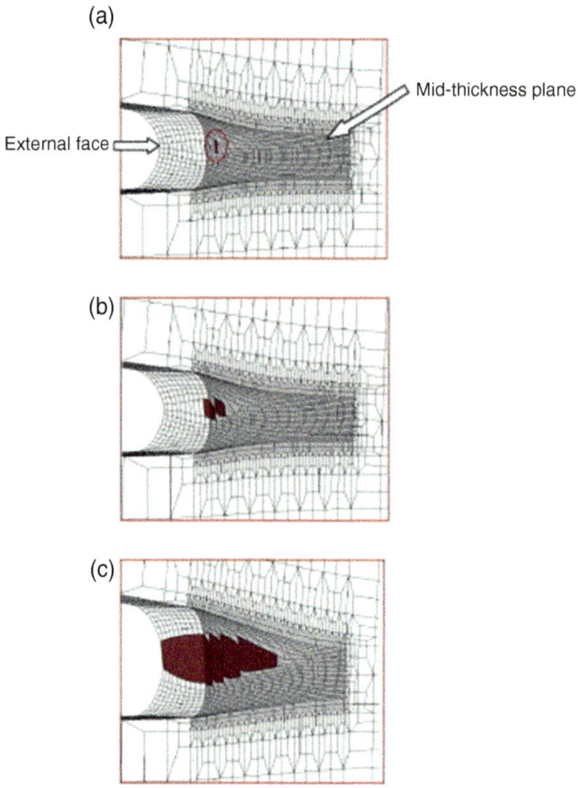

Figure 13.37 FE simulation of crack growth in POM. Red regions denote failed elements. (a) Fracture initiating at the mid-plane position. (b) Crack growth away from the initiation point. (c) Crack has grown to the outside surface. Ricard et al. (2014) / with permission of Elsevier.

Lach and Grellman (2008) have studied the relationship between fracture and viscoelasticity, using impact testing to provide fracture data and DMTA to measure viscoelasticity. They have explored the temperature dependences of fracture energy and of viscoelasticity, the latter in the form of tanδ (see Chapter 5). The examples of chlorinated PVC (PVC-C) and a blend of PVC-C and chlorinated polyethylene (PE-C) are shown in Figure 13.40. Here local maxima in fracture energy are clearly related to the β transition. The maxima were also seen to coincide with high levels of plastic deformation of the fracture surfaces. Similar work on polycarbonate and PMMA showed less definite correlations, but with fracture energy peaks fairly close to the β transition range. For semicrystalline polymers PE and PP, correlations were found between maxima in fracture toughness and the glass (α) transition.

In this work some interesting correlations are seen, but the pattern is not entirely consistent. It is relevant that DMTA produces results for the small strain, linear viscoelastic regime, whereas, as we saw in Section 13.11.1, nonlinear viscoelasticity is necessary for the realistic representation of fracture where there are sizeable plastic zones.

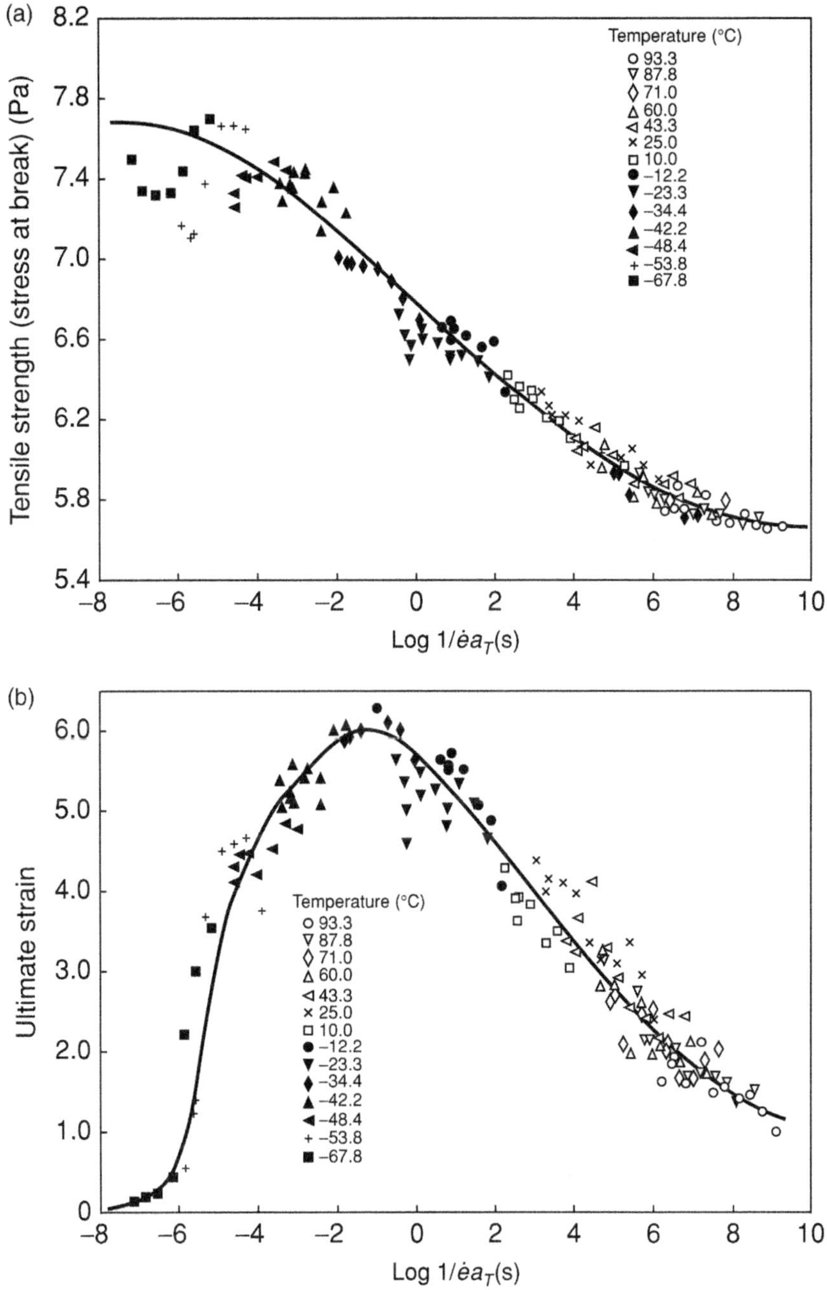

Figure 13.38 Variation of tensile strength (a) and ultimate strain (b) of a rubber with reduced strain rate $\dot{e}a_T$. Values were measured at various temperatures and rates and reduced to a temperature of 263 K. Smith (1958) / John Wiley & Sons.

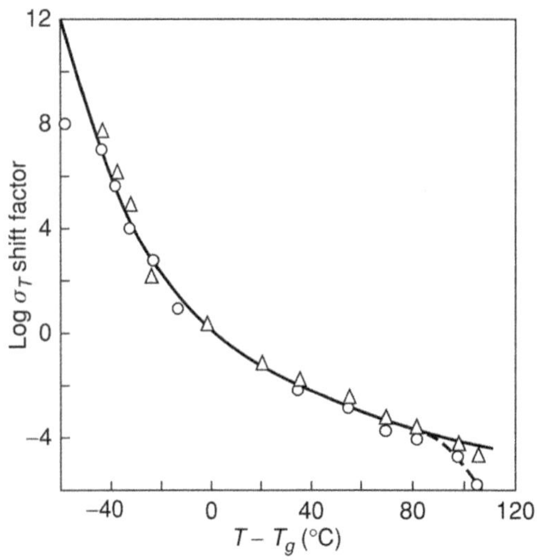

Figure 13.39 Experimental values of log σ_T shift factor obtained from measurement of ultimate properties compared with those predicted using the WLF equation: (Δ) from tensile strength; (O) from ultimate strain; (- - - -) WLF equations with T_g = 263 K. Smith (1958) / John Wiley & Sons.

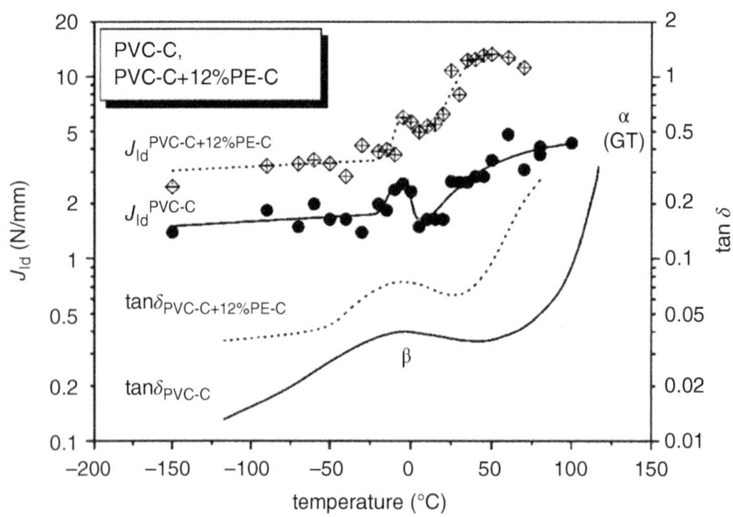

Figure 13.40 Comparison of temperature dependence of fracture energy (J_{1d}) and the viscoelastic loss parameter tanδ for two amorphous polymers. Mahajam and Hartmaier (2012) / American Physical Society.

13.12 Fatigue in Polymers

Materials frequently fail by fatigue due to the cyclic application of stress below that required to cause yield or fracture when a continuously rising stress is applied. The effect of such cyclic stresses is to initiate microscopic cracks at centres of stress concentration within the material or on the surface, and subsequently to enable these cracks to propagate, leading

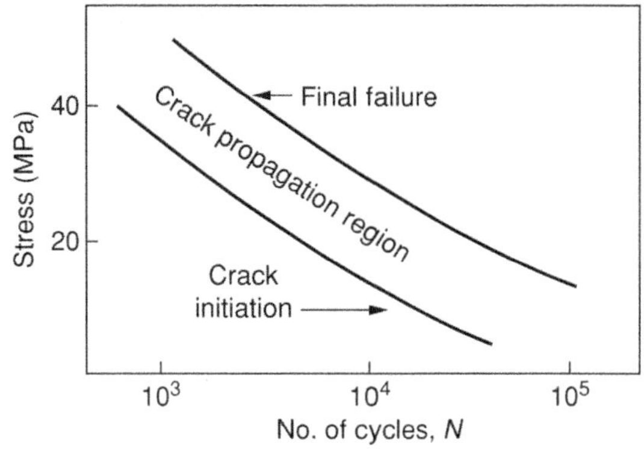

Figure 13.41 Fatigue response of PVC: relationship between applied stress σ and number of cycles to failure N, for both initiation of fatigue cracks and final failure. Adapted from Manson and Hertzberg, 1973.

to eventual failure. Fatigue in polymers is often associated with crazing as the precursor to cracking.

Early studies of fatigue in polymers concentrate on stress cycling of unnotched samples, to produce S versus N plots similar to those that have proved so useful for characterising fatigue in metals (S being the maximum loading stress and N the number of cycles to failure). An example of this type of plot for PVC (Manson and Hertzberg 1973) is shown in Figure 13.41. A major aspect of such a test is the question of adiabatic heating, which can lead to failure by thermal melting. Clearly there will be a critical frequency above which thermal effects become important.

Stress cycling tests on unnotched samples do not readily distinguish between crack initiation and crack propagation. Further progress requires a similar approach to that adopted in fracture studies, namely the introduction of very sharp initial cracks in order to examine crack propagation utilising fracture mechanics concepts.

For glassy polymers, fracture mechanics has been the usual starting point (Borduas *et al.* 1968; Hertzberg *et al.* 1970; Arad *et al.* 1971; Harris and Ward 1973), with the fatigue crack growth rate usually expressed as an empirical relationship

$$\frac{\mathrm{d}c}{\mathrm{d}N} = A(\Delta K)^m, \tag{13.56}$$

where c is the crack length, N is the number of cycles, ΔK is the range of the stress intensity factor (i.e. $K_{\max} - K_{\min}$, where $K_{\min}$ is usually zero) and A and m are constants depending on the material and test conditions.

Equation (13.56) is the most general form of the law proposed by Paris (1964) and Paris and Erdogan (1963) for predicting fatigue crack growth rates in metals. The general situation for glassy polymers is illustrated in Figure 13.42(a), with some typical results shown in Figure 13.42(b). The data differ in two respects from the Paris equation: first, analogous to the case of rubbers, there is a distinct threshold value of ΔK, denoted by ΔK_{th}, below which no crack growth is observed; secondly, as ΔK approaches the critical stress intensity factor K_c, the crack accelerates.

For accurate examination of this latter effect, according to Equation (13.56) the crack growth rate needs to be determined for known values of ΔK. For most fatigue testing the applied force and therefore the applied stresses $\sigma_{\max}$ and $\sigma_{\min}$, rather than $K_{\max}$ and

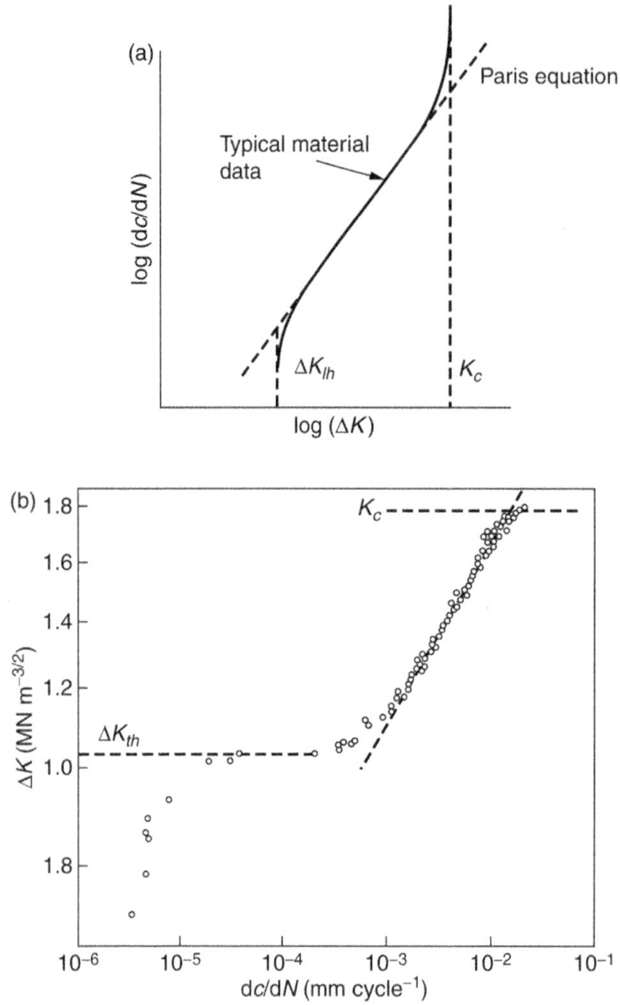

Figure 13.42 (a) Schematic diagram of fatigue crack growth rate dc/dN as a function of the range of stress intensity factor ΔK. (b) Fatigue crack growth characteristics for a vinyl urethane polymer. Adapted from Harris and Ward, 1973.

K_{min}, are controlled. As the crack length c increases during the course of a fatigue test, K_{max} will increase according to Equation (13.5). In most cases, particularly when $K_{min} = 0$, ΔK also increases. This effect has been investigated by Favier *et al.* (2002), who developed a system of feedback from video capture of cracks to produce fatigue testing equipment capable of operating at controlled stress intensity. Crack growth for force control and intensity control conditions is compared in Figure 13.43.

For the force control tests, the maximum and minimum forces were kept constant and at a constant ratio R, so that $R = \sigma_{max}/\sigma_{min}$ to give $\Delta K = K_{max}\ (1 - R)$, increasing as K_{max} increases with crack length in accordance with Equation (13.5). The figure (1) shows an accelerating rate of crack growth under force control, but not noticeably so in (2) for

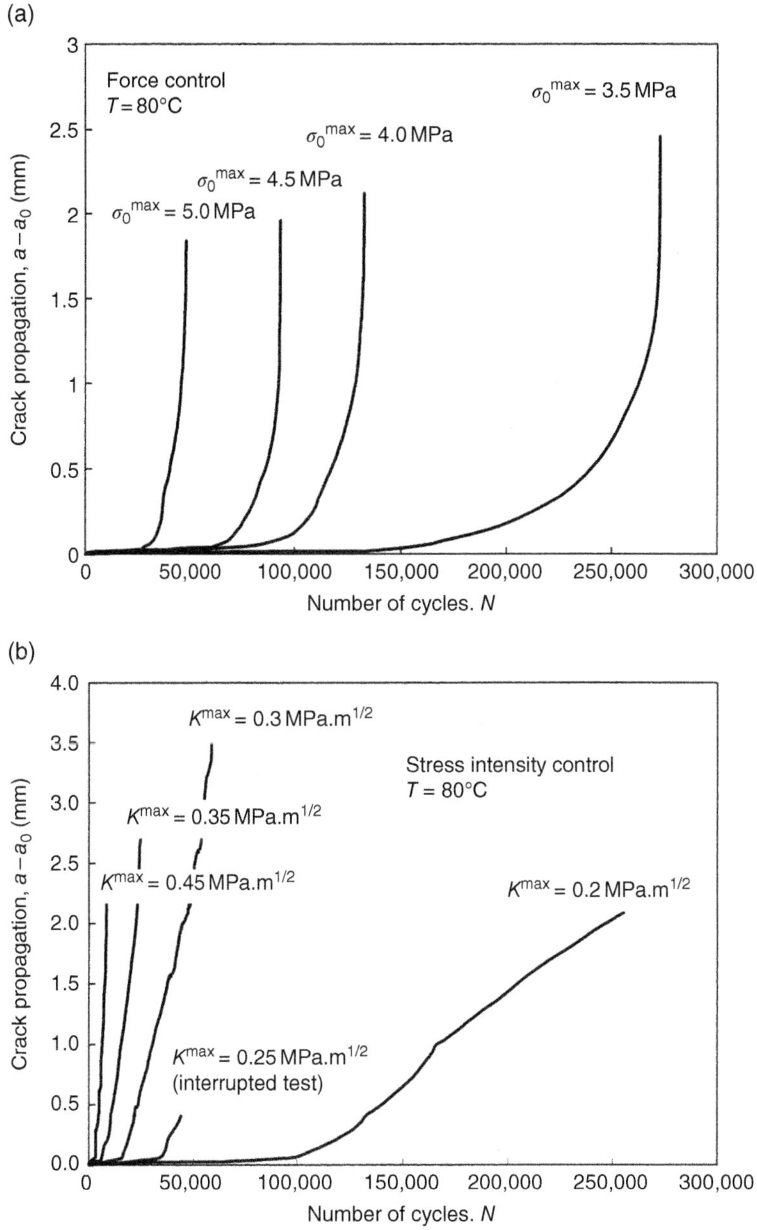

Figure 13.43 Crack growth for polyethylene in fatigue for (a) controlled force and (b) controlled stress intensity. Favier et al. (2002). / with permission of Elsevier.

intensity control. In the case of force control, the correct (increasing) values of ΔK could be calculated as the crack lengths were known. The resulting Paris law plots from these tests were almost indistinguishable from those at controlled intensity. This shows that Equation (13.56) is valid in an instantaneous sense, capable of giving good predictions independent of the crack growth history. This gives the Paris law a degree of robustness.

Equation (13.56) can be criticized on the grounds that it allows for the influence of the stress intensity factor but not for the mean stress. We have already seen that mean stress influences craze nucleation, and the detrimental effect of mean stress on fatigue lifetimes has been reported (Janssen *et al.* 2008; Amjadi and Fatemi 2020a, 2020b). This consideration led Arad *et al.* (1972) to suggest an equation of the form

$$\frac{dc}{dN} = \beta\lambda^n, \tag{13.57}$$

where $\lambda = \left(K_{\max}^2 - K_{\min}^2\right) = \Delta K(K_{\max} + K_{\min})$ is simply related to the mean intensity and therefore the mean stress. In a review of modelling of slow crack growth not restricted to polymers, Chudnovsky (2014) reports a number of variations on the Paris equation to enable more realistic predictions, some of which include the effect of mean stress.

It has become generally accepted (Pitman and Ward 1980; Williams 1977) that it is difficult to assign physical significance to the parameters in the Paris equation. However, it is possible to make use of the assumption that crack growth rate is uniquely dependent on ΔK to perform numerical analyses of fatigue crack growth. Ding *et al.* (2014) began their analysis by relating crack growth rate to dissipated energy W by the relation

$$\frac{dW}{dN} = \left(\frac{dW}{da}\right)\left(\frac{da}{dN}\right) \tag{13.58}$$

so that $\frac{dW}{dN}$ is the plastic work per load cycle and $\frac{dW}{da}$ is the plastic work per unit crack length. This latter quantity is related to the critical strain energy release rate or fracture energy G_c, so that

$$\frac{da}{dN} = \left(\frac{1}{G_c}\right)\left(\frac{dW}{dN}\right) \tag{13.59}$$

In the analysis of Ding *et al.* (2014) elastic–plastic finite element analysis is used to evaluate $\frac{dW}{dN}$, and thence $\frac{da}{dN}$. As the simulation proceeds, the crack grows, leaving yielded zones at the edges of the newly created crack – a 'plastic wake'. The dissipated energy within the plastic wake is evaluated by the finite element program to evaluate $\frac{dW}{dN}$. G_c is accepted as a material property and assumed to be a constant. Ding *et al.* had no available date for this property and so estimated it by matching a single experimental point $\left(\Delta K, \frac{da}{dN}\right)$ with a point from the simulation. Yield stresses as a function of strain rate were required as input for the material constitutive models. Fatigue crack growth predictions were made for a number of polymers with good accuracy. Figure 13.44 shows the example of polycarbonate. A similar approach has been adopted by Wang and Jiang (2022) who studied the effect of mean stress.

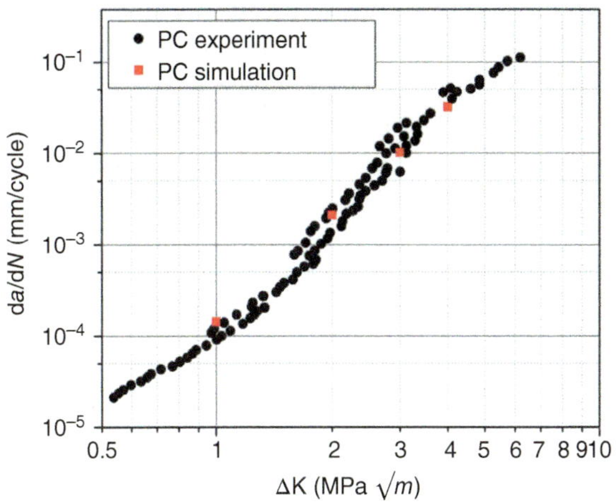

Figure 13.44 Fatigue crack growth observations and predictions for polycarbonate. Adapted from Ding et al. 2014.

Fatigue is a technologically important area for polymers, as applications such as pressure pipes often demand highly reliable performance over many years. This has driven the use of pragmatic approaches where the fundamentals of the phenomena were poorly understood. Fracture mechanics concepts can be deployed to good effect, in some cases with the assistance of numerical methods.

References

Al Laham, S. (1999) *Stress Intensity and Limit Load Handbook*, British Energy Generation Ltd., Gloucester.

Almomani, A., Mourad, A.-H.I., Deveci, S., Wee, J.-W. and Choi, B.-H. (2023) Recent advances in slow crack growth modeling of polyethylene materials. *Mater. Des.*, 227, 111720.

Amjadi, M. and Fatemi, A. (2020a) Multiaxial fatigue behavior of thermoplastics including mean stress and notch effects: experiments and modelling. *Int. J. Fatigue*, 136, 105571.

Amjadi, M. and Fatemi, A. (2020b) Creep and fatigue behaviors of high-density polyethylene (HDPE): Effects of temperature, mean stress, frequency, and processing technique. *Int. J. Fatigue*, 141, 105871.

Anand, L., Kamrin, K. and Govindjee, S. (2022) *Introduction to the Mechanics of Solid Materials*, Oxford University Press.

Andrews, E.H. (September 1966) Fracture mechanics approach to corrosion stress cracking in plastics. in *Proceedings of the Conference on the Physical Basis of Yield and Fracture*, Oxford, (ed. A.C. Strickland), Institute of Physics & The Physical Society London, p. 127.

Andrews, E.H. (1974) Generalised theory of fracture mechanics. *J. Mater. Sci.*, 9, 887.

Andrews, E.H. and Bevan, L. (1972) Mechanics and mechanism of environmental crazing in a polymeric glass. *Polymer*, 13, 337.

Arad, S., Radon, J.C. and Culver, L.E. (1971) Fatigue crack propagation in polymethylmethacrylate – effect of mean value of stress intensity factor. *J. Mech. Eng. Sci.*, 13, 75.

Arad, S., Radon, J.C. and Culver, L.E. (1972) Growth of fatigue cracks in polycarbonate. *Polym. Eng. Sci.*, 12, 193.

Aranda-Ruiz, J., Ravi-Chandar, K. and Loya, J.A. (2020) On the double transition in the failure mode of polycarbonate. *Mech. Mater.*, 140, 103242.

Argon, A.S. (2011) Craze initiation in glassy polymers – revisited. *Polymer*, 52, 2319–2327.

Argon, A.S. and Hannoosh, J.G. (1977) Initiation of crazes in polystyrene. *Philos. Mag.*, 36, 1195–1216.

Argon, A.S. and Salama, M.M. (1977) Growth of crazes in glassy polymers. *Philos. Mag.*, 36, 1217–1234.

Argon, A.S., Hannoosh, J.C. and Salama, M.M. (1977) *Initiation and growth of crazes in glassy polymers. Fracture 1977*, Vol. 1, Pergamon Press, Waterloo, pp. 445–470.

Arzakhova, O., Yu, K.A., Nazaroc, A.I., Dolgova, A.A. and Volynskii, A.L. (2020) "Green" environmental crazing of polymers in oil-in-water emulsions with high water content. *Polymer*, 186, 122020.

Ayatollahi, M.R., Nejati, M. and Ghouli, S. (2022) Crack tip fields in anisotropic planes: a review. *Int. J. Fract.*, 234, 113–139.

Battachary, S.K. and Brown, N. (1984) Micromechanisms of crack initiation in thin films and thick sections of polyethylene. *J. Mater. Sci.*, 19, 2519–2532.

Begley, J.A. and Landes, J.D. (1972) The J integral as a fracture criterion. *ASTM STP*, 514, 23.

Benbow, J.J. (1961) Stable crack propagation in plastics. *Proc. Phys. Soc.*, 78, 970.

Benbow, J.J. and Roesler, F.C. (1957) Experiments on controlled fractures. *Proc. Phys. Soc.*, B70, 201.

Bernier, G.A. and Kambour, R.P. (1968) The role of organic agents in the stress crazing and cracking of poly(2,6-dimethyl-1,4-phenylene oxide). *Macromolecules*, 1, 393.

Berry, J.P. (1961) Fracture processes in polymeric materials. I. The surface energy of poly (methyl methacrylate). *J. Polym. Sci.*, 1, 107–115.

Berry, J.P. (1963) Determination of fracture surface energies by cleavage technique. *J. Appl. Phys.*, 34, 62.

Boisot, G., Laiarinandrasana, L., Besson, J., Fond, C. and Hochstetter, G. (2011) Experimental investigations and modeling of volume change induced by void growth in polyamide 11. *Int. J. Solids Struct.*, 48, 2642–2654.

den Boogaert, V. (1966) *The Physical Basis of Yield and Fracture*, Physical Society, Oxford.

Borduas, H.F., Culver, L.E. and Burns, D.J. (1968) Fracture-mechanics analysis of fatigue-crack propagation in polymethylmethacrylate. *J. Strain Anal.*, 3, 193.

Bowie, O.L. (1964a) Rectangular tensile sheet with symmetric edge cracks. *J. Appl. Mech.*, 86, 208–212.

Bowie, O.L. (1964b) Symmetric edge cracks in tensile sheet with constrained ends. *J. Appl. Mech.*, 86, 726–728.

Bowie, O.L. and Freese, C.E. (1972) Central crack in plane orthotropic rectangular sheet. *Int. J. Fract. Mech.*, 8, 49–57.

Bowie, O.L. and Neal, D.M. (1965) Single edge crack in rectangular tensile sheet. *J. Appl. Mech.*, 32, 708–709.

Boyer, R.F. (1968) Dependence of mechanical properties on molecular motion in polymers. *Polym. Eng. Sci.*, 8, 161.

Bramuzzo, M. (1989) High speed fracture mechanics by photography of polypropylene copolymers. *Polym. Eng. Sci.*, 29, 1077.

Broberg, K.B. (1968) Critical review of some theories in fracture mechanics. *Int. J. Fract.*, 4, 11.

Broutman, L.J. and McGarry, F.J. (1965) Fracture surface work measurements on glassy polymers by a cleavage technique. I. Effects of temperature. *J. Appl. Polym. Sci.*, 9, 589–608.

Brown, H.R. (1973) Critical examination of impact test for glassy polymers. *J. Mater. Sci.*, 8, 941.

Brown, N. (1980), in *Methods of Experimental Physics*, (ed. R.A. Fava), Vol. 16, Part C, Academic Press, New York, p. 233.

Brown, N. and Imai, Y. (1975) Craze yielding of polycarbonate in N2, AR, and 02 at low pressures and temperatures. *J. Appl. Phys.*, 46, 4130.

Brown, H.R. and Ward, I.M. (1973) Craze shape and fracture in poly(methyl methacrylate). *Polymer*, 14, 469.

Brown, N., Metzger, B.D. and Imai, Y. (1978) Equation for creep in terms of craze parameters. *J. Polym. Sci. Polym. Phys. Ed.*, 16, 1085.

Brown, N., Donofrio, J. and Lu, X. (1987) The transition between ductile and slow-crack-growth failure in polyethylene. *Polymer*, 28, 1326–1330.

Bucknall, C.B. (2007) New criterion for craze initiation. *Polymer*, 48, 1030–1041.

Bueche, F. (1955) Tensile strength of plastics above the glass temperature. *J. Appl. Sci.*, 26, 1133.

Bueche, F. (1957) Tensile strength of rubbers. *J. Polym. Sci.*, 24, 189–200.

Bueche, F. and Halpin, J.C. (1964) Molecular theory for tensile strength of gum elastomers. *J. Appl. Phys.*, 35, 36.

Chan, M.K.V. and Williams, J.G. (1983) J-integral studies of crack initiation of a tough high-density polyethylene. *Int. J. Fract.*, 23, 145.

Chudnovsky, A. (2014) Slow crack growth, its modeling and crack-layer approach: a review. *Int. J. Eng. Sci.*, 83, 6–41.

Clarke, P.L. (1982) Tensile yield and fracture of linear polyethylene. Ph.D. thesis, Leeds University.

Crouch, B.A. and Huang, D.P. (1994) The J-integral technique applied to toughened nylons under impact loading. *J. Mater. Sci.*, 29, 861.

Curtis, J.W. (1970) The effect of pre-orientation on the fracture properties of glassy polymers. *J. Phys. D. Appl. Phys.*, 3, 1413–1422.

Dear, J.P. and MacGillivray, J.H. (1991) Strain gauging for accurate determination of K and G in impact tests. *J. Mater. Sci.*, 26, 2124–2132.

Deng, B., Wang, S., Hartquist, C. and Zhao, X. (2023) Nonlocal intrinsic fracture energy of polymerlike networks. *Phys. Rev. Lett.*, 131, 228102.

Ding, G., Karlsson, A.M. and Carlsson, L.A. (2014) Numerical evaluation of Paris-regime crack growth rate based on plastically dissipated energy. *Eng. Fract. Mech.*, 124–125, 155–166.

Dobbs, H.S., Field, J.E. and Maitland, A.H. (1973) Fracture surface energies using laser-induced cracks. *Philos. Mag.*, 28, 33–47.

Doll, W. and Weidmann, G.W. (1976) Interferential optical measurement of craze zone before fracture process in PMMA with different molecular weights. *Colloid Polym. Sci.*, 254, 205.

Donald, A.M. (1997), in *The Physics of Glassy Polymers*, (eds R.N. Haward and R.J. Young), 2nd edn, Chapman Hall, London.

Donald, A.M. and Kramer, E.J. (1981) The mechanism of craze-tip advance in glassy polymers. *Philos. Mag. A*, 43, 857–870.

Dugdale, D.S. (1960) Yielding of steel sheets containing slits. *J. Mech. Phys. Solids*, 8, 100.

ESIS Technical Committee on Polymers and Composites (May 1995) A testing protocol for conducting J crack growth resistance curve tests on plastics.

Estevez, R., Tijssens, M.G.A. and Van der Geissen, E. (2000) Modeling of the competition between shear yielding and crazing in glassy polymers. *J. Mech. Phys. Solids*, 48, 2585–2617.

Evans, R.M., Nara, H.R. and Bobalek, R.G. (1960) Prediction of impact resistance from tensile data. *Soc. Plast. Eng. J.*, 16, 76.

Fasce, L., Bernal, C., Frontini, P. *et al.* (2001) On the impact essential work of fracture of ductile polymers. *Polym. Eng. Sci.*, 41, 1.

Favier, V., Giroud, T., Strijko, E., Hiver, J.M., G'Sell, C., Hellincckx, S. and Goldberg, A. (2002) Slow crack propagation in polyethylene under fatigue at controlled stress intensity. *Polymer*, 43, 1375–1382.

Flory, P.J. (1946) Effects of molecular structure on physical properties of butyl rubber. *Ind. Eng. Chem.*, 38, 417.

Fraser, R.A.W. and Ward, I.M. (1974) The fracture behaviour of notched specimens of polymethyl methacrylate. *J. Mater. Sci.*, 9, 1624.

Fraser, R.A.W. and Ward, I.M. (1977) The impact fracture behaviour of notched specimens of polycarbonate. *J. Mater. Sci.*, 12, 459.

Gearing, B.P. and Anand, L. (2004) Notch-sensitive fracture of polycarbonate. *Int. J. Solids Struct.*, 41, 827–845.

Gent, A.N. and Lai, S.M. (1994) Interfacial bonding, energy dissipation, and adhesion. *J. Polym. Sci. Part B-Polym. Phys.*, 32, 1543–1555.

Girard, D., Castagnet, S., Gacougnolle, J.L. and Hoschstetter, G. (2007) On the relevance of a notch creep test for the comprehension and prediction of slow crack growth in PVDF. *Polym. Test.*, 26, 937–948.

Greensmith, H.W. and Thomas, A.G. (1955) Rupture of rubber. III. Determination of tear properties. *J. Polym. Sci.*, 18, 189–200.

Griffith, A.A. (1921) The phenomena of rupture and flow in solids. *Philos. Trans. R. Soc.*, 221, 163.

Guo, J., Zehnder, A.T., Creton, C. and Hui, C.-Y. (2020) Time dependent fracture of soft materials: linear versus nonlinear viscoelasticity. *Soft Matter*, 16, 6163–6179.

Halpin, J.C. (1964) Fracture of amorphous polymeric solids – time to break. *J. Appl. Phys.*, 35, 3133.

Hammond, M.J. and Fawaz, S.A. (2016) Stress intensity factors of various size single edge-cracked tension specimens: a review and new solutions. *Eng. Fract. Mech.*, 153, 25–34.

Hanson, D.E. (2011) The molecular kink paradigm for rubber elasticity: numerical simulations of explicit polyisoprene networks at low to moderate strains. *J. Chem. Phys.*, 135, 054902.

Hanson, D.E. and Barber, J.L. (2013) The theoretical strength of rubber: numerical simulations of polyisoprene networks at high tensile strains evidence the role of average chain tortuosity. *Model. Simul. Mater. Sci. Eng.*, 21, 075013.

Hanson, D.E. and Martin, R.L. (2009) How far can a rubber molecule stretch before breaking? *Ab initio* study of tensile elasticity and failure in single-molecule polyisoprene and polybutadiene. *J. Chem. Phys.*, 130, 064903.

Harris, J.S. and Ward, I.M. (1973) Fatigue-crack propagation in vinyl urethane polymers. *J. Mater. Sci.*, 8, 1655.

Hashemi, S. and Williams, J.G. (2000) Temperature dependence of essential and nonessential work of fracture parameters for polycarbonate film. *Plast, Rubber Comp.*, 29, 294.

Heijboer, J. (1968) Dynamic mechanical properties and impact strength. *J. Polym. Sci. C*, 16, 3755.

Hertzberg, R.W., Nordberg, H. and Manson, J.A. (1970) Fatigue crack propagation in polymeric materials. *J. Mater. Sci.*, 5, 521.

Hine, P.J. (1981) The fracture behaviour of polyether-sulphone. Ph.D. thesis, Leeds University.

Hoff, E.A.W. and Turner, S. (1957) A study of the low-temperature brittleness testing of polyethylene. *Bull. Am. Soc. Test. Mater.*, 225, TP208.

Hua, W., Li, J., Zhu, Z., Li, A., Huang, J. and Gan, Z. (2023) A review of mixed mode I-II fracture criteria and their applications in brittle or quasi-brittle fracture analysis. *Theor. Appl. Fract. Mech.*, 124, 103741.

Huang, Y.-L. and Brown, N. (1990) The dependence of butyl branch density on slow crack-growth in polyethylene – kinetics. *J. Polym. Sci. Polym. Phys. Ed.*, 28, 2007.

Huang, J.J. and Paul, D.R. (2006) Comparison of fracture behavior of nylon 6 versus an amorphous polyamide toughened with maleated poly(ethylene-1-octene) elastomers. *Polymer*, 47, 3505–3519.

Imai, Y. and Brown, N. (1976) Environmental crazing and intrinsic tensile deformation in polymethylmethacrylate. 1. Mechanical behavior. *J. Mater. Sci.*, 11, 417.

Inglis, G.E. (1913) Stresses in a plate due to the presence of cracks and sharp corners. *Trans. Inst. Nav. Architec.*, 55, 219.

Ippitsu, R., Okuda, T., Ito, A., Nitta, K. and Hiejima, Y. (2023) In-situ observation of falling weight impact damage in isotactic polypropylene sheet. *Polymer*, 283, 126225.

Irwin, G.R. (1957) Analysis of stresses and strains near the end of a crack traversing a plate. *J. Appl. Meek*, 24, 361.

Irwin, G.R. (1958) Fracture. *Handbuch Phys.*, 6, 551.

Irwin, G.R. and Kies, J.A. (1952) Fracturing and fracture dynamics. *Welding Res. Suppl.*, 31, 95–100.

Irwin, G.R. and Kies, J.A. (1954) Critical energy analysis of fracture strength. *Welding J. Res. Suppl.*, 33, 1935.

Janssen, R.P.M., de Kanter, D., Givaert, L.E. and Meijer, E.H. (2008) Fatigue life predictions for glassy polymers: a constitutive approach. *Macromolecules*, 41, 2520–2530.

Jiang, H., Zhang, J., Yang, Z., Jiang, C. and Kang, G. (2017) Modeling of competition between shear yielding and crazing in amorphous polymers' scratch. *Int. J. Solids Struct.*, 124, 215–228.

Johnson, A.E., Moore, D.R., Prediger, R.S. *et al.* (1986) The falling weight impact test applied to some glass-fiber reinforced nylons. 1. Appraisal of the method. *J. Mater. Sci.*, 21, 3153; (1987) The falling weight impact test applied to some glass-fiber reinforced nylons. 2. Some results and interpretations. *J. Mater. Sci.*, **22**, 1724.

Jones, D.P., Leach, D.C. and Moore, D.R. (1986) The application of instrumented falling weight impact techniques to the study of toughness in thermoplastics. *Plast. Rubber, Proc. Appl.*, 6, 67.

Kambour, R.P. (1964) Structure and properties of crazes in polycarbonate and other glassy polymers. *Polymer*, 5, 143.

Kambour, R.P. (1966) Mechanism of fracture in glassy polymers. 3. Direct observation of craze ahead of propagating crack in poly(methyl methacrylate) and polystyrene. *J. Polym. Sci. A2*, 4, 349.

Kambour, R.P. (1973) A review of crazing and fracture in thermoplastics. *J. Pol. Sci.; Macromol. Rev.*, 7, 1.

Kambour, R.P., Gruner, C.L. and Romagosa, E.E. (1973) Solvent crazing of dry polystyrene and dry crazing of plasticized polystyrene. *J. Polym. Sci.*, 11, 1879.

Kambour, R.P., Gruner, C.L. and Romagosa, E.E. (1974) Bisphenol-A polycarbonate immersed in organic media – swelling and response to stress. *Macromolecules*, 7, 248.

Karger-Kocsis, J., Moos, E., Mudra, E. and Varga, J. (1999) Effects of molecular weight on the perforation impact behavior of injection-molded plaques of α-and β-phase lsotactic polypropylenes. *J. Macromol. Sci. Part B Phy. B*, 38, 647.

Karger-Kocsis, J., Varga, J. and Drummer, D. (2002) Instrumented falling weight impact of co-injection-molded multi-polypropylene sandwich plaques containing β-phase polypropylene core. *J. Macromol. Sci. Part B Phy. B*, 41, 881.

Katsamanis, F.G. and Delides, C.G. (1988) Fracture surface energy measurements of PMMA: a new experimental approach. *J. Phys. D. Appl. Phys.*, 21, 79–86.

Kattekola, B., Ranjan, A. and Basu, S. (2013) Three dimensional finite element investigations into the effects of thickness and notch radius on the fracture toughness of polycarbonate. *Int. J. Fract.*, 181, 1–12.

Kinloch, A.J. and Williams, J.G. (1980) Crack blunting mechanisms in polymers. *J. Mater. Sci.*, 15, 987.

Knauss, W.G. (1973) On the steady propagation of a crack in a viscoelastic sheet: experiments and analysis, in *Deformation and Fracture of High Polymers*, (eds H.H. Kausch, J.A. Hassell and R.I. Jaffee), Springer, US, pp. 501–541.

Knauss, W.G. (2015) A review of fracture in viscoelastic materials. *Int. J. Fract.*, 196, 99–146.

Koiter, W.T. (1965) Discussion: "Rectangular tensile sheet with symmetric edge cracks". (Bowie, O.L. (1964) *ASME J. Appl. Mech.*, **31**, 208–212). *J. Appl. Mech.*, 87, 237.

Kojouri, A.S., Rikaee, H.K., Kalteremidou, K.-A. and Van Hemelrijck, D. (2023) Experimental evaluation of J-integral in elastic and elastic–plastic polymers by means of digital image correlation and higher-order eigenfields under mode-I. *Eng. Fract. Mech.*, 291, 109534.

Kramer, H.H. (1983) Microscopic and molecular fundamentals of crazing. *Adv. Polym. Sci.*, 52, 1–56.

Kramer, H.H. and Berger, L.L. (1990) Craze growth and fracture. *Adv. Polym. Sci.*, 91, 1–68.

van Krelelyan, D.W. and Nijenhuis, K. (2009) *Properties of Polymers: their Correlation with Chemical Structure: Their Numerical Estimation and Prediction from Additive Group Contributions*, 4th edn, Elsevier. Chapter 7

Kulmi, U. and Basu, S. (2006) A molecular dynamics study of the failure modes of a glassy polymer confined between rigid walls. *Model. Simul. Mater. Sci. Eng.*, 14, 1071–1093.

Lach, R. and Grellman, W. (2008) Time- and temperature-dependent fracture mechanics of polymers: general aspects at monotonic quasi-static and impact loading conditions. *Macromol. Mater. Eng.*, 293, 555–567.

Lake, G.J. and Thomas, A.G. (1967) The strength of highly elastic materials. *Proc. R. Soc. A*, 300, 108–119.

Landes, J.D. and Begley, J.A. (1979), in *Post Yield Fracture Mechanics*, (ed. D.G.A. Latzko), Applied Science, London.

Leitão, N. and Gilabert, F.A. (2021) A unified J-Integral-based procedure to investigate at different loading regimes the fracture by FEM simulations and image analysis. *J. Mech. Phys. Solids*, 149, 104318.

Liu, J. and Yee, A.F. (2000) Effect of local conformational transition on craze initiation in polyestercarbonates containing cyclohexylene linkages. *Macromolecules*, 33, 1338–1344.

Lu, X. and Brown, N. (1986) The relationship of the initiation stage to the rate of slow crack growth in linear polyethylene. *J. Mater. Sci.*, 21, 2423.

Mahajam, D.K. and Hartmaier, A. (2012) Mechanisms of crazing in glassy polymers revealed by molecular dynamics simulations. *Phys. Rev. E*, 86, 021802.

Mai, Y.W. and Cotterell, B. (1986) On the essential work of ductile fracture in polymers. *Int. J. Fract.*, 32, 105.

Manson, J.A. and Hertzberg, R.W. (1973) Fatigue failure in polymers. *CRC Crit. Rev. Macromol. Sci.*, 1, 433.

Marshall, G.P., Culver, L.E. and Williams, J.G. (1970) Craze growth in polymethyl-methacrylate – a fracture mechanics approach. *Proc. Roy. Soc.*, A319, 165.

Marshall, G.P., Williams, J.G. and Turner, C.E. (1973) Fracture toughness and absorbed energy measurements in impact tests on brittle materials. *J. Mater. Sci.*, 8, 949.

Martinatti, F. and Ricco, T. (1994) High rate testing of toughened polypropylene. *Polym. Test.*, 13, 405.

McCutcheon, C.J., Zhao, B., Jin, K., Bates, S.B. and Ellison, C.J. (2020) Crazing mechanism and physical aging of poly(lactide) toughened with poly(ethylene oxide)-block-poly(butylene oxide) diblock copolymers. *Macromolecules*, 53, 10163–10178.

McMurry, J. (2023) *Organic Chemistry: A*, 10th edn, Openstax, Rice University, Houston.

McNeill, S.R., Peters, W.H. and Sutton, M.A. (1987) Estimation of stress intensity factor by digital image correlation. *Eng. Fract. Mech.*, 28, 101–112.

Moore, D.R. and Prediger, R.S. (1988) A study of low energy impact of continuous carbon-fiber-reinforced composites. *Polym. Compos.*, 9, 330.

Murakami, Y. (1986) *Stress Intensity Factors Handbook*, Pergamon.

Naz, S., Sweeney, J. and Coates, P.D. (2010) Analysis of the essential work of fracture method as applied to UHMWPE. *J. Mater. Sci.*, 45, 448–459.

O'Connell, P.A., Bonner, M.J., Duckett, R.A. *et al.* (1995) The relationship between slow crack-propagation and tensile creep – behavior in polyethylene. *Polymer*, 36, 2355.

Orowan, E. (1949) Fracture and strength of solids. *Rep. Prog. Phys.*, 12, 185.

Osumu, R., Yasui, T., Tanake, R. *et al.* (2023) Impact of strain-induced crystallization on fast crack growth in stretched cis-1,4-polyisoprene rubber. *ACS Macro Lett.*, 11, 747–752.

Oxborough, R.J. and Bowden, P.B. (1973) A general critical-strain criterion for crazing in amorphous glassy polymers. *Philos. Mag.*, 28, 547–599.

Paris, P.C. (1964) *Fatigue, An Interdisciplinary Approach*, Syracuse University Press, New York, p. 107.

Paris, P.C. and Erdogan, F. (1963) A critical analysis of crack propagation laws. *J. Basic Eng. Trans. ASME*, 85, 528.

Pitman, G.L. and Ward, I.M. (1979) Effect of molecular weight on craze shape and fracture toughness in polycarbonate. *Polymer*, 20, 895.

Pitman, G.L. and Ward, I.M. (1980) The molecular weight dependence of fatigue crack propagation in polycarbonate. *J. Mater. Sci.*, 15, 635.

Plati, E. and Williams, J.G. (1975) Determination of fracture parameters for polymers in impact. *Polym. Eng. Sci.*, 15, 470.

Pluvinage, G. (2004) Chapter 3: Stress concentration factor, in *Fracture and Fatigue Emanating from Stress Concentrators*, Springer, Dordrecht. https://doi.org/10.1007/1-4020-2612-9_3

Puttick, K.E. (1978) The mechanics of indentation fracture in poly(methylmethacrylate). *J. Phys. D. Appl. Phys.*, 11, 595–604.

Qi, Y., Li, X., Venkata, S.P., Yang, X., Sun, T.L. and Hui c-Y., Gong J P., Long, R. (2024) Mapping deformation and dissipation during fracture of soft viscoelastic solid. *J. Mech. Phys. Solids*, 186, 105595.

Radon, J.C. and Turner, C.E. (1969) Fracture toughness measurements by instrumented impact test. *Eng. Fract. Mech.*, 1, 411–428.

Ramsteiner, F. (1999) J(0.2)-values by impact testing. *Polym. Test.*, 18, 641.

Razavi, M., Cheng, S., Huang, D., Zhang, S. and Wang, S.-Q. (2020) Crazing and yielding in glassy polymers of high molecular weight. *Polymer*, 197, 122445.

Ricard, J., Guigné, F. and Laiarinandrasana, L. (2014) Damage and fracture mechanisms of polyoxymethylene: Multiscale experimental study and finite element modelling. *Eng. Fract. Mech.*, 115, 270–283.

Rice, J.R. (1968a), in *Fracture – an Advanced Treatise*, (ed. H. Liebowitz), Academic Press, New York. Chap. 3

Rice, J.R. (1968b) A path independent integral and approximate analysis of strain concentration by notches and cracks. *J. Appl. Mech.*, 35, 379.

Rivlin, R.S. and Thomas, A.G. (1953) Rupture of rubber. 1. Characteristic energy for tearing. *J. Polym. Sci.*, 10, 291.

Robeson, L.M. (2013) Environmental stress cracking: a review. *Polym. Eng. Sci.*, 53, 453–467.

Rooke, D.P. and Cartwright, D.J. (1976) *Compendium of Stress Intensity Factors*, HMSO London.

Sakulkaew, K., Thomas, A.G. and Busfield, J.J.C. (2011) The effect of the rate of strain on tearing in rubber. *Polym. Test.*, 30, 163–172.

Schapery, R.A. (1975) A theory of crack initiation and growth in viscoelastic media. I. Theoretical development. *Int. J. Fract.*, 11, 141–159.

Sih, G.C., Paris, P.C. and Irwin, G.R. (1965) On cracks in rectilinearly anisotropic bodies. *Int. J. Fract. Mech.*, 1, 189–203.

Smith, T.L. (1958) Dependence of the ultimate properties of a GR-S rubber on strain rate and temperature. *J. Polym. Sci.*, 32, 99.

Smith, T.L. (1960) Ultimate tensile properties of amorphous polymers. *Soc. Plast. Eng. J.*, 16, 1211.

Smith, T.L. and Stedry, P.J. (1960) Time and temperature dependence of the ultimate properties of an SBR rubber at constant elongations. *J. Appl. Phys.*, 31, 1892.

Snyder, M.D. and Cruse, T.A. (1975) Boundary-integral equation analysis of cracked anisotropic plates. *Int. J. Fract.*, 11, 315–328.

Stearne, J.M. and Ward, I.M. (1969) The tensile behaviour of polyethylene terephthalate. *J. Mater. Sci.*, 4, 1088.

Sternstein, S.S. and Ongchin, L. (1969) Inhomogeneous deformation and yielding of glasslike high polymers. *Am. Chem. Soc. Polym. Prepr.*, 10, 1117.

Sternstein, S.S., Ongchin, L. and Silverman, A. (1968) Yield criteria for plastic deformation of glassy high polymers in general stress fields. *Appl. Polym. Symp.*, 7, 175.

Sumpter, J.D. and Turner, C.E. (1973) Applicability of J to elastic–plastic materials. *Int. J. Fract.*, 9, 320.

Svensson, N.L. (1961) Variation of fracture energy of brittle plastics with temperature. *Proc. Phys. Soc.*, 77, 876.

Sweeney, J. and Ward, I.M. (1988) The fracture behaviour of oriented polyethylene at high pressures. *Proc. R. Soc. Lond. A Math. Phys. Sci.*, 420, 55–80.

Thomas, A.G. (1955) Rupture of rubber. 2. The strain concentration at an incision. *J. Polym. Sci.*, 18, 177.

Thomas, A.G. (1960) Rupture of rubber. VI. Further experiments on the tear criterion. *J. Appl. Polym. Sci.*, 3, 168–174.

Thuy, M., Pedragosa-Rincón, M., Niebergall, U., Oehler, H., Alig, I. and Böhning, M. (2022) Environmental stress cracking of high-density polyethylene applying linear elastic fracture mechanics. *Polymers*, 14, 2415.

Torabi, A.R. and Shahbhazian, B. (2020) Notch tip plastic zone determination by extending Irwin's model. *Theor. Appl. Fract. Mech.*, 108, 102643.

Trofimchuk, E.S., Moskvina, M.A., Nikonorovaa, N.I. *et al.* (2020) Hydrolytic degradation of polylactide films deformed by the environmental crazing mechanism. *Eur. Polym. J.*, 139, 110000.

Truss, R.W., Duckett, R.A. and Ward, I.M. (1983) A novel technique for measuring the impact properties of a tough polyethylene. *Polym. Eng. Sci.*, 23, 708.

Tsunoda, K., Busfield, J.J.C., Davies, C.K.L. and Thomas, A.G. (2000) Effect of materials variables on the tear behaviour of a non-crystallising elastomer. *J. Mater. Sci.*, 35, 5187–5198.

Van Melick, H.G.H., Govaert, L.E. and Meijer, H.E.H. (2003) Localization phenomena in glassy polymers: influence of thermal and mechanical history. *Polymer*, 44, 3579–3591.

Venkatesan, S. and Basu, S. (2015) Investigations into crazing in glassy amorphous polymers through molecular dynamics simulations. *J. Mech. Phys. Solids*, 77, 123–145.

Vincent, P.I. (1960) The tough brittle transition in thermoplastics. *Polymer*, 1, 425.

Vincent, P.I. (1971) *Impact Tests and Service Performance of Thermoplastics*, Plastics and Rubber Institute, London.

Vincent, P.I. (1974) Impact strength and mechanical losses in thermoplastics. *Polymer*, 15, 111.

Wada, Y. and Kasahara, T. (1967) Relation between impact strength and dynamic mechanical properties of plastics. *J. Appl. Polym. Sci.*, 11, 1661.

Wang, J. and Jiang, W. (2022) Numerical analysis of fatigue crack growth using a plastically dissipated energy factor based model. *Proc. IMechE Part C J. Mech. Eng.*, 236, 4232–4252.

Williams, J.G. (1977) A model of fatigue crack growth in polymers. *J. Mater. Sci.*, 12, 2525.

Williams, J.G. (1980) *Stress Analysis of Polymers*, 2nd edn, Ellis Horwood, Chichester.

Williams, J.G. and Marshall, G.P. (1975) Environmental crack and craze growth phenomena in polymers. *Proc. R. Soc. Lond. A Math. Phys. Sci.*, 342, 55.

Williams, J.G. and Rink, M. (2007) The standardisation of the EWF test. *Eng. Fract. Mech.*, 74, 1009–1017.

Wu, J. and Mai, Y.-W. (1996) The essential fracture work concept for toughness measurement of ductile polymers. *Polym. Eng. Sci.*, 36, 2275.

Wu, P.D. and Van der Giessen, E. (1996) Computational aspects of localized deformations in amorphous glassy polymers. *Eur. J. Mech. A. Solids*, 799–823.

Yarysheva, A., Strel'tsov, D., Malakhov, S., Arzakhova, O., Yarysheva, L. and Volynskii, A. (2020) Surface modification of poly(tetrafluoroethylene) and poly(ethylene terephthalate) films via environmental crazing. *Polym. Int.*, 69, 627–634.

Yarysheva, A., Khavpachev, M., Bagrov, S., Bakirov, A., Efimov, A., Trofimchuk, E. and Chvalun, S. (2021) Breathable materials and hybrid nanocomposites with antimicrobial activity based on porous poly(ε-caprolactone) obtained via environmental crazing. *Macromol. Mater. Eng.*, 306, 2000636.

Zhou, Z. and Brown, N. (1996) A new automated method for recording and predicting failure by slow crack growth in polyethylene. *Polym. Test.*, 15, 549–558.

Zimmermann, C., Klemm, W. and Schönert, K. (1984) Dynamic energy release rate and fracture heat in polymethylmethacrylate (PMMA) and a high-strength steel. *Eng. Fract. Mech.*, 20, 777–782.

Further Reading

Kinloch, A.J. and Young, R.J. (1995) *Fracture Behaviour of Polymers, SpringerScience+Business Media*, B.V, Dordrecht.

Index

Mechanical Properties of Solid Polymers, Fourth Edition. John Sweeney and Peter Hine.

d

e